ENCYCLOPEDIA OF PHYSICS

EDITED BY

S. FLÜGGE

VOLUME XLVI/1

COSMIC RAYS I

WITH 150 FIGURES

SPRINGER-VERLAG

BERLIN · GÖTTINGEN · HEIDELBERG

1961

HANDBUCH DER PHYSIK

HERAUSGEGEBEN VON

S. FLÜGGE

BAND XLVI/1

KOSMISCHE STRAHLUNG I

MIT 150 FIGUREN

SPRINGER-VERLAG
BERLIN · GÖTTINGEN · HEIDELBERG
1961

ISBN 978-3-642-45966-5   ISBN 978-3-642-45964-1 (eBook)
DOI 10.1007/978-3-642-45964-1

© by Springer-Verlag OHG. Berlin · Göttingen · Heidelberg 1961
Softcover reprint of the hardcover 1st edition 1961

Contents.

The Origin of Cosmic Rays.

By

P. MORRISON.

With 15 Figures.

A. Summary of the properties of cosmic rays.

1. Introductory. For more than a generation cosmic rays have been studied primarily as a natural beam of high-energy particles. It is unnecessary to detail here how fruitful that "laboratory" has been, how much of our knowledge about fundamental particles and their interactions, from the cascade process of electromagnetic radiation through the nature of hyperons and tau-mesons, has come from cosmic ray studies. Most of the subsequent articles in this volume are devoted to an exposition of such material. But the study of the physics of the "beam" has thrown little light on the long-discussed question of the origin of cosmic rays themselves.

The unique nature of this stream of high-energy particles has led to a wide set of speculations concerning their origin. Many of these were all but metaphysical, assigning the cosmic rays to the original pre-stellar state postulated for the universe as a whole[1], or to some unimagined nuclear events of spontaneous decay. The problem of cosmic-ray origin became regarded as one of fundamental theory, rather than of the identification of astrophysical processes which might account for their presence. It is the goal of this article to expound exactly the contrary view. While we cannot specify precisely the history of the cosmic rays, we propose to sketch the general outline of a model of that history, sufficiently in detail so that its predictions can be tested, yet with freedom enough so that the rapid growth of the subject to be anticipated in the next years will not wholly invalidate the discussion.

The experimental physicists in at least several of the best-equipped high-energy laboratories now can dispose of a beam of protons at almost the mean cosmic-ray energy, some five or ten Gev/nucleon. Their work tends more and more to restrict the efficient use of the cosmic-ray beam to the highest energy region. But in return for this severe competition, the machine designers have given cosmic-ray workers more and more confidence in an understanding of the processes of cosmic-ray origination. For the natural process, as so often, appears to involve just the phenomena which have proved decisive in the design considerations of the synthesizers of cosmic rays.

Four developments of the last decade may be regarded as the key to the establishment of our model. Two of these arise from cosmic-ray studies:

(i) The demonstration that the primary cosmic-ray beam is grossly a sample of the nuclear abundances familiar in the stars, and that electrons and gammarays of Gev energies, at least, are all but absent. The presence of heavier nuclei seems directly to imply a more or less adiabatic energy gain process, for the kinetic energies observed are at least thousands of times greater than the disassociation energies of these structures. No plausible nuclear process known

---

[1] G. LEMAÎTRE: Rev. Mod. Phys. **21**, 357 (1949), and references therein.

could be expected to yield such a range of heavy nuclei. The near-absence of particles with strong electromagnetic interaction points to the importance of electromagnetic processes in controlling the particle history.

(ii) The detailed study of solar-connected time variations in cosmic rays more and more clearly demonstrates the importance of magnetic fields in the solar atmosphere and in interplanetary space. Cosmic-ray particles are the best probes for such weak but extended magnetic fields. The origin of some particles of low cosmic ray energies at or near the sun is now sure.

The two other developments arose in a more purely astrophysical context:

(iii) The development of hydromagnetics, with recognition that many astrophysical phenomena, from radio emission to the polarization of starlight by galactic absorption, are controlled by hydromagnetic interactions.

(iv) Direct optical and radio-astronomical observation of the presence of large densities of highly relativistic particles in specific objects[1].

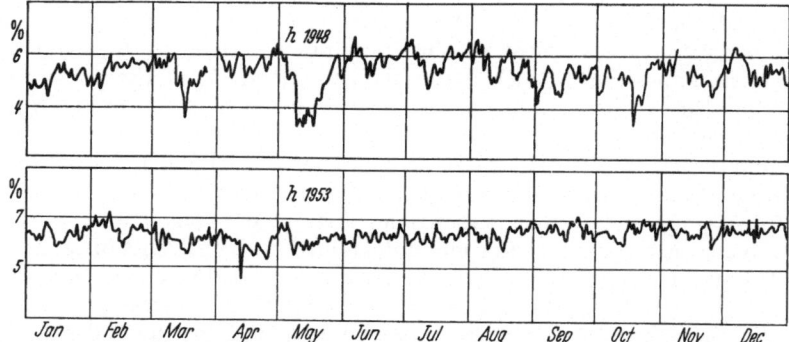

Fig. 1. Time variations of cosmic-ray intensity. Cosmic-ray ionization as recorded at Huancayo for the years 1948 and 1953. Comparison of the two years gives an illustration of the long-time variations. This refers to primaries with energies >15 Gev. From S. E. FORBUSH: J. Geophys. Res. **59**, 525 (1954).

With these results, it seems that the study of the cosmic ray primaries should take its place with optical and radioastronomy as a means of gaining important information about the structure of the world in the large. The cosmic rays no longer represent a separate and rather mystical phenomenon, but become a key part of the astronomical description of the world. Cosmic-ray astronomy, and indeed the still undeveloped branches, gamma-ray and neutrino astronomy, will one day belong properly to the astrophysical volumes of future *Handbooks*. To put the question of how the cosmic rays originate will appear as naive as it does today to ask how starlight originates. Both questions elide a wealth of information.

The present situation is not so clear. Our model of cosmic-ray origin is still both qualitative and ambiguous. This want follows both from the poor state of understanding of the essentially non-linear behavior of hydromagnetic processes and from our still primitive information about the physical conditions in the spatial regions of most importance for our theory: the space between stars and galaxies, and the neighborhood of certain energy-rich objects, like supernovae. But it is the contention of this article that apart from these admittedly serious lacunae no essentially new physics is needed for a full understanding of cosmic ray origin. This implies that the general considerations of cosmology are not relevant for certainly they are still no secure part of physics. Here we commit

---

[1] Early recognition: H. ALFVÉN and N. HERLOFSON: Phys. Rev. **78**, 616 (1950).

ourselves then to the view that the cosmic rays are phenomena of no larger scale than the scale of galaxy clusters, within which ordinary time-space concepts may be taken to work well. Moreover, we postulate matter which contains no unfamiliar particles, and supports no new or very rare elementary processes. In how far we must depart from this conservatism for cosmic rays of the very highest energies, beyond 10^8 Gev/nucleon, we defer to the final section of the article.

2. Time variations. Certain general properties of the cosmic rays must find explanation in the model. Subsequent articles in this volume describe the experimental results and their foundation in detail; here we propose simply to state the facts in approximate form, giving only a sketch of the supporting evidence. Most of the present information is far from secure, and considerable uncertainty attaches to the properties of the primaries we wish to employ. In particular, many of the statements are merely plausible extrapolations, say, to energy ranges or to past times about which very little empirical information is at hand. It appears at present that no qualitative changes in the picture will result from more extensive and more precise measurements.

The first property to be cited is the *time constancy* of the cosmic-ray primaries. It is evident that this is a gross abstraction from the data, as Figs. 1 and 2 will demonstrate.

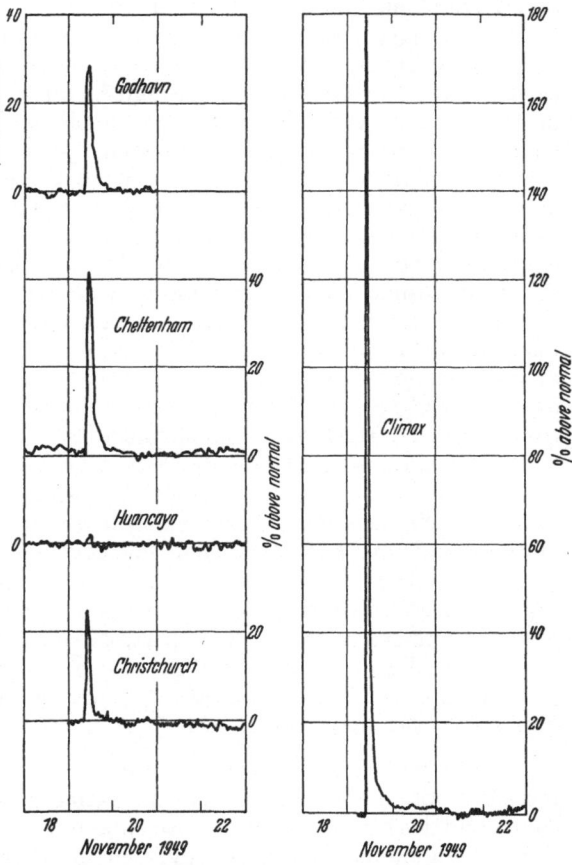

Fig. 2. A large solar flare. Increase of cosmic-ray intensity on Nov. 19, 1949. Observations by S. E. FORBUSH, T. B. STINCHCOMB and B. SCHEIN: Phys. Rev. **79**, 501 (1950),

But it is a general feature of the cosmic ray beam that if we look apart from variations of a few percent the beam above ~ 10 Gev remains time-independent to a remarkable degree. Only during rare great solar flares (Fig. 2) are there major intensity changes for incoming particles above a few Gev/nucleon. It is to be observed that the lowest energies (below a total energy of two or three Gev/nucleon) are exceptional in this respect. Particles below such an energy may be all but lacking in years of maximum solar activity[1], and present in full intensity for years near the minimum of the eleven-year solar-activity cycle. All these changes, together with the shift in the equatorial plane of the effective dipole magnetic field of the earth as seen from the cosmic-ray latitude effect compared with its orientation as found from geomagnetic measurements at the

---

[1] H. NEHER: Phys. Rev. **103**, 228 (1956).

earth's surface[1], demonstrate the presence of magnetic conditions in the interplanetary space capable of modulating the low-energy beam. They here serve
but to emphasize the constancy of the cosmic rays of energies from, say, 30 Gev/nu-
cleon up to the highest known primary energies, about 10^9 Gev/nucleon. No changes
above statistical fluctuations are seen for these energies. The limit is probably
as good as a part per thousand in the daily mean, for times up to some years,
and energies up to 10^4 Gev/nucleon. For higher energies, the accuracy is poorer.

The primary beam seems to remain constant for times of the order of ten
thousand years, as judged from the agreement between documentary and radio-
carbon dating of old historical events[2]. Such measurements give information
on the particles with energy from a couple of Gev up to 10 Gev or so, within
which range the bulk of the incoming energy lies. It is probably correct to assert
that the intensity of this major component averaged over five thousand years
or more has not varied more than ten percent from the present value. Measurements on cosmic-ray produced isotopes in meteors are subject still to uncertainties
of interpretation of serious nature, but indicate that within one order of magnitude
the mean cosmic-ray flux at these typical energies has remained unchanged for
hundreds of millions of years[3]. From all these indications we take it to be plausible, and a fundamental feature of the phenomenon, that the cosmic-ray flux,
at least that portion not influenced by the sun or the earth, is roughly constant
over times of geological span. Such a point of view does not exclude variations
from the mean which do not much affect its value when averaged over long times.

Almost all measurements are made at such rates that statistical fluctuations
mask any but the most exceptional intensity variations over times of an hour
or less. Such variations, if they exist, must refer to rather local inhomogeneities
of the cosmic-ray beam; we shall not consider them.

Our conclusions then are that the primary beam—outside the special earth-
sun region where we must observe it—is constant in time for periods up to hundreds of millions of years, with possible small or short-duration changes only.
For energies below 30 Gev the effect of sun and perhaps of earth grows, becoming
predominant below a couple of Gev, where the phase of solar activity cycle
determines the observed intensity.

3. Isotropy. The rotation of the earth on its axis and its revolution about
the sun imply a connection between spatial anisotropy and time variations. Of
course, intrinsic time variations are superimposed upon any periodic time variations which follow from anisotropy. The presence of sharp time variations during
great solar flares also makes possible the detection of sharp anisotropies in the
first hour or so of such events. But all diurnal or annual variations which can
so far be ascribed to the primary beam anisotropy (and not to effects of atmospheric origin) refer to the energies below 30 Gev. If we define the anisotropy δ
as the ratio

$$\delta = (\varphi_{max} - \varphi_{min})/\tfrac{1}{2}(\varphi_{max} + \varphi_{min})$$

between the flux φ_{max} in the direction of maximum intensity and that, φ_{min},
in the direction of minimum intensity we can set limits as follows for various
energies:

$E = 10-30$ Gev	$E = 10^4$ Gev
δ: varying, near 1%	$\leq 0.1\%$

[1] See, e.g., J. Simpson, K. B. Fenton, J. Katzman and D. C. Rose: Phys. Rev. **102**,
1648 (1956).
[2] J. L. Kulp and H. L. Volchok: Phys. Rev. **90**, 713 (1953).
[3] F. Singer: Phys. Rev. **105**, 765 (1957). — Nuovo Cim. **8**, 539 (1958).

No observation reliably setting δ greater than zero has been made for any energy beyond about 30 or 40 Gev. Below that energy the values are high, strongly time-dependent, and clearly solar-connected.

The high degree of isotropy for all energies above the manifestly solar-influenced ones is crucial, for it implies that no strong sources of cosmic rays are "visible", except rarely the sun. Since on Copernican grounds it is unlikely that the true sources of the rays are isotropically situated around the earth, the overall gross isotropy implies that the rays propagate in such a way as to obscure the direction of their sources[1].

4. Nature of the incoming beam. The primary particles are mainly protons, but there are also found alpha-particles, and heavier nuclei up at least as far as iron, $Z = 26$, have been identified.

The number of shower-producing particles, electrons or photons, with energies above 1 Gev, is below one percent of the total flux. Again excepting the lowest energies, the charge distribution of the incoming beam appears to be roughly independent of energy up to some hundred Gev/nucleon. A few individual heavy nuclei have been seen with much greater energies, suggesting at least that the charge spectrum remains grossly similar up to 10^4 Gev/nucleon. Beyond that energy the air shower experiments have at least given no sign (as by multiple cores of the great showers) that protons do not still predominate up to the highest energies.

There is some evidence that the heavy components ($Z = 2$ and up) become relatively less prominent as the energy per nucleon increases from a few to some tens of Gev. We comment on this energy-dependence in the next section.

Table 1. *Atomic abundances.*

	Cosmic-ray beam (mainly below 10 Gev)[2]	Adjusted solar system[3] composition (so-called "cosmic" abundances)
H	100	100
He	15	15
Li		4×10^{-7}
Be	0 to 0.4	1×10^{-7}
B		1×10^{-7}
C		0.037
N	1.2 ± 0.4	0.010
O		0.10
F		
Ne	0.2	0.003
Mg	0.09	0.003
Si	0.07	0.004
Fe (plus Co and Ni)	0.06	6.10^{-4} up to 0.004
beyond Ni[4]	less than 10^{-5}	10^{-6}

It is rather well established from the observed effects of the geomagnetic latitude on the incoming beam that the nuclei entering the earth's neighborhood are completely ionized, within the accuracy of the measurements, which might show up an electron cortege of a few electrons on the heavier nuclei. There is evidence for the presence of nuclei of charge 3 through 5 (Li, Be, B) in the beam. These are rare in stellar and nebular gases. It seems likely that these elements are indeed relatively less rare in the cosmic rays than in the stellar atmospheres.

We present a summary table (Table 1) showing the element abundances in the cosmic ray primary beam, mainly between a few and 50 Gev/nucleon energy, and a comparison with typical spectroscopic analyses of astronomical objects.

---

[1] But compare Y. SEKIDO: Nature **177**, 35 (1956), and [*17*], for a contrary view, which we regard as statistically not convincing. He has presented still further data in Phys. Rev. **113**, 1108 (1959).

[2] For cosmic rays, see Table 10.

[3] For adjusted solar values, A. G. W. CAMERON, Astrophys. J., 1959, in press.

[4] L. V. KURNOSOVA *et al.*, in Satellite Symposium, CSAGI, Moscow, 1958.

The overall similarity is striking; it appears that the cosmic-ray beam may be relatively richer than the sun in the heavy elements, but the indicated effect is at best a factor under 10. Since there is strong reason to believe that the astronomical abundances are not universal, but vary from object to object, depending on the thermonuclear evolution of the material, the cosmic ray beam may be regarded as a particular sample of a different kind, with some marks of its special history, but probably no striking ones (compare Sect. 26 below).

The observations of the incoming beam composition are made difficult by the overlying atmosphere, since until now they have been made from balloons. There remains above the detectors about five or ten grams/cm² of atmosphere. Interactions within this layer are important for the heavy elements, at least, and impose a necessary correction to the data. Qualitatively, the error in the observed spectrum could be represented by saying that the amount of material through which the beam has passed is uncertain by some fraction of the residual atmospheric layer above. This implies an error of a couple of grams/cm² of matter. But the experiments suggest that the total matter traversed has been of that same magnitude. It cannot be large compared to the mean free path for iron, say, which is only some five gm/cm² of hydrogen, for then the whole beam would have had to be iron at injection. Indeed, this has been suggested, but appears to fail quantitatively; there is too much hydrogen for that. The presence of the elements with Z from 3 to 5, which are so perishable from thermonuclear reactions that they are really rare in the sun, is generally ascribed to interactions, and used to measure the amount of matter traversed. The point is still unsettled experimentally, though the weight of evidence lies on the side of some incoming Li, Be, and B, say, $(\mathrm{Li+Be+B})/(\mathrm{C+N+O})={}^{1}/_{10}$. Taking the total traversed matter since injection as between about one and up to five or eight grams/cm² at most cannot be far wrong. We know no cosmic densities so well that better data will be completely convincing as to the history of the beam. This result rests upon only the qualitative picture of the beam, and can hardly be wrong.

5. Energy spectrum. The energy spectrum of the cosmic rays is in some ways their most striking property. By a wide variety of methods, the cosmic ray primaries have been shown to extend in energy from particles with a kinetic energy less than a tenth of their rest mass, up to particles, very likely single protons, with energies as high as a billion times their rest mass. Over such a wide range of energies it is plain that high accuracy or even high resolution is not to be expected, but the qualitative position is clear: the incoming flux per unit energy is a grossly smooth and monotone decreasing function of particle energy, independent of time, approximately given for total energies E of the incoming proton by the expression:

$$n(E) = 0.3/E^{2.5\pm0.2}, \qquad E \gtrsim 10 \text{ Gev}, \qquad (5.1)$$

where $n(E)$ is the number of incoming protons per unit solid angle per second per square centimeter per Gev. We can write, of course,

$$E = (T+1)\, m_0 c^2 = \gamma\, m_0 c^2$$

where T is the kinetic energy, and γ the familiar dimensionless value for the total energy in terms of the rest mass. Fig. 3 indicates the validity of the formula. Below $T \approx 2$ Gev, down to the lowest identifiable energies (merging into the auroral particles), the formula (5.1) is not useful. In the low-energy region there appears a so-called "low-energy" cut-off during the years of high solar activity. In times of low solar activity, as in 1954, the low-energy spectrum tended to rise with decreasing energy as fast as or perhaps even faster than the formula (5.1). No cut-off was seen at that time.

At high energies, only the study of the giant air showers appears capable of giving a reasonable energy spectrum. If the cosmic rays are to receive any rational explanation, there must be some energy beyond which their intensity falls off more rapidly than the simple power law predicts (compare Sect. 7). So far no such high-energy cut-off has been seen.

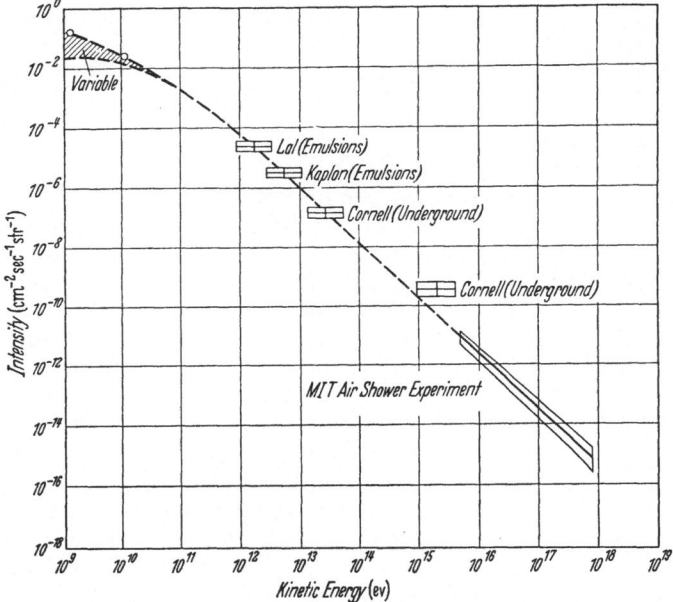

Fig. 3. General nature of the energy spectrum of cosmic rays. The plot shows the integral spectrum: all those particles which have energy greater than the indicated value of E contributing to the intensity. Detailed accounts of these studies are found in the article in this volume by Schopper. The plotted data are by no means definitive; the solar-variable region is shaded. The high-energy tail is not well located in the plane of the graph. But it is clear that a fairly satisfactory fit can be obtained by the use of a simple power law over a very wide range of energies. The line dotted in is a more complicated function, but a straight-line fit is as good as the data. The "best slope" is not well known, but cannot be much less than 1.4 or more than about 1.9. This curve is modified from a paper by G. Clark, J. Earl, W. Kraushaar, J. Linsley, B. Rossi and F. Scherb: Nature 180, 406 (1957).

The differential energy spectrum of (5.1) is often cited in integral form:

$$N(> E) = \int\limits_{E}^{\infty} n(E)\, dE$$

where $N(>E)$ is the number of cosmic rays per unit solid angle, square centimeter, and second, with total energies at least E Gev. This integral spectrum is generally more directly obtained from measurement. It is often interesting to know what fraction of the total energy in cosmic rays is contained in particles lying within a certain energy range. We define the fraction $f(E)$ by the relation:

$$f(E) = \int\limits_{E}^{\infty} n(E)\, E\, dE \Big/ \int\limits_{E_0}^{\infty} n(E)\, E\, dE$$

and tabulate the result in Table 2. Here E_0 is chosen to exclude the variable region; α denotes the exponent in the power-law spectrum $n(E) \sim E^{-\alpha}$.

Table 2.

E (Gev)	$f(E)$		
	$\alpha = 1.3$	$\alpha = 1.5$	$\alpha = 1.8$
2	1	1	1
10	0.62	0.45	0.28
10^2	0.31	0.14	0.04
10^3	0.16	0.045	6.9×10^{-3}
10^4	0.078	0.014	1.1×10^{-3}
10^5	0.039	4.5×10^{-3}	1.7×10^{-4}
10^7	9.8×10^{-3}	4.5×10^{-4}	4.4×10^{-6}
10^9	2.5×10^{-3}	4.5×10^{-5}	1.1×10^{-7}
10^{11}	0.6×10^{-3}	4.5×10^{-6}	2.8×10^{-8}

$E_0 = 2$ Gev.

6. Absolute energy of cosmic rays. An important guide to the theory has been a comparison of the cosmic ray energy density as observed on earth with other energy densities of astrophysical interest. A table (Table 3) of these values gives an immediate physical insight into the nature of the cosmic ray origin problem.

Table 3. *Energy densities in the galaxy.*

Type of energy	ev/cm³
Cosmic rays near sun	1
Starlight	1
Turbulent gas motion	1 — 10
Magnetic energy at 3 microgauss	1
Kinetic energy of rotation of galaxy as a whole	1000

Most of these values are of course order-of-magnitude estimates only. They have been chosen of course with the ultimate model in mind; yet they are highly suggestive in their own right. It is plain that the relation between starlight, which is the natural degradation product of every nuclear energy source, and cosmic rays for which we have no such general theory, must be explained in any acceptable model. It is also clear, as it has been for decades of thought on this problem, that the cosmic-ray flux observed on earth cannot be considered as extending throughout galactic and extra-galactic space without promoting cosmic rays to truly cosmological importance. *It is here that the modern viewpoint most radically departs from that of the workers of previous decades:* they took it as likely that on a cosmic scale very much of the world's energy was in these rays. We instead, following Fermi, seek processes which can maintain a density gradient in the cosmic ray flux, and assert that the total energy in cosmic rays is not vast, because at the earth we sample a density appropriate to a spiral arm, but space as a whole contains a much smaller mean density of cosmic rays. The spatial magnetic fields can act, as later chapters will make clearer, to maintain this kind of inhomogeneity. It should be clear also tbat it is easier to tolerate a large volume filled with cosmic rays of higher mean energy than one for the bulk of the rays, for the total energy assignable goes down rapidly as the energy per particle increases. This, too, is consistent with the behavior of magnetic fields, which cannot contain equally well rays of arbitrary energy.

B. Structure of the model

7. From the point of view here taken, the problem of cosmic ray origin becomes a problem of identifying certain regions in phase space, from which samples of gas are transferred to the neighborhood of the earth, now typically in higher momentum ranges. Equally necessary is an account of the processes by which this motion is brought about. The problem is essentially a statistical one; the rays arriving here bear little mark of a place or time of origin individually, and a statistical theory is fully satisfactory at the present time. The logical structure of such a theory can be presented succinctly, and will serve to guide the reader through the rest of the article.

The steps of the process are of course generally not intrinsically differentiated; the motion in phase space and the time sequence of events is essentially a continuous one. But the analysis into distinct phases serves as a guide to the approximations required and the physical assumptions implied. All current versions of this general model do present an equivalent for each of the steps here outlined. (The influence of electronuclear machine design on this analysis ought to be clear.)

α) *Injection.* Since the cosmic ray particles are a sample of ordinary matter, some feature of the state of the particular particle must have determined its entry into the process. Ionization is a possible requirement; it is very likely necessary, or nearly so, but it is far from sufficient. Most of the matter which can at all take part is already highly ionized. The initial step is probably determined by the position in phase space: for example, the particle may have belonged to the tail of the thermal distribution at some point in space at the time the process began. Any such distinguishing feature might do: position, velocity, charge. The process by which this feature enabled the start of the long path in phase space may be described as the injection step of the process leading in the end to cosmic rays.

β) *Acceleration.* Continuous with the act of injection, but perhaps distinguished from it is the important step of acceleration. A cosmic ray particle becomes one by somehow being injected into an accelerating process. Where and how the great energy gain takes place is the essence of any model. This act may be located close in space and time to injection, or very remote from it.

γ) *Stirring.* The observed cosmic-ray beam is nearly isotropic. Some process has deflected the particles so that statistically they strike the earth from all directions, even though the galactic position of the earth is by no means central. Only if all extra-galactic space sped particles to us could this isotropy be a natural consequence of our location. Since this is untenable, for the whole beam, as the sequel will demonstrate and the previous Sect. 3 suggests, some process must be considered which will stir the particles in momentum space sufficiently so that the high isotropy can arise. This stirring process may of course be part and parcel of the acceleration process, but the distribution in momentum direction as well as the distribution in momentum magnitude must follow from the model.

δ) *Storage and loss.* Not all particles attain the same energy; not all follow the same path. This implies that different particles take differing times to arrive to us. The stirring step alone makes this most natural. The time of storage, or the mean delay time between injection and detection, is clearly an important parameter of the model. Moreover, the cosmic-ray particles can be lost to the process, either by nuclear collisions which can break up the heavy nuclei and produce observable electromagnetic radiation and electron secondaries, or by entering some unfavorable part of phase space. Neither the beam composition nor the absolute energy densities can be accounted for by any model which does not specify the time of storage and the processes by which particles are lost to the observable beam.

It is clear that the steady-state assumption made natural by the properties of the rays gives the relation

$$n/\tau = \bar{q} \tag{7.1}$$

between $\bar{q}$ the mean source intensity per unit volume and τ the storage time. Long storage implies a small source intensity for the observed density; this point immediately makes it understandable how the isotropized cosmic rays can be present in energy density comparable with the starlight, which has much stronger sources, but a storage time determined wholly by straight-line propagation among the stars of the galaxy.

ε) *Cut-off.* The cosmic ray properties are not very precisely known. But there is at least superficially a continuous and smooth variation with energy of the intensity, which drops by many powers of ten over the known range. No other property shows much change: neither the isotropy nor the beam composition. On the most general grounds, the processes which produce cosmic rays ought to vary with the energy produced. There must be some energy beyond which the process is no longer governed by the same statistical parameters. We may call this energy the cut-off. Beyond the cut-off energy either the nature of the beam or the isotropy or the smooth variation of the intensity ought to change. If such a cut-off is found, or indeed, a succession of such cut-off energies, the properties of the rays beyond the cut-off would throw much light on the main and on subsidiary processes. If such a cut-off does not exist, the cosmic ray phenomenon could not be ascribed to any plausible set of physical conditions, but rather some fundamental or cosmological law would determine the simple behavior extended over so vast an energy range. Any physical model, then, ought to predict a cut-off energy, beyond which the properties of the rays would show some decisive change. Experimental search for the cut-off, looking at the rarer and rarer events of higher and higher energy, is an essential part of the study of cosmic-ray origin, and has been carried on for some years (cf. Ref. [*13*]).

C. Propagation of the cosmic ray beam in space.

Every galactic theory assumes a more-or-less prolonged wandering of the individual particles in interstellar space. Matter is always present in this space, either as free electrons and nuclei, or in heavier aggregates up to stars; and with it is radiation, varying in frequency from the radio region through the visible to the unknown gamma-ray components (which we have strong reason to expect). The cosmic-ray particles interact with all these as they travel. The spatial density of the cosmic rays themselves is certainly less than about 10^{-10} particles/cm^3, using the local density; collisions between particles of the primary beam are thus negligible compared to interactions with the material or photons in space in all known regions (compare Table 3). This remains true over such a wide range of cosmic ray intensity that we can neglect interactions between beam particles throughout the discussion. But interaction with the matter and radiation of the medium is crucial. Since every study of such propagation must make use of the properties of such interactions, we present in this chapter a summary of the processes which appear relevant at this time.

But in addition to this local, particulate sort of interaction, the cosmic rays may interact with matter which remains *at a distance* from the cosmic-ray particle. This interaction is of course mediated by fields, gravitational and electromagnetic, so that it may be regarded as local interaction again, this time not with matter but with field energy. The forces of gravity do not appear relevant, certainly not until it is necessary to consider the behavior of cosmic radiation moving over distances comparable to or larger than the "radius of the universe", the reciprocal of the Hubble constant, ca. $10-15 \times 10^9$ lightyears. This case does not arise, and we shall neglect all gravitational effects, as well as any which may be implied by certain cosmological theories implying the change of physical laws with time or distance. The case is quite otherwise with the electro-magnetic field. The charged cosmic-ray particles interact with electric and magnetic fields very effectively. It appears that space cannot very well contain electric fields of sizeable magnitude, for the high conductivity of such tenuous and rather well-ionized matter soon shorts them out. But the very act of shorting the electric fields implies the flow of currents, whose time duration is long in virtue of the large dimensions and low ohmic losses. These currents necessarily imply magnetic fields, which serve to couple the motion of large and widely extended masses of gas with the motion of the individual particles of the cosmic rays. This coupling is the most important physical phenomenon involved in the whole process; it seems probable today that on one or another scale it determines the processes of injection, acceleration, stirring, storage and loss. It induces a strong energy loss by radiation as well; direct astronomical observations of the consequent emitted radiation are among the most important pieces of evidence in the whole logical chain. The second portion of this chapter is therefore devoted to a summary of the nature of the motion of cosmic-ray particles in cosmic magnetic fields.

8. Interaction of the cosmic-ray particles with particles of the medium. α) *Ionization and Coulomb scattering.* The cosmic-ray particles are very probably born as ions. Even if they were not, their high energy implies a stripping of the electron cortege away from the nucleus in a rather short space. No precise theory of the capture and loss of electrons by heavy atoms passing through gas is available. But it is well known that the electrons will be lost very readily provided the atom of nuclear change $+Ze$ has a velocity V in the medium such that $V \gg V_e$, the orbital electron velocity. The stripping cross section for any electrons whose velocity in their bound orbits is smaller than the critical value V_e is about πa_e^2, with a_e their orbit radius. This relation implies that the light atoms, say from H to O, are completely stripped in passing through from 10 to 100 mg/cm² of matter, if their energies are above some tens of Mev. Even the heaviest atoms will be stripped of all but the most tightly-bound electrons in a similar layer of stopping matter; the K electrons, however, will in general not be easily stripped from heavy atoms. For the electron velocity of K electrons is just $\alpha Z c$, with $\alpha = e^2/\hbar c$, the fine structure constant, and the K shell radius is about a_0/Z, with a_0 the Bohr orbit, $a_0 = 0.56$ Å. To strip iron of every electron requires an energy above 15 or 20 Mev/nucleon for the atom, and a matter layer many grams/cm². Even at the extreme relativistic energies, then, such heavy ions may be accompanied by a few bound electrons; the small reduction in specific charge from the stripped case can safely be neglected for most purposes.

β) *Energy loss by ionization.* The stopping of fast particles by the loss of energy to ionization and excitation of the medium in which they move is a classical part of the study of high-energy particles. It is reviewed in such a work as that of BETHE and ASHKIN [23]. The application to the cosmic-ray propagation problem presents one novel feature. The medium is invariably dilute, so that the problem of the influence of neighboring atoms (the Fermi effect [23]) does not arise. Unlike the usual stopping material, however, the intragalactic medium is typically wholly or partially ionized. The effect of the free electrons must be taken into account. (The same phenomenon occurs in metals, but less markedly since the conduction electrons are relatively few.) A number of authors have considered this question. We follow the discussion of HAYAKAWA and KITAO[1].

The total contribution is the sum of three components: the neutral atoms present, if any, contrbute according to the standard calculations; the ions enter also in the ordinary way, but with reduced number of electrons; finally, the free electrons must be included. These electrons adjust themselves to the passage of the fast particle, moving to screen its Coulomb field, reducing its influence on the medium at large distances from the trajectory. But this screening fails for distances smaller than the Debye distance D, within which the direct energy transfer by Coulomb interaction may take place. For larger impact parameters, the transfer of energy is due to the excitation of plasma oscillations. These may be treated by the classical procedures, using the dielectric constant of the medium expressed in terms of the plasma frequency ω_p. The plasma frequency is just

$$\omega_p^2 = 4\pi n_e \frac{e^2}{m_e} = \frac{\langle v_e^2 \rangle}{3 D^2}; \quad \nu_p = \frac{\omega_p}{2\pi} = 9.0\, n_e^{\frac{1}{2}} \quad \text{kilocycles} \tag{8.1}$$

where D is the Debye length (written here with a rather unusual definition).

We cite formulae obtained in this way, written in forms appropriate to the present problem. For a nuclear particle of charge z and mass a, moving in hydrogen of density ϱ gm/cm^3 with a degree of ionization α, where

$$\alpha = \frac{\text{number of singly-ionized atoms}(= \text{no. of free electrons})}{\text{numbers of singly-ionized atoms} + \text{neutral atoms}}$$

we get:

$$-\frac{dE}{d(\varrho x)} = \frac{0.15 z^2}{\beta^2} \left[\left(21 + 2 \ln \frac{\beta^2}{1-\beta^2} - \beta^2 \right) + \alpha \left(\ln\left(1-\beta^2\right) + \beta^2 - \ln\alpha + 54 \right) \right] \left. \begin{array}{c} \\ \text{Mev/g-cm}^{-2}, \\ \\ \end{array} \right\} \tag{8.2}$$

$$E = \text{energy of fast particle} = \gamma\, a c^2 = \frac{1}{(1-\beta^2)^{\frac{1}{2}}}\, a c^2.$$

For electrons, the bracketed term in (8.2) is reduced by about $\frac{1}{3}$.

The effect of ionization in the medium is quite generally to increase the rate of energy loss by fast particles. This increase is appreciable in typical cases. For example, in the passage of protons through hydrogen gas, for proton energies near $E/Mc^2 = 3.5$, at the minimum of the energy loss curve, the comparison is:

in neutral hydrogen   4.2 Mev/g-cm^{-2},

in completely ionized hydrogen   12 Mev/g-cm^{-2}.

For higher energies, the difference remain about constant; for lower energies, the difference slowly increases. A rough estimate of the effect of ionization may be taken by multiplying the familiar energy loss rate by $1 + \alpha \left(2 - \frac{\ln\alpha}{25} \right)$ for a medium of charge Z, once ionized.

The importance of hydrogen in the medium and in stellar atmospheres allows the use of these simple formulae in most cases. A straightforward extension of the equations to include a sum over multiply-ionized atoms is adequate in more general cases.

γ) *Coulomb multiple scattering.* The energy loss accompanying passage of fast particles implies a related momentum transfer. This leads to multiple scattering, by the target nuclei, which slowly causes a parallel beam to diverge. In the

---

[1] S. HAYAKAWA and K. KITAO: Progr. Theoret. Phys. **16**, 132 (1956).

absence of any other effects, this phenomenon would set a limit to the angular definition with which a distant point source of cosmic rays would be "seen". For small angles of scattering, which is the only simple case, the familiar result may be taken over:

$$\langle \vartheta^2 \rangle = z^2 \left(\frac{21 \text{ Mev}}{m \gamma \beta^2 c^2} \right)^2 t, \quad t \text{ in radiation lengths}, \tag{8.3}$$

with the limitations that $\beta \approx 1$, $t > \frac{7 A^{\frac{1}{3}}}{(Z/137)^2}$ g/cm². For fully ionized media, the mean square angle should be increased by the factor $\frac{\log D/a}{\log 183\, z^{-\frac{1}{3}}}$, with D the Debye length, and a the atomic radius. This is an increase by a factor of the order of ten in typical cases.

The thickness t in (8.3) is measured in units of the so-called radiation length, which is familiar in cascade theory. We list here a few values of the radiation length for various elements, after [23]. These values include the effects of the atomic electrons. The quantity is defined as:

$$\frac{1}{x_0} = 4 N_{\text{at}} Z (Z+1) \frac{1}{137} \left(\frac{e^2}{m_e c^2} \right)^2, \quad N_{\text{at}} = \text{atoms/g}. \tag{8.4}$$

Also included are the so-called critical energies, at which the radiation loss by electrons equals the ionization loss (in un-ionized matter).

Table 4. *Radiation lengths and critical energies.*

Element	H	He	C	O	Si	Fe	Pb
X, g/cm²	58	85	43	34	23	14	6
E_c, Mev	340	220	103	77	44	24	7

δ) *Bremsstrahlung.* Fast charged particles lose energy in collisions by other means than by the transfer of momentum to their collision partners. For they are coupled to the electromagnetic field, and the acceleration during such collisions disturb the "virtual photons", setting some of them free as real photons, to carry electromagnetic energy infinitely far away. With the related process, pair production by photons, this phenomenon forms the heart of the cascade theory of cosmic-ray showers. For detailed treatment, see other chapters of this volume. We will here cite a few relevant results.

The energy loss by radiation for a particle of mass m and charge z moving through a medium of a definite atomic number Z and mass A is given by the formulae:

$$\left. \begin{array}{l} -\dfrac{1}{W} \dfrac{dW}{d(\varrho x)_{\text{rad}}} \approx \dfrac{z^4}{(m/m_e)^2} \dfrac{1}{X_0(Z)}, \text{ for un-ionized medium}, \quad W \gg \dfrac{137 m_e c^2}{Z^{\frac{1}{3}}} \\[2ex] \text{same} \times \dfrac{\ln(D/\hbar/mc)}{\ln(183 Z^{-\frac{1}{3}})}, \text{ fully-ionized medium}, \quad W < W_{\text{max}}. \end{array} \right\} \tag{8.5}$$

The latter formula holds up to a certain high cut-off energy, W_{max}, given by $\frac{W_{\text{max}}}{m c^2} = \frac{D}{\lambda_c}$ where D is the Debye length, and $\lambda_c = \hbar/m_e c$ is the Compton wavelength of the electron. Beyond that energy W_{max}, the radiation loss is constant.

Here W is the energy of the fast particle, ϱx the thickness of medium traversed, in g/cm², and $X_0(Z)$ is the radiation length, defined above (Sect. 8γ).

It is plain from (8.5) that this phenomenon is unimportant in the very tenuous media of interstellar space, except possibly for electrons. The nuclear components have almost no loss at all to bremsstrahlung.

To a reasonable approximation, the energy spectrum of bremsstrahlung is flat; the power radiated per unit frequency is constant from the lowest frequencies up to the maximum, given by $W = h\omega_{max}$. This may be written:

$$P(W, \nu)\, d\nu = \text{const}\, d\nu. \tag{8.6}$$

$P(W, \nu)$ is the power radiated per unit frequency, and the constant is found by setting $\int_0^{W/\hbar} P\, d\nu$ equal to the total loss rate given in (8.5). The particle velocity can be put equal to c. A tenth of the radiated energy goes into quanta with energies up to a tenth of the initial energy, and so on. This is good enough for most estimates.

ε) *Nuclear collisions*. Finally, we come to the direct nuclear interactions between the nuclear components of the primary beam and the nuclei of the medium.

Table 5. *Inelastic interaction cross-sections for protons against nuclei.*

Nucleus	H	He	C	O	Si	Fe	Pb
Cross-section (millibarns)	inelastic: 25 elastic: varies from 20 down to 5	150	300	350	430	650	1800
Interaction mean free path in millions of light years[1]	80	15	7	5.5	4.5	3	1.2

This is of course a complex problem, involving much of the subject matter of another part of this volume. It is well-known that in the relativistic region both inelastic and elastic collisions are possible. The inelastic collisions transfer a fair fraction of the relative kinetic energy into collision products, mainly mesons of various kinds, and into some antinucleon-nucleon pairs. The mesons will all decay in a short time to electrons, positrons, and neutrinos; any neutrons or anti-neutrons will decay also by beta-decay, and the only constituents of the beam which can persist for more than a very few years are electrons, protons, neutrinos, photons, anti-protons, and nuclei. Complex nuclei, from helium on up, which participate in nuclear collisions, will very nearly always lose at least one nucleon, and often many more, from their structure. From this point of view, the nuclear collisions may be thought of essentially as removing any given complex nucleus from the beam, and restoring to the beam some fraction of its energy as lighter nuclei, down to the proton. A proton-proton collision is a little different; from this, two protons are certain to emerge in the end, only with the original energy divided between them and several other products. Thus, nuclear collisions are an actual cause of loss to the heavy components of the beam, and only an energy-degrading process to the protons. We will later see how these notions may be used to discuss the history of the observed beam. Here it is appropriate to collect some estimates of the nuclear-cross-sections, and the mean free path for interaction to which they correspond. Any purely elastic collisions represent an energy-degrading process for all nuclei; they will be disregarded here as relatively unimportant, except perhaps for proton-proton collisions.

We tabulate (Table 4) a representative sample of inelastic collision cross-sections for protons against nuclei, and the mean free paths to which they correspond in a medium of given hydrogen density. The data are obtained both

---

[1] Computed with a medium density of $\frac{1}{2}$ proton/cm³, or 10^6 light years equals 0.8 g/cm².

from direct measurement with machine-produced beams[1], and by inference from cosmic-ray penetrating shower studies[2]. There is good reason to believe that the cross-sections stay roughly constant from 1 or 2 Gev up to 50 or 100 Gev at least, and some evidence that they do not change much for the rest of the energy range, up to the highest energies.

From the rapid decrease of mean free path with mass of nucleus, it follows that the observation of a distinct peak in the Fe region of the mass distribution implies the passage of the beam through a good deal less than one free path of hydrogen. For even the assumption that the initial beam was all iron would result in more secondary light nuclei than are observed, after traversing 80 million light years of space at the nominal density above. This extreme assumption is perhaps admissible for very high energies, but we actually see the injection of particles of energies near 10 Gev from the sun which from their geomagnetic behavior must be mainly protons. It is not likely that the initial beam is so free of hydrogen, and therefore the most probable mean thickness traversed is one which could not reduce by an order of magnitude the iron/hydrogen ratio: say, not above 5 or 10 grams/cm². This topic will be discussed in more detail when specific models are considered (see Sect. 26).

9. Collisions with photons. To a relativistic cosmic-ray particle, the electromagnetic radiation in space has appearance very different from what we see looking out from the earth. In the reference frame in which the fast particle is at rest, the radiation field is strongly Doppler-shifted, with an energy spectrum very much dependent on the direction of observation. Looking in the forward direction, optical and even radio photons are much more energetic, while in the opposite direction every radiation is soft. The photons can collide with the cosmic-ray particle. They may elastically scatter, make particle pairs, or induce photo-disintegration, depending on the circumstances of the collision. The general galactic radiation field contains something like one optical photon of a few electron volts energy per cm³, as well as weak photons in the radio region, with energy of about a microvolt per photon, and a density also around one per cm³. Since this photon density is similar to the particle density, cosmic-ray collisions with photons might be expected to have an effect on the beam not very different from the collisions with the proton gas of space, discussed in Sect. 8.

The effect of collisions with the photon gas in space was first worked out be FEENBERG and PRIMAKOFF[3] in considerable detail, and has been treated also by DONAHUE[4].

It turns out that the only effect of any significance in galactic space is to degrade the energy of very fast electrons, rather more slowly than the magnetic field interactions described later in Sect. 13. Electrons stored for a long time in strong light, near the surface of a star, will lose energy rapidly, while nuclei can be reduced in charge by photo-disintegrations. Since in the most important cases the effects are not large, we shall here give only a sketch of the theory; details will be found in the references just cited.

If a particle of rest mass m moving through space with total energy $\gamma\,mc^2$ collides with a photon of energy k, the collision can be described by giving the angle ϑ included between the velocity of the particle and the reversed velocity of the photon before collision. Then the Lorentz transformation from the frame

---

[1] CHEN, LEAVITT and SHAPIRO: Phys. Rev. **103**, 220 (1956). COYLE, WENZEL and CANSEY: Phys. Rev. **107**, 859 (1957).

[2] A. BRENNER and R. WILLIAMS: Phys. Rev. **106**, 1020 (1957).

[3] E. FEENBERG and H. PRIMAKOFF: Phys. Rev. **73**, 449 (1948).

[4] T. M. DONAHUE: Phys. Rev. **84**, 972 (1951).

of the galaxy to that in which the cosmic-ray particle is stationary supplies the relation:

$$k' = \gamma k (1 + \beta \cos \vartheta)$$

with the photon energy in the galactic frame written as k and in the cosmic-ray rest frame as k'. As usual $\beta = v/c \approx 1$. It is adequate to replace $\beta \cos \vartheta$ by unity, and to ignore those directions in which $\cos \vartheta \approx -1$. Then for most collisions we have roughly:

$$k' \approx \gamma k.$$

The process most likely to affect the heavy nuclei of the cosmic-ray beam is the photo-emission of single nucleons, mainly of neutrons. This process takes place chiefly in the energy region between 10 and 20 Mev, except for the very lightest nuclei. The typical optical photon will be capable of such a process only when the nucleus has energy such that $\gamma \approx 10^7$. For such a collision, the cross-section is very roughly given by

$$\sigma(\gamma, n \text{ or } p) \sim \tfrac{1}{3} A^{\frac{2}{3}} \text{ millibarns.}$$

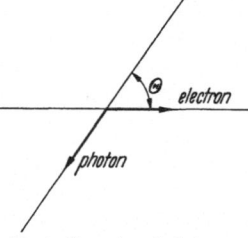

Fig. 4. Geometry of photon-electron collision.

Taking a photon density about $1/cm^3$, which is appropriate in the galactic disc, the mean free path for reducing by one unit the mass (and usually charge by subsequent beta-decay) of a nucleus like iron is around 2×10^7 lightyears. This is rather longer than the distance between collisions with the interstellar gas. In extragalactic space, the photon density is not well-known; in the visible, it is perhaps an order of magnitude smaller than that in the galaxy. This sets an upper limit on the free path against photon collisions which might be smaller than that due to nucleon collisions, since the nucleon density in that portion of space is still less well-known. It seems probable that an iron nucleus of very high energies cannot wander more than a few times 10^8 years in space anywhere without losing nucleons. The effect might extend to much smaller values of γ and hence much less rare particles, if there is any appreciable density of soft x-ray quanta in the extragalactic space.

The photon collisions are relatively more important for the fate of any electrons in the cosmic-ray beam. Here the effect is elastic scattering in the rest frame of the electron, which in the frame of the sun appears like a considerable energy transfer from fast electron to soft quantum. We can divide the events into two classes:

(a) If the photon energy k' in the electron rest frame is small compared to mc^2, then the Thomson process occurs, and the photon after collision also has the energy k'. But in the rest frame of the sun, the photon will appear to have the energy $k'_f = \gamma k'$; the electron has supplied this energy, losing the amount

$$\Delta U \approx \gamma k'_f = \gamma^2 k.$$

The cross-section for this process is roughly the Thomson cross-section σ_T (a small reduction occurs because not all angles of scattering will result in energy loss). We can write the energy loss with time:

$$-\frac{dU}{dt} \approx (\Delta U/\text{collision}) \times c\, \sigma_T \varrho = \frac{8\pi}{3} c\, r_0^2 \gamma^2 k\, \varrho(k)$$

and from this extract the useful expression:

$$-\frac{1}{U}\frac{dU}{dt} = \frac{8\pi}{3} c\, r_0^2 \gamma \frac{k}{mc^2} \varrho, \qquad \gamma \gg \frac{mc^2}{k}, \tag{9.1}$$

giving the lifetime against energy loss by photon collision, where $\varrho(k)$ is the density of photons with energy k in the rest frame of the sun.

(b) The other case occurs when the photon is very energetic before the collision in the rest frame of the electron, so that the Compton process, and not the Thomson scattering, applies. For this $k' \gg mc^2$. Then the energy loss becomes, after averaging over angles of collisions,

$$\Delta U \approx U\left(1 - \frac{1}{\ln \dfrac{2\gamma k}{mc^2}}\right)$$

and the cross-section is given roughly by:

$$\sigma_c \approx \pi r_0^2 \left(\frac{mc^2}{\gamma k} \ln \frac{2\gamma k}{mc^2}\right).$$

For this case, one obtains:

$$-\frac{1}{U}\frac{dU}{dt} = \pi r_0^2 c \varrho(k) \frac{mc^2}{\gamma k}\left(\ln \frac{2\gamma k}{mc^2} - 1\right) \tag{9.2}$$

under the restriction that $\gamma \ll mc^2/k$.

Cases intermediate between (a) and (b) can best be treated by interpolation.

These formulae have been applied by DONAHUE in particular, to study the fate of electrons stored near the earth in the high flux of solar photons. Electrons of $\gamma \approx 10^6$ can survive in daylight only some months. In stellar atmospheres, where the photon density may reach 10^7 or 10^8 times the value for daylight, this process may represent an important energy drain on any mechanisms for accelerating electrons; this is one more, though surely not the most important, reason to expect the weakness of electron sources, especially for high cosmic-ray energies, $\gamma \gtrsim 10^5$. These scattering processes are less probable for protons, by a factor $(m_e/m_p)^2$.

In Ref. [16] the effect of photons on stripping electrons from atoms has been shown to be small in the cases at hand. This is also true for nuclear photo-disintegration. In stellar atmospheres or other regions of unusually high optical flux these effects might become important.

10. Interaction of particles with the magnetic field. *α) Uniform and nearly-uniform magnetic fields.* The most important influence on the motion of the charged particles of the cosmic ray beam through galactic space is the magnetic field. It will be seen (Sect. 11) that electric fields are in general small. The magnetic fields are responsible for the processes of stirring and acceleration: for the latter, only when they vary in *time.* We will here present the elements of the theory of motion of charged particles in uniform, and in nearly uniform, magnetic fields. The treatment will of course be relativistic throughout; any results which apply only in the non-relativistic domain of energies will be specially indicated.

The Lorentz force forms the foundation of the study; it may be written

$$\frac{dp}{dt} = ze\,(E + v \times B) \tag{10.1}$$

where p is the particle momentum (relativistic), satisfying the relations:

$$p = mv; \quad m = \gamma m_0; \quad \gamma = (1 - v^2/c^2)^{-\frac{1}{2}}, \tag{10.1a}$$

where m_0 is the particle rest mass. In (10.1) the MKS system has been used; we shall often quote results, however, in other convenient units.

We here consider the case with $\boldsymbol{E}=0$ and $\boldsymbol{B}$ constant in time and space. Plainly:

$$\frac{dp^2}{dt} = 2\boldsymbol{p} \cdot \frac{d\boldsymbol{p}}{dt} = 0; \quad E = m_0 c^2 \gamma, \quad \text{with} \quad E^2/c^2 = p^2 + m_0^2 c^2,$$

and the energy E is a constant of the motion. The velocity magnitude $|\boldsymbol{v}|$ is constant, and $\boldsymbol{p}$ may be written as the sum of two components:

$$\boldsymbol{p} = \boldsymbol{p}_\| + \boldsymbol{p}_\perp = m(\boldsymbol{v}_\| + \boldsymbol{v}_\perp) \quad \text{and} \quad d\boldsymbol{p}_\|/dt = 0$$

with $\boldsymbol{v}_\|$, $\boldsymbol{v}_\perp$ respectively parallel to and normal to the magnetic field $\boldsymbol{B}$. The motion now takes the familiar helical form, with the particle describing circular orbits centered on a field line, while the plane of these orbits drifts along the field line with constant velocity $\boldsymbol{v}_\|$. This is the familiar helical motion, with the circular frequency in the orbit plane ω_c (the cyclotron frequency), and the radius of the orbits R_L (the Larmor radius), given by:

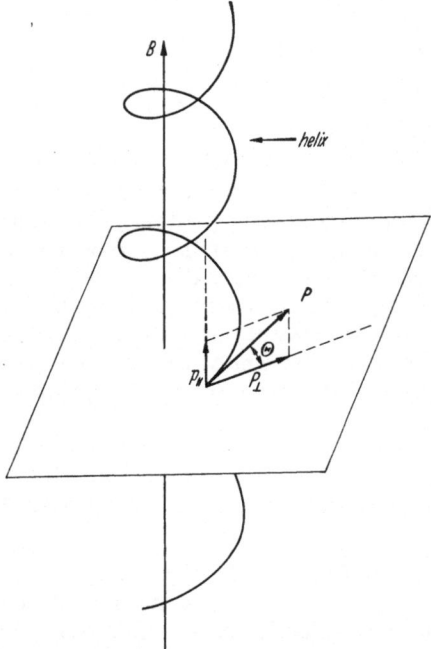

$$\left.\begin{aligned} \omega_c &= 2\pi\nu_c = v_\perp/R_L = z\,e\,B/m, \\ R_L &= m\,v_\perp/|z\,e|\,B. \end{aligned}\right\} \quad (10.2)$$

For positive ze, the rotation is in the direction of a left-handed screw advancing along the positive direction of the field; for negative charge, a right-handed screw. Numerical evaluation of the formulae (10.2) gives:

for electrons:

$$\nu_c = \frac{2.80}{\gamma} B_{\text{gauss}} \quad \text{Megacycles/sec,} \quad (10.2a)$$

for particles of charge ze; mass, A atomic mass units:

$$\nu_c = \frac{1.53\,z}{\gamma\,A} B_{\text{gauss}} \quad \text{kilocycles/sec} \quad (10.2b)$$

and in general:

Fig. 5. Geometry of the orbit helix of a charged particle in a constant magnetic field.

$$R_L = 1/300 \,(v_\perp/c\,|z|)\, E_{\text{ev}}/B_{\text{gauss}} \text{ cm.} \quad (10.2c)$$

In quite different, astronomically appropriate, units, we obtain as well, for $v/c \sim 1$:

$$R_L = 3.52 \times 10^{-6} E_{\text{GeV}}/|z|\, B_{\text{microgauss}} \text{ lightyears} \quad (10.2d)$$

where the units are appended as subscripts.

In Fig. 5 the geometry of the orbit helix is described in the familiar way. The pitch angle (ϑ), formed by the plane normal to the magnetic field and the momentum $\boldsymbol{p}$, determines the momentum components:

$$p_\| = p \sin \vartheta; \quad p_\perp = p \cos \vartheta.$$

The distance moved along the field line in one gyration is the pitch distance of the helix; it is just $2\pi\varrho \sin \vartheta$, while the projected distance around a single gyration circle is $2\pi\varrho \cos \vartheta$, where $\varrho = \dfrac{R_L}{\cos \vartheta} = \dfrac{p}{|z\,e\,B|}$.

This general form of motion is retained even when the magnetic field is no longer strictly uniform. For small departures from uniformity, the orbit changes slowly as the particle moves. If the scale of field variations is such that within the distance ϱ there is little change, the orbit can be quantitatively estimated.

First, we compute transverse accelerations of the particle. Initially, as an aid to calculation, we consider these as produced by an electric field, $\boldsymbol{E}$, which is uniform, and has a component $E_{\parallel}$ parallel to $\boldsymbol{B}$, and one, $E_{\perp}$, normal to $\boldsymbol{B}$. If the motion is observed in another inertial frame, one moving with velocity $\boldsymbol{v}_D$, perpendicular to the plane containing $\boldsymbol{E}$ and $\boldsymbol{B}$, we may write $\boldsymbol{v}_D = \varkappa \boldsymbol{B} \times \boldsymbol{E}$, $\varkappa$ a constant. Now, in this frame, a Lorentz transformation produces an electric field component normal to $\boldsymbol{v}_D$, which is:

$$\boldsymbol{E}'_{\perp} = \gamma(\boldsymbol{E} + \boldsymbol{v}_D \times \boldsymbol{B}_{\perp}).$$

If we choose $\varkappa$ such that $\boldsymbol{E}'_{\perp} = 0$, there will be *no* transverse electric field in the moving frame, and the particle orbit will again be a helix directed along the magnetic line of force. (The pitch will increase if there is a finite $E_{\parallel}$, but this axial motion evidently can be separately treated.) If we write $\boldsymbol{E}'_{\perp} = 0$, we obtain the relation:

$$\boldsymbol{E}_{\perp} \varkappa [(\bar{\boldsymbol{B}} \times \bar{\boldsymbol{E}}) \times \bar{\boldsymbol{B}}]_{\perp} = \varkappa(-B^2 \boldsymbol{E} + (\boldsymbol{B} \cdot \boldsymbol{E}) \boldsymbol{B})_{\perp}$$

so that $\varkappa = -1/B^2$, and $\boldsymbol{v}_D = (\bar{\boldsymbol{E}} \times \bar{\boldsymbol{B}})/B^2$. From such a moving frame, then, the orbit will be seen as a simple helix along the magnetic field line. In the original frame, then, a transverse electric field simply produces a drift of the helical axis, which moves with velocity $\boldsymbol{v}_D$. It is to be noted that $\boldsymbol{v}_D$ is independent of the charge or the mass of the moving particle.

We may generalize this drift to other forces, provided the effect is not so great as to require relativistic treatment of the applied forces. Suppose the transverse force is such to cause the acceleration $\boldsymbol{a}$. Then we can replace the $\boldsymbol{E}$ field by its equivalent:

$$\boldsymbol{E} \to \boldsymbol{F}/z\,e = m\,\boldsymbol{a}/z\,e$$

and then

$$\boldsymbol{v}_D = \boldsymbol{a} \times \boldsymbol{B}/|B|\,\omega_c,$$

now mass and charge dependent. We apply this now to the inertial acceleration produced by a slow plane curvature of the field lines which form the guiding axes of the helical motion. If the particles move parallel to the magnetic field with velocity $v_{\parallel} = v \sin \vartheta$, their orbit helices will drift normal to the plane of the curving field lines with velocity v_D:

$$v_D|_{\mathrm{curv}} = v^2 \sin^2 \vartheta / R\,\omega_c = v\,\frac{\varrho}{R} \sin^2 \vartheta$$

where R is the radius of curvature of the lines of force. A similar, slightly more complicated case arises when the field varies in magnitude, normal to the lines of force, but not in direction. ALFVÉN [1] has developed the first-order theory of this drift, which is also worked out for a special case by ROSSI [10]. The drift velocity here becomes:

$$\boldsymbol{v}_D|_{\mathrm{grad}} = \frac{1}{Z} \frac{v_{\perp}^2}{\omega_c} \frac{V_{\perp} |B|}{|B|}$$

where $V_{\perp} |B|$ is the gradient of the magnitude of $\boldsymbol{B}$ in the direction normal to $\boldsymbol{B}$, and $v_{\perp}$ is the velocity component in the same direction. Combining the two drifts, the particle will move in a slightly non-uniform magnetic field in such a way that it has always a helical motion, with the appropriate local value of

R_L and ω, but with the helical axis drifting away perpendicular to the line of force with the drift velocities given above. If the currents flowing within the medium are unimportant, then

$$\nabla \times \boldsymbol{B} \approx 0$$

and in this case $\boldsymbol{B}$ can be written as a vector field normal to the radius vector $\boldsymbol{r}$ measured from the center of curvature. Then the transverse variation $|\nabla_\perp B/B| \approx 1/R$, and the two drifts are in the same direction, given by $(\boldsymbol{B} \times \nabla B)\, e$. The resultant velocity of drift is:

$$v_D|_{\text{total}} = \frac{v\,\varrho}{R}\left(\frac{1}{z}\cos^2\vartheta + \sin^2\vartheta\right).$$

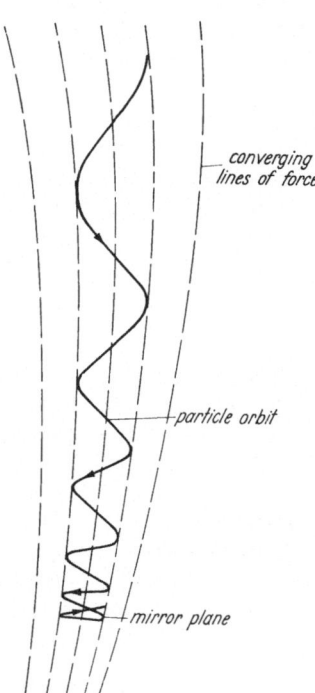

converging lines of force

particle orbit

mirror plane

Fig 6. Magnetic mirror. Reflection of a particle from a region of converging magnetic field.

The last case of motion in a nearly-uniform magnetic field is the most interesting. Here we consider a field which changes slowly in magnitude, but retains about the same direction, as the particle progresses along the line of force. If the magnitude is increasing, the field lines tend to converge; if decreasing, to diverge. The Lorentz force is:

$$\frac{d\boldsymbol{p}}{dt} = z\,e\,\boldsymbol{v}\times\boldsymbol{B} = z\,e\,\frac{\boldsymbol{p}\times\boldsymbol{B}}{m}$$

and we form the expression:

$$\boldsymbol{r}\cdot\frac{d\boldsymbol{p}}{dt} = \frac{z\,e}{m}\,(\boldsymbol{r}\times\boldsymbol{p})\cdot\boldsymbol{B}.$$

Now again we resolve the momentum into parallel and transverse components relative to the field $\boldsymbol{B}$:

$$\boldsymbol{p} = \boldsymbol{p}_\perp + \boldsymbol{p}_\|$$

and $\dfrac{d\boldsymbol{p}_\|}{dt} = 0$, while $\dfrac{d\boldsymbol{p}_\perp}{dt} = \dfrac{p_\perp^2\,\boldsymbol{n}}{m\,R_L}$, since the particle moves with constant velocity along the helical axis while it rotates around it. Here $\boldsymbol{n}$ is a unit vector tangential to the orbit circle, and normal to $\boldsymbol{B}$. We can write, therefore:

$$\boldsymbol{r}\cdot\frac{d\boldsymbol{p}_\perp}{dt} = \frac{z\,e}{m}\,(\boldsymbol{r}\times\boldsymbol{p}_\perp)\cdot\boldsymbol{B} = \frac{p_\perp^2}{m}$$

and from this $p_\perp^2/m = 2\mu B$, where we call μ the orbital magnetic moment of the particle; $\mu = \dfrac{z\,e}{2m}\,\boldsymbol{L}_\|$ where $\boldsymbol{L}_\| = \boldsymbol{r}\times\boldsymbol{p}_\perp$ is the component of angular momentum along the guiding field line. From the familiar adiabatic theorem, we expect the angular momentum (an action variable) to vary only by an amount depending on the change of B not over the whole of the motion but over a single rotation. In the limit of slow variation of the field, we get:

$$p_\perp^2/B = 2m\mu = z\,e\,L_\|, \tag{10.3}$$

an adiabatic invariant. Introducing the familiar pitch angle, ϑ, we can write:

$$p^2 \cos^2\vartheta/B = \text{const.} \tag{10.3a}$$

Noting also that $p^2 =$ const, we obtain the result:

$$\mu = (z\,e\,B\,R_L)^2/2m\,B \propto \pi\,R_L^2\,B = \text{const} \tag{10.4}$$

which states that the flux Φ enclosed by the orbital circles also remains constant through adiabatic changes in field.

Thus the particle spirals along slowly converging or diverging lines of force, expanding its orbital circles in weak fields, and contracting them in strong ones, to maintain a constant enclosed flux. Plainly the pitch angle also changes during such a motion, for we require

$$\cos^2 \vartheta \propto B; \qquad \frac{\cos^2 \vartheta_i}{\cos^2 \vartheta_f} = B_i/B_f$$

as the motion carries the particle from an initial field B_i to some final position with field B_f.

Evidently the orbit cannot enter any region where the magnitude of B exceeds a value $B_{\max}$:

$$B_{\max} = B_i/\cos^2 \vartheta_i.$$

In fact, the particle is reflected from such regions of strong field, and returns along its previous path. From the relations:

$$p_\perp = p \cos \vartheta; \qquad p_\| = p \sin \vartheta; \qquad p^2 = p_\|^2 + p_\perp^2$$

the conservation of energy guaranteeing the constancy of p^2, we can write the result

$$p_\|^2 (f) = p_\|^2 (i) + p_\perp^2 (i) \left(\frac{B_f}{B_i} - 1 \right) \tag{10.5}$$

which holds for arbitrary p, relativistic or not. The only proviso is the constancy of B over the distance R_L.

It is useful to examine the nature of this turning point along a line of force moving into a stronger field. We can use the appropriate cylindrical coordinates, writing z for the axis distance, and r, ϑ for the transverse coordinates. We take $\boldsymbol{B}$ azimuthally symmetrical. Now

$$\boldsymbol{V} \cdot \boldsymbol{B} = 0; \qquad \frac{1}{r} \frac{\partial}{\partial r} (r\,B_r) + \frac{\partial B_z}{\partial z} = 0$$

so that we have the integral

$$B_r = - r/2 \frac{\partial B_z}{\partial z}$$

where the constant of integration can be set equal to zero if the field at remote z values is taken as uniform. Then we can write:

$$\frac{dp_z}{dt} = \frac{e}{m}\, p_\vartheta\,B_r = - \mu\, \frac{\partial B_z}{\partial z}$$

which may be expressed as:

$$d\boldsymbol{p}_\| / dt = - \mu\,V_\|\,B$$

where $V_\|\,B$ is the gradient of the magnitude of the field along the lines of force (compare with v_D). It is instructive to consider the case with restoring force approximately linear, slope fixed by the length D:

$$\partial B_z/\partial z = B_0 z/D^2; \qquad B_r = \frac{1}{2}\,B_0\,(z/D)^2.$$

Then the circular frequency $\omega_\parallel$ of the "oscillation" near the turning point is:

$$\omega_\parallel/\omega_c = R_L/D, \qquad D \gg R_L \tag{10.6}$$

again showing the nature of the slow field variations assumed.

Another simple adiabatic invariant of the motion of a charged particle in a slowly-varying field is known. It is a function of the longitudinal component of the momentum, the component along the line of force which guides the helix, rather than of the transverse momentum. The invariant is defined by the integral:

$$J = \int_1^2 dl\, p_\parallel. \tag{10.7}$$

Here the line integral is carried out along any line of force. The end points of the integral may refer to the same point in space, for the rare case of a strictly periodic motion, or to two distinct turning points, say mirror points, along any line of force. The asymptotic nature of this definition is clear: the motion is, of course, never exactly along any line of force. The obvious resemblance to the action integral is a strong sign of the invariance of J; it has been proved to be a proper adiabatic invariant of the motion[1]. The constancy of J forces the particle to move in a helix of steeper pitch if it reaches a region where the field lines have shorter runs between mirror points. But if the shorter lines are also in a stronger field, as will usually be the case, the near-constancy of the magnetic moment, μ, will require the pitch to be flatter. The interplay of the two means that the orbits are rather more stably fixed to a given region of the field than one might have expected, at least if shorter field lines go with strong fields. The roughly dipolar field of the earth presents a good example of this case. The invariant J can also lead to useful conclusions about slowly time-varying fields, as did the magnetic moment.

We have finally a complete picture of a particle of any energy moving in a complicated static magnetic field, provided only that nowhere do the fields vary in magnitude or direction by an appreciable amount of their own local magnitude in the distance R_L. In this case particles will advance spiralling on the surface of a single tube of force, passing on to other tubes by drifting transversely, with the very slow drift velocities given above, as they encounter either simple bends in the tube of force, or actual gradients in the field magnitude transverse to the local field direction. It will take many revolutions to cross from one tube of force to another. As the orbit enters a region where the lines of force tend to diverge, it will simply become steeper in pitch and larger in radius; as it enters converging regions, it becomes flatter in pitch, and tighter in radius. Strong convergence will turn back the orbit whence it came, except for those entering particles whose original pitch was sufficiently steep.

This general picture may be expected to represent at least roughly the behavior in any static field. Where the changes are not slow in the sense indicated, of course, the quantitative estimates will fail. But something like this picture will be useful unless the field quantities vary strongly within the local Larmor radius. The approach to a dipole, like that of the earth, is such a case for low energies. Here the tendency to be reflected from strong fields, and to spiral in more and more tightly along converging lines of force, as into the earth's poles, is clear. For cosmic-ray particles of even rather high energies, say 10^5 Gev, and fields like those found in large spatial regions, say from a microgauss up to a milligauss,

---

[1] See the discussion by M. ROSENBLITH and C. LONGMIRE, Ann. of Phys. **1**, 210 (1957).

the Larmor radius generally remains small compared to the scale of the phenomenon. This fact lends great usefulness to the adiabatic theory of the foregoing
sections.

β) *Departures from adiabatic invariance.* Since its introduction by ALFVÉN
(see [1]) the adiabatic invariant of the motion of a charged particle in a magnetic
field, which we have written both in terms of the angular momentum L

$$z e L = p^2 \cos^2 \vartheta / B \qquad (10.8)$$

and of the magnetic moment of the orbital motion about the guiding center

$$\mu = \frac{z e}{2 m} L = p^2 \cos^2 \vartheta / 2 m B \qquad (10.9)$$

has been extremely useful. But its invariance is after all only approximate; it
remains constant only in the limit where the spatial gradients, for example,
vanish entirely. It was already shown by HELLWIG[1] that the variations in L
or μ were only of third order at most in the time and space derivatives of the
field. A more detailed study has been made by HERTWECK and SCHLÜTER[2].

These authors could not consider the general case of deviations from uniform
and constant field. Instead, they studied a particular case which seems very
likely to be typical of most possibilities. They treated the motion of a particle
which moved in a helix about a guiding line of force in a uniform field. Then
they allowed the field to increase in magnitude, remaining homogeneous in space
and constant in direction. They examined the behavior of the adiabatic invariant
with varying time rates of increase of the field. The calculation is complex, and
in part numerical. For their general importance, we here review only the results;
the methods can be consulted in the original paper.

The time variation of the field can be characterized by a quantity

$$\tau = B(t) \Big/ \frac{dB}{dt}$$

and this time compared with a time τ_0 characteristic of the field magnitude:

$$\tau_0 = \frac{1}{\omega_0} = m c / e B.$$

We use the subscript i or f to indicate the initial or final values of any quantity.
For slow field changes, satisfying the condition

$$\tau \gtrsim 1.5 \, \tau_0$$

the change in the adiabatic invariant is given by the expression

$$\frac{\Delta L}{L} = e^{-a \tau / \tau_0}$$

where a is a numerical constant not far from unity. (Its exact value will presumably depend somewhat upon the form of the assumed time variation.) For these
relatively slow changes, the adiabatic invariant remains quite constant, the
variation falls exponentially with increasing slowness of the field variation. For

---

[1] G. HELLWIG: Z. Naturforsch. **10**a, 508 (1955).
[2] F. HERTWECK and A. SCHLÜTER: Z. Naturforsch. **12**a, 844 (1957).

somewhat faster rates of change, their result becomes roughly:

$$\frac{\mu_f}{\mu_0} \approx \frac{1}{10} \frac{(\omega_f + \omega_i)^2}{\omega_f \omega_i}$$

for

$$\tfrac{1}{2}\tau_0 \lesssim \tau \lesssim 2\tau_0$$

and in the limit of sudden change, with $\tau \ll \tau_0$, they find

$$\frac{\mu_f}{\mu_i} = \frac{1}{4} \frac{(\omega_f + \omega_i)^2}{\omega_f \omega_i} . \tag{10.10}$$

It is interesting to examine the case of sudden change with completely flat helical motion: i.e., with the motion in a plane circular orbit around the line of force. Then the change in field must mean a change in energy of the particle. From the relation (10.10) we can find the two results:

$$\frac{T_f}{T_i} = \frac{1}{4}\left(\frac{B_f}{B_i} + 1\right)^2 \tag{10.11}$$

when the motion is non-relativistic, $T \ll m_0 c^2$, and

$$\frac{E_f}{E_i} = \left(\frac{B_f}{2B_i}\right)^{\frac{2}{3}} \tag{10.12}$$

when the motion is highly relativistic, with $E \gg m_0 c^2$. It is interesting to collect the results of the adiabatic (betatron) acceleration with these. We see:

	Slow (betatron) acceleration rates	Sudden
Relativistic energies . . .	$E \sim B^{\frac{1}{2}}$	$E \sim B^{\frac{2}{3}}$
Non-relativistic energies .	$T \sim B$	$T \sim B^2$

It appears clear that high relativistic energies cannot well be reached with only modest changes of magnetic field. Repetitive events, or some quite different mechanisms, must be relied upon.

11. Acceleration by magnetic fields. In a static magnetic field, no work is done on a charged particle. The energy is thus a constant of the motion; this property of static fields is independent of the degree of spatial non-uniformity. All velocity changes are *transverse* to the velocity, and the energy E and total mass γm_0 are constants of the motion. Unless electric fields are present, no energy can be gained or lost.

This property of course fails once the magnetic field varies in time. Then there are electric fields, arising from induction, and the particles can gain or lose energy. It is primarily such mechanisms which have been suggested as means to produce cosmic ray energies from injected particles in various astronomical settings. No doubt many such mechanisms are possible, and indeed many may occur. We here discuss a few typical, simple, and effective ones; it is still unsure which are the true sources of cosmic ray energy. To establish that, it will probably be necessary to have a sufficiently deep knowledge both of the theory of such magnetic plasma and of the actual conditions present in each suspected region. We can only propose mechanisms at present which are likely to work under the general conditions of the objects under study, which do not demand either improbable field motions and structures, or precisely-adjusted values of the field parameters.

α) *Betatron mechanisms.* This class of mechanisms arises from the postulate that at a given point in space the local value of B changes with time. The original suggestions of SWANN[1] and the detailed examination by RIDDIFORD and BUTLER[2] may be consulted for detailed dynamics. A satisfactory estimate of the energy gain possible follows directly from the adiabatic invariance of $p_\perp^2/B$. If now $\dot{B}$ changes in time slowly, so that $\dot{B}/B \ll \omega_c$ (a very plausible condition for our situations, cf. the frequencies of (10.2), we know that the pitch must change, since the electric field is transverse, and $p_{\shortparallel}$ is constant. Then

$$p_{\shortparallel} = p \sin \vartheta = \text{const},$$

$$p_\perp^2/B = p^2 \cos^2 \vartheta/B = \text{const}$$

so that

$$B \tan^2 \vartheta = \text{const}$$

$$\frac{p_\perp^2}{B} = \frac{p^2 - (p \sin \vartheta)^2}{B} = \frac{p_0^2 \cos^2 \vartheta_0}{B_0},$$

$$\frac{p^2}{p_0^2} = \frac{B}{B_0} \cos^2 \vartheta_0 + \sin^2 \vartheta_0,$$

and for any important increase in energy, $B(t)/B_0 \gg 1$. The total momentum then varies with time according to the relation:

$$\frac{p^2(t)}{p_0^2} \approx \frac{B(t)}{B_0}$$

so that for non-relativistic kinetic energies T:

$$\frac{T(t)}{T_0} = \frac{B(t)}{B_0} \quad \text{N.R.} \tag{11.1a}$$

and for relativistic cases:

$$\frac{E(t)}{E_0} = \left(\frac{B(t)}{B_0} \right)^{\frac{1}{2}} \quad \text{R.} \tag{11.1b}$$

Such a mechanism is not very likely to take particles through a wide range of relativistic energies; it is not easy to imagine a state of affairs in which the initial field can be multiplied by a factor of 10^{10} or so. As an important mechanism in the early stages, it cannot be excluded. It is to be noted that particles will escape the "betatron" as soon as their Larmor radius exceeds the dimensions of the region within which the field is coherently changing with time. It is also to be observed that the energy stored in the particles is taken from the currents which are causing the field increase. Since all the ionized particles in a region can take part, it is clear that this is usually a very severe load on the producing currents. It seems probable that if such field changes do occur at a point in space, they act to sweep the space clear of the ions rather quickly; not many particles can remain to receive high energies. The large magnetic field changes in sun-spots are so slow that it appears unlikely that they can account for any of the cosmic rays seen to arise in the solar flares. Some more or less steady production of particles from sun-spots in this way is not at all unlikely; but it is at present unverified. The energy drain by collision tends to become important in the solar layers where these large field changes occur, and nuclear collisions as well may arise. It is very likely that such processes do produce fast particles of energies well below the cosmic ray energies, but possibly in the region of Mev, where they may induce

[1] W. F. G. SWANN: Phys. Rev. **43**, 217 (1933).
[2] L. RIDDIFORD and S. T. BUTLER: Phil. Mag. **43**, 447 (1952).

nuclear reactions of some importance in the formation of rare constituents of stellar atmospheres. Apart from sun spots (or star spots, often much larger), there seem to be few places where large $B(t)/B_0$ ratios may be found. Smaller but repetitive events of this kind may arise in hydromagnetic wave patterns [20][1].

β) *Moving magnetic mirrors.* One of the most general of mechanisms has been suggested by Fermi [11]. This operates by use of the longitudinal magnetic forces which serve to turn the particles back from a region of converging lines of force. Such a region may be thought of as a reflecting wall or mirror. Another form of reflection (perhaps less like a mirror than like a reentrant light pipe) arises when a line of force bends around until it returns parallel to itself. In either case, a particle will come back from the field, which so far is static, with reversed momentum in the direction of its motion, and unchanged energy (see Fig. 6). But now imagine the field pattern to move as a whole with respect to the inertial frame of observation. Now the particle sees a varying magnetic field, hence an electric field, and energy is transferred between field pattern and particle. The energy change may be computed without detailed knowledge of the field pattern. It is necessary only to assume that the pattern moves as a whole, and that the spiralling direction is completely reversed. Either of the mechanisms above will do.

Consider a moving frame of reference, in which the field pattern is stationary. In such a frame the collision of particle and moving "mirror" transfers no energy, and simply reverses the momentum of the particle. If we first transform by special relativity from the frame in which the pattern is moving to that in which it is at rest, and then back again after the elastic reversal of momentum, we can compute the energy change observed in the reference frame of observation. Let us take the mirror to have the velocity V along the direction of approach in the observing reference frame; the particle, velocity $+v \cos \alpha$ along V, initial energy E_i and final energy E_f. The pitch angle of the particle is $\Theta = \dfrac{\pi}{2} - \alpha$ as usual. Then in the (primed) frame moving *with* the mirror:

$$p'_i = - p'_f; \quad E'_i = E'_f$$

and we may use the transformations:

$$p = \gamma \left(p' - V \cos \vartheta \, E/c^2 \right); \quad \gamma = \frac{1}{(1 - v^2/c^2)^{\frac{1}{2}}}$$

and

$$E = \gamma \left(E' - V \cos \vartheta \, p' \right).$$

The result of performing the two transformations is:

$$E_f = E_i \gamma^2 \left(1 + \frac{2 V v}{c^2} \cos \alpha + \frac{V^2}{c^2} \right),$$

and it is seen that maximum energy is gained when the collision is head-on, so that $\cos \alpha = +1$ and maximum energy lost by the particle when the collision is overtaking, with $\cos \alpha = -1$.

Such a moving magnetic mirror can thus accelerate or decelerate particles. The energy is transferred between the kinetic energy of the material which bears the magnetic field pattern and the particle; in the present context, the energy change of the field pattern is small. With the fastest-moving gas masses seen, having velocities perhaps a hundredth of that of light, a particle may gain a percent or two in energy in a single favorable collision. This transfer of energy

---

[1] L. Davis: Phys. Rev. **96**, 743 (1954).

to many fast particles will in the end slow down the motion of the field patterns. Of course, the assumption that they move as rigid is reasonable only so long as the total energy content of the cosmic ray particles present is small compared to that of the matter; in some of the contexts considered, this approximation is no longer valid. Then a much more complicated energy exchange between the field patterns maintained by currents flowing in the gas plasma, and the cosmic rays, must be considered.

Two simple types of repeated collisions between particles and moving "mirrors" may be considered. They form limiting cases. The first of these is a so-called trapping mechanism. Here a particle moves between two "mirrors" which may be approaching or receding from each other, say with uniform velocity. The particle advances, is reflected from one mirror, and recedes to strike the other, from which it is in turn reflected back to the first. This continues until something stops the process. Receding mirrors will remove energy from the particle, for all collisions are of the overtaking type; approaching mirrors raise the particle energy until it becomes so high that the "mirror" no longer reflects it. As the particle gains energy, $p_{\parallel}/p_{\perp}$ increases and the pitch increases. But a mirror of the converging force-line sort cannot reflect particles beyond a maximum pitch angle given by (10.3), and one of the reentrant line-of-force sort cannot operate when the Larmor radius is large compared to the radius of curvature of the line of force. Within these limits, which depend upon the strength and shape of the field pattern, the energy transfer rate may be estimated. Consider the mirrors, or trap jaws, approaching with relative velocity V. We treat the case $V/c \ll 1$, $v_{\text{particle}}/c \approx 1$. Taking one jaw as stationary, the energy change ΔE upon reflection from the moving jaw, assuming the simplest parallel collisions with $\cos \alpha = 1$, is just

$$\frac{E_f - E_i}{E_i} = \frac{2V}{c} = \frac{\Delta E}{E}.$$

The number of such reflections per second is $c/2L$, where $L(t)$ is the changing distance between the jaws. Then the energy transfer rate is

$$\frac{d \ln E}{dt} = \frac{c}{2L} \cdot \frac{2V}{c} = -\frac{d \ln L}{dt}, \tag{11.2}$$

and this rate will continue until the particles escape the trap, or the jaws come so close that the field patterns superimpose, or, in the receding case, the particle loses all longitudinal velocity. In the relativistic case, then, the energy of the particles is inversely proportional to the distance between the trap jaws:

$$E \propto \frac{1}{L}$$

and a similar estimate can be made for the non-relativistic case, giving:

$$T \propto \frac{1}{L^2}$$

where T is the kinetic energy.

It is interesting to compare these results with the adiabatic compression of a perfect gas [2]. In the case just computed, the density ϱ varies like $1/L$, for the compression was one-dimensional only. If the relative velocities were randomized during the process, we would expect that ϱ would instead satisfy $\varrho \propto 1/L^3$. This leads exactly to the result of the adiabatic compression of a perfect gas. In general, for such a process

$$T \propto \varrho^{\gamma - 1}$$

with the ratio γ of the specific heats for a monoatomic perfect gas

$$\gamma = 1 + \varrho V / \langle T \rangle = \begin{cases} 1 + \frac{2}{3} & \text{non-relativistic} \\ 1 + \frac{1}{3} & \text{relativistic} \end{cases}$$

using the statistical result. The longitudinal adiabatic invariant can be made to yield the same result.

We can expect acceleration or deceleration of cosmic ray particles, then, in the uniform adiabatic compression or expansion of a chaotically-magnetized plasma. Such a field pattern will have many regions where opposing mirrors exist, and the overall effects of compression or expansion will be as estimated above, i.e.

$$\left.\begin{array}{l} E \propto \varrho^{\frac{1}{3}} \quad \text{relativistic,} \\ T \propto \varrho^{\frac{2}{3}} \quad \text{non-relativistic.} \end{array}\right\} \tag{11.3}$$

The limit to the degree of acceleration may be set by the escape from the trap discussed above, or by other deviations from the idealized considerations here set down.

A related result can be stated for the mean magnetic field within such a turbulent magnetized plasma. In the approximation in which the field lines are "frozen" into the matter, the flux is preserved in any circuit. Thus $BL^2 = \text{const}$, where L is a typical dimension, and if the isotropy of the turbulent fluid is conserved, we have that $\varrho \propto 1/L^3$, where ϱ is the plasma density. Therefore the magnetic energy density, or magnetic pressure, will vary as

$$\langle B^2 \rangle \propto U_{\text{mag}} \propto \varrho^{\frac{2}{3}}$$

exactly like the non-relativistic case of (11.3). This tendency for the magnetic energy to follow the particle energy is of course a special case of the general theorem of equipartition, which surely holds in the limit of thermal equilibrium: i.e. if transfer of energy is possible among all the motional degrees of freedom of field, plasma currents, and cosmic-ray particles. How far such transfers are inhibited by the dynamical conditions of the system in particular configurations, and over the times available of course can never be generally stated, but must be computed for each distinct case. It is possible to regard the whole acceleration problem as the problem of the rate of approach to energy equilibrium between the particles of the cosmic-ray beam and the magnetized gas with which they interact. As in the ordinary theory of turbulence, there is no reason to anticipate that equilibrium will in fact be attained. As long as turbulence itself persists, it is sure that actual thermal equilibrium has not been reached. But such equilibrium considerations often provide useful orienting estimates; they must be used only tentatively.

12. Hydromagnetic waves. The description of the fields carried by plasma which we have so far used is of course wildly naive; neither isolated rigid clouds of field, nor uniformly-approaching jaws of field can be thought of as anything but idealizations which might be approximated in some very special cases. Rather, the mutually-dependent fluid motion and field equations must be solved in a more general way to represent any real state of affairs. This is a program plainly more difficult than the description of a real compressible but non-conducting fluid; patently, we cannot even begin to sketch the general features of this still far from developed program here. It is perhaps enough to point out that many types of motion exist in which the fluid flow and the magnetic field are strongly coupled. Most—but by no means all—of such motions can be regarded as the superposition of hydromagnetic waves of various types, wave motions in which fluid and field have related periodic changes in time. The most familiar waves of this sort are the Alfvén waves [1], [5] in which the lines of force act as loaded strings, transmitting transverse waves which deflect the force lines and their attached fluid, with a velocity

$$v^2 = \frac{B^2}{4\pi\varrho}.$$
                    (12.1)

These waves are discussed in detail in Ref. [1] and [20]. They propagate quite linearly in the rather unrealistic limit of an incompressible gas; in more realistic approximations, they show the dispersion typical for all waves of finite amplitude, and can develop something like the steep shock fronts familiar for strong sound waves. Longitudinal modes of wave motion also exist, with a still stronger tendency to develop steep fronts. The minimum thickness of such fronts may be estimated as about the Larmor radius of the plasma protons, which is in most astrophysical contexts a good deal smaller than the collision mean free path.

Now if the cosmic-ray gas is immersed in a medium bearing a complex set of hydromagnetic waves, the fields it encounters are evidently varying in time. This will give rise to various betatron-like accelerations by induction, as well as the accelerations produced by the Fermi mechanism, in which the particles are turned around by an oncoming wave front. DAVIS[1] and FAN[2] have examined the coupling between betatron and Fermi-like collisions, and have argued that in some configurations at least there will be a strong tendency for acceleration by catching particles between approaching hydromagnetic waves, a version of FERMI's jaws. W. B. THOMPSON[3] and J. H. PIDDINGTON[4] have described quite similar situations. In all of these discussions, it seems fair to say, the key question is how to prevent those wave-particle collisions which accelerate from being cancelled by those which decelerate. This is exactly the problem Fermi saw so presciently when he described the two mechanisms, the rigid clouds, which give no acceleration in the first order, $\frac{\Delta E}{E} \propto \left(\frac{v}{c}\right)^2$, and the jaws, which give $\frac{\Delta E}{E}$ of the first order, proportional to (v/c), but only if regions of receding jaws are overlooked.

There are three approaches to using the repetitive collisions for a rapid energy gain, preferably one with $\frac{\Delta E}{E} \propto \frac{v}{c}$:

1. Taking account of the growth of the field as waves approach each other (DAVIS, FAN). Then deceleration is less effective than acceleration.

2. Maintaining isotropic collisions to the highest energies (PIDDINGTON, THOMPSON, FERMI). This means that the adiabatic invariant $B/\sin^2\vartheta$ must be broken. Unless it is so broken, the particles will be accelerated only until they are moving in spirals of such steep pitch that they penetrate freely through the advancing wave. On the other side they will see it as receding, and lose all the energy have gained. The usual scheme for such isotropization is to invoke the cresting up of the waves of finite amplitude into shock fronts, or their analogues, in which the field gradients are no longer small, and the particles cannot follow

[1] L. DAVIS: Phys. Rev. 101, 351 (1954).
[2] C. T. FAN: Phys. Rev. 101, 315 (1954).
[3] W. B. THOMPSON: Proc. Roy. Soc. Lond., Ser. A 233, 402 (1955).
[4] J. H. PIDDINGTON: Austral. J. Phys. 10, 515, 530 (1957).

the new direction of the field lines. Then the spirals may be relatively flat again, and net acceleration takes place.

3. Parker [20] has described a modification of this process in which the approaching waves, crested to maintain isotropic scattering, produce a net acceleration in the first order simply because the particles spend more time between approaching waves than between receding ones. This crowding of particles between approaching waves results from tendency to spend a longer time in the flatter spirals before penetration of the wave front than in the spirals after penetration, even assuming the re-distribution of pitches by the sharp field gradients is isotropic in pitch angle.

It is no criticism of all these authors to say that the matter is still left rather unclear. They have produced models which somewhat mitigate the special nature of the jaws of Fermi, but which are not close enough to any real situation to improve the details. They do confirm strongly the idea of the transfer of energy from a varying and complex hydromagnetic wave pattern to the cosmic-ray gas, and the possible high importance of local kinks in the field lines sharp enough to break down the adiabatic invariance.

It seems physically rather more attractive to invoke special geometry, and not to depend upon the generality of the phenomenon. We see a few systems—the solar flare, the supernova—where an efficient acceleration does take place. In these cases, there is present not an isotropic expansion (which would cool the cosmic-ray gas), nor general criss-crossing of waves, nor a guided jaw motion like that of Fermi, but a kind of radiation of a few or of many strong hydromagnetic pulses, all travelling outward from a single center. Around that center, there is generally present a more or less complex field pattern with lines generally transverse to the motion of the hydromagnetic pulse. It is tempting to look at this asymmetric situation as the natural extension of Fermi's approaching jaws. The particles are repeatedly compressed between the strong, damped hydromagnetic pulses and the stationary pattern surrounding the whole active center. One might estimate the energy gain per pulse to be given roughly by the simple Fermi relation for scattering by a cloud:

$$\frac{\Delta E}{E} \propto \frac{v}{c}.$$

Of course, eventually the particles will diffuse out of the whole structure. Related ideas have been described by Davis [17a] and Woltjer [22].

It is important to observe that even if repetitive processes lead to zero mean net gain in energy, but merely to fluctuations in energy, the particles present will be drawn out into a long energy spectrum by mere random walk in the energy space, without any one-sided driving force. This is considered fully in Sect. 16.

Finally, we must mention the other side of the coin. If cosmic rays gain energy from the gas, the gas loses energy to the cosmic rays. It seems pretty sure that the transfer is rapid under the relevant conditions. This suggests that the near-coincidence between cosmic ray energy densities and those of galactic gas motions is not at all special. Wherever these processes can go on for a suitable length of time, (i) because collision losses are small, (ii) because of the size or the shape of the enclosing field patterns, and (iii) because there is a source of injection there will be a tendency to build up cosmic-ray energy until it equals roughly the energy of the ordinary supersonic gas motion. At that point, further cosmic rays would either escape the region at its weak points, or act as a heavy load on the gas motions themselves. Parker has very persuasively argued in this same way in [20], his Sect. 9.

13. Radiation by charged particles in a magnetic field. Although the Lorentz force alone does no work on a charged particle moving in a time-constant magnetic field, the self-force does do work. The result is a loss of energy: it leaves the particle entirely, radiated away into space. This radiation is referred to as *synchrotron* radiation, from the fact that it can be observed coming from the electron beam in such machines, or as *magnetic bremsstrahlung* (to distinguish it from the familiar case of radiation caused by accelerations of charges in the Coulomb field in particle collisions).

It was I. POMERANCHUK[1] who first called attention to the significance of this mechanism for cosmic-ray electrons approaching the earth through the terrestrial magnetic field. They must radiate, and hence reach the atmosphere with energy reduced from the asymptotic value. Indeed, one may consider only the radiation reaction as acting to fix the particle's orbit, and obtain in this way lower limit for the energy radiated. The calculation is carried through, for example, in the book of LANDAU and LIFSCHITZ [23]. It turns out that in this way a limiting energy can be set such that no charged particle can arrive to an observer within a magnetic field bearing energy higher than the limit. If we call this limiting energy E_B, and the initial and final energies for the particle traversing the field E_i and E_f respectively, one obtains the relation

$$\frac{1}{E_f} = \frac{1}{E_i} + \frac{1}{E_B}.$$

The observed energy must be less than E_f, for only a lower limit to the energy loss has been evaluated.

The limiting energy E_B can be computed in terms of the transverse magnetic field encountered by the particle on its path of length L. This relation is:

$$E_B = \frac{3}{2} \frac{(m_0 c^2)^4}{(z e)^4 \langle B^2 \rangle L}. \tag{13.1}$$

No particle can cross a magnetic field and carry a higher energy than E_B. This sets indeed a high limit to the energy of cosmic rays, but it is conceptually important to have *some* limit to their energy. We tabulate the limiting energies for protons and electrons, assuming appropriate values for $\langle B^2 \rangle$ and L.

	$\langle B^2 \rangle$ (gauss)^2	L (cm)
For entry into the earth's field .	0.1	10^9
For entry into the solar system .	10^{-8}	10^{14}
For traversing the galaxy . . .	10^{-11}	10^{23}

and the limiting energies become:

	Earth	Solar system	Galaxy
Electron	10^9 Gev	10^{11} Gev	10^3 Gev
Proton	10^{22} Gev	10^{24} Gev	10^{15} Gev

For galactic electrons, the limits are not outside of present observations. The absence of high-energy electrons is very likely due to just this process; a more detailed account appears below. For protons, the limit is not in sight; probably no processes actually originate nucleons for which this process is dominant.

[1] I. POMERANCHUK: J. Phys. USSR. **2**, 65 (1940).

The equation of motion for relativistic particles, including the radiation reaction, has been examined with some care by Wentzel[1] who shows that the radiation reaction does not affect the equation of motion in any important way for energies below the limitting energy E_B. Of course, the slow decline of the total energy E does influence the orbit, but the approximations of Sect. 11 are still valid. In particular the magnetic moment, $\mu = \dfrac{m_0 V_\perp^2}{2B}$ remains a constant of the motion, and the guiding centers drift just as without the radiation effects. The orbit radius slowly shrinks down onto the line of force in a uniform field as the energy declines, while the velocity components $v_\perp$ and $v_\parallel$ remain nearly constant in magnitude. The mass is what is mainly radiated away, without much change in the velocities of the motion.

The details of this radiation in a uniform magnetic field have been known for a long time (Schott[2]). Detailed studies have been carried out by Schwinger[3] and by Pomeranchuk and co-workers [22], with numerical computations given by Ginzburg [14a] and Oort [21] in forms closely adapted to our needs, and by Vladimirsky[4]. The Soviet workers have made a very thorough study[4] of the limits of the approximations used, showing that there are only minor effects of quantum theory, spin, and so on, within the energy region of interest[5]. In particular, if the particle energy E remains below the value given by

$$E < \left(\frac{R}{\lambda_c}\right)^{\frac{1}{3}} m_0 c^2$$

where R is the Larmor radius and λ_c the Compton wavelength, and the emitted radiation studied satisfies the condition $\dfrac{\hbar \omega}{E} \ll 1$, the classical theory is fully satisfactory. Tomboulian and Hartman[6] have experimentally verified the classical theory for the spectral distribution of the radiation, and Corson[7] that for the total power radiated.

Some insight can be gained into the process by which a relativistic particle radiates as it moves in its circular orbit in a uniform magnetic field. It is plausible that the radiation intensity has a polar diagram concentrated about the forward (tangential) direction. For in the frame co-moving with the electron, one would expect dipole radiation, and the transfer to the laboratory frame will strongly concentrate the radiation in direction, into a forward cone, with opening half-angle $\bar{\vartheta} = 1/\gamma$, where $\gamma\, m_0 c^2$ is the particle energy. Into this cone the particle will send what is actually a superposition of harmonics of the cyclotron frequency $\omega_0 = c/R$. But these frequencies will be strongly shifted upward by the approach of the electron to the observer. The electron will emit radiation during a short time only, as its forward cone sweeps across the observer. This time may be estimated as

$$\frac{2R}{c}\, \bar{\vartheta} = \frac{2R}{c} \cdot \frac{1}{\gamma}$$

the time required for the direction of motion to move through $\bar{\vartheta}$. But the time interval for the reception of that pulse is very much shorter. The retardation

[1] D. G. Wentzel: Astrophys. Journ. **126**, 559 (1957).

[2] G. Schott: Electromagnetic Radiation, Chap. VII. Cambridge: Cambridge Univ. Press 1912.

[3] J. Schwinger: Phys. Rev. **75**, 1912 (1949).

[4] V. Vladimirsky: Ž. eksp. teoret. Fiz. **18**, 393 (1948).

[5] A. Sokolov and A. Matveev: Ž. eksp. teoret. Fiz. **30**, 126 (1956).

[6] D. Tomboulian and P. Hartman: Phys. Rev. **102**, 1423 (1956).

[7] D. R. Corson: Phys. Rev. **90**, 748 (1953).

connects the time of emission and the time of reception by the relation:

$$t_{\text{rec}} = t_{\text{em}} + \frac{r}{c}$$

where r is the vector extending from source to detector. Then, with $v = -\dfrac{dr}{dt}$,

$$\Delta t_{\text{rec}} = \left(1 - \frac{r}{r} \cdot \frac{v}{c}\right) \Delta t_{\text{em}}$$
$$= (1 - \beta \cos \vartheta) \Delta t_{\text{em}}.$$

Thus the pulse is detectable during the time interval

$$\Delta t_{\text{rec}} \approx \left[1 - \beta\left(1 - \frac{\bar{\vartheta}^2}{2} + \cdots\right)\right] \Delta t_{\text{em}} \sim \frac{R}{c} \frac{1}{\gamma^3}$$

so that the Fourier analysis of the pulse will contain frequencies up to a maximum which is of the order of

$$\omega_m \sim \frac{1}{\Delta t_{\text{rec}}} \sim \gamma^3 c/R = \gamma^3 \omega_0.$$

If the motion is strictly periodic, all Fourier components are harmonics of ω_0,

$$\omega_0 = \frac{c}{R} = \frac{z l B_\perp}{m_0 c \gamma}.$$

The full calculation confirms this picture: the emitted radiation contains a wide spectrum from low frequencies up to a maximum in the neighborhood of ω_m, and then rapidly falls.

It is evident that the varying current which radiates the energy flows strictly in the plane of the orbital motion. This then must contain the electric vector of the emitted radiation, which is thus completely plane-polarized, with the electric vector transverse to the magnetic lines of force.

For a particle of charge e and rest mass m_0, the total power radiated in a uniform magnetic field of strength B, with the orbital plane inclined to the force line so that $B_\perp = B \sin \vartheta$, where ϑ is the angle between the orbital plane and the force line, is given by

$$P_{\text{tot}} = -\frac{dE}{dt} = \frac{2e^4}{3 m_0^2 c^3} B_\perp^2 \left(\frac{E}{m_0 c^2}\right)^2. \tag{13.2}$$

In a useful set of units, this may be written *for electrons:*

$$-\frac{dE_{\text{Gev}}}{dt} = 3.79 \times 10^{-6} E_{\text{GeV}}^2 B_\perp^2 \text{ sec}^{-1}, \quad \text{with } B_\perp \text{ in gauss}, \tag{13.3a}$$

and the fractional loss rate is given by the relation:

$$-\frac{1}{E_{\text{Gev}}} \frac{dE_{\text{Gev}}}{dt} = b E_{\text{Gev}} B_\perp^2, \quad b = 3.79 \times 10^{-6} \tag{13.3b}$$

also for electrons. This differential relation (10.3) can be integrated to give:

$$\frac{E_{\text{Gev}}}{E_0} = \frac{1}{b B_\perp^2} \cdot \frac{1}{(E_0 t + 1/b B_\perp^2)} = \frac{1}{(1 + t/T_{\frac{1}{2}})}$$

where E_0 is the initial energy in Gev at time $t = 0$, and the time $T_{\frac{1}{2}}$, by which time the energy has dropped to half E_0, is given by

$$\text{(electron)} \quad T_{\frac{1}{2}} = \frac{2.64 \; 10^5}{B_\perp^2 E_0} \text{ sec}, \quad E_0 \text{ in Gev.} \tag{13.4a}$$

Expressed in these terms, the lifetime against radiation energy loss varies with the inverse fourth power of the charge of the particle, and directly with the cube of the rest mass. For protons, the lifetime becomes

$$\text{(proton)} \quad T_{\frac{1}{2}} = \frac{5.2 \times 10^7}{B_\perp^2 E_0} \text{ years}, \quad E_0 \text{ in Gev.} \quad (13.4\text{b})$$

These loss rates are simply specific examples of the process estimated in Eq. (13.2) above; an electron stays within the earth's field for something like a tenth of a second at tenth-gauss mean field, very roughly, even coming straight in. Except for orbits along the lines of force into the poles, then, the transverse field will cause a radiation loss adequate to halve the energy in transit when the electron has a value of E_0 about 10^9 Gev, so that $T_{\frac{1}{2}}$ equals about 0.1 sec. This is the same order of magnitude given above. Heavy particle energies are unaffected in such fields at all observed energies.

Perhaps the most unexpected and one of the most important results of recent times in this whole problem is the direct observation of the radiation so emitted by relativistic particles in various astronomical objects (not in the field of the earth). These results can be understood on the basis of the known spectral distribution of the power radiated in synchrotron radiation. We may write

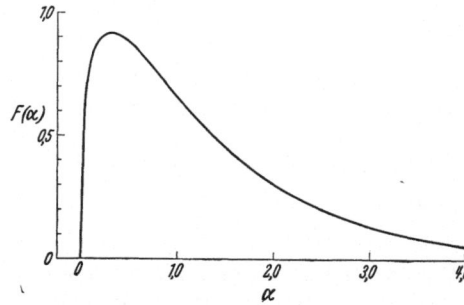

Fig. 7 a. Radiation of an electron in a magnetic field. The radiation function $F(\alpha) = \alpha \int_\alpha^\infty K_{\frac{5}{3}}(\eta)\, d\eta$ of an electron as a function of $\alpha = \nu/\nu_c$. After J. H. Oort and Th. Walraven: Bull. astronom Inst. Netherl. 12, 285 (1956).

$$-\frac{dE}{dt} = P(\nu)\, d\nu,$$

with ν the frequency in cycles/sec, and the spectral distribution becomes [14a], [21]:

$$P(\nu, E/m_0 c^2) = \sqrt{3} \frac{(z|e|)^3}{m_0 c^2} B_\perp F\left(\frac{\nu}{\nu_c}\right) = 2.34 \times 10^{-22} B_\perp F \quad \text{for electrons} \quad (13.5)$$

where we have used the relation for the Larmor radius

$$R = m_0 c^2 \gamma / z e B_\perp$$

and the usual notation for charge e, rest mass m_0. The function F is defined by:

$$F\left(\frac{\nu}{\nu_c}\right) = \frac{\nu}{\nu_c} \cdot \int_{\nu/\nu_c} K_{\frac{5}{3}}(y)\, dy \quad (13.6)$$

where $K_{\frac{5}{3}}(y)$ is the usual Bessel function of the second kind with imaginary argument. This integral is plotted in Fig. 7a. (The literature contains various slightly discrepant numerical values; ours follow Oort [21]. The parameter ν_c is

$$\nu_c = \frac{3 z e}{4 \pi m_0 c} B_\perp \gamma^2 = \frac{3 c}{4 \pi R} \gamma^3.$$

The flat maximum of $F(\nu/\nu_c)$ occurs at $\nu_{\max}$, where the function

$$F\left(\frac{\nu_{\max}}{\nu_c}\right) = 0.92 \quad \text{with} \quad \nu_{\max} = 0.30\, \nu_c.$$

It is handy to compare the frequency ν_{max} (in cycles/sec) with the circular frequency, called the cyclotron frequency,

$$\omega_0 = \frac{c}{R} = e\,B_\perp/m_0\,c\,\gamma$$

and therefore $\nu_{max} = 0.72\,\gamma^3\omega_0$. From Sect. 10$\alpha$ we can take the numerical relations which yield:

electrons: $\nu_{max} = 1.26\ B_\perp\,\gamma^2\,\mathrm{Mc/sec}$ ⎫
protons: $\nu_{max} = 7.1\ \ B_\perp\,\gamma^2\,\mathrm{kc/sec}$ ⎬ $B_\perp$ in gauss.
⎭

The function $F(\nu/\nu_c)$ shows a rather asymmetrical peak. It falls to one-half its maximum value at $\nu = 0.01\,\nu_c$ on the low-frequency side of the maximum, and at $\nu = 1.47\,\nu_c$ on the high-frequency side. Good approximations to the behavior of F are given by

$$F\left(\frac{\nu}{\nu_c}\right) = 2.15\left(\frac{\nu}{\nu_c}\right)^{\frac{1}{3}} \tag{13.7a}$$

for $\nu/\nu_c \ll 1$, which is within a few percent for frequencies up to $\nu = 0.01\,\nu_c$, and by

$$F\left(\frac{\nu}{\nu_c}\right) = 1.26\left(\frac{\nu}{\nu_c}\right)^{\frac{1}{2}} e^{-\nu/\nu_c} \tag{13.7b}$$

good to the same limit for frequencies beyond $\nu = 10\,\nu_c$.

It is useful to give an equivalent band width $\overline{\Delta\nu}$, defined by

$$P(\nu_{max})\,\overline{\Delta\nu} = P_{tot} = -\frac{dE}{dt} \tag{13.8}$$

with the value of

$$\overline{\Delta\nu} = 1.76\,\nu_c.$$

Thus the synchrotron radiation spectrum is roughly like a uniform intensity sent in all frequencies ranging from $\sim 0.1\,\nu_c$ to $\sim 1.7\nu_c$, with $-\dfrac{dE}{dt}$ given by (13.3), but with a tail extending to lower frequencies, and weakly up to higher ones.

The correctness of the intensity formulae above depends upon the absence of coherent effects. If the radiating particles are too close in space—essentially within a wavelength of the radiation in question—important modifications could occur. Electrons immersed in a plasma may also radiate differently because of plasma interactions producing absorption and refraction. GINZBURG [14a] has estimated these effects, and concludes that they remain unimportant, provided the plasma is not too dense.

The radiation formulae are also inaccurate whenever the radiating particles move in their helix with a velocity nearly parallel to the direction of the line of force. The angle α between the velocity of the particle and the line of force must exceed the opening angle $1/\gamma$ of the cone of synchrotron radiation. Otherwise, the emitted radiation is much reduced, but it may still be of interest in the context of solar radio emission.

It is often necessary to integrate the spectral emission over a distribution of particle energies and directions. Let us assume that the particle density spectrum is given by the power law

$$N(>E) = \int_E^\infty \varrho(E)\,dE, \qquad \varrho(E)\,dE = \varkappa\left(\frac{E_0}{E}\right)^n dE \tag{13.9}$$

3*

where $\varrho(E)$ is the number of particles in the energy range dE per unit volume. Take the simplest case, in which the magnetic field directions are distributed isotropically with respect to the observer's line of sight, and the inclination angle ϑ is random also. Then the effective transverse magnetic field will be:

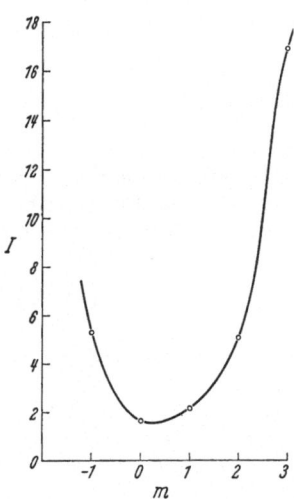

$$B_\perp = \langle B \rangle \langle \sin \vartheta \rangle_{\text{hemisphere}} = \frac{\pi \langle B \rangle}{4}.$$

We compute $q(\nu)$ the source power density, in ergs $\text{sec}^{-1} \text{cm}^{-3} (\text{cycle/sec})^{-1}$, for light of frequency ν in $d\nu$, if K is given in $\text{cm}^{-3} \text{erg}^{-1}$:

$$\left. \begin{aligned} q(\nu)\, d\nu &= \int P(\nu, E)\, \varrho(E)\, dE\, d\nu \\ &= \frac{\sqrt{3}}{2} (z|e|)^3\, B_\perp\, \varkappa \left(\frac{\nu_0}{\nu}\right)^{\frac{n-1}{2}} I\left(\frac{n-3}{2}\right) d\nu \end{aligned} \right\} \quad (13.10)$$

with

$$\nu_0 = \frac{\nu_c}{\gamma^2} = \left(\frac{3z|e|}{4\pi m_0 c}\right) B_\perp.$$

The notation is as in (13.5) above. Here the integral

$$I(m) = \int_0^\infty dy\, y^m\, F(y)$$

Fig. 7b. Radiation of an electron in a magnetic field. The integrated function $I(m) = \int_\infty^0 y^m F(y)\, dy$, plotted as a function of m. After H. J. Oort and T. Walraven: See Fig. 7a.

and is plotted in Fig. 7b (from numerical integrations by J. Leonard). Note especially that the radiation spectrum follows a power law

$$q(\nu) \sim \nu^{-\frac{n-1}{2}}$$

while the source particle spectrum follows the steeper power law $1/E^n$. For n in the neighborhood of two or three, as observed for the cosmic-ray beam of heavy particles near the earth, the emitted radiation falls off only like $\nu^{-\frac{1}{2}}$ or ν^{-1}. This indeed is observed in the spectrum of the Crab Nebula and a few other sources. (Compare the following sections, 24 and 27).

Table 5. *Electron energies for optical and radio emission.*

$B_\perp$ (gauss)	Optical $\lambda = 5000$ Å $\nu_{max} = 6 \times 10^{14}$ cycles/sec	Radio $\nu_{max} = 100$ Mc/sec
	$E(\nu_{max})$ (ev)	$E(\nu_{max})$ (ev)
10^{-2}	1.12×10^{10}	4.6×10^6
10^{-3}	3.55×10^{10}	1.44×10^7
10^{-4}	1.12×10^{11}	4.6×10^7
10^{-5}	0.55×10^{11}	4.6×10^7
10^{-6}	1.12×10^{12}	4.6×10^8

In every case the bulk of the radiation at a given frequency is emitted by those particles whose critical frequency of emission ν_c lies within a factor of a few of the observed frequency. A table of values for optical and radio emission follows.

A simpler estimate of the electron density needed to produce any observed radio flux is given by the use of the power formula (13.2) and the equivalent band width of Eq. (13.8). The intensity of a radio signal (or of an optical continuum) may be expressed by giving $J(\nu)\, d\nu$, the energy received per square centimeter per second in a band-width $d\nu$ around the frequency ν. If a source of mean power density $q\, d\nu$ and total volume V is found at distance R from the earth, the received flux is given by

$$J(\nu) = q(\nu)\, V/4\pi R^2.$$

Now it is easy to estimate the energy radiated by the use of the equivalent band width of (13.8). We regard the radiation as arising from electrons which have the density $\varrho_e\big(E = E(\nu_{\max})\big)$, and using the equivalent band bandwidth we have

$$J(\nu) = 0.128 \frac{(z\,|e|)^3}{m_0\,c^2} B_\perp \frac{\varrho_{\mathrm{eV}}}{R^2}$$

which can be written for electrons

$$J(\nu) = 1.76 \times 10^{-23} B_\perp \frac{\varrho_e(E_{\max})\,V}{R^2} \quad \mathrm{erg\ sec^{-1}\ cm^{-2}\,(cycle/sec)^{-1}} \qquad (13.11)$$

with $B_\perp$ in gauss, R in cm, and V in cm³. The energy $E_{\max}$ is given by the relation

$$E_{\max}/m_0\,c^2 = \left(\frac{\nu}{1.26\,B_\perp}\right)^{\frac{1}{2}} \quad \text{with } \nu \text{ in Mc/sec, } B_\perp \text{ in gauss}$$

and typical values are found in Table 5.

BURBIDGE[1] has made a similar estimate from the total rate of radiation, but taking into account the power-law distribution appropriate for electron energies. His formulae, or the accurate one of (13.10) are sometimes preferable to the estimate of (13.11).

The polarization of the emissions, which has been the most important clue for detecting synchrotron radiation, depends of course upon the coherence of the magnetic directions within the observed volume. This polarization would be reduced if the field directions were chaotic, or if the angular resolution is imperfect, allowing light to reach the detector from regions of non-parallel magnetic field. Even if the source emitted light with electric vector in a single plane, the observed polarization might be reduced if the light traverses optical paths through a magnetized plasma. For the plane waves moving through such a region undergo Faraday rotation of the plane of polarization. This can be estimated by using a relation familiar in ionosphere studies (cf. [21]), which gives the rotation $\Delta\varphi$ of the plane of the electric vector in passage along the line of sight, r, through a medium of electron density n_e, for radiation of wavelength $2\pi\,\lambdabar$, in a magnetic field $B(r)$ making an angle ϑ with the line of sight. The relation is:

$$\Delta\varphi = \left(\frac{e^2}{m_0\,c^2}\right)\left(\frac{e}{m_0\,c^2}\right)\lambdabar^2 \int B(r)\cos\vartheta\, n_e(r)\, r^{-1}\, dr \quad \text{radians.} \qquad (13.12)$$

Whenever the detector admits light from paths along which the rotation $\Delta\varphi$ differs significantly, the polarization will be reduced. This is plainly the more important the lower the frequency detected. Radio polarizations are small both because of the factor $\lambdabar^2$ in (13.12) and because the angular resolution is poor by optical standards.

The application of these formulae to the problem of the cosmic rays and their origin is best discussed after the astronomical findings have been presented (Sects. 24, 25 below), but it is worth anticipating them here in brief summary.

1. The isotropy and time-constancy of cosmic radiation can be explained as the effects of a more-or-less chaotic, more-or-less ubiquitous magnetic field in the galaxy. The field strength anticipated is in the neighborhood of a few to ten microgauss. Such a field will cause electrons present, with energies and densities of cosmic-ray order, to radiate radio emission about equal in intensity and spectrum to the observed radio radiation from the galactic halo in our own and in other galaxies. Such electrons will lose most of their energy in such radiation, and their ultimate energy spectrum will steepen up, so that they lie below the observed limits for primary cosmic ray electrons.

---

[1] G. BURBIDGE: Astrophys. J. **124**, 416 (1956).

2. A polarized optical continuum has been seen in a few objects. Two such anomalous objects are known to emit both polarized continuous light and strong radio waves; in one at least radio polarization has been found. All this is convincing evidence of synchrotron radiation. These sources must contain large numbers of electrons of very high γ, and probably, but not certainly, cosmic-ray protons and heavier nuclei as well. While the protons do not radiate, it is hardly possible to doubt their presence, and then these same objects must be sources of cosmic rays. If we extrapolate the few observed sources to all the other possible sources which seem physically like them, the number of sources would very plausibly account for most cosmic rays from a few tens of Gev up to the highest observed energies. Most of the rest of this article is devoted to elaborating the relations between the cosmic-ray beam and the sources of synchrotron radiation in the visible and in the meter-wavelength region. This general connection was first worked out by Shklovski and Ginzburg [4], [15].

D. The diffusion of the cosmic-ray particles.

14. Elements of the theory. The general processes which serve to make the cosmic rays have been described. Insofar as our ideas are correct, the cosmic-ray gas is simply that portion of the interstellar matter which has been subjected to a more or less random walk about the magnetic fields located at stellar neighborhoods or far out in space. The random walk carries the particles in some statistical way to regions of phase space far from their original location, and may as well contain steps in which collision between cosmic-ray particles and the particles of the medium serves to modify the particle type. The task of the theory is to describe this random walk as well as possible, and to identify from the properties of the beam those more or less particular space-time and momentum regions from which the particles originally come. The statistical nature of the process itself, no less than of the sampling measurements which are all we have to go on, plainly implies some version of the theory of Brownian motion, or more specifically, of diffusion, for we cannot expect to derive much from a more detailed treatment. The generally isotropic and time-constant nature of the beam confirms the hope that a rather approximate treatment, emphasizing the multiplicity of the steps in the history of a single particle, will be adequate. Davis[1] has examined the only deviation from diffusion theory which appears to be of possible importance: he has looked at the nature of the angular distributions to be expected if indeed the magnetic field outside the solar system is homogeneous for a distance of many light years. We shall simply neglect any such local correlation; no specific data are available for comparison with the model in any case (most of the data he treated have been by now found unreliable). Let us consider first in general, and later in more detail, the diffusion of cosmic rays.

In the introductory Sect. 7 we have already set out the main steps of the process. Here we shall review the main ideas of the Fermi theory [11] in the most elementary and transparent form[2].

The simplest point arises from the near-isotropy of the beam (above 30 Gev). If we write as in Sect. 3

$$\frac{\varphi_{max} - \varphi_{min}}{\frac{1}{2}(\varphi_{max} + \varphi_{min})} = \delta$$

where δ is the anisotropy, and φ_{max}, φ_{min} the observed maximum and minimum fluxes for a given energy, we can use the conservation of particles to write the

---

[1] L. Davis: Phys. Rev. **96**, 743 (1954).
[2] G. Cocconi: Phys. Rev. **83**, 1193 (1951).

relation:

$$A(\varphi_{max} - \varphi_{min}) = \overline{Q} V \qquad (14.1)$$

where A measures the area bounding the region of volume V in which the cosmic rays are made, with a source strength $\overline{Q}$ per unit time and per unit volume, averaged over V. Now, if we define the mean lifetime of a particle of the energy and type we include in φ as just τ, then the average particle density ϱ is

$$\varrho = \overline{Q}\tau$$

as before, Eq. (7.1). Without any further assumption, we can write for τ the quantity L/v, defining the mean path length L by that relation. Then we have

$$\varphi_{max} - \varphi_{min} = \frac{\varrho v V}{LA}. \qquad (14.2)$$

This may be expressed in terms of a distance R from the sun to the centroid of the region in which the rays are made, and we have, neglecting small geometrical factors, the rather general result: (cf. [13], and COCCONI [18]).

$$\delta \approx \frac{V}{LA} \approx \frac{R}{L}. \qquad (14.3)$$

Now experimentally an upper limit is put on δ, which ranges from less than 10^{-3} in the region from a few tens of Gev up to 10^6 Gev, and is uncertain, but definitely less than 10^{-1} at the highest observed energies, say 10^8 Gev or more. On the other hand, it is clear that the charge spectrum of cosmic rays set an upper limit on L, measuring it not by centimeters, but by the mass traversed/cm^2. For the nuclear collisions would eliminate heavy nuclei from the beam if L grows arbitrarily. Such collisions would moreover overstock the rare Li, Be, B portion of the spectrum. Since the density of the regions concerned can be estimated at least roughly, this column mass limit can be translated provisionally into a length. The length becomes, as we shall see in more detail later, a length of a few grams. At the density of the galactic disc, perhaps 0.5 H atom/cm^3, this amounts to a few times 10^7 light years. Of course, the path may end not by collision but by leakage into a space of differing properties. So L cannot exceed the distance given, and may be much less. This places an upper limit on R, which can be used to help fix the model. Moreover, L cannot be less than a few times the Larmor radius appropriate to the fields involved. For the higher energy cosmic rays, the condition appears to exclude the possibility of a purely solar-system origin and confinement. But this general criterion does not seem to demand much more than that a large spatial volume be accessible. It works strongly against a disc or spiral-arm confinement, for it then requires (since R is small) a rather small value of L, much below the upper limit.

It is clear that the removal of particles by collision or by leakage can be described roughly by the relation for the total lifetime in the region:

$$\frac{1}{\tau} = \frac{1}{\tau_{coll}} + \frac{1}{\tau_{leak}}, \qquad (14.4a)$$

where τ_{coll} and τ_{leak} are partial lifetimes against collision and leakage. This suggests a simple estimate of τ_{leak} from random-flight theory. In random-walk in three dimensions $d^2 = 2N\lambda^2$, where d^2 is the mean square distance reached after N magnetic scattering (*not* nuclear) collisions, mean free path between such collisions being λ. If the distance d approaches the dimensions of the confining region, the particle will leak out. Thus the lifetime for confinement in a region with smallest dimension h is estimated as $\tau_{\text{leak}} = N_{\text{leak}} \lambda/c \approx h^2/\lambda c$, and the relation

$$\frac{1}{\tau} \approx \frac{1}{\tau_{\text{coll}}} + \frac{\lambda c}{h^2} \tag{14.4 b}$$

follows. The leakage may also be expressed through the probability of loss of particles having suffered N collisions, estimated by

$$P = e^{-N/\bar{N}} \tag{14.5}$$

where $\bar{N} = c\tau/\lambda$, limited by leak, since $\tau_{\text{leak}} < \tau_{\text{coll}}$. This leads immediately to the heart of the theory [11]: the energy is taken as a smooth function of the number of collisions N since injection into the process. If we write the simple relation of Fermi (cf. Sect. 15)

$$\frac{dU}{dN} = \alpha(U)\, U - \beta(U) \tag{14.6}$$

we get $\ln \dfrac{U}{U_0} = \int\limits_0^N \alpha\, dN'$ for any energy well above the critical energy U_c, where $dU/dN = \alpha U - \beta$ vanishes. Now this can be written as

$$\ln \frac{U}{U_0} = \bar{\alpha}(N)\, N$$

where $\bar{\alpha}$ is a suitable average. Then we have the energy spectrum:

$$\varphi(U) = \varphi_0\, P\big(U(N)\big) = \varphi_0 \exp\left(- N/\bar{N}\right) = \varphi_0 \left(\frac{U_0}{U}\right)^{+1/\bar{\alpha}\bar{N}} \tag{14.7}$$

where φ is the scalar flux of particles with collision number between N and $N + dN$. If the combination $\bar{\alpha}\bar{N}$ is roughly independent of energy, the power-law arises naturally. Non-equilibrium processes such as cosmic-ray production and turbulent eddies seem naturally to have a power-law distribution, as equilibrium ones seek the canonical exponential distribution. This scheme presents the origin of a power law in this natural way; no proposal has been made for any other distinctly different way to produce the power-law spectrum *without* introducing an ad hoc power-law dependence on some parameter of the theory[1]. On this view, the exponent becomes $\bar{\alpha}\bar{N} \approx \alpha h^2/\lambda^2 \approx 1$ or 2 experimentally. This is a relation between certain magnetic properties of a medium and its overall extent; it is not fully clear that it should be present, but it at least does not involve such disparate quantities as nuclear collision cross-sections, which would be the case if collision, and not leakage, were taken to be the cause of the particle's demise.

Note also the storage feature of this viewpoint: the lifetime determines the density. The long lifetimes possible, compared to the straight-line transit time, suggest that the processes which free the cosmic-ray particles from thermal equilibrium need be much less copious than those which emit photons, if the energy density of the two are to be comparable, as we see them.

Finally, the occurrence of cut-offs from the smooth spectrum is an essential part of the theory. For very high energy, magnetic confinement must fail, and the magnetic scattering mean free path λ approaches nuclear collision values $\lambda_{\text{nuc}} \approx 1/(n_{\text{medium}} \sigma_{\text{nuc}})$. For low energies, particles with low values of N may be absent, if sources are not uniformly distributed with respect to the observer. Of course, we regard the time variability of the low-energy cut-off, with its clear dependence on solar activity, to be a demonstration that the low-energy cut-off is a feature of the solar neighborhood, not of the cosmic rays in general.

---

[1] Cf. Alfvén and Åström: Nature, Lond. **181**, 330 (1958).

15. The diffusion equation. Most of the studies of cosmic-ray transport are done with the diffusion equation, an approximation adequate for nearly-isotropic beams, especially in the present rough state of the theory. Here the fundamental idea is, of course, the use of the net current $\boldsymbol{J}$ as given by the linear relation

$$\boldsymbol{J} = -D\boldsymbol{V}\varrho = -\frac{\lambda v}{3}\boldsymbol{V}\varrho = -\frac{\lambda}{3}\boldsymbol{V}\varphi, \qquad \varphi = \varrho v \tag{15.1}$$

where ϱ is the particle density, and λ the transport mean free path. The ordinary diffusion equation follows at once from the equation of continuity:

$$\boldsymbol{V} \cdot \boldsymbol{J} + \frac{\partial \varrho}{\partial t} = -\boldsymbol{V} \cdot (D\boldsymbol{V}\varrho) + \frac{\partial \varrho}{\partial t} = 0. \tag{15.2}$$

The angular distribution of the beam in this theory is that of an isotropic main part on which a small distribution proportional to $\cos \vartheta$ is superposed. The anisotropy parameter δ is given as:

$$\delta = \frac{4}{3}\lambda \left| \frac{\boldsymbol{V}\varphi}{\varphi} \right|.$$

Local measurements (see DAVIS [17]) do not really bear on the goodness of the approximation, which ought to be applied only in the mean over regions large in diameter compared to the mean free path. In a steady state, the density ϱ or the scalar flux φ, which is more often used, will approximately satisfy the relation $|\boldsymbol{V}\varphi| \approx \varphi/R$ with R the diameter of the region. This recovers the more general relation of Eq. (14.3), $\delta \approx \lambda/R \approx R/L$ since $N \sim R^2/\lambda^2$, $L \sim N\lambda$.

Let us build up one by one the various cases of diffusion theory. The overall description can be obtained by including all the effects in one single balance equation but this is rarely necessary.

α) *Spatial diffusion and sudden loss processes.* The simplest application is the determination of the density distribution as a function of position, which may vary with respect to distance from the source of particles or to boundaries beyond which the conditions of the random walk differ. The equation is of course:

$$\frac{\partial \varrho}{\partial t} = \frac{\lambda}{3}\Delta\varphi + q(\boldsymbol{r}, t) \tag{15.3}$$

where φ is the scalar flux of particles at time t, and the source density is q. This theory has a large literature from its importance for thermal neutrons; we will make very little use of its geometrical complexities. The simplest boundary condition is to set the density zero at the interface against a vacuum. (This corresponds to neglecting reflection.) For the contact of differing regions, 1 and 2, the equation of current continuity requires:

$$j_{1 \to 2} = j_{2 \to 1}, \qquad \boldsymbol{j} = \lambda \frac{\boldsymbol{V}\varphi}{\varphi}.$$

In a steady state, with some localized source present, it is clear that φ must be a solution of LAPLACE's equation satisfying the boundary conditions, and quite generally will satisfy the estimate

$$\boldsymbol{V}\varphi \sim \varphi/R$$

of its gradient. If we introduce any processes of loss, such as nuclear collision, we obtain at once:

$$\frac{\partial \varrho}{\partial t} = \frac{\lambda}{3}\Delta\varphi - \frac{\varrho}{\tau_{\text{coll}}} + q \tag{15.4}$$

so that the steady state in an infinite region yields

$$\varrho = \bar{q}\,\tau_{\text{coll}}$$

as we saw in general, and in a finite region from which leakage is important,

$$\tau = \frac{L}{v}, \qquad \frac{1}{L} = \frac{1}{L_{\text{coll}}} + \frac{1}{L_{\text{leak}}},$$

where the mean distance diffused before leaking out is given by

$$L_{\text{leak}} = k\,R^2/\lambda.$$

The numerical constant k is fixed by the geometry of the problem.

It is possible to go still further, taking account of the nuclear collisions, which feed the lighter components of the beam with the secondaries from collisions of heavier components. This problem is important in the interpretation of the observations of primary interactions in the earth's atmosphere, discussed elsewhere in the volume. Ginzburg and Fradkin[1] have applied the equation to the diffusion of the beam throughout the galactic sphere, with a lifetime for protons of several hundred million lightyears, less by a factor of five or ten than that appropriate to the disc of the galaxy. There is not much effect of collisions even in this model. The authors cited would want to interpret the probable richness of the cosmic-ray beam in heavy particles ($Z \geqq 10$) as a reflection of the source richness, little modified by collisions. The general absence of nuclear collisions seems pretty well established, especially at the higher energies. The diffusion equation takes the form:

$$\frac{\partial \varrho_i}{\partial t} = \boldsymbol{V} \cdot (D_i \boldsymbol{V} \varrho_i) - \frac{\varrho_i}{\tau_i} + \sum_{j > i} p_{ij}\frac{\varrho_j}{\tau_j} + q_i \tag{15.5}$$

where the index i indicates the type of particle, and the probabilities p_{ij} represent the mean number of particles of type i, lighter than type j, produced in one collision between a type j particle and interstellar matter. The usual scheme is to consider at most five types: protons, alphas, "light" nuclei, which are Li, Be, and B; medium ones, which are C, N, O, F; and heavy ones, with Z greater than 10, neon and beyond. The data are still not beyond criticism (see the Bombay work[2]); all we need at the moment is the general result: the absence of many collisions, with the path traversed certainly below a few grams for the bulk of the cosmic rays, even those with energies up to 10^5 Gev. Nothing is known for rays above that energy.

β) *Acceleration processes.* In addition to the processes of sudden change, there are processes of continuous or near-continuous nature. It is evident that many "collisions" between the particles and the moving magnetic fields of some region of space change the energy more or less gradually from thermal energies to somewhat beyond to cosmic ray values. It was these processes which were discussed in Sect. 11. What this section undertakes is to study the effects of such repetitive processes, whatever their nature.

The simplest process is one in which the energy is a function of the number of collisions with diffusing magnetic elements. This is the analogue of the "age" theory of neutron diffusion. If we regard the scalar flux φ now as a function of N, the number of scattering collisions since injection, we obtain the diffusion

[1] Y. L. Ginzburg and M. Fradkin: Astronom. Zhurn. **33**, 579 (1956).
[2] M. Appa Rao, S. Biswas, R. Daniel, K. Neelakantan and B. Peters: Phys. Rev. **110**, 751 (1958).

equation now in the form:

$$\frac{\partial \varrho}{\partial t} = \frac{\lambda}{3} \Delta \varphi - \frac{1}{\lambda} \frac{\partial \varphi}{\partial N} - \frac{\varphi}{L_c} + q(\boldsymbol{r}, t, U_0)\, \delta(N) \tag{15.6}$$

where $\varphi = \varrho v$, with $\varrho(\boldsymbol{r}, t, N)$ the number of particles per unit volume at position $\boldsymbol{r}$ and time t, which have undergone between N and $N + dN$ collisions since injection from a source with energy U_0, and initial $N = 0$. The source density is written as a function of U_0, and refers to injection in energy range dU_0 at U_0. If there is a spread of injection energies, one needs to add up the result for each; there is implied a connection between U and N. This needed connection is given for example by the integration of FERMI's result, in which each collision on the average increased the particle energy by the amount αU, with

$$\langle \Delta U \rangle = \alpha U$$

where α (discussed in Sect. 11) is a property of the medium. It is reasonable to include the losses from plasma interaction, or such processes as synchrotron radiation, in a loss term. A differential equation easily results (14.6):

$$dU/dN = \alpha(U)\, U - \beta(U). \tag{15.7}$$

Evidently there is a critical energy: for $U \leq U_c = \beta/\alpha$, particles cannot be accelerated. Only particles which reach U_c are said to be "injected" into the cosmic-ray region.

The steady-state solution of the diffusion equation (15.6) is easy to describe in general, though its evaluation is of course dependent upon the geometry. The boundary conditions are furnished by the relation for $N = 0$:

$$\varphi(\boldsymbol{r}, t, 0, U_0) = \lambda q(\boldsymbol{r}, t, U_0)$$

and by the condition imposed at the spatial boundaries of the diffusing region. Nominally this can be taken to be $\varphi \to 0$ on the boundary; reflecting boundaries will make some difference (cf. [13]). Now the solution is found in general by writing a Fourier-Bessel series, using functions $u_i(r)$ appropriate to the geometrical shape of the region involved. Quite generally for the steady state, one obtains the series

$$\varphi(\boldsymbol{r}, N) = \sum_{\infty}^{i=1} A_i\, u_i(\boldsymbol{r}) \exp(-N/N_i) \tag{15.8}$$

where the expansion coefficients A_i are fixed by the distribution of the sources, and are proportional to the Fourier-Bessel coefficients in the expansion of the source spatial distribution in terms of the functions u_i. More interesting are the mean numbers of collisions in each mode i:

$$\frac{1}{N_i} = K_i\, \lambda^2/h^2 + \lambda/L_{\text{nuclear}} \tag{15.9}$$

where K_i is a numerical constant depending of course on the index i and on the geometry. For low N, small enough so that $\lambda/\overline{N}$ is small compared to h, the smallest dimension (or a mean dimension) of the diffusing volume, the higher "modes" of the series may contribute, and the geometry of the source distribution will tend to determine the functional dependence of φ on N. If the sources are far enough away so that their distance d_q implies a random walk $d_q \sim \sqrt{N_q}\,\lambda$ with numerous enough collisions $N_q \gg h^2/\lambda^2$, the spectrum will be depleted below $U(N_q)$. C. Y. FAN[1] has used this scheme with sources concentrated at the galactic

---

[1] C. Y. FAN: Phys. Rev. **82**, 209 (1951). — Astrophys. J. **123**, 491 (1956).

center, and diffusion throughout a galactic sphere, to give an account of the spectrum below 100 Gev. The time variations of the low-energy spectrum seem nowadays to imply the solar-system origin of the low-energy cut-off. For $N > N_1$, that is, $N > h^2/\lambda^2$, and sources more or less uniform, the higher modes of the Fourier expansion are negligible, and the spectrum depends in a simple exponential way upon the collision number N, like $\exp(-N/N_1)$. irrespective of geometry of the diffusing volume or of the source distribution. This is of course the heart of the simple theory of the spectrum described in Eq. (15.7) above.

It is easy to extend the formalism to the case where the transport mean free path λ varies with the collision number. One need only replace λN and $\lambda^2 N$ in the exponential of Eq. (15.8) by the integrals

$$\int_0^N \lambda \, dN', \quad \int_0^N \lambda^2 \, dN'.$$

If the relation (15.6) between collision number and energy involves a variable $\alpha(U)$, the spectrum is given by

$$\varphi \frac{dN}{dU} \sim \frac{\text{const}}{U \alpha(U)} \exp\left(-\frac{k}{h^2} \int_{U_0}^U \frac{\lambda^2(U')}{\alpha(U')} \frac{dU'}{U'}\right)$$

whose main dependence is of course given by the exponential. The asymmetry is determined by $\lambda(U)$ at the given value of U. If one examines the resolution of the detectors used, it becomes fairly clear that as long as the acceleration is gradual, near-isotropy will persist to very high energies, and the cut-off in the energy spectrum will enter before the beam becomes strongly anisotropic. The whole question is experimentally much vexed (cf. [13]) but no marked departure from isotropy or from the spectral shape has yet been seen up to the highest energies, near 10^8 or perhaps even 10^9 Gev.

There is still another source of energy change, which is better described by a somewhat different formalism. That is deceleration by magnetic bremsstrahlung, or synchrotron radiation. As we have seen in Sect. 13, this effect is much the strongest for the electrons whose absence from the primary beam is such a conspicuous property of the cosmic rays. The electron radiation loss is given by:

$$dU/dt = -b\,U^2, \quad U = \text{total energy}/m_e c^2$$

where the coefficient

$$b = \frac{ze}{3}\left(\frac{e^2}{mc^2}\right)^2 \langle B^2 \rangle \approx 10^{-3}\,(B\,\text{gauss})^2\,\text{ev/sec}$$

depends on the magnetic fields present. Neglecting the spatial diffusion, we may write for an electron beam density of ϱ_e at energy U in range dU, the continuity equation along the energy axis:

$$\frac{\partial \varrho_e}{\partial t} + \frac{\partial}{\partial U}\left(\varrho_e \frac{dU}{dt}\right) = q_e(U)$$

where the "current" along the energy axis is just $\varrho_e \, dU/dt$. If we introduce ordinary nuclear bremsstrahlen as well, in which a single collision may cause the radiation of a large fraction of the energy of the electron, and represent its probability by the lifetime $c\tau_r = L_r$, where L_r is the radiation length in the medium (see Sect. 8), we have

$$\frac{\partial \varrho_e}{\partial t} = -\frac{\varrho_e}{\tau_r} + \frac{\partial}{\partial U}\left(\varrho_e(U) \frac{dU}{dt}\right) + q_e(U). \tag{15.10}$$

Here $q_e(U)$ is the source density for electrons of energy U. The steady-state solution, as given by GINZBURG [15 a], is simply the result of integration:

$$\varrho_e = \frac{1}{b\,U^2}\exp\left(-\frac{1}{b\,U\,\tau_r}\right)\int\limits_U^\infty dU'\,q_e(U')\exp\left(\frac{1}{b\,U'\,\tau_r}\right).$$

Let us write this for a delta-function source spectrum, injecting at U_0. Applying this relation to electrons in typical galactic space, where an ambient field of several to ten microgauss is present, while $\tau_r \sim 10^{16}$ sec, we see that for $U > 10^{3-4}$, the electron spectrum is strongly impoverished by the exponential factor. Below a few Gev, the spectrum is pulled out into a power-law spectrum, with the exponent governed entirely by the rate of synchrotron radiation. This type of energy dependence is characteristic for such processes. GINZBURG[15] and BURBIDGE[1] (and others) have sought to place the origin of the fast electrons partly or wholly in the decay of the mesonic products of collisions between cosmic-ray protons and the protons of the medium. Such collisions in fact lead to a yield of electrons amounting to some five or ten percent of the cosmic-ray energy for each collision, and thus to a steady source of electrons. The electron source spectrum q can be estimated very roughly from a simplified model of such nuclear collisions. One may take the secondary products of collision to have roughly the energy $E_2 = \delta E_1$, where δ varies little with primary energy E_1. Then the electron source becomes just

$$q_e(E_e) = \frac{f\,\varrho_N}{\tau_N}\,(E_N/\delta)$$

where N indexes nuclear properties and f is the fraction of secondary energy appearing in electrons and positrons, after all transitory products have decayed.

For a power-law primary spectrum $\varrho_N\,dE_N = \dfrac{\varkappa}{E_N^\gamma}\,dE_N$, we have

$$q_e\,dE_e \approx \frac{f}{\delta}\,\frac{\delta^\gamma}{\tau_r}\,\frac{\varkappa}{E_N^\gamma}\,dE_N \tag{15.11}$$

and from this then

$$\varrho_e \propto \int \frac{\varkappa\,dE_N}{E_N^{\gamma+2}},\quad \text{with } E_N \gtrsim \text{ few Gev}.$$

Thus the electron spectrum for secondaries would fall off with a power-law dependence, perhaps one power of E more steeply than does the primary cosmic ray beam. Such a spectrum is plausible in radio sources, as we have seen. The full radio emission may or may not be the consequence of such secondary electrons; it is easier to ascribe to them the radio emissions of such a weak source as our galactic halo than it is to explain the strong sources. As we have earlier seen (Sect. 11), the generation of relativistic electrons by the processes which produce the cosmic rays themselves is quite probable. But the absence of electrons from the primary beam is naturally explained by the same synchrotron radiation which does account rather well for the numerous sources of radio emission.

16. Fluctuations in energy: the Fokker-Planck equation. The third formulation of the acceleration problem is the most general one. The first two depended upon a unique connection between time, or collision number, and the energy of the particle. Now we shall consider as well the energy changes which result from a random walk along the energy axis, in which the energy change in a collision

---

[1] G. BURBIDGE: Phys. Rev. **103**, 264 (1956).

may be either a gain or a loss, and a final spectrum in energy may result merely from the fluctuations of a random energy-changing process.

Let us consider a simplified model in which the particle suffers an energy-changing collision every τ seconds. — In each collision the fractional change in energy may be either positive or negative:

$$\Delta \mathscr{E} = \Delta E/E = \pm \alpha.$$

We set the probability of a gain in energy equal to p, a loss, to q, with $p+q=1$. Now the expected value of the change in the logarithm of the energy $\langle \Delta E/E \rangle = \langle \Delta \mathscr{E} \rangle_1$ in one collision, is given by the definition:

$$\langle \Delta \mathscr{E} \rangle_1 = \langle \Delta E/E \rangle = p \alpha - q \alpha = (p - q) \alpha.$$

We assume that a time interval Δt is large enough to contain many collisions, but small enough so that only a small change in E occurs. Then the mean change in N collisions will be given by:

$$\langle \Delta \mathscr{E} \rangle_N = N \langle \Delta \mathscr{E} \rangle_1$$

which we shall write in terms of the time interval Δt,

$$\langle \Delta \mathscr{E} \rangle_N = \bar{D} \Delta t = \frac{\Delta t}{\tau} \alpha (p - q).$$

This defines an energy drift rate,

$$\bar{D} = \alpha (p - q)/\tau.$$

The variance of $\mathscr{E}$ in a single collision is found from its definition:

$$\langle (\Delta \mathscr{E})^2 \rangle_1 - \langle \Delta \mathscr{E} \rangle_1^2 = p \alpha^2 + q (-\alpha)^2 - \alpha^2 (p - q)^2$$
$$= \alpha^2 \big(1 - (p - q)^2\big) = 4 p q \alpha^2$$

and extending the process over a set of N independent collisions yields:

$$\langle (\Delta \mathscr{E})^2 \rangle_N - \langle \Delta \mathscr{E} \rangle_N^2 = 4 p q \alpha^2 N = D_2 \Delta t$$

where again we define an energy-gain coefficient, D_2, this time for the fluctuations in energy:

$$D_2 = 4 p q \alpha^2/\tau.$$

If the probability of gain in energy in a single collision is equal to that of loss, then $p = q$. The mean rate of energy gain $\bar{D}$ vanishes, but the fluctuations still occur, and the value of D_2 remains finite.

This formulation derives from the treatment of Brownian motion. Just as there, we can write the Fokker-Planck equation, a differential equation sufficient to describe the random walk in cases of sufficiently numerous and not too rapidly-varying collisions (see Chandrasekhar[1], p. 31 ff.). This procedure was first adapted in a rather special case to the cosmic-ray problem by Terletski and Luganov[2], and more generally by Davis [19], whose procedure we follow.

It is appropriate, just as in the neutron age theory [13], to introduce as we did above the logarithmic measure of energy:

$$\mathscr{E} = \ln \frac{U}{U_0}.$$

[1] S. Chandrasekhar: Rev. Mod. Phys. **15**, 1 (1945).
[2] I. P. Terletski and A. A. Luganov: Ž. exsp. teor. Fiz. **21**, 576 (1951); **23**, 682 (1952).

We introduce a density $\varrho(\mathscr{E}, t)$ to represent the number of particles per unit volume with the value $\mathscr{E}$ in $d\mathscr{E}$ at time t (and position $\boldsymbol{r}$, if spatial diffusion is to be included). If there is a catastrophic collision loss, governed by a lifetime τ_n, and a slow energy-dependent loss (perhaps by ionization, for the particles of cosmic rays, or by radiation, as for electrons), then the continuity equation may be written:

$$\frac{\partial \varrho}{\partial t} = -\frac{\varrho}{\tau_n} + \frac{\partial}{\partial \mathscr{E}} (l\,\varrho) - \frac{\partial}{\partial \mathscr{E}} (\overline{D}\,\varrho) + \frac{1}{2} \frac{\partial^2}{\partial \mathscr{E}^2} (D_2\,\varrho) + q(\mathscr{E}, t) \qquad (16.1)$$

where the parameter l is the smooth rate of loss of $\mathscr{E}$, $-l = d\mathscr{E}/dt$. q is the source density, and it is plainly easy to add a spatial diffusion term $\dfrac{\lambda}{3}\,\varDelta\,\varrho$ if needed.

The most useful solution of Eq. (16.1) is the simple steady-state case, with injection constant at energy U_i. Then for all $\mathscr{E} > \mathscr{E}_i = \ln\dfrac{U_i}{U_0}$ the spectrum becomes

$$\varrho = \varrho_0 \exp(-\gamma\,\mathscr{E})$$

with

$$\gamma = -\frac{\overline{D} - l}{D_2} + \left[\left(\frac{\overline{D} - l}{D_2}\right)^2 + \frac{2}{D_2\,\tau}\right]^{\frac{1}{2}}.$$

Transforming to the energy scale, one obtains

$$\varrho(U)\,dU = \varrho_0\,U_0^\gamma\,\frac{dU}{U^{1+\gamma}}, \qquad (16.2)$$

the familiar power law which we saw to follow in the "age" theory, with $dU/dN = \alpha U$. Indeed, if we set $D_2 = 0$, $l = 0$, and write for $\overline{D}$ just the value given in the simple Fermi theory, $\overline{D} = \beta^2/\tau_{\text{coll}}$, we recover the familiar result:

$$\gamma = \frac{\tau_{\text{coll}}}{\tau_{\text{nuc}}}\,\frac{1}{\beta^2}.$$

The fluctuations ought not to be disregarded. If the collisions according to the Fermi mechanism yield $\overline{D} = \beta^2/\tau_{\text{coll}}$, then we may easily identify $p \approx \frac{1}{2} + \beta$, $q \approx \frac{1}{2} - \beta$, and moreover $\beta \ll 1$. Then $D_2 \approx \beta^2/\tau_{\text{coll}}$ and roughly at least $\overline{D} \approx D_2$. In this case, the exponent

$$\gamma = -1 + \left[1 + \frac{2\tau_{\text{coll}}}{\beta^2\,\tau_{\text{nuc}}}\right]^{\frac{1}{2}}.$$

But since the spectral shape we are looking for demands that γ be of the order of magnitude of unity, it cannot much affect the result, whether or not we consider fluctuations represented by D_2. Indeed, even if we had set $\overline{D} = 0$ and thus depended entirely upon fluctuations for the acceleration process, the required parameters would not be much changed. The inference is clear: mechanisms which accelerate in the mean, or which simply provide fluctuations in energy, will be more or less equally available for the production of cosmic rays with something like the observed spectrum. As we saw in Sect. 11, the most plausible types of accelerations show some tendency for both contributions, but are surer to induce energy fluctuations.

Davis[1] has examined the approach to the steady-state condition. It appears that the random walk fluctuations approach the steady state value about as rapidly as a mean energy gain with the same resulting power law would accelerate particles. The logarithmic range of energy over which the steady state would be

---

[1] L. Davis: Phys. Rev. **96**, 743 (1954).

nearly reached grows linearly with the time, and not merely with the square root of the time.

17. The mean free path in magnetic diffusion. Perhaps the key parameter in the treatment of cosmic ray diffusion is the transport mean free path. This is the statistical measure of the tendency of the particle to make progress through the magnetic field. Hardly anyone has been brave enough to try to treat in any detail the motion of the fast particles through some complex magnetized region. Simple and plausible models must be made to serve.

As we have seen in Sect. 15, the simplest and likeliest motion would be diffusion with a mean free path independent of energy. It is useful to define the familiar transport mean free path λ_t by the relation

$$\lambda_t = \lambda/(1 - \langle \cos \vartheta \rangle)$$

where λ is the mean distance between deflections in the field, and $\langle \cos \vartheta \rangle$ is the mean cosine of the angle of deflection. This simple kinetic point of view already suggests the model which might admit of such a description: If a magnetized region consists a random array of tight bundles of lines of force, and of weak fields straying between the bundles, the motion of a charged particle would be very much like that of a molecule moving in a field-free space between collisions with hard spheres. If the bundles of force lines, which could also be called knots or clouds of magnetic field, satisfy the condition that the Larmor radius of the diffusing particle is small compared to the diameter of the cloud, this hard-sphere analogy can be realized. The angle of deflection in each cloud is large compared with 2, and the mean cosine vanishes. The particle "forgets" which way it was going on each collision. Then the mean free path will be just the geometric one defined by the diameter d and the mean density of the clouds in space, N:

$$\lambda_t = \frac{1}{\pi d^2 N} \cdot$$

In particular, it will not depend on the energy of the charged particle. Of course, this independence upon energy cannot be maintained to arbitrary energies. For some suitably high energy, the Larmor radius will begin to approach the cloud diameter, and the mean free path will begin to rise. For an energy such that the Larmor radius is small compared to the diameter of the cloud, the other extreme condition will apply. The mean angle of deflection will be given by $\langle \vartheta \rangle = \frac{d}{R_L}$, and the transport mean free path will become

$$\lambda_t \approx \frac{1}{\pi d^2 N} (R_L^2/d^2)$$

where R_L is the Larmor radius, d the mean diameter of the scattering cloud, and N the number of such clouds per unit volume. The two limiting cases then are given by

$$\lambda_t = \frac{1}{\pi d^2 N} \qquad \text{for "hard" clouds,}$$

$$\lambda_t \approx \frac{1}{N d^2} \frac{R_L^2}{d^2} \qquad \text{for "soft" clouds.}$$

In the latter case, the mean free path evidently increases with energy like E^2 in the relativistic region. The energy of passing over from predominantly hard to predominantly soft collisions is given by the condition

$$\frac{\gamma m_0 c^2}{e B} = R_L \approx d.$$

Since there is little doubt that even if this very naive view is anywhere applicable, the clouds present have a wide distribution of diameters and field strengths, there is really no very good argument for how this transition takes place. It is usual to imagine a sudden change-over at a critical energy, below which the mean free path is constant and relatively small compared to the diffusing volume, and above which the mean free path suddenly jumps to large values. This suggests a cut-off at high energies in any spectrum maintained by the trapping action of diffusing regions. In [13] some effort is made to discuss observational possibilities for detecting such a cut-off. A suitable model which introduces the limiting relations in a simple if arbitrary way has been used by PARKER[1] to fit some observed data on diffusion in the solar system. He simply put:

$$\lambda_t = \frac{1}{\pi \, d^2 \, N} \left(1 + \frac{R_L^2}{d^2}\right).$$

The boundary condition on the particle density in ordinary diffusion theory is, of course, to set the density at zero on any bounding surface against the vacuum. The vacuum acts as a complete absorber. In the present context, it is not perfectly clear how to treat the boundary of a magnetically diffusing volume. It does not appear very likely that there is a region beyond which the magnetic fields really drop to zero, even in the space between clusters of galaxies. But it is fairly clear that there are regions of generally low magnetic field, and hence of long mean free path, which might roughly be represented by vacuum. On the other hand, there is a considerable disposition to consider magnetic fields bounding various diffusing regions as being better represented by a reflecting boundary. If any boundary is penetrated by lines of force, so that at least in some areas field lines extend into the space of lower mean field, the boundary cannot be very hard to penetrate. For the particles will move along the field lines as guiding centers. This may be expected to occur whenever the field energy density is not sizeably larger than the energy density of diffusing particles. For then particles which here or there escape the region will tend to drag out the field lines with them. If this is not the case, and the boundary is maintained by more or less smooth lines of force running predominantly in the interface itself, and never crossing it, the situation is not so clear. The particles will tend to use the lines of force as guiding lines for their precession circles, and thus not leave the region of higher field strength. But they must drift along the interface, across the lines of force, because of the field gradient normal to the surface. More than that, if there are field gradients anywhere which are not exactly normal to the surface, the particles will drift off the lines of force and move normally to the general interface. This normal drift will usually be outward into the region of weaker field. The reflectivity of a surface will not be large unless there is some special symmetry of the fields-not to be expected for the cases at hand. A force-free field, which may be ex, pected to bound some diffusing volumes (see Sect. 24 on the Crab Nebula) will usually have rather large gradients, which make it less reflective than otherwise might be expected. The whole topic is unsettled, but it seems probable that the special nature of boundaries will not much influence those situations in which the diffusing volume really has a field pattern sufficiently irregular to allow treatment anywhere within the volume by the diffusion approximation.

A similar remark applies to diffusion into a boundary from a region of generally weaker field. This is clearly slow, in the sense that any particle entering is likely to appear again. The albedo is high. If absorption is neglected, the albedo of a semi-infinite plane is unity from simple diffusion alone. The albedo of a large

---

[1] E. PARKER: Phys. Rev. **103**, 1518 (1956).

finite volume is close to unity, reduced from it by a quantity of the order of λ/D, where D is the characteristic diameter of the diffusing region, and λ the typical mean free path. This is small if diffusion theory is anywhere useful. It takes a long time for the reflected particle to emerge. In general, the steady state solution will describe the state of affairs reasonably well. Once a particle from outside has penetrated several mean free paths, it will have no particular difference from a typical particle generated within the region. Again the special problem of entering a region of stronger mean field will be unimportant if the field is suitably complex, having either special symmetries or almost-closed lines of forces.

E. The astronomical setting for cosmic rays.

18. The sun and the solar system. The sun is the only sure source of cosmic rays which we know. It is also by an enormous factor the closest possible source. It is the burden of this whole work to demonstrate that in spite of these facts, the sun is *not* the major source of cosmic rays whose spread in energy is so marked. It may be that most of the energy in cosmic rays comes from the sun originally, most of the energy being carried by particles just above, or even below, the few Gev/particle energy at the typical observed cut-off. A large part of Simpson's contribution to this volume is devoted to the solar cosmic rays; but we need here to recapitulate a few of the solar facts relevant to the discussion of non-solar sources[1].

For reference we include a summary table of some properties of the sun and of the solar system [5]:

Table 6.

Sun's Spectral class: $dG2$, a weak-line star of the galactic disk population
  Radius of sun (to photosphere) 6.96×10^5 km.
  Mass of sun $\qquad 1.99 \times 10^{33}$ g.
  Solar power (in visible region, plus u.v. + i.r.) $(3.92 \pm 0.2) \times 10^{33}$ erg/sec.
  Solar radio emission power: quiet sun $\qquad \sim 10^{24}$ erg/sec ($\lambda \geqq 1$ cm)
  $\qquad\qquad$ extreme disturbances $\sim 10^{26}$ erg/sec ($\lambda \geqq 1$ cm).
Mean Radius of earth's orbit: 1 a.u. $= 1.495 \times 10^8$ km. Of Jupiter's: 5.2 a.u.
  $\qquad\qquad\qquad$ Of Pluto's: 40 a.u.
  Approximate depth to photosphere (to optical depth $\tau = 1$) ~ 3 g/cm^2.
  Particle density: at photosphere $(R = R_\odot) \sim 10^{17}$/cm^3
    at lower chromosphere $(R = R_\odot + 3000$ km) $\sim 10^{11}$/cm^3
    in corona, at $\;1.1\,R_\odot \qquad 10^8$/cm^3
      at $\;2.0\,R_\odot \;\; 6 \times 10^6$/cm^3
      at $10 \quad R_\odot \;\; 2 \times 10^4$/cm^3

$\alpha)$ *Solar emission of matter.* These clouds or beams—their geometry is not clear, and both types may occur—extend out to planetary distances. At distances near 1 a.u. (the earth's orbit), it can be concluded from studies of the behavior of comet's tails, of magnetic storms, etc. [5], that typical properties are:

Streaming velocity	Particles/cm³ at 1 a.u.	Diameter
For beams emitted from normal dipolar regions on the sun:		
300 km/sec	$\sim 10 - 100$	$\leqq 0.1$ a.u.
For beams emitted from unusual events, like strong solar flares:		
2000 km/sec.	$\sim 10^5$	0.3 a.u.

---

[1] See also C. de Jager's contribution to Vol. LII of this Encyclopedia, pp. 215—222 and passim.

The magnetic fields are not directly known. They cannot be larger than some 10^{-3} gauss in even the most unusual events, for the variations in the earth's surface magnetic field are rarely so large, and these fields are not so rapidly-varying as to be screened out entirely by the ionosphere. From equipartition arguments, which are *not* reliable for such relatively transient phenomena, one might estimate that $\varrho v^2 \sim 10^{-3}$ ergs/cm^3, and hence have fields up to 10^{-2} gauss. But the study of the directional and time variations of the cosmic rays produced from the sun give evidence that the fields are, at some times at least, of magnitudes near 10^{-4} gauss, over distances up to 1 a.u.

How these clouds and their fields are arranged is not known, and is surely complex and changing. It is the presence of this magnetic system, with its constant change and renewal, which is certainly responsible for the time variations and isotropies found in the cosmic-ray beam, particularly at low energies. All these matter are discussed in other parts of this volume; we give the present account to serve as a model for the even less well known conditions in other parts of space, far from the sun, and to make it plain that the low-energy portion of the cosmic ray beam may be solar-modulated to a great extent. These are particles whose Larmor radii are less than some a.u. in fields of some 10^{-4} gauss, so that they have energies below some hundreds of Gev. Above that region, it is hard to see how the sun and the solar system can much affect the trajectories of cosmic ray particles, much less actually set them into motion. At even lower energies, say below 2 Gev, the sun and its solar system of magnetic-field-bearing clouds may be responsible for the main features of the spectrum. The variation of the cut-off at low energies with the solar activity cycle of eleven years seems to demonstrate this fact directly. The low-energy cut-off is not a problem in the study of the galactic origin of the cosmic rays.

β) *Cosmic ray emission by the sun.* The close connection in time between the few great sudden increases of cosmic ray intensity in the region below some 5 Gev, and the occurrence of great solar chromospheric eruptions, or solar flares, demonstrates that cosmic rays can be made by the sun. The direction from which these incremental rays originally come makes it pretty sure that they are in fact accelerated by processes near the solar surface, certainly within a very few solar radii away. These events are fully discussed by Professor SIMPSON in this volume. Here we wish to emphasize three points; they will all serve as arguments by analogy in our later work[1].

1. The solar-flare produced cosmic rays have a steeply-falling energy spectrum, and contain no important contribution above say 20 Gev, even though they make a very slight increase there in the most extreme of the known events, that of 23 Feb. 1956.

2. The observed energy stored in relativistic protons has been estimated to be about 20 ev/cm^3 during the great cosmic flare of 23 Feb. 1956. The low-energy spectrum was not observed, and is likely to contain several times more energy than that. The volume filled by the cosmic-ray burst was certainly as large as 10^{-3} of the sphere circled by the earth's orbit; it was much less than the whole sphere in angular opening, and not enormously larger in radius. It seems probable that the total energy sent out to regions beyond the sun's near field—much energy may have remained near the sun—was of the order of 10^{30} ergs. It may have been larger. The total visible light emission from this same event was several times 10^{32} ergs (not counting the "reddish-blue fan" of the Tokyo observers); the emission in the u-v and x-ray region may be guessed from the rocket

---

[1] See [*17*], and E. N. PARKER: Phys. Rev. **107**, 830 (1957).

data of Friedman *et al.* on other occasions[1] as 10^{30-31} ergs; the radio emission escaping from the corona, about 10^{27} ergs.

The efficiency of conversion of energy into relativistic protons at the earth is therefore about 10^{-3} or higher; there are indications that even more energy was converted into fast particles, including electrons and low-energy protons, which were not detected on the earth. This is an efficiency well beyond that of most man-made accelerators. High efficiencies are characteristic of cosmic-ray acceleration processes.

3. Synchrotron radiation from such events has perhaps been observed in the so-called Type IV radio outbursts of the Meudon group[2]. In these events, interferometer records at 200 Mc/sec show a strongly-emitting area, diameter perhaps a third the solar disc, which moves away from the disc as far as several solar radii. The area moves with a velocity of some hundreds of kilometers per second, and radiates as it goes with an intensity which may be up to 10^{-21} watts/cm²/cps at the earth, about as strong as any solar emissions whatever at these frequencies. They occur after solar flares, and apparently were present in all known cases when cosmic rays found the earth. They may last an hour or so, during which time their intensity first grows and then slowly declines. But unlike the other types of solar radio emission, the intensity has no rapid variation with time; records taken with time constants of seconds show perfectly smooth course, whereas the other bursts present very jagged plots. It is very tempting to identify these emissions as synchrotron radiation from great balls of trapped relativistic electrons moving out in the solar fields. They suggest that solar flares frequently produce cosmic rays, but that only when special propagation conditions hold in the region between earth and sun do we receive the sharp bursts described in detail by Simpson. (Compare Cocconi *et al.* in [*17*]) Moving areas emitting in the visible have also been seen crossing the solar disc after some flares[3].

In the special event of 23 Feb. 1956 the Tokyo observers[4] saw a fan-shaped patch of light which may be regarded as a blend of the reddish H_α with the blue continuum of synchrotron emission. It too was beyond the disc; it was not reported as moving (Fig. 8).

All of these results, while yet inconclusive, strongly suggest that the sun emits cosmic-ray particles up to some ten Gev in energy, many more than we can directly relate to individual solar events. Most of them simply diffuse out to space, to join the beam already present.

γ) *The cut-off.* The constancy of cosmic rays in time is now known to be a kind of artifact of observation. What one ought to say is that the rays seen near sea-level and at moderate geomagnetic latitudes are constant in time. But below a few Gev/particle the incoming beam is far from constant; it shows a nearly-complete cut-off below $\frac{1}{2}$ to 1 Gev/particle at times of solar activity maximum, but at times of solar inactivity (1953.5, for example, and nearly at eleven-year intervals from that date), the cut-off is absent. The particle number spectrum rises with decreasing energy, about like $1/E^2$, so that the total energy input above a kinetic energy of a few hundred Mev varies from peak to trough of the solar cycle by about 20%. It is here enough[5] to say that this seems surely due to a general modulation of the cosmic-ray beam by magnetic field conditions in the solar system, or perhaps around the earth. In no other way does it seem possible

[1] F. Chubb, H. Friedman, R. Kneplin and J. Kuperian: Nature, Lond. **179**, 861 (1957).
[2] See A. Boischot: C. R. Acad. Sci. Paris **244**, 1326 (1957).
[3] R. Bray, R. Longhead, V. Burgess and M. McCabe: Austral. J. Phys. **10**, 319 (1957).
[4] M. Notaki, F. Hatanaka and W. Unno: Publ. Astro. Soc. Japan **8**, 52 (1956).
[5] For details see Simpson's article in this volume.

to connect the change so closely with the solar cycle, and yet to ensure that the greatest cosmic ray influx to earth occurs when the solar output of cosmic ray particles is at a minimum. The sun controls the cosmic ray beam, presumably by gas emission which either brings with it magnetic fields, or generates them say around the earth.

We take these observations as a guide to the wider regions here to be considered. What the sun does, so also other objects do in differing degree, depending on the appropriate scale of the physical factors concerned. We shall present evidence for each of these solar phenomena in remote objects, often on prodigious scale. All of this will fit into a general scheme, according which the cosmic rays

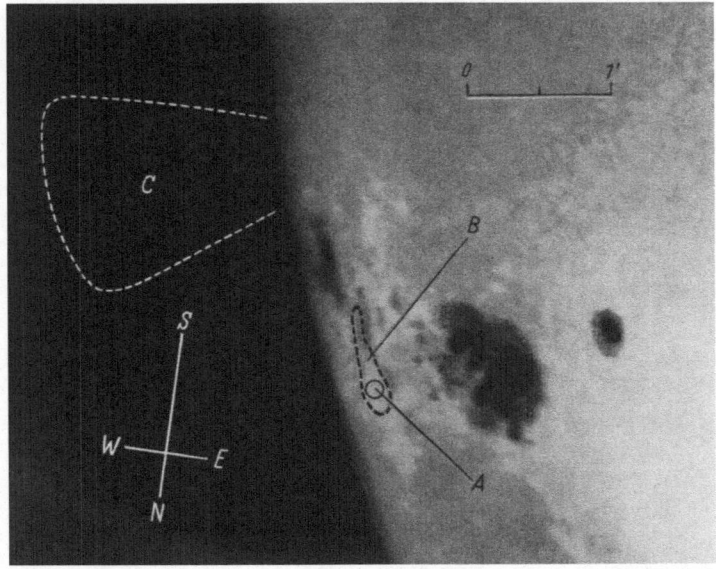

Fig. 8. The region C indicates approximately the extent of the fan of visible light seen outside the limb of the sun during the unusual flare of February 23, 1956. From M. NOTUKI, T. HATANAKA and W. UNNO: Publ. Astr. Soc. Japan 8, 52 (1956).

beyond ten or a hundred Gev reflect not solar origin or modulation, but phenomena of far wider extent in time, space and energy. It is noteworthy that no important variations (above 1%) in cosmic rays above 100 Gev energy are time-correlated either with the solar cycle or with any particular known event on the solar surface. On the other hand, as described by SIMPSON, the numerous variations in isotropy, intensity, and spectrum which are measured in tens of percent for rays of a few Gev are nearly all correlated with specific manifestations of solar activity, perhaps mediated by geomagnetic effects. It is natural to ascribe all these effects to the solar and solar system magnetic climate, even though the details are still far from understood. But the cosmic rays of low-latitude cut-off energies and beyond tend toward time constancy, and represent processes insensitive to local solar system conditions. We shall return to these points in Sect. 29.

Finally, we shall estimate the cosmic-ray output of the sun. Using the few known cosmic-ray flare events, we find that the visible energy output, in cosmic-ray particles between a few tenths of Gev and say ten Gev, with a very steeply falling energy spectrum, amounts to about 10^{31} ergs in a solar cycle, or about 10^{23} ergs/sec on the average. But this is very likely a lower limit. If we take it that many, or indeed most flares, produce cosmic-rays, but that the earth does not receive any well-defined pulse unless special conditions hold, we would be inclined to increase that limit by two orders of magnitude, for there are very many more optical flares than the cosmic-ray events indicate, though admittedly only the largest flares contribute much to the total. It seems plausible to attribute to the sun an injection of cosmic-ray energy in the low energy range of about 10^{25} ergs/sec.

19. The solar magnetic field[1]. The sun has a complex and varying magnetic field. It is probable that the many hydromagnetic phenomena we see, from the

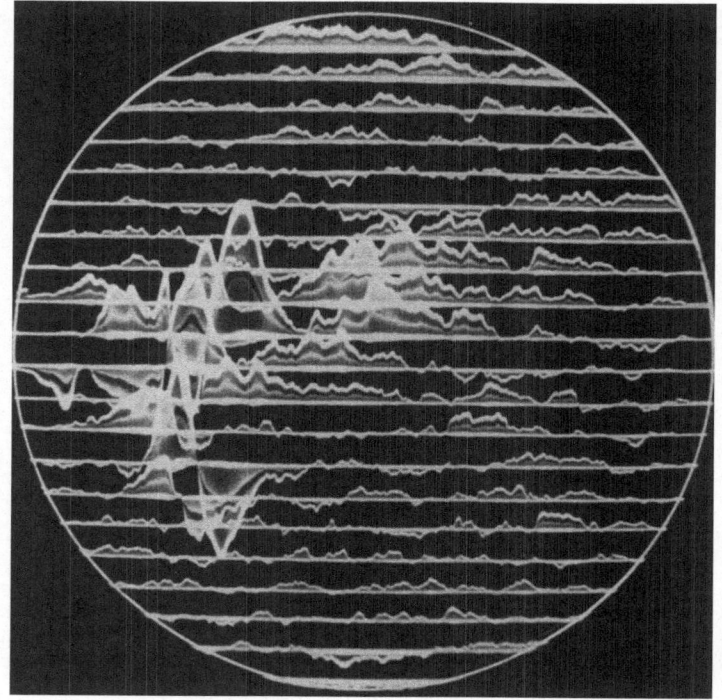

Fig. 9. The sun's magnetic field. Solar magnetograph measurements, made on 19 Aug. 1953, by H. W. and H. D. Babcock, Astrophys. J. **121**, 349 (1955), using the Zeeman splitting of one of the FeI lines. The magnetograph scans along parallels of latitude, shown as straight horizontal lines, and the deflection of the oscillograph trace is proportional to the magnetic field intensity in a narrow strip centered on the appropriate parallel. A deflection equal to the line spacing corresponds to a field of about one gauss; an upward deflection indicates a magnetic vector whose radial component (in the line of sight) is toward the observer, and vice versa.

intense local fields some (thousands of gauss) in the spots to the smooth large-scale fields which surely determine the shape of the tufts of coronal polar streamers seen in eclipses, are themselves secondary to large-scale and more or less intense internal fields about which we know little. But on the solar surface, the classical work of Hale at Mt. Wilson in the spots, and the later work of Thiessen and of the Babcocks, have given us a fair picture of the sun's field (Fig. 9).

It appears that the polar regions of the sun support a grossly dipole-like field, whose force lines are somehow made visible as the polar coronal tufts. This more or less smooth, but in detail fluctuating both in intensity and in direction, amounts to about one or two gauss at the poles. The field lines leave the north pole and enter the south pole (opposite to the earth's field); it is conceivable that this polarity changes over many solar cycles. In the equatorial regions, say

---

[1] Cf. also Vol. LII of this Encyclopedia.

fifty or sixty degrees north and south of the equator, there is a much more complex field. During the absolute minimum of solar activity (as in early 1954) this region is free of any fields of as much as 0.2 gauss, resolved along the line of sight. But at other times, this region is more disturbed. There are usually present a number of discrete regions of higher field strength, whose shape and strength may change from day to day, but which tend, like sunspots to show a recognizable identity for weeks or months. These regions range from the smallest, like single spots, diameters about 10^4 km, but fields of some three thousand gauss, to larger spots and groups of spots, to areas much larger than spots, covering up to a tenth of the whole hemisphere, with maximum field intensities of five or ten gauss. Spots tend to develop out of the larger magnetic areas. Most of these areas, like spot groups, contain poles of opposing sign, so that the field lines arch from one to the other, extending presumably out from the photosphere a tenth or more solar radii. But some regions occur which, like the rather more permanent polar caps, show at least a transient—lasting weeks or months—of magnetic field of a single sign, covering an area up to a couple of tenths of the disk, with a field strength of a few gauss.

Coronal streaming pictures confirm this general view. The more or less dipole field of the whole sun, which wavers and fluctuates, is dominant mainly in the polar caps. Near the equator, with great and frequent effect at times of solar maxima, diminishing at times of solar activity minima, but not vanishing, there are disturbed regions. These appear to be places where field lines leave the sun to arch out into space, closing back nearby in some cases. In others the field lines simply extend radially outward (or come in to the disk), with no evident area of return flux. Typical fields are measured in some few gauss over large areas, increasing in strengths as the area involved decreases, until some huge spots may maintain thousands of gauss over areas as large as 10^{-3} of the disk area. This sketch is much simplified, and the full complexity of the phenomenon is neither understood nor within our scope. But it is hard to avoid the conclusion that the solar field is carried bodily out into space by the outpouring of ionized gas, such as the stuff seen on the limb in prominence observations. The equatorial region of the sun must be the source of great clouds of magnetized conducting gas, first connected at their roots with the solar vicinity, later perhaps freed.

20. Stars with activity of solar type. There are no stars about whose surface events we can gain as detailed information as we have about the sun's. But even from the little information we do obtain, it has been possible to conclude that there is a numerous class of stars which show activity somewhat resembling solar flares. Of course, normal dwarf main-sequence stars like the sun might on general grounds be assumed to behave more or less as the sun does. But there are two classes of objects which display unmistakable activity: we can detect it in the integrated light of the star either because (i) the star's main radiation is much weaker than is the sun's, so that the flare disturbance, which is comparable with a solar flare, is much more conspicuous, or (ii) because the actual flare-like event is on a scale far exceeding the solar one.

The two classes of objects are:

α) *Flare stars.* Since the work of JOY and HUMASON[1] and of LUYTEN[2], it has been clear that among the reddish low-luminosity dwarf stars, to the right of the sun along the main sequence, a special kind of variability is not uncommon.

[1] A. H. JOY and M. L. HUMASON: Publ. Astronom. Soc. Pacific **61**, 133 (1949).
[2] W. LUYTEN: Astrophys. J. **109**, 532 (1949). — P. ROQUES: Publ. Astronom. Soc. Pacific **67**, 34 (1955).

Some dozen stars have been observed to display an irregular variability, with outbursts taking from minutes to hours, which considerably augment the total light from the star, by as little as ten percent or so up to a factor of fifty in one case (Luyten). Most of the increase in light can be ascribed to emission lines, chiefly H_α, but some also to the familiar Ca emission lines, exactly as in the optical light from the ordinary solar flare. These stars are of low luminosity, say from under $10^{-2}L_\odot$ to as low as $10^{-4}L_\odot$, and of spectral type dM or nearby. They are objects with masses about 0.1 or $0.2M_\odot$, and radii about $0.3R_\odot$. The light power output in such a stellar flare is from 10^{29} ergs/sec to a few times 10^{30} ergs/sec, compared with a big solar flare power of $\sim 10^{32}$ ergs/sec. There is very little information on how often such flares occur; but this class of stars is numerous. There are about ten times as many such stars in the galactic neighborhood of the sun as there are stars of higher luminosity, and about a tenth of them appear to show flaring at least from time to time. It is plausible that at least some of these stellar flares, like some solar flares, are cosmic-ray sources. It is hard to estimate the total cosmic-ray ejection even roughly; it is not likely to be much different from the solar output in absolute amount, though it is relatively larger, compared to the small light output of such dwarf stars. From the sizes of such objects, it is also plausible to infer that the cosmic rays produced will not be of energies exceeding those of solar cosmic rays; probably most of the energy injected is in particles below a few Gev.

β) *T Tauri stars.* Several hundred stars are known[1] which are also rapid, irregular emission-line variables, showing wide emission lines against a background absorption spectrum not very different from the main sequence. Their emission spectra are reminiscent of the sun's chromosphere, including a strong background continuum, well into the ultra-violet. Moreover, some forbidden lines are seen, implying emission from very tenuous gas. These stars are dwarfs, but not so under-luminous as the flaring variables; their outputs range from about the solar value down to some 10^{-2} solar, with spectral type F and G, plus, of course, the characteristic emissions. They are characteristically associated with galactic nebulosity, of fluorescent and dusty type, and with dark matter. Near some of the stars, for example the prototype, T Tauri, a nebular fan can be seen closely associated with the star.

The close relation between these stars and the great nebular complexes, for example, in the region Orion-Auriga-Taurus-Monoceros, is evidently responsible for their distribution in direction, which shows peaks both toward the galactic center and away from it. Presumably these stars belong to the spiral arms, and we lie more or less in a position to inspect the arms most closely along a radius, looking either way from the center. Such stars are not rare in a suitable region: in the diffuse gas of the Orion nebula, a region perhaps 10^5 cubic lightyears in volume, and some 1500 lightyears away, hundreds of them have been noticed (Herbig, Haro[2]). Now, T Tauri stars have been seen in dynamical association with very luminous and hence young stars. Nor have they been found elsewhere. This strongly suggests their great youth, which may be measured in millions of years only. From their color-luminosity spread, it appears that they may evolve sizably in this short time, a point difficult to understand, since they send off rather little visible light. They do radiate abnormally in the near ultra-violet; in that spectral region they typically send off more light than a normal star of the same color by a factor of three to five. This ultraviolet light fluctuates as well, but the total output and the ultra-violet continuum do not show markedly

[1] A. H. Joy: Astrophys. J. **102**, 108 (1945); **106**, 288 (1947); **110**, 427 (1949).
[2] G. Haro: Astrophys. J. **117**, 73 (1953).

correlated fluctuations. These stars must lose much energy in novel ways. Polarized light has been reported[1], but (alas) not confirmed[2].

21. Mass ejection by stars. The emission of cosmic-ray particles is demonstrated for solar flares, all but certain for supernovae, and by analogy at least very likely for the novae, the magnetic variables, and the flaring variables. But there is good evidence that the sun, for example, not only emits rays during flares, but steadily emits particles of lower energies, in the range up to tens of kilovolts, (velocities near a thousand kilometers per second). The most direct evidence is so far that of BIERMANN[3], who discussed the observed brightness variations and recoil motions of the gas in comets' tails. Even if no real explosive process is seen, it is rather probable that such steady emission also contributes at least some flux of cosmic-ray energies, perhaps whenever certain magnetic instabilities arise in the emitting region. If these particles possess a steeply falling energy spectrum, they will not show up in terrestrial observations as a large solar component of cosmic radiation. Other stars, built on altogether larger scale, may contribute cosmic rays even in the Gev region as concomitants of a steady particle emission[4]. Several types of stars are held to give evidence of steady particle emission; we will describe them, with the plausible assumption in mind that at least some of this efflux of energy and matter can appear in the energy region of cosmic-ray importance.

α) *Supergiant stars of various spectral types.* Stars which are more luminous than the sun by a factor of from 10^4 to 10^5 are called supergiants. They are found in all temperature ranges. At the high-temperature end they merge into the high-luminosity end of the main sequence; at the red end of their range they resemble the red giants, these latter being typically less luminous by a factor of a hundred. Such great stars cannot survive very long; they are not numerous, although they contribute a major part of the total light of the stars. It is fairly sure that they represent evolutionary stages in the life of massive stars which arise after nuclear reactions are pretty well advanced. From direct luminosity measures ABT[5] has shown that many, perhaps all, of these stars are variable, demonstrating some sort of surface instability. Asymmetrical absorption line profiles have been interpreted as showing the likelihood that the gas has an outward mass motion. This is clear for a red supergiant, α Herculis, studied by DEUTSCH[6], who ascribes to this star a mass loss of 10^{42} particles/sec, with a mass velocity of 10 km/sec. But supergiants up to the early spectral class A probably share this property in greater or lesser degree, even though ABT felt that the profiles could be interpreted as due to rotation alone. In the case of the cool supergiant α Herculis, it is clear that non-gravitational forces are acting to support the outflowing gas, which seems to be non-uniform in density. It seems very plausible to allow such unstable and highly-luminous stars the property of emitting some cosmic rays, perhaps in the region of tens of Gev. We will make a guess as to the total output in the close of this section.

A few unusual supergiants, like the nova of 1600, P Cygni, show the presence of very definite and large emission, an outward moving shell surrounding this star which may take off as much as 10^{44} particles/sec at some hundreds of

[1] K. HUNGER and G. KRON: Publ. Astronom. Soc. Pacific **69**, 347 (1957).

[2] W. HILTON and B. IRIARTE: Astrophys. J. **127**, 510 (1958).

[3] L. BIERMANN: Z. Astrophys. **29**, 274 (1951). — Z. Naturforsch. **7**a, 127 (1952).

[4] A. UNSÖLD: Phys. Rev. **82**, 857 (1953).

[5] H. ABT: Astrophys. J. **126**, 138, 503 (1952); **127**, 658 (1958).

[6] A. DEUTSCH: Astrophys. J. **123**, 210 (1956). — Publ. Astronom. Soc. Pacific **68**, 308 (1956).

kilometers/sec. This star loses more mass by ejection than by radiation. Few such objects are known, though they may be thought of as merging into the planetary nebulae. These too are stars which are either now or were originally highly luminous, which are now surrounded with an expanding shell 0.1 to 1 lightyear across, typically fluorescent, moving out with velocities in the tens of km/sec. There may be 500 of these objects in the galaxy; the mass sent out is something like $2 \cdot 10^{53}$ particles/sec total.

β) *Shell stars on the main sequence.* (a) Wolf-Rayet stars. These objects are not infrequently the origins of the planetary nebula. They form some of the class of stars of O type at the high-luminosity end of the main sequence, or perhaps just below it, and show very wide characteristic emission lines, with the asymmetrical absorption which suggests an ejected shell. The velocities are high, in the hundreds of km/sec, and the mass ejected may be 10^{44} particles/sec.

(b) Emission-line stars of the main sequence. A good many B stars, perhaps one in ten, show emission lines with the asymmetrical absorption. They resemble the prototype P Cygni, though they are generally less luminous. These, too, are probably surrounded by outgoing shells, velocities in range of 20 or 30 km/sec. The mass ejected is not easy to estimate; evolutionary arguments have been applied to it in somewhat circular fashion. These stars are of course much more numerous than the unusual objects like the P Cyg ni supergiants.

(c) Tidally disrupted stars. Stars with very high rotation, or stars which are members of close binaries are very often surrounded by shells which must be ascribed to tidal disruption of the stellar surface by inertial and gravitational forces. These motions would not in themselves generate any cosmic rays; such mass motions might very well couple energy into magnetic fields, and begin, at least in regions where the density became suitably low, the emission of cosmic rays.

Nearly all of the processes listed are characteristic of stellar population I (see Sect. 25). Most of them are even restricted to the patchy spiral-arm kind of distribution which is called extreme population I. In any case, these processes are plentiful in the solar galactic neighborhood. It is interesting to make what is patently a very rough estimate of the cosmic-ray source these processes might represent. If energy emitted as kinetic energy of ejected mass is called E_m, we may set the postulated cosmic-ray emission as equal to $\varepsilon_{CR} E_M$, with ε_{CR} the energy efficiency of cosmic-ray emission from such objects. Summing up all the emission processes discussed under 1 and 2 (a) and (b) above (not counting tidal losses), we arrive at a very rough figure (see Biermann[1]) of $E_m \approx 10^{39}$ ergs per sec in the whole galaxy. To correct this for the concentration of these objects toward the galactic plane, we may estimate they fill a cylinder of galactic diameter and thickness a thousand lightyears. The result for cosmic-ray emission is then $10^{28} \varepsilon_{CR}$ ergs/sec in 100 (lightyears)^3, giving the source power in a volume like that we can assign to a single star like the sun. These processes then, if they exist at all, could be more important than the contribution of stars like the sun if the efficiency ε_{CR} exceeds the value of 10^{-5}. For the sun, the total energy in steady particle emission is 10^{30} ergs/sec, and in cosmic rays something like 10^{23} ergs/sec up to 10^{25} ergs/sec (see Sect. 18). Thus the solar value for ε_{CR} is from 10^{-7} to 10^{-5}. The value appropriate for the unstable and turbulent objects here described is not plausibly put any lower. It follows that there is a sizable cosmic-ray emission from these objects. We emphasize again that these are all merely arguments from analogy; only from the sun can we yet be certain cosmic rays come.

---

[1] L. Biermann, in: Gas Dynamics of Cosmic Clouds, IAU Symposium Z. New York: Interscience Publ. 1955.

22. Magnetic variables. H. W. Babcock[1] has found that many stars show strong coherent magnetic fields within their atmosphere, with field strengths ranging up to thousands of gauss, or even more, and showing large and rapid fluctuations, some even reversing coherently in a matter of several days. It seems almost certain that a fair fraction of this changing magnetic energy will induce the sorts of hydromagnetic shocks which can generate cosmic-ray particles. These magnetic variables can be regarded as members of a series which begins with the turbulently magnetic flaring dwarfs, goes through the sun, with its flares and related complex magnetic disturbances of large scale and continues into the stars of spectral type B, a few of which show strong magnetic fields. The fields are measured by the ingenious application of the Zeeman effect: observations of the polarization of the Zeeman-split wings in strong iron lines of the spectrum demand this interpretation. While under a hundred stars of this type have been found, it seems probable that nearly all stars of class A are magnetic, except some which by chance have negligible angular momentum.

Though we have no direct evidence, it is probable that these stars in fact do emit very many cosmic rays. Estimating the volume and the field changes involved, and comparing this, say, with a solar flare, one might easily guess that such a star emits from 10^5 to 10^7 times as many cosmic rays as does the sun. These A stars may also generate some rays up to energies scaled up for diameter of field region and for field strength from the values appropriate to solar flares. This might mean a spectrum extending up to 10^{12} ev.

Such stars are less common than the dwarfs and near-dwarfs like the sun or smaller. They belong to the intermediate population I (see Sect. 25), and in our galactic neighborhood are perhaps an order of magnitude less numerous than stars of solar and near-solar type. Their overall light output is an order of magnitude greater than the total light output of all dwarfs of solar type and below. Whether on the basis of a constant fraction of light output, or on the even more impressive estimate of magnetic energy stored, these stars might as a whole outweigh the cosmic-ray contribution of dwarfs by several orders of magnitude, not yet approaching what is needed for overall cosmic-ray generation in the galaxy. It is quite possible, though, that in the patch of a spiral arm where we live with the sun, perhaps in that loose assembly of diameter ~ 2000 lightyears called the "local system" or "Gould belt" which is defined by population I objects, there is a considerable trapping of cosmic rays generated by stars like the supergiants, the magnetic variables, and the flare-type dwarfs, and thus a higher mean density of cosmic rays. We shall return to this general picture in the final Sect. 29.

23. Explosive stars. $\alpha)$ *Novae* [7]. A couple of hundred events have been seen, most within our galaxy, but scores in near-by galaxies as well, which are called *novae*. A nova is a star which suddenly rises in visible luminosity, increasing its light output power by a factor of greater than 10^4 in from one to a few days. The pre-nova state of the star appears to be not very different from that of the sun in mass or luminosity, perhaps somewhat less luminous. From this state it sharply rises in output power, reaching a rather uniform maximum luminosity of about $10^5 L_\odot$. During this time it ejects a thickish irregular shell of glowing gas, often merely a set of more or less orderly jets, which can be observed by their specific emission spectra, showing complex Doppler shifts and other forms of broadening, and has in a few cases been seen as an expanding nebular disc (especially in Nova Aquilae 1918, or V 603 Aquilae). The gases move out with a velocity of around

---

[1] H. W. Babcock: Phys. Rev. **109**, 2210 (1957).

1000 km/sec, with no very great tendency to spherical symmetry. The light output begins to fall off strongly after a few days to a few hundred days. The light curves, like the spectra, are neither unique nor simple, often showing periodic

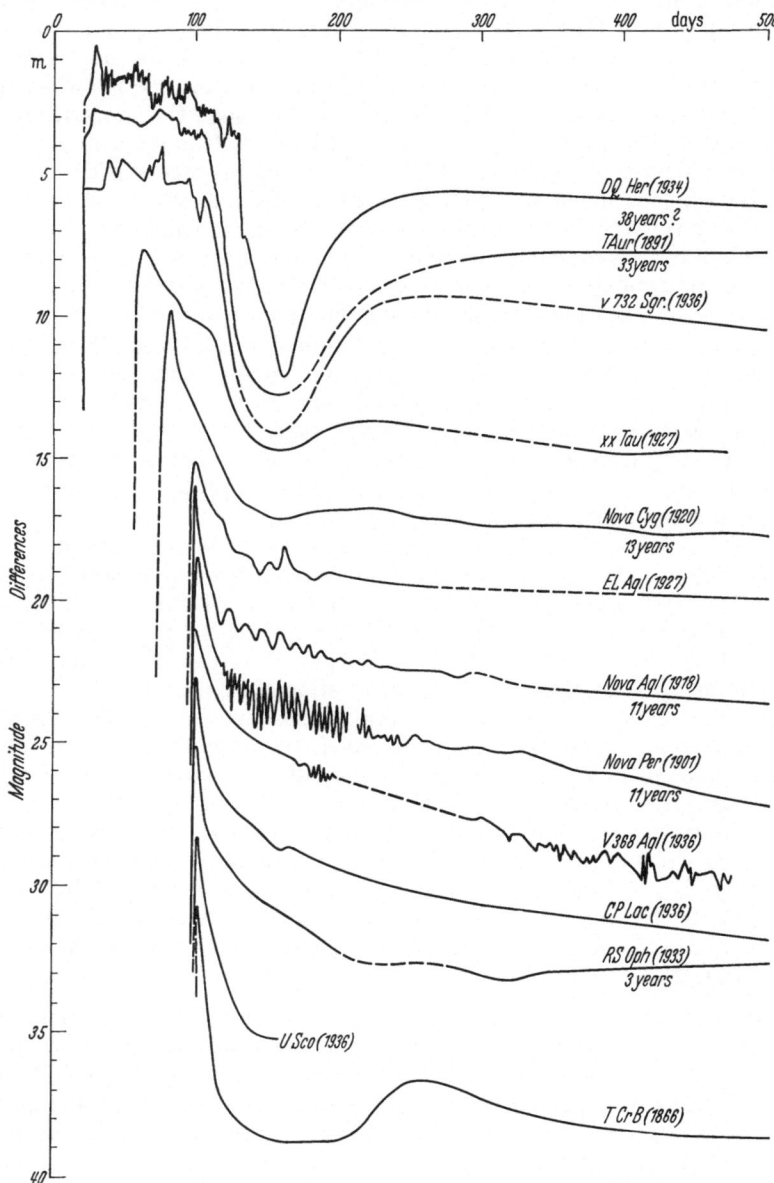

Fig. 10. Light curves of some novae. Time given for some curves is the time required for the nova to return to normal brightness. From L. Campbell and L. Jacchia: The Story of Variable Stars, p. 126. Philadelphia 1941.

changes, fluctuating maxima and minima in light output around the general decline, etc. After some years the star steadies down, presumably with the shell no longer emitting very much, and so far away that changes are slow. Estimates of the ejected mass run around $10^{-3} M_\odot$.

While each nova is different from the others, and a whole series of light curves and spectra may be established, there is a strong family resemblance (Fig. 10). The maximum power is rather uniform, the integrated light output somewhat less so, ranging from about $6 \cdot 10^{44}$ ergs for a typical strong nova, down by a factor of ten or so to the weaker events. There is even a kind of periodicity; some novae have been seen to erupt more than once, and there are irregular variable stars which appear to be quite weak, quasi-periodic "novae". But the main class, for all its detailed variations, is plainly recognizable.

The event is clearly spectacular, but it is not really catastrophic. The mass lost is a small part of the star's total mass, and even the light output amounts only to some 10^5 years' total, out of the billions of years the star presumably is capable of shining. The cause is unclear, but it seems likely to involve the passage of a strong shock wave through the star's outer layers, possibly initiated by some thermonuclear instability. Novae appear to occur with a frequency of about fifty per year in our galaxy, showing a mild tendency to take place more frequently in the star-rich nucleus of the galaxy, but less concentrated there than some other objects, such as planetary nebulae, which are not unlike fossil "weak" novae. They occupy a flattened disc intermediate in shape between the overall star population of the galactic framework and the rotating flattened disc of the spiral arms and their stars. (See the next section for further comment.)

There is no proved connection between cosmic rays and novae. The signs of cosmic-ray production—strong radio emission, or polarized visible light—are not known for any nova so far examined. But their sudden emission of masses of ionized gas is reminiscent of solar flares, and cannot be ignored. SHKLOVSKY[1] [4] in particular has expected novae to be strong cosmic-ray sources, emitting radio and polarized light. Perhaps it will yet prove to be so; but among a good many novae[2] no such source has been found.

$\beta)$ *Supernovae.* More than fifty events, including three in our own galaxy, are known of a related but extreme kind. These were named "supernovae" by BAADE and ZWICKY (1934). They are genuinely catastrophic explosions of stars, in which the star is deeply and irreversibly modified, losing some percent of its mass as an ejected shell, and sending out an integrated amount of light equal to a billion years output of the sun.

Even in this small sample of events, there have been recognized two distinct types. Supernovae of Type I have been seen perhaps twenty times, including the three supernovae in our own galaxy. They are a remarkably uniform class; they reach in a day or less a peak magnitude, with an output in the visible of about $2 \times 10^8 L_\odot$, and begin to fade at once, dropping smoothly by a factor of twenty or more in some three or four weeks. Then they begin a strictly and simply exponential decline in luminosity, remarkably regular and uniform, which carries their output down with a half-life of about 55 ± 5 days, until after six or seven hundred days, they fade from sight (in the distant galaxies, for which alone we have such late data). This remarkable light curve has given the clue to the nature of the event[3] (Fig. 11).

Supernovae of Type II are more frequent, perhaps by as much as ten times. They are less luminous at peak by a factor of about ten, but they show a wider dispersion of peak output. Their light curves are less uniform, and do not have the simplicity of type I. Typically they display a rapid rise, a slow fall, and

---

[1] I. S. SHKLOVSKY: Astronom. Zhurn. **32**, 215 (1955).

[2] B. MILLS, A. LITTLE and K. SHERIDAN: Austral. J. Phys. **9**, 84 (1956).

[3] E. BURBIDGE, G. BURBIDGE, W. FOWLER and F. HOYLE: Rev. Mod. Phys. **29**, 547 (1957). — Contains account of element formation in general Bibliography.

finally a rather swifter decline, setting in after a few months. The "shoulder" in the light curve (logarithm of luminosity vs. time) is taken as typical. None have been seen in our galaxy; it is possible to regard Type II supernovae as merely extreme examples of the nova phenomenon, but this seems less likely than their close relation to the Type I supernovae.

The total light output for a Type I supernovae is about 10^{49-50} ergs, for a Type II, ten to a hundred times less. It appears that supernovae of Type I ought to occur in the galaxy perhaps once every one to two hundred years, and those of Type II several times more frequently. These estimates are very doubtful, for they are based on a small sample from a wide variety of galaxies. Indeed,

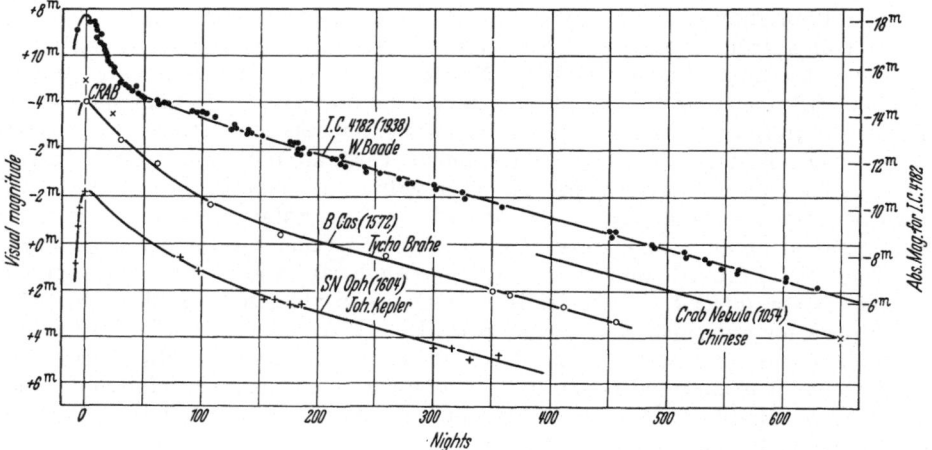

Fig. 11. Light curves of supernovae by BAADE. Measures for SN IC 4182 are by BAADE; those for B Cassiopeiae (1572) and SN Ophiuchi (1604) have been converted by him to the modern magnitude scale from the measures by Tycho Brahe and Kepler. The three points for the supernova of 1054 are uncertain, being taken from the ancient Chinese records. The abscissa gives the number of nights after maximum; the left-hand ordinate gives the apparent magnitude (separate scale for each curve); the points for the Crab Nebula belong on the middle scale, i.e., that for B Cassiopeiae. The right-hand ordinate gives the absolute magnitude for SN IC 4182 derived by using the current distance scale. From G. R. BURBIDGE, E. M. BURBIDGE, W. A. FOWLER and F. HOYLE: Rev. Mod. Phys. 29, 547 (1957).

three supernovae of Type I have been seen within the last thousand years in a volume of only about 1% of our own galaxy; in rather more galaxies than would be expected on a purely random basis, more than one supernovae has been observed. It is generally said, on rather thin evidence, that the Type I supernovae are to be found in the galactic centers, and in a smooth flattish distribution along the galactic plane (Population II, so-called, see Sect. 25), while the Type II supernovae are confined to the spiral arms. This is very likely to be modified as statistics improve. Type II, in particular, may turn out to be an inhomogeneous class.

The supernovae, like the novae, eject material at high speed into a shell of gas, more or less spherical in shape. For Type II supernovae, a very blue continuum is seen at maximum, and later strong emission lines, which resemble the spectrum of "a gigantic ordinary nova", and are broad enough to allow the assignment of expansion velocities of some five thousand km/sec. For Type I supernovae, the matter is less clear. The spectrum is "enigmatic", resembling nothing else seen by astronomers, it consists of wide overlapping bright lines, none identifiable, and hardly distinguishable from a modulated continuum. In the Type II spectrum, hydrogen lines can be identified, and some others, like N III.

Because of the violent nature of the event, the supernovae have often been implicated as sources for cosmic rays. Not much can be said about this from the distant supernovae, but the study of the few remnants which are found in our near galactic neighborhood has given real strength to that point of view. The discussion which follows outlines what is known about the nearby supernovae.

Three or four galactic supernovae are known. Three were seen visually to explode: the Crab Nebula in 1054 A.D.; Tycho's nova, B Cassiopeiae, in 1572, and Kepler's nova, in 1604. Only the Crab is at all striking today. Tycho's star, like Kepler's, does appear to be a weak radio source. Tycho's star cannot be seen, but it lies in a direction heavily obscured by dust. Kepler's star can be seen, much reddened and obscured, as an expanding nebular wisp, much less bright than the Crab Nebula, and without any discernible blue continuum. (Compare the next section for details of the Crab.) Cassiopeia A, the brightest of radio sources (see Table 11) can also be seen as a rapidly-diverging set of narrow nebular wisps. They can be interpreted as pieces of an expanding shell arising from a galactic supernova which is unknown in the records, though only two or three hundred years old. (W. Baade, unpublished but widely cited.) All four of these supernovae lie close to the galactic plane. The Crab lies on an extension of that galactic radius which passes through the sun, but further out from the center, probably within the next spiral arm. The Cassiopeia A source seems to be rather closer to us, and perhaps lies within our own spiral arm.

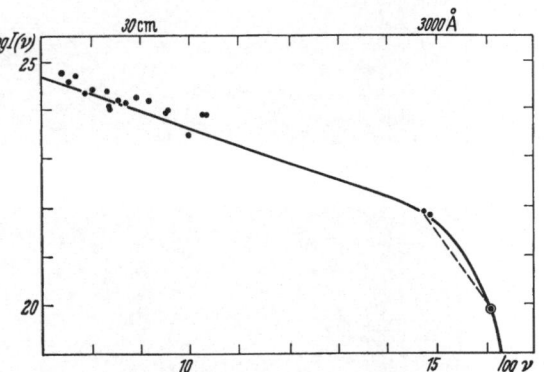

Fig. 12. The spectrum of the continuous emission from the Crab Nebula. The units of ν are (cycles/sec) and those of $I(\nu)$ are erg-sec^{-1} — (c/sec)$^{-1}$. From L. Woltjer: Bull. astronom. Inst. Netherl. **14**, 39 (1958).

24. The Crab Nebula. This unique object was the first in Messier's list, and is NGC 1952. It is known to be the remnant of a supernova which became visible in June of 1054 A.D.[1]. The angular displacement of features of its boundaries can be extrapolated backwards in time. The reversed motions collapse the present object into a small "circle of confusion", at a time about 100 years later than the observed date. This small acceleration appears real. A measured Doppler splitting implies a radial velocity, i.e., in the line of sight, of 1150 km/sec. Taking this with the angular motions, assuming the shape remained similar throughout the motion, the distance becomes about 6000 lightyears. (The usual value of about 3000 lightyears is modified by Woltjers [11] because it is based on a model of the motion which forces a curiously anisotropic emission of light. Such a phenomenon is unlikely, though possible; the distance must still remain in doubt.) The Crab appears as a tangled filamentary shell surrounding an amorphous cloud within (Figs. 13a, b). The central cloud emits an unusual optical continuum while the filaments outside show strong emission lines, not unlike other nebulae. The shell measures roughly six lightyears by about four lightyears as projected, and is about six lightyears deep. The analysis of intensity of emission lines leads to a visible mass of the filamentary shell about 0.1 solar masses, one third hydrogen. The total mass may be 10 times higher.

---

[1] Hsi Tse-tsung: (Translation), Smithson. Contrib. Astrophys. **2**, 109 (1958).

a

b

Fig. 13a—e. The Crab Nebula, M1, NGC 1952. (a) Photographed in the light of Hα (6300—6700 Å filter). (b) Photo
graphed in a portion of the spectrum without emission lines, seen in the light of its continuum only (7200—8400 Å filter).

About nine-tenths of the optical emission of this object is in the continuum.
This amounts to about 5×10^{36} erg/sec. The object is a strong radio-source as
well, and emits in the radio-spectrum some few times 10^{33} ergs/sec. The complete

c

d

Fig. 13 c and d. In (c) and (d) the continuous emission is again seen, but now through a polarizing filter in two different planes of polarization. The white arrows mark the direction of the electric vector passed by the filter in each case. See W. Baade: Bull. astronom. Inst. Netherl. **12**, 312 (1956).

spectrum of the continuous emission from radio to ultra-violet is shown in Fig. 12, plotted on log-log scales. There is evidence for a cut-off beginning around 3000 Å.

   Some changes are still observed to take place in the nebula. A few times each year "light-ripples"—small bright wisps—appear near the supposed central star and move away with a velocity of about one-tenth that of light, merging into the amorphous glowing mass when they reach its outer regions. Largerscale changes are seen to occur over several years; it seems likely that the nebula is the sum of many such structures, the old ones having merged into the nearly uniform background. A single light-ripple contributes about 10^{-4} of the whole emission.

It was the brilliant prediction of SHKLOVSKI that this continuous radiation was strongly polarized, rapidly confirmed both at radio and optical frequencies which began to explain the enigmatic emissions of this object. The observed

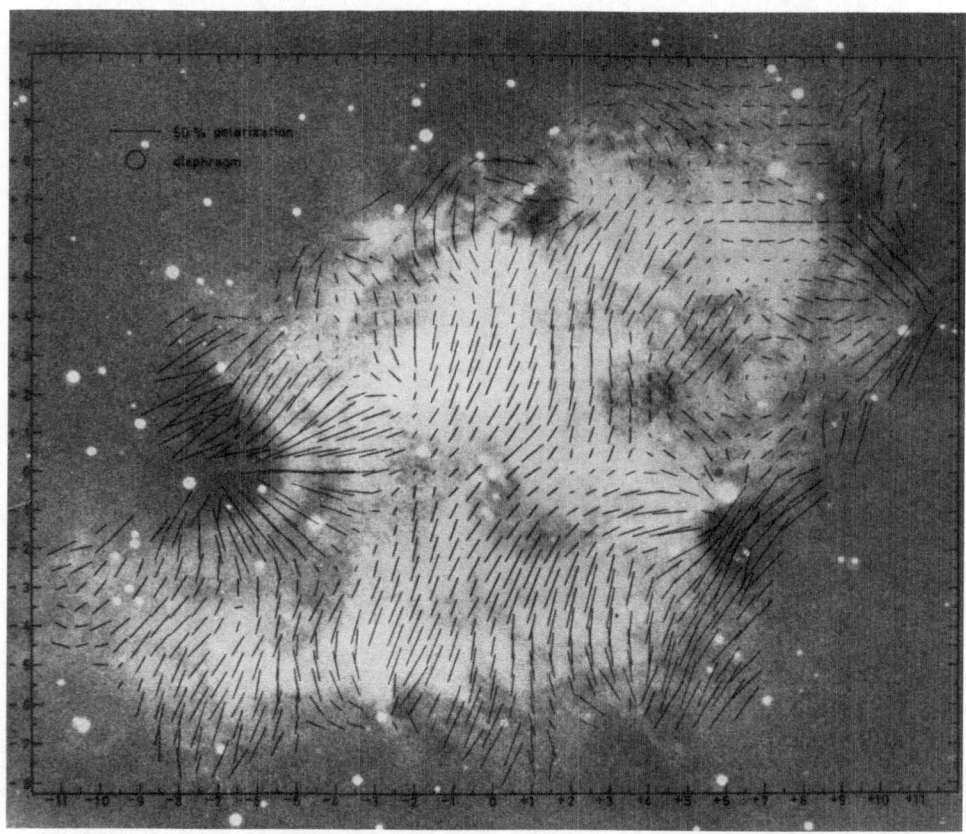

Fig. 13 e. The measured polarizations, projected against a photograph of the Crab in the light of [S II] lines (4068.6−4076.3 Å filter). From L. WOLTJER: Bull. astronom. Inst. Netherl. 14, 39 (1958).

polarizations are shown in Figs. 13 c, d, and e. There can be little doubt that we see here the electron synchrotron radiation described in Sect. 13. From the physics of this process, we can explain the polarization structure as implying the presence of magnetic fields, whose force lines run everywhere normal to the observed plane of the electric vector. Using this clue and the Doppler and angular velocities, WOLTJERS has shown the field to be a roughly "force-free" field, (i.e. $\nabla \times \boldsymbol{B}$ proportional to $\boldsymbol{B}$, so that $\boldsymbol{j} \times \boldsymbol{B} = 0$) nearly tangential everywhere around the boundary of the nebula. The emission filaments which wrap the outside are concentrations in a sheet of current around the surface of the whole object, which is the source of the field. The dark bay at the lower left of Fig. 13 e is a point at which we look down the axis of current flow, the lines of force encircling the current. The polarization directions are plainly arranged just as they ought to be around some great wire, while the darkness must imply that electrons cannot easily diffuse into this strong and regular field with its locally almost closed lines of force. There is indication, from the comparison of radio with optical brightness distributions, that the field is somewhat stronger in the center, by a factor under five, and that centrally the electron spectrum is richer in high energies as well. The moving ripples are all but completely polarized, and move perpendicular to the local direction of the magnetic lines of force. The structure of the field, while complicated, is not of very small scale; the field appears to remain roughly uniform over a fair fraction of a light year.

These qualitative features are so striking and fit so well together that they fully confirm the general model of a great expanding magnetized region filled with relativistic electrons radiating polarized continuous radiation over a very wide range of frequencies. Beyond that, to a more quantitative account, it is as usual not at all easy to go. Using a power-law energy spectrum for the electrons, one may fit the observed continuous radiation, deriving an energy density for electrons of all energies up to a maximum energy, for those which radiate light near the ultra-violet cut-off. This density depends on the assumed field strength. The higher B, the less electron energy is needed. But for high B, the total magnetic energy becomes large. A table of values, formed following WOLTJERS with some modifications, from the material of Sect. 13 on synchrotron radiation, makes the point clearly.

τ is the mean lifetime against energy loss by radiation for an "optical" or a "radio" electron. The minimum total energy lies between the limits given. If this criterion is disregarded, it is still clear that too high field strengths will so shorten the lifetime against energy loss by radiation that the uniformity of the structure would be impossible, while too low values rapidly increase the energy demands on the original event and might allow the fast particles to "break through" the field. It is useful to list a few other properties of the region in appropriate units:

gas density: less than 10 atoms/cm^3,

energy loss by ionization (relativistic particles): 40 ev/year.

The low density of the gas implies that collision losses are negligible compared to the acceleration gains for all relativistic particles, at least.

The key question is of course how these fast electrons are related to cosmic rays. We know they are not themselves important components of the general cosmic ray beam. But they imply such efficient acceleration mechanisms that it is very hard to see why protons should not be accelerated as well, either to yield the same spectrum of rigidity, and hence the same total energy, or even to yield the same values of E/mc^2, and thus 10^3 times as much energy in cosmic rays as in electrons. BURBIDGE[1] has even suggested that the electrons are all injected at high energies by the decay of mesons secondary to proton collisions in the gas. This places a very heavy total energy requirement on the system, and seems rather unlikely. But it is to be stressed that the assignment of large cosmic ray source strengths to supernovae is a plausible argument from analogy only. It appears to be a very plausible one; it plays the major role in the discussion of the Moscow workers (see GINZBURG [16] and SHKLOVSKI[2] [4]).

The supernovae of Type I have been broadly explained by BURBIDGE and others who make the remarkable exponential decay curve the main clue to their

---

[1] G. B. BURBIDGE: Astrophys. J. **127**, 48 (1958).

[2] First in Dokl. Akad. Nauk SSSR. **90**, 983 (1953). Also Astronom. Zhurn. **30**, 15 (1953); **32**, 215 (1959).

argument. That curve reflects, they say, the radioactive decay of the heavy isotope californium 254, which has just that lifetime and is unique among radioactive substances in that it is the lightest nucleus which decays primarily by spontaneous fission. Thus its energy release will dominate any mixture in which it occurs with an abundance a few percent of the total. The supernova explosion is the result of the implosion of a star whose core has consumed all its nuclear energy sources, and is too massive to be held stable against collapse by the pressure of a degenerate electron gas. The internal collapse starts an implosion of the outer layers, which are supposed to contain enough hydrogen and other thermonuclear fuel to generate a great explosion upon collapse. Neutrons are made in great numbers, possibly even from the whole of the core (Cameron[1]), and their capture in rapid succession yields elements of every mass up to such transuranics as the californium. The energy available must be at most something like that obtained by converting a few solar masses from hydrogen to helium. This is a maximum of about 10^{52} ergs. The kinetic energy of the shell expanding now amounts to 10^{49} ergs, the magnetic and electron energy are quoted in Table 7

Table 7.

Magnetic field gauss	E_{max} Gev	E_{radio} Mev	Total electron energy ergs	Magnetic energy ergs	Electron energy density ev/cm³	τ years opt.	τ years radio
10^{-3}	600	200	$5 \cdot 10^{47}$	$3 \cdot 10^{48}$	1500	3	10^{4}
10^{-4}	1800	600	$15 \cdot 10^{49}$	$3 \cdot 10^{46}$	4500	100	$3 \cdot 10^{5}$

above. The visible energy emitted in the initial stages of the supernova must come out in other than visible form, the total emission lies around 10^{51} ergs, only one order of magnitude below the plausible upper limit. Since the development of the shell into something like its present size, say for the past 10^{10} seconds, it is probable that the object has emitted at least 10^{-4} of the total energy emitted in the event.

Whether any of this energy appears as cosmic rays is not certain. It is hard to doubt that at least in the early stages, much did come out in such form. Now it appears that the acceleration process takes place by virtue of collisions between the charged particle and the moving ripples. This appears to be an efficient means of acceleration. Relativistic electrons are constantly being injected from beta-decay, as are protons from neutron decay, following spontaneous fission. The electrons are rather more abundant than the protons, and have a much lower ionization loss to be accelerated. The energy supply of the object may now come from the radioactive debris, which should liberate about enough energy to supply the losses in radiation and in stretching the magnetic field, which amounts to some 10^{37} ergs/sec. The injection of beta-rays is a much better way to start the process of acceleration than anything we know of for ions; on the other hand, the synchrotron radiation forms a much higher loss burden on fast electrons than ionization or radiation forms for fast protons. If the solar flares manage injection of protons, it is hard to doubt that at least at one period the supernova managed it also, and probably still does it in the form of the light-ripples. In any case, it looks reasonable to impute to a supernova somewhere between 10^{47} and 10^{50} ergs in cosmic ray source energy. This amounts to a mean source power of between 10^{37} and 10^{40} ergs/sec, since the supernovae occur perhaps once in 300 years. Of course, not all such objects may behave exactly in

[1] A. G. W. Cameron: Astrophys. J. 1959 (in press).

the same way. The other remnants do not seem to be so remarkable, though they are heavily obscured in the visible. There is evidence that the strongest radio source, the Cassiopeia A source (see Table 11) consists of the remains of a supernova shell, assigned to an explosion some 300 or 400 years ago. This might suggest the Crab properties are at least not unparalleled.

WOLTJERS [22] has discussed the origin of the field. It seems difficult to admit that it all expanded from an original field possessed by a star; the required field strengths at origin would have exerted much too great a pressure. The field-making processes must have magnified the original magnetic energy. If they do so, they probably manage it by folding the lines of force back on themselves at the expense of a high kinetic energy of turbulent mass motion. The field scales by a factor depending on the turbulent velocity and the size of the region. But the maximum energy of particle which can be trapped is proportional to the product of field and diameter, whereas the total energy in magnetic field scales like $B^2 R^3$. Thus the maximum particle energy satisfies the relation:

$$E_{max}^3 / B = \text{const}.$$

If, as Woltjers suggests, the magnetic energy might have increased by a factor of 10^6, while the radius grew by a thousand and the mean field went down by less than a factor of 100, E_{max} could never have been an order of magnitude different from its present value. That value is about 10^{17} or 10^{18} ev. If repetitive processes are needed for large energy gain, the maximum energy is much less than that. It seems plausible that supernovae would not contribute in any case particles much above 10^{15} ev, even in early stages. The present electron spectrum cuts off near 10^{12} ev; this might imply the same energy limit for protons, or might imply a $\gamma \sim 10^6$, $E_{max} \sim 10^{15}$ ev.

We shall assume that supernovae can supply cosmic rays at the power of some 10^{37} ergs/sec in a galaxy. It could well be more, by some fair factor. No one can be sure that the number is not zero. But such a remarkable event seems on general grounds sure to produce cosmic rays. They would probably have a large admixture of heavy elements, with a z_E approaching 20 or even 50 or more, since in the explosion much hydrogen is consumed. This conclusion is by no means very strong; the neutrons made do in the end become hydrogen in large part by decay (see CAMERON, footnote above).

25. The Galaxy [8]. We live with the sun in an arm of a well-developed spiral galaxy. The plane of the Milky Way, nearly a great circle, proves that we live near but a little north of the equatorial plane of the Galaxy itself. The center of the complex but fairly symmetrical distribution of stars, gas, and dust lies about 27000 ($\pm$ 1500) lightyears away from us, in the direction of the constellation Scorpio, not far from the sun's apparent postion at winter solstice. (Galactic coordinates, about 328° longitude, and about 1.5° south latitude.)

Around this center, more or less concentric, are distributed the various great classes of objects which comprise the galaxy. Far outside, in a kind or roughly spherical halo, are found the globular clusters. About a hundred of these objects are known, each itself a tight condensed spherical aggregate of stars, with a diameter of around 200 lightyears, and a population of say 10^6 stars. Even the globular clusters are not spherically distributed; they form a mass distribution such that the surfaces of equal density are distinctly ellipsoidal, the more flattened the closer to the center. But the most outlying clusters suggest a nearly spherical boundary for the whole system, perhaps 55000 or 60000 lightyears in radius. Radio emission makes it plain that there is gas as well in such remote regions—we will consider this later (Sect. 27) — and some thousands of stars are identifiable as belonging to the halo, and not within any cluster. A few stars are known as far as 70000 lightyears from the center. Coming in along a polar radius and thus towards the center, towards the equatorial plane, there is first reached the galactic nucleus, represented by a flattened globe of typically faint stars, spectral types from F to M and below, which form the great majority of all stars. They are distributed in a flattened and more or less centrally condensed solid of revolution, with a polar radius of 5000 or 6000 lightyears to the edge (say, where the density is down by a factor of 100 or more from the central density), and a full thickness, between planes of half the central density, amounting to about 1000 lightyears. Along the equatorial plane, this distribution of stars extends much further from the center, falling to half the central density at a radius of about 4000 lightyears. At this radius, the full thickness between planes of half-density is not much less than at the center. This distribution of stars is relatively low in gas, and does not appear to have much rotation, the stars far away from the center moving in highly eccentric orbits, more or less radially. All the stars we have so far described comprise around 90% of the whole galactic mass; that mass is centrally concentrated in the galactic nucleus, but has a well-marked "flange" extending out in the galactic equatorial plane to the sun's position and beyond.

Most of the mass is thus contributed by this smooth, little-flattened, central and faint population of stars. But a little of the mass and much of the complexity of gas, dust, and irregularity of feature is contributed by Population I, which consists of clouds of gas and dust, the spiral arms recognizable in the arrangement of the gas and dust, and the bright, short-lived stars of spectral types O, A, and B, which are so closely associated to the gas out of which they form and into which they again dissipate their mass. Population I extends from the edge of the galactic nucleus, at a radius of a couple of thousand lightyears or perhaps even much less, out to the extreme edge of the galaxy, at a radius of 50000 lightyears or more. The sun lies on a radius two-thirds of the way out to that radius at which the density has fallen a factor of about a hundred from the central density.

Everyone is familiar with the pictures of our galaxy, or of external galaxies of similar type, as a flattened disk surrounded by the halo of faint stars and of globular clusters. It is this gross distinction which gave rise to the distinction between Populations I and II. Further study has tended to show that these divisions are really more likely to be parts of a continuum, in which the stars and interstellar matter can be classified along a single parameter scale. The classification can be based on the three quite distinct properties; distribution within the great galactic sphere, the kinematics of the motion with respect to the galactic nucleus, and the physical type of the object involved. A recent summary by Oort is extremely helpful[1]. We present it as Table 8. It is evident that this is rather a working classification than a hard and fixed set of divisions; there is no full agreement on just how to chop up the continuum. But the orderly correlation of diverse properties does exist. It is almost certain that it reflects primarily the distribution in time of the formation of the various subsystems; the oldest part of the galaxy is the halo Population II (perhaps excepting its gas content), and the youngest surely the very bright O- and B-stars of extreme Population I, with the formative T Tauri stars. None of these objects is likely to last in its present form for more than some 10^7 years, five percent of the time of a galactic rotation, and perhaps a thousandth of the age of the halo Population II.

---

[1] J. Oort, in: Mc Connell, ed., Stellar Populations. New York: Interscience Publ. 1959.

It follows from this point of view that the present distribution among populations is by no means the distribution for all epochs of galactic history. There is a series of arguments, ranging from the helium and iron content of the disk popu-

Table 8.

Name of Population	Halo population II	Intermediate Population II	Disk Population	Intermediate Population I	Extreme Population I
Typical Members	most globular clusters, subdwarfs, red giants (less luminos then supergiants), very dilute gas	most stars with high velocity relative to sun	most stars near the sun, the sun other weak-line stars, planetary nebulae, normal novae, Type I supernovae (?), unobserved faint stars	A-type stars and strong-line stars	interstellar gas and dust, T Tauri stars, galactic clusters, the very bright O- and B-stars associated with the spiral arms, Cepheids, all supergiants, Type II supernovae
Approximate density near sun (units: a.m.u./cm³: overall density ≈ 6 particle/cm³)	0.06/cm³	$< 10^{-2}$/cm³	~ 2.5/cm³	~ 0.05/cm³	interstellar matter ~ 1/cm³ all stars ~ 0.1/cm³
Total mass in Galaxy (units of solar mass)	$\geqq 1.5 \cdot 10^{10}$	$\sim 5 \cdot 10^{10}$		6.5×10^{10}	$\sim 0.2 \cdot 10^{10}$
r.m.s. velocity (km/sec)	130	50	30	~ 20	$5-15$
Mean height from equatorial plane of Galaxy (light-years)	7×10^3	$\sim 2.5 \times 10^3$	~ 1200	500	400
Axial ratio of representative ellipsoid	< 2	5	~ 25		~ 100
Abundance for elements of $Z \geqq 6$:	0.3%	1%	2%	3%	4%
Distribution	smooth, strongly concentrated (perhaps by 10^4) towards nucleus	intermediate but concentration probably smooth		in spiral arms, patchy, little concentrated to center	in spiral arms very patchy, little central concentration, or even less dense in center

lation to the relatively small amount of gas remaining for further star formation, which suggests that the rate of star formation, the rate of nuclear transformation, and the rate of rapid expulsion of matter from various forms of stellar instability were all much greater in the past. This would be to say that the role of Population I was much larger then, for these phenomena mainly belong to that group, the two or perhaps the three right-hand columns of Table 8. It is in just this population that cosmic rays are presumably formed, and the suggestion of stronger cosmic-ray formation in the past is strong. Whether this is still observable to-day depends on the presence of a region for long-time storage, storage over several times 10^9 years. The enhancements of rate of the various processes mentioned could have been at least a factor of about ten.

Magnetic fields in the Galaxy. There are two lines of evidence for the presence of magnetic fields in the galaxy, that is, of fields whose spatial extent is so large that they cannot be thought of as associated with any single star or even cluster of stars.

The first of these suggests fields in the spiral arms, fields whose lines of force run mainly along the arms, and have magnitudes of several to ten or even twelve microgauss. These fields are suggested by an observed weak polarization of starlight which affects the redder, more distant stars preferentially, and which shows a strong correlation in planes of polarization among stars in a particular direction. The whole picture is consistent only with a polarization in transit, and not with polarized sources (this is sharply to be distinguished from the strong, bluish polarized continuum due to synchrotron radiation from a few special sources.) The polarization is ascribed to preferential absorption of the electric vector which lies along the long axis of dust grains aligned with the field by paramagnetic relaxation processes. FERMI and CHANDRASEKHAR[1] in particular estimated the strength of the field on the grounds of its dynamic influence on the stability of spiral arms, (not much different from simply assuming energy equipartition) and they found probable its overall alignment with the arms. Whether there are not places in the field where a more chaotic field pattern sets in is not known; on the whole it seems probable that there are patches of more or less uniform field and patches of frayed and tangled field lines. The motion of hydromagnetic waves guided by the field lines, and the influence of the differential rotation of the galaxy itself on these pulses have been invoked (by DAVIS for example, and by L. MARSHALL) to employ the general galactic motion as an energy source for cosmic-ray acceleration.. That some such effects take place is rather likely; but it is difficult to know how important they are quantitatively, or how much of a drain they in fact place on the galactic mass motion. The presence of other sources of cosmic rays in number makes the invocation of overall galactic mechanisms unnecessary, but it cannot be excluded.

The second line of evidence is the radio emission observed in the meter wavelengths to come more or less uniformly and isotropically from the spheroidal galactic halo. In the absence of any other plausible isotropic sources out there, the radio emission has been ascribed (by SHKLOVSKI [4] and BURBIDGE[2]) to synchrotron emission from relativistic electrons moving in large-scale magnetic fields within the halo. The halo gas, its fields, and the electrons are all postulated from this one line of argument. Several sources for such ejected gas have been proposed (PIKELNER[3]); there is at least partial evidence for the presence of neutral hydrogen in the halo from 21-centimeter observations. A consistent picture of the halo gas can be given if its magnetic field is taken to be a few microgauss, its particle density about 10^{-2} or 10^{-3} cm^{-3}, its extent some fifty or a hundred thousand lightyears, and the energy density of relativistic electrons about 0·1 to 1 ev/cm^3 at an energy of 10^8 or 10^9 ev. These last follow of course from the observed emission at ~ 100 Mc/sec, of some 10^{37} ergs/sec, and the formulae of Sect. 13. Whether there are cosmic-ray protons present as well is of course only conjecture, but extremely probable. Indeed, it is likely that some of the rela-

---

[1] E. FERMI and G. CHANDRASEKHAR: Astrophys. J. **118**, 113 (1953).
[2] G. R. BURBIDGE: Phys. Rev. **103**, 214 (1956).
[3] S. PIKELNER: Rev. Mod. Phys. **30**, 935 (1958).

tivistic electrons whose radio emission we can see are only secondary to relativistic protons. For the lifetime of such electrons is fixed by emission, and here turns out to be some 10^8 or 10^9 years. The lifetime against collision of cosmic-ray protons in the 10^{10} ev energy region is some 10^{10} or 10^{11} years. These collisions will convert a few percent of their energy finally into electrons and positron of ten times less energy or so. Thus if the energy density in cosmic-ray protons were a hundred times that in electrons, the protons could be responsible for the observed radio emission by the production of electrons (and positrons) in the halo. This implies an energy density of cosmic-ray protons in the halo which is a good deal greater than that near the earth, perhaps by a factor of ten or a hundred. It does not appear likely that this can be true, for unless the external rays are of lower energy than 10^{10} or 10^{11} ev, they should diffuse to the solar region. It is possible, as BURBIDGE believes, that acceleration processes within the halo maintain the energy of the secondary electrons. Or it may be that sufficient electrons are injected into the halo with high energy to meet the observed emission. The total energy involved is not great compared to the observed cosmic-ray energies, for while the halo volume is large, the lifetime against either collision or leakage out of its dilute and large volume is perhaps a hundred times greater than that for cosmic-rays amounting to, in the disk, some 10^9 years or more. It is unnecessary to invoke any new mechanism of acceleration (though it may be present); we need to say only that as much cosmic-ray energy enters the halo per unit time as goes into any other part of the galaxy. Then the halo volume —say a hundred times the rest of the galactic volume—fills with cosmic rays to perhaps a tenth of the solar region energy density, with very likely a higher mean energy, and a smaller flux, with relatively rather more electrons. This is consistent with the fact that we view the cosmic rays from the solar position well within the disk, with its larger magnetic fields and gas density. Whether the loss to the extragalactic space by diffusion from the halo limits the lifetime of halo protons too severely is unknown. If the mean free path is like the Larmor radius for particles of 10^{11} ev in a few microgauss field, say some 10^{14} cm, there is plenty of room for random walk, and leakage need not be appreciable. The random walk lifetime is given by $\tau \approx \dfrac{R^2}{L C}$, and is enormously longer than the collision lifetime for energies up to about 10^{17} to 10^{18} ev, when the Larmor radius becomes comparable with the halo radius R. Of course, the field is not likely to be so obliging in shape in this dilute gas, but there is no evidence on what its form actually is; we could expect that it is not so simple as to allow easy leakage, and we have a good deal of time to spare. On the view we have just sketched, the cosmic-ray sources belong to Population II region of the halo. Low-energy rays, in the neighborhood of 10 Gev and lower, may be patchily distributed throughout the arms and disk, and may be proper members of the Population I.

26. Chemical composition of various stars. The chemical composition of the cosmic-ray beam is an obvious and hotly pursued clue to their sources. The whole problem of stellar chemical evolution[1] is connected with this argument and it appears very easy to overinterpret the few numbers which are known. In Table 9 are collected analyses of various stars, which seem to show that the cosmic-ray beam cannot, without some special sorting at least, be a fair sample of the stars which lie at type A or later on the main sequence. On the other hand, the giants and supergiants are objects in which hydrogen burning has been carried out at high temperature, and even helium burning has proceeded for a

---

[1] BURBIDGE et al.: Rev. Mod. Phys. 21, 547 (1957).

long time. Such stars may well be rich in the elements of the CNO group, and even in Si and its neighbors. Since just such stars together with those which have intrinsically unusual surface conditions, capable of modifying even nuclear abundances, have been suggested as likely to eject cosmic rays, this much of the story is complete. Large amounts of iron, and even of the heavier elements, have been sent out by the supernova processes, as described above. Many objects and types of stars are known in which a wide diversity of abundance anomalies are present; it seems premature to attempt too close a fit between the complex and poorly known astrophysical data and the rather uncertain cosmic-ray information. All one wants to say is that the high values of z seem to be connected with stars in which evolution has been carried well along. Such stars were within

Table 9. *Relative abundances in various stars and in cosmic rays.*

Element	H	He	Li	Be	B	C	N	O	F	Ne	Mg	Si	"Fe" (Cr to Ni)	z
Sun (disk population I)	100	20	$5 \cdot 10^{-12}$	10^{-10}	—	0.035	0.01	0.10	—	—	0.002	0.004	$7 \cdot 10^{-4}$	
Vega[1] (inter. population I)	100	14 (5—50)					0.15				0.01	0.006	$8 \cdot 10^{-3}$	2—
τ Scorpio[1] (B0 star, extr. population I)	100	17				0.025	0.04	0.12	—	0.05	0.008	0.008		≲
HZ 44[2] (an unusual O-type subdwarf, probably late in chemical evolution	100	25					0.5—1				0.1—0.2			~
Cosmic rays[3]	100 (± 10)	15±4	~0.1	(0—0.4)		1.2 ± 0.4				0.2	~0.09	~0.07	~0.06	~

the last millions of years young stars of Population I, in general. Of course, such stars were also found in early galactic history, billions of years ago, and if the rays have been stored in the halo for times comparable with the galactic age, they may very well represent a chemical composition no longer reproducible in the galaxy. This forms part of the very strong argument of Biermann and Davis[4] emphasizing the importance of galactic youth for typical cosmic ray processes. For such cosmic-ray particles as may have been stored for less times, of course, we must look for living stars with a high z.

Table 9 contains an entry for HZ 44, which possesses perhaps the highest z for any star yet analyzed spectroscopically. Estimates have been made for the values of z expected for supergiants which have carried on helium burning, and for supernovae, both on very rough grounds. These figures are:

$$\text{supergiants} \quad z \approx 5$$

$$\text{supernovae} \quad z \sim 20 \text{ to } 50.$$

It is this high value of z which makes the supernovae attractive as major sources of cosmic rays. It is of course very poorly known. But this number, with the

---

[1] L. Aller: Element Abundances. New York: Interscience 1959 (in press).
[2] Burbidge et al.: Rev. Mod. Phys. **29**, 547 (1957).
[3] Ref. [17], and Koshiba et al.: Nuovo Cim. **9**, 1 (1958).
[4] L. Biermann and L. Davis: Z. Naturforsch. **13a**, 909 (1958).

visible evidence of the Crab, support strongly the importance of the supernovae. On the other hand, really high Z atoms are *not* found, in spite of their manufacture in Type I supernovae. It seems premature to dismiss the many other sources ejecting material of high conductivity with high velocities. The mixture we receive may by no means represent a universal sample of cosmic radiation. In Table 10a, b we summarize a possible synthetic model which attempts to include all the processes which appear to contribute.

More detailed analyses have been made (Ref. 3 of Table 11, and [16]), but they seem indecisive. Errors in Z of one are easily possible, tending to wash out odd-even differences, which do remain observable. Moreover, solar system magnetic sorting of the heavies introduces an unknown error, though this perhaps mainly affects the ratio of all above hydrogen to the proton flux. It seems very likely that the over-abundance of the elements beyond helium, especially the CNO group and a few of the others, is real, and may amount to a factor 10. This does not

Table 10. *Nearest galaxy clusters.*

	Mean distance lightyears	Mean diameter of cluster lightyears	Number of galaxies
Local group		ca. 2×10^6	ca. 20
M 81 group (Ursa Major W) . .	$4.6 \cdot 10^6$	$1-2 \times 10^6$	> 9
M 101 group (Ursa Major E) . .	$5 \cdot 10^6$	$1-2 \times 10^6$	> 6
Virgo cluster	$45 \cdot 10^6 \, (\pm 50\%)$	$8-10 \times 10^6$	$> 2.5 \times 10^3$

after E. HOLMBERG, Medd. Lund. Observ., No. 128, Ser. II, 1950.

continue beyond the iron region: Sputnik III data[1] show a very low flux beyond $Z = 31$. We have introduced the quantity z_E, defined by the relation:

$$z_E = \frac{\text{amount of total energy in nuclei with } Z > 2}{\text{total energy}}$$

and we normally take this quantity as representative of the chemical composition of the beam. Whether z_E is really energy-independent is unclear; as we pointed out above, there is evidence that it does not change in order of magnitude up to 10^3 Gev/nucleon or more. Beyond some 10^6 Gev/nucleon, nothing is known for sure about z_E.

27. External galaxies. *α) Normal external galaxies.* We know much less about external galaxies in general than we do about our own. This general statement is perhaps inaccurate when applied to the most conspicuous galaxies which are immediate neighbors, the two Magellanic Clouds, and the Great Nebula in Andromeda, Messier 31. M 31 seems in most ways to be a good mirror for us to watch: it is a spiral of the very same type as our own, and of about the same size. In our own galaxy, as BAADE put it, it is difficult to see the forest for the trees. It was M 31 in which he first made out the two great stellar populations. But even the sister galaxy, M 31, is not exactly like our own. It seems to be about four times more massive than ours, and less than half as luminous. The two Clouds, on the other hand, are each about two percent of the mass of our galaxy, and under a percent as luminous. Moreover, the Clouds seem to contain as much as about half their total mass in interstellar form, while our galaxy has only a percent or so. The Clouds are predominantly extreme Population I; the Andromeda nebula seems to be a mixture about like our own.

---

[1] L. KURNOSOVA, L. RAZORENOV and M. FRADKIN: CSAGI Satellite Symposium, Moscow, 1958.

The Clouds lie about 2×10^5 lightyears from the center of our own galaxy, the Andromeda, about 1.5×10^6. There are a dozen or so other objects of galactic size (mostly small spheroidal types) which lie inside a sphere of a couple of million lightyears' radius about us. These are often said to form the Local Group of galaxies, and they are weakly associated dynamically, with some bridges of matter connecting one with another. Whether there are fields and gas clouds intervening is unknown; the mean density appears to be about 10^{-4} particles per cm^3 averaging over the whole system; any material between the galaxies must be more tenuous than that. Mutual motions of the galaxies do not appear to be likely to produce field changes and hence acceleration in the ionized plasma. But cosmic-ray storage is possible.

The picture of the Local Group—which contains objects more diverse than the four we have described—stands very well for external galaxies as a whole. They tend to cluster in large assemblies often of very diverse type, sometimes more uniform. Galaxies appear to range from little irregular clouds of extreme Population I, like the Clouds of Magellan, or small spheroids (like the Andromeda companions), up to giant spheroidal collections of clusters and stars entirely free from Population I, with no bright blue stars and no dust. Their luminosity varies over a range of some 200; their mass, by even more. It is probable that there is little interstellar gas within spheroidal galaxies. The small companions of the Andromeda Nebula are of this type, and seem to show (m gas/m total) $\approx 10^{-3}$.

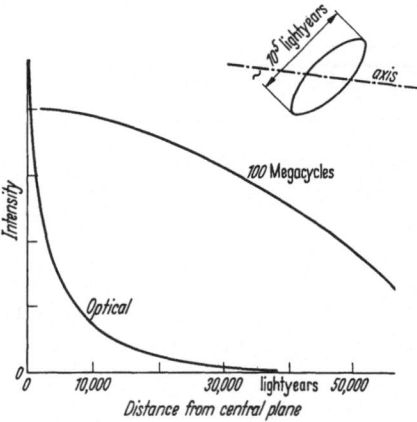

Fig. 14. Intensity distributions of radiation from the Andromeda galaxy, NGC 224. The upper curve shows the radio source intensity at 100 Mc/sec, plotted as a function of the distance from the central plane. The lower curve shows the light intensity calculated from the optical observations of stars, in a similar graph. The inset shows the axis along which the intensity was measured. After J. E. Baldwin: Nature, Lond. 174, 320 (1954).

The spiral galaxies are numerous and bright; it is plausible that they all contain cosmic rays developed more or less as our own. Indeed, the Andromeda shows a strong halo radio emission, extending out perhaps as far as half a million lightyears; it was on this object that the radio halo was first unambiguously demonstrated (Baldwin[1], Fig. 14). It is hard to see that they could contribute much to the cosmic radiation we observe. If their cosmic radiation is like our own, their great distance, and the difficulty of entering the galactic halo imply that the galaxy tends to be isolated from the cosmic-ray point of view. This remains true for all particles below some energy so high that it enters readily. But this is near 10^{18} ev or more; normal galaxies make very few such particles, if any at all, by any processes we know of.

The galaxies, or the very frequent clusters of galaxies, seem to show a random velocity dispersion, perhaps due to dynamical interaction, amounting to some 300 km/sec. More distant galaxies or clusters show the proportionate increase of recession velocity with distance, the famous "red shift". This amounts to somewhere between 150 and 300 km/sec for a distance of ten million lightyears (Sandage[2]), and is linear to very large distances. Distance measurements beyond a few tens of millions of lightyears are still unreliable except for the red shift

[1] J. Baldwin: Monthly Notices Roy. Astronom. Soc. London 115, 684, 690 (1955).
[2] A. Sandage: Astrophys. J. 127, 513 (1958).

itself. It is instructive to tabulate the nearest galaxy clusters. Beyond these, of course, some 10^8 galaxies can be detected as small, rather poorly-resolved images on emulsions, out to a limiting red shift of $\Delta\lambda/\lambda \approx 0.2$, a distance of perhaps 2 to 4×10^9 lightyears.

β) *Unusual galaxies.* Among the sample of only a few thousand galaxies about which we have pretty good information, some few show properties which we must regard as unusual even compared to the diversity of galaxies in size, shape, mass, luminosity, and population mixture. Whether these far objects are truly unique and singular, or simply so rare that our small sample by chance contains no other examples, we cannot be sure.

Two of these rare objects are of high relevance for cosmic-ray origin. The first of these is the great spheroidal galaxy M 87, one of the most luminous members of the Virgo cluster. This huge galaxy displays almost a pure Population II, with strong central condensation, no flattening to speak of, and a great halo of globular clusters. It is remarkable in two respects:

1. M 87 is the seventh brightest discrete radio source known. It is an extended source, with a diameter at 100 megacycles a couple of times greater than the diameter of its visible halo. From its distance and angular aperture [the distance comes from measured luminosities of globular clusters (BAUM[1]), the radio halo must be 200 or 300 thousand lightyears in radius. The radio power emitted is prodigious. It amounts to about 2×10^{41} erg/sec, about ten thousand times the emission of our own galaxy, or of a typical large spiral. Its optical emission is high, making it one of the brightest of known spheroidal galaxies, but it is in visible light approximately like our own or the Andromeda. It is thus really a radio galaxy, emitting in that region about a percent as much as its visible emission. Since all the radio power goes into a frequency range of 1000 Mc/sec instead of into the optical band of 2×10^8 Mc/sec width, its power per unit frequency is a thousand times greater in the radio region. From what we have seen already, it is easy to conclude that this galaxy is surrounded by a halo rich in relativistic electrons, with a magnetic field and particle density large enough to emit synchrotron radiation. This proposal was considered by BURBIDGE[2]. Various combinations of particle flux and magnetic field will, of course, yield the same emission; another relation is needed, as we saw in the discussion of the Crab Nebula (Sect. 24). The lifetime considerations used there are not really available here, but rough equipartition may be used to guess the values. They turn out to imply fields of some ten microgauss and particle energy densities around 10 or 100 ev/cm^3 in electrons alone. If, as is rather likely, protons are also present with high energy, and the electrons are secondary to the proton flux, produced by collisions in the galactic halo of M 87, then the energies are higher, about 10^4 ev/cm^3. The whole galaxy is bathed in cosmic rays with a flux ten thousand times what we have here.

2. The radio emission is not the only peculiarity of M 87. Near the center of the neatly spherical mass of stars, there is a unique linear feature, called the *jet* (see Fig. 15 a, b). This portion emits a bright blue continuum, and is strongly polarized, showing uniformity of the plane of polarization over considerable fractions of its whole dimension. This jet has been known for a generation; its polarization was predicted by SHKLOVSKI from the knowledge that M 87 was a strong radio source. We can hardly fail to follow his argument: the whole appearance is analogous to that of the Crab Nebula, and can be explained at a stroke

---

[1] W. BAUM: Publ. Astronom. Soc. Pacific **67**, 328 (1955).
[2] G. BURBIDGE: Astrophys. J. **124**, 416 (1956).

if we postulate a strong magnetic field, indicated by the polarization of the light, and ascribe optical and radio emission as well to the presence of fast electrons giving synchrotron radiation. Again the presence of heavy cosmic-ray particles

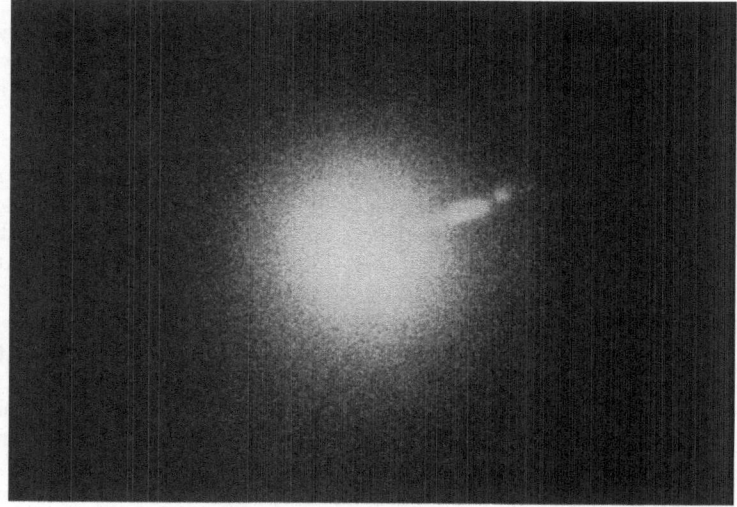

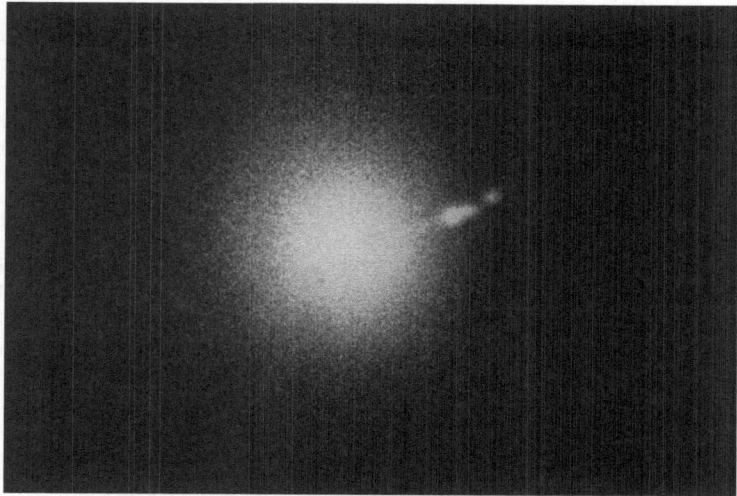

b

Fig. 15 a and b. The bright central portion of the spherical galaxy M 87 or NGC 4486, in the Virgo cluster. Note the broken linear structure extending up and right from the center. It is called the jet of M 87. The two photographs are taken with differing directions of the electric vector, showing a marked polarization of the light from the different knobs within the jet. From W. BAADE: Astrophys. J. **123**, 550 (1956).

is unproved but plausible. The radio emission of the halo strongly suggests diffusion of the particles away from their presumed source in the jet out into the galaxy's halo. But the magnitude of this phenomenon is impressive; the dimensions of the jet are not spanned by a few lightyears, as is the Crab Nebula, but instead it is four or five thousand lightyears long, and about a tenth of that distance across its axis. There are a couple of "knobs" along the jet which seem particularly bright and show uniform polarization over distances of almost a thousand lightyears. The light intensity and the total radio emission imply that this jet is something like the Crab remnant, but larger in volume by a factor of 10^7. Its energy content again depends on the model, but considerations like those mentioned in discussing the halo can be used. It is not easy to impute to the jet an energy density less than about 10^8 ev/cm³ if the magnetic field is one microgauss, and about 10^3 ev/cm³ if the magnetic field reaches 10^{-2} gauss. On the first of these assumptions, the electrons present need energies reaching 10^{13} ev, on the latter, only 10^{11} ev (cf. Table 5). The total energy involved does not vary so much as the particle energy, for the magnetic energy must be large when the particle energy is taken small. The duration of the phenomenon is of course unknown; it might be guessed that the length of the jet represents some continuing consequence of a collision of fields, and if one took velocities of the order of 300 km/sec, which is indicated by a Doppler shift of a curious oxygen emission line seen in the object, the jet may have been forming for 10^6 or 10^7 years. The total energy present in the jet is certainly as much as 10^{56} ergs, which would account for the emission observed for that length of time. There is no way to exclude an energy content fully a thousand times higher than that, up to 10^{59} or 10^{60} ergs. A supernova releases about 10^{50} ergs. The jet is larger in volume by 10^7, in specific energy density by up to 10^3, and in duration by up to 10^5 or so, all compared to the present state of the supernova of 1054 A.D. A nuclear source for this energy appears all but out of the question; annihilation of a star of solar mass with its anti-star (protons *vs.* anti-protons) would release only 4×10^{53} ergs. Such a process seems highly implausible (even though perhaps then only fast electrons need be postulated, no cosmic-ray protons, and the event could just supply the energy then required), and can almost surely be disregarded.

The minimum energy required is about the kinetic energy of collision of two large gas clouds moving with galactic velocities near 300 km/sec and of a size and mass about like those of a large globular cluster with a mass some 10^{-3} of the whole galaxy. Whether such collisions are likely, and how a sizeable fraction of the collision energy can be placed into cosmic rays, are questions still unanswered. But the presence and nature of the M 87 jet are facts.

The second unusual object is in fact the brightest discrete radio source in the sky, Cygnus A. BAADE and MINKOWSKI (Table 11, Ref. [3]) have shown that this object is in fact a pair of distant galaxies in collision. Two galactic nuclei can be roughly resolved; they lie in a rich cluster of galaxies; the spectrum shows a nearly uniquely strong emission-line spectrum, largely in forbidden lines of N and O. All these facts are consistent with the intense interactions expected upon the highly supersonic collision of two gas-filled galaxies, with a relative velocity of several hundreds of km/sec. The red shift is large, about 16.8×10^3 km/sec, putting the object between 0.5 and 1.0×10^9 lightyears away. The angular diameter observed yields a geometrical diameter of around a hundred thousand lightyears, reasonably confirming the idea that this is a collision of galaxies in progress. That such a remote object dominates the radio night sky is evidence enough for its remarkable nature. The radio power emitted is about 10^{44} ergs/sec or more, perhaps twice the total optical emission.

Neutral hydrogen has been observed in this object by its 21 cm emission line, and confirms the assignment of galactic mass to the emitter. The radio source appears to be double, with its two centers separated perhaps three times their width, which is the same as the overall optical width. SHKLOVSKI [4] and M. SAVE-DOFF[1] have suggested that what is seen here is the result of the collision of two galactic halos, the stars continuing to move together while shock waves move out in the halo gas, finally causing the halos to separate while the star masses continue in collision. The strong shocks produce magnetic fields and fast electrons, with the accompanying synchrotron radio emission. Some such picture seems likely, but this event too is far from understood. This object has radiated at least 10^{56} ergs, and must contain a great deal more energy in fast particles than that. There is no blue optical continuum of importance, suggesting that the strong fields required for synchrotron radiation in the visible are not now present. The total energy available in a collision of galaxies is around 10^{60} ergs.

Table 11.

Constellation	Position [2,3]		Approximation [3] angular size	Flux density [3,6] at 100 Mc/sec $10^{-24} \dfrac{\text{Watts}}{m^2-(c/sec)}$	Identity [3,4]	Distance [4] (lightyears)	Approximation radio power emitted (ergs/sec) $(\Delta f = 500$ Mc/sec$)$
	R.A.	Dec.					
Virgo A	12.4 h	12.7°	4′	12	M 87[4] NGC 4486	5.2×10^7 *	1.3×10^{41}
Cygnus A	20.0 h	40.6°	1′	125	Colliding[4] galaxies	5.5×10^8 *	1.5×10^{44}
Taurus A	05.5 h	22.0°	5′	18.5	M 1 NGC 1952 Crab Nebula Supernova (1054)	6×10^3	2.3×10^{33}
Centaurus A	13.4 h	−42.7°	4°	50	NGC 5128[4] Two nebulae (Colliding galaxies ?)	3×10^6	1.7×10^{39}
Cassiopeia A	23.4 h	58.6°	4′	190	Supernova Type II (~1600) ? ?	1.6×10^3	1.8×10^{33}
B Cassiopeiae	00.4 h	~64°		≦10 ?	Supernovae (1572)		
Puppis A[6]	08.3 h	−42.5°	50′	~15	Supernova Type II ? ?	1.6×10^3 ?	1.5×10^{32}
Pictor[3,6]	05.3 h	~−45°		18	"Extended source"		
Perseus	03.3 h	41.3°	≦1′	≳0.9	NGC 1275[4,5] Colliding galaxies	1.6×10^8 *	8.7×10^{40}
Andromeda[2,3]	00.7 h	~41°	3.5°	~1	M 31 NGC 224	1.5×10^6	1.2×10^{37}

Two more strong radio sources, Centaurus A and Perseus A, have been identified with the unusual extragalactic optical objects NGC 5128 and NGC 1275 respectively. These appear to be similar to the Cygnus A source, looking like superimposed galaxies, and showing unusually strong emission lines. Quite probably these two are also pairs of colliding galaxies, but possibly they represent evolutionary processes of unusual sort, related to the M 87 jet. Their distances are not well known, but seem to imply radio powers less by a couple of orders of magnitude than the Cygnus A source. NGC 5128 may have the power of the M 87 jet, and NGC 1275 not much more. Still another source, the Fornax A source, appears to be of rather similar kind [7,8].

---

* Distances based on red-shift, using 100 km/sec per 10^6 psc. for the Hubble constant.
[1] See Rev. Mod. Phys. **30**, 1074 (1958).
[2] C. W. ALLEN: Ref. [25].
[3] J. L. PAWSEY: Astrophys. J. **121**, 1 (1955).
[4] W. BAADE and R. MINKOWSKI: Astrophys. J. **119**, 206, 215 (1954).
[5] G. R. BURBIDGE and E. M. BURBIDGE: Astrophys. J. **125**, 1 (1957).
[6] K. V. SHERIDAN: Austral. J. Phys. **2**, 400 (1958).
[7] G. BURBIDGE und E. BURBIDGE: Astrophys. J. **125**, 1 (1957).
[8] For spectra: G. R. WHITFIELD: Monthly Notices Roy. Astronom. Soc. London **117**, 680 (1957).

If we suppose that these unusual objects represent a reasonable sampling of what can occur, and if we impute the radio and optical emission to cosmic-ray processes as we have implied, the total cosmic-ray production does not receive a large contribution from these events. The large number of supernova explosions in the history of a normal spiral galaxy outweights the jet event of a rare galaxy. Allowing for the poor observations we can make of distant galaxies, the jet contributes to the total cosmic-ray generation of the Virgo cluster certainly not more than the same order of magnitude as the summed normal galaxies. It is concentrated within a single galaxy, not thousands, and in a time probably rather short compared to the galactic lifetime. But it will not affect distant galaxies, like our own, in any very important way, save possibly one. It is plausible that the energy limit to which cosmic rays can be accelerated in an object like the jet will rise with the scale of the fields involved. The jet is hundreds or thousands of times as large across as the supernova shell. Even at the same magnetic field strengths it could contain rays of a thousand times greater energy. It is not implausible to suggest that such an event as the M87 jet may produce cosmic rays up to 10^{19} ev energy at least, and possibly much higher. The energy placed in these rays may be a small fraction of the total energy. But these rays spread in space, not much affected by the fields of normal galaxies, and perhaps only in time rendered rather isotropic by weak but large-scale fields in the space between the galaxies of a great cluster.

28. Cosmological topics. The cosmic rays have been pushed up to the scale of clusters of galaxies by more or less direct inference from astronomical data. Beyond that scale there is little present need to go. The overall expansion which solves the paradox of OLBERS about the darkness of the night sky also prevents the really distant galaxies from making a large contribution to cosmic rays. Indeed, the likelihood that the charged cosmic rays spend much more time on devious paths through extragalactic space than does light in its straight-line motion implies that the effective cut-off for cosmic ray propagation falls earlier in time than a few times the Hubble constant[1].

One possible intermediate scale between the cosmological and the cluster of galaxies has been suggested by DE VAUCOLEURS[2]. This is the existence of a super-galaxy of which we and the Virgo cluster form parts. This "cluster of the second order", with a weak dynamical connection among its members, might include a few thousand clusters of galaxies, and extend some hundreds of millions of lightyears around the Virgo cluster. Statistical evidence for it seems present if weak; there are certainly some unusual correlations of velocities and even of orientations of planes of spiral galaxies. If this system is present, it could very well contain rays up to the highest energies, supposing that fields exist at least in some of this volume.

A second possible cosmological point relevant to our problem is the hypothetical existence of large-scale samples of anti-matter. Conceivably matter-antimatter collisions might supply some of the energy for the very energetic sources of radio emission. By supplying fast electrons and positrons in the hundred Mev range, without the presence of any fast nucleons at all, such processes might reduce the energy imputed to these sources. This view has been put forward by G. BURBIDGE and F. HOYLE[3]. It is verifiable probably by searching for gamma-rays

---

[1] Only Farley has tried to give an account of cosmic rays which makes their extragalactic nature of importance, apart from the early theories which ascribed the rays to processes no longer current in the universe. F. FARLEY in [17].

[2] G. DE VAUCOLEURS: Nature, Lond. under press (1959).

[3] G. R. BURBIDGE and F. HOYLE: Nuovo Cim. **4**, 558 (1956).

from such objects. Such searches are still in a very rudimentary state[1]. It does not appear yet necessary to postulate such revolutionary processes, which would still prove inadequate for the M 87 optical source without further acceleration.

F. Synthesis of a model.

29. A heterogeneous model. The recognition of a connection between cosmic radiation and radio astronomy, the brilliant prediction by Shklovski and Ginzburg that an analogous optical connection might be found, and the remarkable confirmation of those views which we have described, place the problem of the origin of cosmic rays on a new footing. It has been merged with the general astrophysical problem of the nature and history of magnetized plasma. From being an unusual phenomenon, cosmic-ray production has become a general feature of plasma on astronomical scale. There is no single source of the rays, the sun and the Crab Nebula are all but certainly two very diverse examples of such sources. In this article, we have tried to outline the main physical processes which take place, and which govern the production and propagation of cosmic rays. In so doing, we have reduced the cosmic ray origin from a striking and novel process to a natural part of the evolution of stars and of galaxies. We remain unconvinced that a unique process which produces all or the bulk of the cosmic rays will be found. In our view, there is a wide variety of sources, each contributing to cosmic ray flux. That flux itself depends upon the galactic position of the observer. The lower the cosmic-ray energy, the more widespread its sources, for modest processes of stellar atmosphere or interstellar gas motions can produce them. The lower the energy also, the more likely is the variability of the cosmic ray flux, in time and in space. For the particles which are physically continuous with cosmic rays, but experimentally distinct, the particles which cause the aurora and the Van Allen radiation bands near the earth, protons and electrons of energies in the Mev range and below, the picture is fairly sure. They originate in solar events of commonplace sort, and their density varies both with time in the scale of days and with position on the scale of the earth's magnetic field diameter. They are stored in the earth's field to reach densities which, at least at times, far exceed the densities of the interplanetary space. Let these serve as the model for the cosmic rays. Those particles whose energies lie in the range of some Mev can be produced by the sun, and very probably by a variety of stellar-envelope processes, as in flare stars, novae, and magnetic variables. Their spatial distribution is controlled by diffusion among the fields of the more or less dense plasmas of the spiral arms. They may diffuse over distances large compared to the earth-sun distance, but small compared to the diameter of the galaxy. They will vary in density depending on the local magnetic conditions, reading "local" as measured in lightyears. Above these energies, particles with energies ranging from tens of Gev up to 10^4 or 10^5 Gev can be generated in supernovae, and wander rather easily out of the narrow confines of Population I into the wider and less dense plasmas of the galactic halo, spending a long time there, but with few collisions and slow leakage out of its great magnetized volume. They superimpose their flux upon the local flux, continuing the spectrum up to the highest energies which it seems plausible to generate by repetitive events within the relatively narrow confines of an expanding supernova shell. Beyond these energies, say above 10^7 or 10^8 Gev per particle, even the galactic halo is not a strong diffuser. The Larmor radius in fields of some microgauss has risen to dimensions comparable with those of the halo itself. Now the unusually powerful

---

[1] P. Morrison: Nuovo Cim. **7**, 859 (1958).

processes seen in M 87 for example can be invoked. Here the great scale permits the repeated acceleration of particles up to the highest observed energies. The total flux of such particles is not well known—it could possibly be zero—but whatever it is, it will have a value which is much the same for a region between galaxies and for the location of the sun, deep in a spiral arm. The higher the particle energy, the larger the volume occupied by such particles. But with the large volume goes the power-law decrease of flux density, so that the total energy stored in the particles of each energy region remains at the most constant for the three groups we have described, or very probably even falls as particle energy grows. It is clear that the division of the groups is arbitrary, depending upon the various recognized sources. In reality, there is a smooth gradation from one group to another, which reflects the changes of the relative importance of the sources with position of observation and with time. Through it all the processes of magnetic stirring guarantee the absence of strong anisotropy or of strong time dependence, except for those energies near and below 10 Gev where the earth and the sun are objects competent to modify the flux in the times available for our study. The energy input to cosmic radiation is on our model nuclear up to the highest energies, where strong collisions of clouds may possibly bring gravitational energies into play. Of course, some energy transfer from the large-scale field motions of the galaxy is possible; such energy would also be gravitational in origin for the most part.

This hierarchical theory seems most natural. It leaves the power-law energy spectrum rather poorly explained; on this view, it is the statistical summation of a number of distinct power laws, each one covering a wide range up to some cut-off. There is a suggestion that the exponents of the several components cannot widely differ, because of the cosmic ray drain on the magnetic energies of the several classes of objects involved and because the energy cut-off is fixed by the physical scale. But the facts as they are here interpreted do not seem to permit regarding as a real phenomenon the single straight line of the logarithmic plot of the primary spectrum, which is observed nearly unchanged over some eight decades. Part of this regularity arises from the poor energy resolution which we are forced to observe, for what is actually measured at high energies is not the primary itself, but the myriad particles of a complex and fluctuating multiple event. It is rather likely also that the galactic stirring processes do tend to smooth out the otherwise less regular course of the energy spectrum, adding some energy to the cosmic-ray beam in its transit.

This model is a little iconoclastic. It suggests that the chemical composition of the rays will not be specially characteristic of a single source, single but of a mixture; and that it may vary somewhat over wide energy ranges. It suggests that the identification of single processes for injection or acceleration will not exhaust the real situation: many such processes can take part. Most of the authors who have suggested such processes—not all—will turn out to be right, but each one only in part. Wherever hydromagnetic processes go on in large scale and in tenuous gas, there some cosmic rays are made. Interpretations which rely exclusively upon single objects, like supernovae, are in the opinion of this author premature. No doubt these play a large role; it is hard to see why they should do everything. Indeed, they cannot well make the rays of highest suspected energy. And they do not make those which come to us during solar flares. The cosmic rays are general samples of varying regions of space and time; they sample the bigger domain, the higher is their energy. Van Allen particles are less than some months or years old, and have never been further than the sun. The highest energy cosmic rays are from distant galaxies, and have been in transit to us for times comparable with the age of the galaxy. That is our picture; it has its grandeur, even it has somewhat diluted the naive enthusiasm with which the present author and many others thought to describe "the origin of the cosmic rays".

We conclude this account by presenting a tabular (Tables 12a, 12b) description of the model here proposed. The numbers which enter are of course both

Table 12a.

	Trapping Region		Cosmic rays					Source	
	diameter	$\langle B \rangle$ gauss	lifetime	typical energy (Gev/ nucleon)	mean free path in field	flux (cm^{-2} sec^{-1})	ev/cm^3	power ergs/sec	z_{II}
I Earth's field (Van Allen band)	10^9 cm	0.1	10^{6-8} sec	<0.1	10^7 cm	10^3	100	10^{10}	2 (?
II Spiral-arm region of sun (local system near galactic plane)	$\lesssim 10^3$ l.y.	10^{-5}	10^{6-7} yr.	$10-100$	10^{13} cm	0.1	~ 1	10^{35}	20 (
III Halo of the galaxy	$\sim 10^5$ l.y.	$\leqq 10^{-6}$	10^{9-10} yr.	10^{3-5}	$>0.1-1$ l.y.	10^{-6}	10^{-2}	10^{38}	~ 2
IV Volume of clusters of galaxies	$\sim 10^7$ l.y.	$<10^{-8}(?)$	$>10^{10}$ yr.	$>10^{7-8}$	$>10^4$ l.y.	10^{-10}	$<10^{-5}$	10^{41}	?

l.y. = lightyear; m.f.p. = mean free path.

Table 12b. *Nature of the sources of cosmic rays.*

Region	Source	Required efficiency ε_{CR}
I	Streams of gas from sun filled with low-energy CR particles	about 10^{-5} for whole output; only a small part strikes earth
II	Stars with mass ejection and magnetic instabilities. Probably flare stars, T Tauri stars, supergiants, even B type and earlier giants. A-type magnetic variables may play a role. All these are in the spiral-arm region not far from the sun. All these are contemporary stars.	10^{-5} to 10^{-3}
III	All the types of region I, plus supernovae and possibly unknown explosive stars. These stars may be extinct, the time of storage of the rays being long enough to go back to early stages of the galaxy.	$\geqq 10^{-3}$
IV	Unusual events like the jet of M 87, postulated to be reasonably frequent within large clusters of galaxies.	10^{-2} to 10^{-3}

tentative and subjective; within the wide ranges of uncertainty, they have been chosen for consistency. But the whole table is meant to suggest how rich a field we have been forced to implicate as a source for cosmic rays by the irresistible results of the astronomers. No one would, I think, have wanted to postulate such an unlikely object as the M 87 jet to help solve the problem of cosmic-ray origin. But it exists. In reading the table, observe that the conditions refer to very different volumes of space. The energy densities will suggest the local importance of each of the regions mentioned. It is unnecessary to remind the reader of the entirely schematic nature of the table; no claim is made for anything but one rough and by no means unique structure which fits together all the diverse sources of data and ideas which we have. Smooth merging of one region into another is of course to be expected; the divisions here taken are expected to be natural but not sharply marked. In particular, the neighborhood of the sun may be expected to involve longer diffusion times than generally to be found in region II; this may imply that a typical region would have smaller power needs than tabulated. Moreover, some effects of general galactic motion in inducing magnetic acceleration may be expected to smooth out the energy spectrum, and somewhat to lessen the power needs from the ultimately nuclear and gravitational sources both for regions II and III. The rather large values of ε_{CR} suggest the degree by which our proposals are all uncertain. No one has shown that even very unusual objects do really place so much of their total power output into cosmic-ray form. This leaves us with a plausible but still conjectural model for the final picture. A great need in this whole field is some definite test of the postulated ubiquity of the cosmic rays.

30. Problems to be solved. There are many problems left untouched. Indeed, this picture, for all that it now appears more sophisticated than the theories of the past, which separated cosmic rays from astrophysics as a whole, is still only qualitative. Any sure quantitative numbers are still lacking. We do not know the energy spectrum or the charge spectrum or the directional distribution except in the most rough and general of ways. Any detailed information on these points, especially at a variety of energies above the solar-influenced range, would be very helpful. They are all experimentally hard to learn. The same remark goes for the presence of photons and electrons in the primary beam. What we have is a model which is fitted to rough and low resolution data. Any gain of resolving power is sure to show either the small effects which this model predicts or their absence, which might serve to eliminate our highly statistical and composite point of view. The search for energy cut-off and for directional preferences in the highest energy range is bound to continue, however extensive the apparatus and patient the experimenter must become. The demonstration that the charge spectrum remains roughly constant to the highest energies, which might be sought by looking for a sign of multiple cores in the greatest giant showers, is an important undone task. None of this adds anything new to the ideas already current among experimenters.

On the astronomical side, we have the same story. Distances and fluxes and magnetic fields are poorly known, and often really only surmised. The rough model is consistent, but it is nowhere sharp. Could we even demonstrate the presence in some radio source of cosmic-ray protons, say by gamma-ray studies, as we have found fast electrons by synchrotron radiation, this would at least confirm one of our most plausible surmises. Any other measure of the distribution or strengths of magnetic fields, in the disk or the halo, would fill in one more gap. Whether a better theoretical treatment of the relation between, say, a source of hydromagnetic waves and the acceleration of fast particles would help is not so clear. For we have not much detailed knowledge of the conditions of the Crab, for example, which would enable us to test such a detailed theory. Perhaps study of the solar events is the most important path to follow along this line; we can expect to find out more about the hydromagnetic processes of solar flares than of any other such process in the galaxy.

The great growth of understanding of stellar and galactic evolution will help. The changing nature of the galaxy in time may become clear enough to give us a better picture of the role of cosmic rays in that evolution, and with that a surer idea of the processes of their origin and long-time storage either in the more distant halo, the nucleus, or the local spiral arm itself. Above all, it is sure that the cosmic-ray problem is no longer a separate one; it is part and parcel of the general study of stellar and of galactic evolution.

It is appropriate to close this article with a remark which is surely already plain to the reader: the material here set forth falls into two very distinct parts. One part, the enumeration of the physical processes which might give rise to cosmic rays, and which surely drain and modify them, is more or less firm. It will not change. The behavior of particles in space, the production of light by synchrotron motion, and so on, are physical problems rather well-solved. They could properly belong in an Encyclopedia. But the specific history of the rays, the enumeration and study of objects from which they in fact come, is ephemeral and primitive. The maturing of evolutionary astrophysics and the widening scope of astronomical studies is likely to change that portion gravely. It is the hope of the author that the model here outlined will at least be recognizable in the future development of our knowledge. It seems that for the first time, thanks to the astrophysicists and to their radio dishes, a cosmic-ray theorist may reasonably, if guardedly, express that hope.

General References.

[1] ALFVÉN, H.: Cosmical Electrodynamics. London: Oxford University Press 1950. — The original, and still very useful, account of particle motion in magnetic fields. The applications are somewhat out-of-date.

[2] SPITZER, L.: Physics of Fully-ionized Gases. New York: Interscience Publ. 1956. — A very useful account of plasma physics.

[3] PAWSEY, J., and R. BRACEWELL: Radio Astronomy. London: Oxford University Press 1955.

[4] SHKLOVSKI, I. S.: Kosmicheski Radioisluchinie. Moscow: Gos. Techniko. teoret. Lit. 1956. — These two are the best books on the topic. They are just a little dated in this rapidly-moving subject.

[5] KUIPER, G. P.: editor: The Sun. Chicago, Ill.: Chicago University Press 1953. — This volume is a compendium of the best kind. The chapters on the photosphere, the corona, solar activity, and solar radio emission are the sources for much of the data used above.

[6] UNSÖLD, A.: Physik der Sternatmosphären, 2. Aufl. Berlin-Göttingen-Heidelberg: Springer 1955. — The best reference on the topic, very widely conceived.

[7] PAYNE-GAPOSCHKIN, C.: Galactic Novae. New York: Interscience Publ. 1957.

[8] SHAPLEY, H.: The Inner Metagalaxy. New Haven: Yale University Press 1958. — These two are excellent references, though rather emphasizing conventional approaches of astronomy.

[9] BIERMANN, L.: Annual Rev. Nucl. Sci. 2, 395 (1953). — An early review of the origin of cosmic rays, still useful reading. Excellent bibliography of early work.

[10] ROSSI, B.: Nuovo Cim. Suppl. 2, 275 (1953). — A similar review.

[11] FERMI, E.: Phys. Rev. 75, 1169 (1949). — The pioneer piece on diffusion of cosmic rays.

[12] FERMI, E.: Astrophys. J. 119, 1 (1954). — Improvements on [11], taking loss into account.

[13] MORRISON, P., S. OLBERT and B. ROSSI: Phys. Rev. 94, 440 (1954). — Similar to [12].

[14] GINZBURG, V. L.: Dokl. Akad. Nauk. USSR. 92, 1133 (1953).

[14a] GINZBURG, V. L.: Fortschr. Phys., Berlin 1, 659 (1954). — Early statements of the present general point of view.

[15] GINZBURG, V. L.: Nuovo Cim. Suppl. 3, 38 (1954).

[15a] GINZBURG, V. L.: Progr. Cosmic Ray Physics 4 (1958). — The latest review of the position of the origin problem, from the point of view of the originator of the present theory. Emphasizes the role of supernovae.

[16] HAYAKAWA, S., K. ITO and Y. TERASHIMA: Progr. Theoret. Phys. Suppl. 6, 1 (1958). — Carries the work of Ref. [15] still further. This review, which is very like the present article in feeling, gives some importance to other than supernova sources. The two [15] and [16], are complete and valuable reviews, differing in conclusion from the present one mainly in the degree of optimism shown about the sufficiency of particular

models. All three reviews really stem from the work of GINZBURG and SHKLOVSKI, but go successively further towards complicating their early ideas. References in both [15] and [15a] span the Soviet work very well.

[17] Nuovo Cim. Suppl. **8**, 126 (1958). — The whole journal contains the proceedings of the (Varenna) meeting of 1957 which forms a very fine source of data and theory on origin and related problems.

[18] COCCONI, G.: Nuovo Cim. **3**, 1433 (1956). — Extragalactic sources.

[19] DAVIS, L.: Phys. Rev. **101**, 351 (1956). — The applicable transport theory.

[20] PARKER, E. M.: Phys. Rev. **109**, 1328 (1958). — A review of acceleration methods, not much changed by the later work of PARKER, see Phys. Rev. **112**, 1048 (1958).

[21] OORT, J., and T. WALRAVEN: Bull. astronom. Inst. Netherl. **12**, 285 (1956).

[22] WOLTJER, L.: Bull. astronom. Inst. Netherl. **14**, 39 (1958). — These two are definitive and brilliant studies of the Crab Nebula. [22] is later and more complete, but [21] has many general points of high interest.

[23] LANDAU, L., and E. LIFSCHITZ: Classical Theory of Fields. Boston: Addison-Wesley 1954. — Derivation of and early references on synchrotron radiation.

[24] BETHE, H., and J. ASHKIN, in: Experimental Nuclear Physics, E. SEGRÈ, ed., Vol. I. New York: Wiley 1953. — Interaction of charged particles with matter.

[25] ALLEN, J.: Astrophys. Quantities. London: University Press 1955. — Tabulated data of very wide scope; useful directly, and as guide to literature. Very valuable for physicists unused to astronomical language.

Theory of the Geomagnetic Effects of Cosmic Radiation.

By

MANUEL SANDOVAL VALLARTA.

With 53 Figures.

1. The discovery of the latitude effect of cosmic radiation simultaneously by CLAY [1] and by COMPTON [2], proving that there are charged particles in the primary radiation, gave very considerable impetus to the study of the motion of such particles in the earth's magnetic field and, in the first approximation, in the field of a magnetic dipole, a problem of very considerable difficulty to which STÖRMER [3] devoted a great deal of attention since the beginning of this century, originally in connection with BIRKELAND's theory of the aurora borealis. Today we know a good deal about the composition of primary radiation, which includes predominantly protons and bare nuclei from helium to iron, all positively charged particles. It is only in the last few years that attempts have been made to work out the motion of such particles in the second approximation to the geomagnetic field, by introducing both the dipole and quadrupole terms.

To begin with the first approximation, the dynamical trajectories of a charged particle in the field of a magnetic dipole are in general solutions of a set of non-integrable differential equations. It is this feature of non-integrability which gives to this, as well as to other dynamical problems, its most interesting and difficult characteristics. In the following we shall attempt, first, to give a summary of the theory of such motions, particularly as far as it concerns the directions from which charged particles of a given energy may, or may not, arrive at a given point of the earth, due regard being paid to the fact that the earth itself acts as an impenetrable body. Such allowed and forbidden regions are of very considerable physical interest and their knowledge is fundamental to understand experimental observations of the intensity of cosmic radiation on the earth.

2. As will be shown in the sequel, the allowed directions at any point of the earth, for particles of any given energy, fill up a cone of many sheets, generally of very complicated shape [4]. This cone has been called by LEMAÎTRE and me the *allowed cone*. The allowed cone in general consists of three regions: first there is the *main cone*, within which all directions are allowed; surrounding the main cone there is a second region in which certain bands or patches of directions are allowed and the rest forbidden, which we have named the *penumbra*; lastly there is a third region, the *shadow cone*, outside of which all directions are excluded. An absolute limit for all allowed directions is the *Störmer cone*, which has the property that outside of it the region of space containing the earth and that containing all trajectories coming from outside the earth are disconnected. This a right circular cone with vertex at the observer and axis along the east-west line. The penumbra lies in general between the main cone and the shadow cone. As a rule it does not reach the Störmer cone, which is only exceptionally touched by the main cone.

There is a fundamental property of all allowed directions to which we wish to call attention at the outset. From LIOUVILLE's theorem on the conservation of volume element in phase space it follows that the intensity of cosmic particles in any allowed direction, defined as the number of particles of given energy crossing unit solid angle per unit time, is the same as it is at their starting point. Therefore, if the distribution at infinity is isotropic, the intensity is the same in all allowed directions for any given energy. This important feature of the dynamical problem, which FERMI and ROSSI and independently LEMAÎTRE and the author pointed out [5] in 1932, results in a major simplification in the physically important problem of calculating intensities. If the distribution at large distances from the earth is isotropic, then it is only necessary to find the allowed cone and multiply the subtended solid angle by the intensity in any allowed direction, for any given energy.

Let $I(x, y, z, v_x, v_y, v_z)$ be the number of particles [6] crossing a point (x, y, z) with velocity v_x, v_y, v_z at time t. At a time $t + dt$ these particles will be in a region in the neighborhood of $x', y', z', v'_x, v'_y, v'_z$ and their number will be

$$I(x', y', z', v'_x, v'_y, v'_z)\, d x'\, d y'\, d z'\, d v'_x\, d v'_y\, d v'_z.$$

These two intensities will be equal if the Jacobian determinant

$$J = \frac{d(x', y', z', v'_x, x'_y, v'_z)}{d(x, y, z, v_x, v_y, v_z)} = 1. \tag{2.1}$$

For this to be so, it is only necessary to show that $dJ/dt = 0$, that is, that first-order terms in the expansion of J in powers of dt all vanish. From $x' = x + v_x\, dt...$, ..., $v'_x = v_x + j_x\, dt ..., ...$, the main diagonal of J is

$$1, 1, 1, 1 + \frac{\partial j_x}{\partial v_x}\, dt, \quad 1 + \frac{\partial j_y}{\partial v_y}\, dt, \quad 1 + \frac{\partial j_z}{\partial v_z}\, dt \tag{2.2}$$

and the terms outside the main diagonal contain dt as a factor. The expansion of the determinant is therefore

$$J = 1 + \left(\frac{\partial j_x}{\partial v_x} + \frac{\partial j_y}{\partial v_y} + \frac{\partial j_z}{\partial v_z} \right) dt + \cdots \tag{2.3}$$

and the intensity is constant if

$$\frac{\partial j_x}{\partial v_x} + \frac{\partial j_y}{\partial v_y} + \frac{\partial j_z}{\partial v_z} = 0, \tag{2.4}$$

which is clearly true in an electromagnetic field for j does not depend on v. In particular, in a pure magnetic field the speed is constant. For particles with velocity between v and $v + dv$

$$d v_x\, d v_y\, d v_z = v^2\, d v\, d\omega \tag{2.5}$$

where $d\omega$ is the element of solid angle. The distribution is characterized by

$$I(x, y, z, \omega)\, d x\, d y\, d z\, d\omega \tag{2.6}$$

and I is constant along any allowed direction.

3. The theory of the allowed cone may be divided into two distinct parts. First, one has to consider the motion of charged particles in the earth's magnetic field, but without any reference to a solid earth, and distinguish three kinds of trajectories: bounded trajectories (which always remain within finite distance from the dipole), semi-bounded trajectories (bounded in the past, but not in the future; or in the future, but not in the past), and unbounded trajectories (attaining infinite distance from the dipole both in the past and in the future). To begin with it is clear that, of cosmic rays come from outer space, only unbounded and semi-bounded orbits are possible cosmic trajectories. Thus the first problem of the theory of the allowed cone is to discover and rule out bounded orbits and orbits semi-bounded in the past. The second, and the only remaining problem, is to introduce a solid impenetrable earth and to determine, among all possible unbounded orbits and orbits semi-bounded in the future, those which are blocked by the earth and those which are not. Only those unbounded and semi-bounded trajectories which do not penetrate the earth are allowed trajectories, and they fill up the totality of the allowed cone. In order to formulate and treat our problem more fully we must now proceed to set up and solve the dynamical equations of motion.

In the first approximation the earth's magnetic field is that of a magnetic dipole [7].

Let us introduce spherical coordinates ϱ, λ, φ (distance, latitude, longitude, respectively), with origin at the dipole. Latitude will be taken positive in the northern hemisphere and longitude positive westwards. Further let us introduce the three unit vectors i_ϱ, i_λ, i_φ which, at a given point denote the direction of the progressive tangents to the coordinate curve along which ϱ, λ, φ increase, respectively. If, for purposes of illustration, the magnetic dipole is assumed to be at the earth's centre, i_ϱ is vertical (towards the zenith), i_λ and i_φ determine the horizontal plane with i_λ pointing northwards and i_φ westwards (Fig. 1).

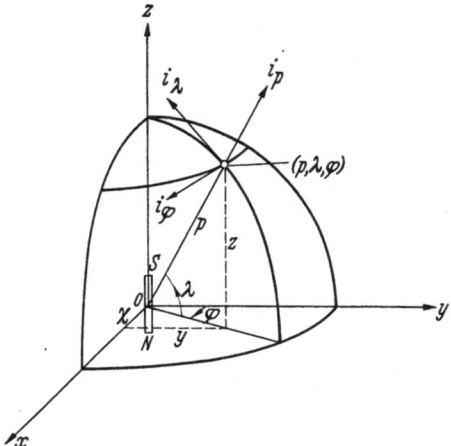

Fig. 1. Coordinate system.

The equations of motion of a particle carrying a charge e can now be written down at once, by equating the ϱ, λ and φ components of the Lorentz force $e\,(\boldsymbol{v} \times \boldsymbol{H})$, where $\boldsymbol{H}$ is the dipole field, to the corresponding components of the product of the relativistic mass (mass at speed v relative to the observer) and the acceleration. To justify this elementary procedure one need only recall that, since the Lorentz force is perpendicular to the particle's path, the work done by it is zero. Hence the kinetic energy of the particle is conserved and so the relativistic mass is constant. We readily obtain

$$m\,(\ddot\varrho - \varrho\,\dot\varphi^2 \cos^2 \lambda - \varrho\,\dot\lambda^2) = - e\,\frac{M\,\dot\varphi}{\varrho^2}\cos^2 \lambda, \tag{3.1}$$

$$m\,(\varrho\,\ddot\lambda + 2\dot\varrho\,\dot\lambda + \varrho\,\dot\varphi^2 \sin\lambda\cos\lambda) = - e\,\frac{2M\,\dot\varphi}{\varrho^2}\sin\lambda\cos\lambda, \tag{3.2}$$

$$\frac{m}{\varrho\cos\lambda}\,\frac{d}{dt}\,(\varrho^2\,\dot\varphi\cos^2\lambda) = e\,\frac{2M\,\dot\lambda}{\varrho^2}\sin\lambda + e\,\frac{M\,\dot\varrho}{\varrho^3}\cos\lambda, \tag{3.3}$$

where M is the dipole's magnetic moment and the dots as usual indicate derivatives with respect to time. To these equations we should add the conservation of kinetic energy

$$\dot\varrho^2 + \varrho^2\,\dot\lambda^2 + \varrho^2\,\dot\varphi^2 \cos^2\lambda = v^2. \tag{3.4}$$

By a change of variables, originally due to Störmer, one may remove at once all the physical quantities (m, M, e) entering in these equations. This is accomplished by using as a unit of length

$$l = \sqrt{\frac{M\,|e|}{m\,v}} \tag{3.5}$$

and then introducing new dimensionless variables r, s defined by

$$\varrho = l r, \qquad v \, dt = l \, ds. \tag{3.6}$$

Making the required transformation in Eqs. (3.1) to (3.3) one obtains:

$$r'' - r \lambda'^2 - r \varphi'^2 \cos^2 \lambda = - \frac{\cos^2 \lambda}{r^2} \, \varphi', \tag{3.7}$$

$$r \lambda'' + 2 r' \lambda' + r \varphi'^2 \sin \lambda \cos \lambda = - \frac{2 \sin \lambda \cos \lambda}{r^2} \, \varphi', \tag{3.8}$$

$$\frac{1}{r \cos \lambda} \frac{d}{ds} (r^2 \varphi' \cos^2 \lambda) = \frac{2 \sin \lambda}{r^2} \lambda' + \frac{\cos \lambda}{r^3} r', \tag{3.9}$$

where dashes indicate derivatives with respect to s. To this must be added, from (4.4)

$$r'^2 + r^2 \lambda'^2 + r^2 \cos^2 \lambda \cdot \varphi'^2 = 1. \tag{3.10}$$

The advantages of STÖRMER's transformation are obvious. A further argument in its favour is the following: the quantity l, Eq. (3.5), for a given particle moving in a given magnetic field, only depends on the momentum of the particle and hence on its energy. The quantity l, as a matter of fact, is, as we shall see, simply the radius of the circular orbit in the dipole's equatorial plane. If now the radius R of the earth is measured in terms of l, r becomes a measure of the energy of the particle. If this energy is measured as usual in electron-volts then

$$r = R \sqrt{\frac{m v}{|e| M}} = R \left(\frac{V}{300 \, M Z} \right)^{\frac{1}{2}} (1 + 600 \, m_0 c^2 / \varepsilon V)^{\frac{1}{4}} \tag{3.11}$$

where V is the voltage which would impart kinetic energy equal to the particle's energy, ε is the absolute value of the electronic charge, $Z = |e|/\varepsilon$, m_0 is the rest mass of the particle, and c is the velocity of light in free space.

Table 1.

r	Protons (Bev)	$_2$He^4 (Bev)
0.1	0.1722	0.1842
0.2	1.618	2.308
0.3	4.49	7.60
0.4	8.61	15.64
0.5	13.79	26.25
0.6	20.50	39.28
0.7	28.23	54.6
0.8	37.19	72.5
0.9	47.29	92.7
1.0	58.50	115.2

The quantity r plays a double role. First it is the radial coordinate to specify the position of a particle and, second, it may be used to fix the radius of the earth in our system of coordinates and in that case it becomes a measure of the particle's energy. As a fitting memorial to the discovery of the transformation (3.6) and of its use, LEMAÎTRE and the author [8] have suggested the name "Störmer unit of length" for the quantity l and for the corresponding specification of the energy given by (3.11) the name "Störmer". Henceforth the energy of a particle will always mean its energy in Störmers. Table 1 gives the equivalence between energy in Störmers and in electron-volts for protons and α-particles.

We now return to the equations of motion (3.7), (3.9). The last of them may be immediately integrated, yielding

$$r^2 \varphi' \cos^2 \lambda + \frac{1}{r} \cos^2 \lambda = \text{const} = 2 \gamma_1. \tag{3.12}$$

Here $2 \gamma_1$ is manifestly a constant of integration which may be evaluated by placing in (3.12) the initial conditions of the motion at infinity. It is readily shown that then the second term of (3.12) vanishes and the first is the component along the dipole axis of the moment of momentum of the particle with respect to the origin. Therefore the arbitrary constant $2 \gamma_1$ is the axial projection of the momentum at infinity and may have all values from $-\infty$ to $+\infty$. Störmer's parameter γ_1 plays a dominant role in the theory we are developing.

An immediate consequence of the existence of the integral (3.12) is that the motion of the particle in space can be resolved into two component motions: one is the motion in the meridian plane where its coordinates are (r, λ); the other is a motion of rotation of the meridian plane around the dipole moment as an axis. This rotation is, from (3.12), governed by the equation

$$\frac{d\varphi}{ds} = \frac{2\gamma_1}{r^2 \cos^2 \lambda} - \frac{1}{r^3} \tag{3.13}$$

and substituting φ' from (3.12) in (3.7) and (3.8) we obtain the equations of motion in the meridian plane

$$r'' - r\lambda'^2 = \frac{(2\gamma_1)^2}{r^3 \cos^2 \lambda} - \frac{6\gamma_1}{r^4} + \frac{2\cos^2 \lambda}{r^5}, \tag{3.14}$$

$$r\lambda'' + 2r'\lambda' = \frac{\sin \lambda \cos \lambda}{r^5} - \frac{4\gamma_1^2 \sin \lambda}{r^3 \cos^2 \lambda}; \tag{3.15}$$

the conservation of kinetic energy (3.10) becomes

$$r'^2 + r^2\lambda'^1 = 1 - \frac{4\gamma_1^2}{r^2 \cos^2 \lambda} + \frac{4\gamma_1}{r^3} - \frac{\cos^2 \lambda}{r^4} \tag{3.16}$$

and it may be observed that (3.16) is not independent of (3.13) and (3.15). Eq. (3.16) thus gives the kinetic energy of the particle in the meridian plane. A number of important consequences follow from (3.16) which are further discussed below.

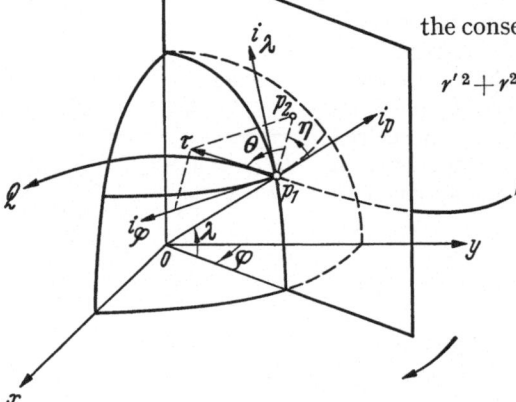

Fig. 2. Angles θ and η.

To specify completely the direction of a trajectory in space, once its motion in the meridian plane, obtained by solving (3.14) and (3.15), is known, we need to know two angles. These may be taken to be the angle θ between the progressive tangent τ (Fig. 2) to the space trajectory and the meridian plane M, and the angle η which the projection of τ on M makes with the zenith direction (direction of i_ϱ). Evidently (Fig. 2)

$$\sin \theta = \cos(\tau, i_\varphi) = r \cos \lambda \cdot \varphi' \tag{3.17}$$

and from (3.13)

$$\pm \sin \theta = \frac{2\gamma_1}{r \cos \lambda} - \frac{\cos \lambda}{r^2}, \tag{3.18}$$

where the positive sign refers to positive particles and the negative sign to negative particles. The angle θ is then counted positive eastwards.

Formula (3.18) is another of Störmer's fundamental contributions. From (3.16) we see at once that

$$Q = r'^2 + r^2\lambda'^2 - 1 - \sin^2 \theta = \cos^2 \theta. \tag{3.19}$$

The angle η is immediately seen to be given by

$$\tan \eta = \frac{r\, d\lambda}{dr}. \tag{3.20}$$

Thus, once the trajectory in the meridian plane is known, its space direction at any point is given by (3.18) and (3.20). This fact enables one to study bounded and unbounded orbits (allowed and forbidden directions) without actually solving (3.13) to find space trajectories. It is only exceptionally that we must integrate (3.13) in addition to (3.14) and (3.15) [9]. As a matter of fact, no use of (3.13) will be made in the sequel.

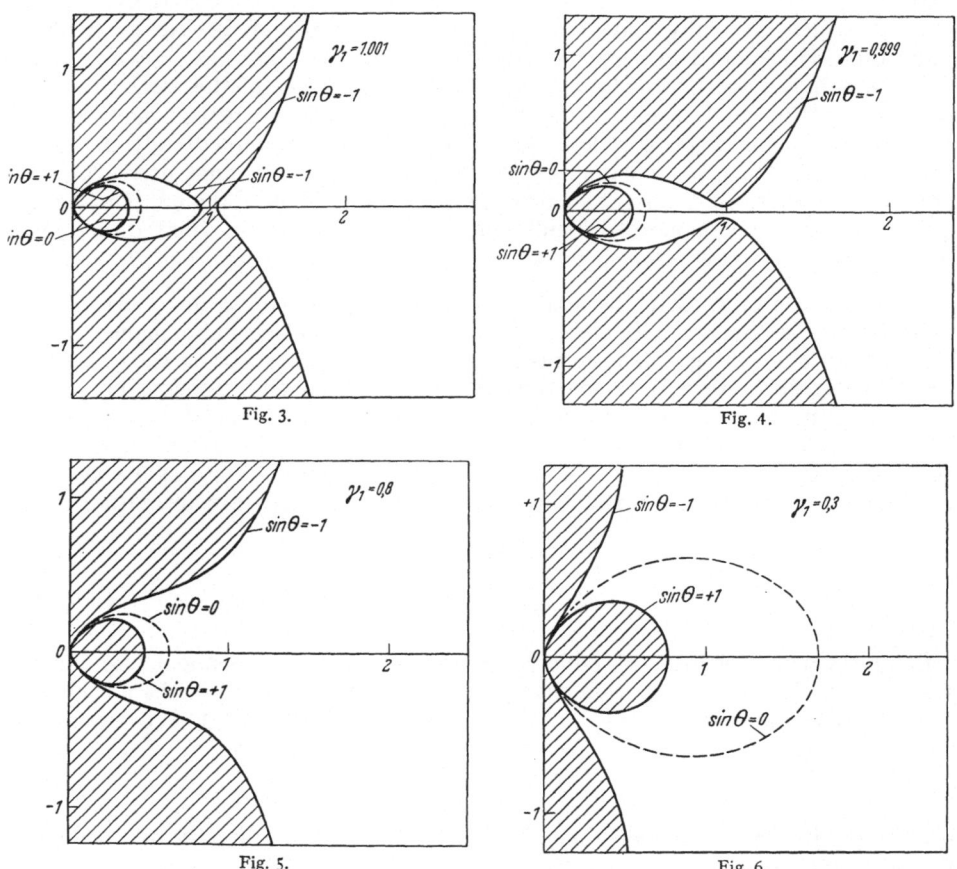

Fig. 3 to 6. Allowed and forbidden regions in r, λ for different γ_1's.

4. The fundamental nature of STÖRMER's formula (3.18) is already brought out by observing that from its very nature $\sin \theta$ must satisfy the condition:

$$-1 \leqq \sin \theta \leqq 1 \tag{4.1}$$

which in conjunction with (3.19) gives the boundary between allowed and forbidden regions of motion in the meridian plane. Remembering that Q, given by (3.19), is essentially positive, being the sum of the squares of two real numbers, it is seen that Q can only exist in the region bounded by and satisfying the condition (4.1).

Figs. 3 to 6, reproduced here from STÖRMER's publications, illustrate the forbidden (cross-hatched) and allowed regions in the meridian plane. The curve $\sin \theta = 0$ along which the velocity vector of the space trajectory lies in the meridian plane, has also been drawn. It is seen that (Fig. 3) when $\gamma_1 > 1$, the internal allowed region containing the earth (for particles of energies less than $r = 1$) has no connection with the external allowed region extending to infinity. Thus all particles for which $\gamma_1 > 1$, coming from infinity, can never reach the earth. It is also obvious that for all such values of γ_1 all motions in the internal region are bounded; it will be proved that all motions in the external region are unbounded. For $\gamma_1 = 1$ a hyperbolic double point appears at $r = 1$ on the equator and for $\gamma_1 < 1$ the inner and the outer allowed regions are connected by a pass (Fig. 4), which gradually widens as γ_1 decreases (Fig. 5 and 6). No particular importance is attached to negative values of γ_1 and these are not shown. Thus it is clear that all particles whose energy places the surface of the earth at $r < 1$ can arrive at the earth, coming from infinity, only if their moment of momentum at infinity, specified by γ_1 is less than unity. It is also immediately obvious that particles of a given energy ($r < 1$) and a given $\gamma_1 < 1$ cannot arrive at all points of the earth, but only at such points as lie in an allowed region. Our next question, and a much more difficult one, is to find out from what directions particles of all energies, from 0 to ∞, may arrive at a given point of the earth.

5. Before we proceed to the analysis of our main problem, a few more preliminaries are necessary. The equations of motion in the meridian plane (3.14) and (3.15) are not integrable in terms of known functions (except on the equator, $\lambda = 0$, where the integrals depend on elliptic functions) and recourse must be taken to numerical or mechanical methods of integration. For this purpose it is most convenient to introduce a transformation which rejects the singularity at $r = 0$ away from the finite region of the meridian plane. Here again we are indebted to Störmer for introducing Goursat's conformal transformation

$$e^x = 2\gamma_1 r, \tag{5.1}$$

$$ds = \frac{e^{2x}}{(2\gamma_1)^3} d\sigma, \tag{5.2}$$

defining the new dependent variable x instead of r and the independent variable σ instead of s. This transformation changes polar coordinates (r, λ) in the meridian plane to Cartesian coordinates (x, λ). It is readily seen that (3.20) becomes

$$\tan \eta = \frac{d\lambda}{dx}, \tag{5.3}$$

and the zenith angle of a trajectory at a given point is given directly by its slope at that point. Eq. (3.18) becomes

$$\pm \sin \theta = 4\gamma_1^2 e^{-x} (\sec \lambda - e^{-x} \cos \lambda) \tag{5.4}$$

from which it is clear that the locus $\sin \theta = 0$ is given by

$$x = 2 \log \cos \lambda. \tag{5.5}$$

The equations of motion (3.14) and (3.15) now become

$$\frac{d^2 x}{d\sigma^2} = a e^{2x} - e^{-x} + e^{-2x} \cos^2 \lambda, \tag{5.6}$$

$$\frac{d^2 \lambda}{d\sigma^2} = e^{-2x} \sin \lambda \cos \lambda - \frac{\sin \lambda}{\cos^3 \lambda}, \tag{5.7}$$

where $a = (1/2\gamma_1)^4$. Finally (3.19) becomes

$$P = \left(\frac{dx}{d\sigma}\right)^2 + \left(\frac{d\lambda}{d\sigma}\right)^2 = a e^{2x} - \left(e^{-x} \cos \lambda - \frac{1}{\cos \lambda}\right)^2, \tag{5.8}$$

and we note that the transformation (5.1), (5.2) requires that $P = Q \, e^{2x}/(2\gamma_1)^4$. It may also be noted that (5.6) and (5.7) may also be written

$$\frac{d^2x}{d\sigma^2} = \frac{1}{2}\frac{\partial P}{\partial x}, \qquad \frac{d^2\lambda}{d\sigma^2} = \frac{1}{2}\frac{\partial P}{\partial \lambda}. \qquad (5.9)$$

6. Corresponding to the boundaries $Q=0$ given by (3.19) we have the boundaries $P=0$ given by (5.8) (Fig. 7), whose significance has already been discussed

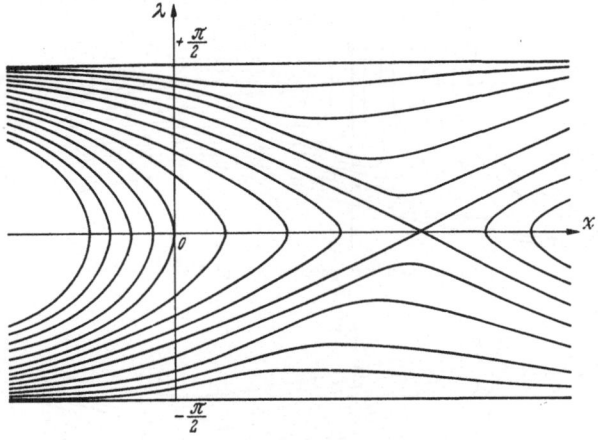

Fig. 7. Boundary lines $\varrho = 0$ for different γ_1's.

above. From (5.9) it follows that, as noted by STÖRMER, the motion in the meridian plane can be thought of as that of a particle in a potential field given by

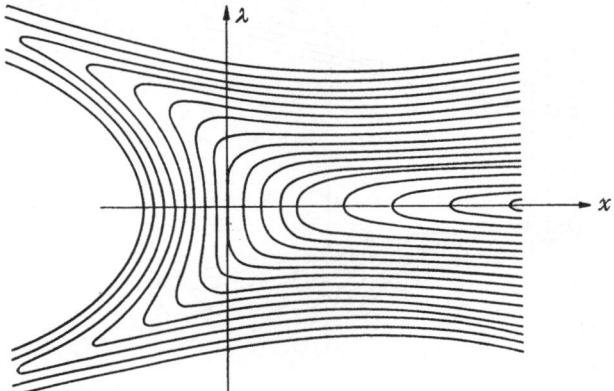

Fig. 8. Level lines $\varrho = \text{const}$ for $0 < \gamma_1 < 2/3^{\frac{3}{4}} = 0.877\,383$.

$-P(x, \lambda, \gamma_1)$ in such a way that its kinetic energy is also given by $+P(x, \lambda, \gamma_1)$, Eq. (5.8), thereby allowing motions of zero total energy only. The boundaries of the regions of allowed motion are the curves $P=0$, and these are also the loci of zero velocity in the meridian plane. From this it follows that if a trajectory in the meridian plane ever reaches the line $P=0$ it must return back on itself. The level lines $P=\text{const}$ are shown in Figs. 8 to 10. The saddle pass to infinity and the elliptic bowl are clearly seen. Other important loci are the lines along which the acceleration components $d^2x/d\sigma^2$ and $d^2\lambda/d\sigma^2$ vanish. On either side of these lines the acceleration has opposite sign. From (5.6) the first is given by the equation

$$a\,e^{2x} + e^{-2x}\cos^2\lambda = e^{-x}, \qquad (6.1)$$

$$e^x = \cos^2\lambda. \qquad (6.2)$$

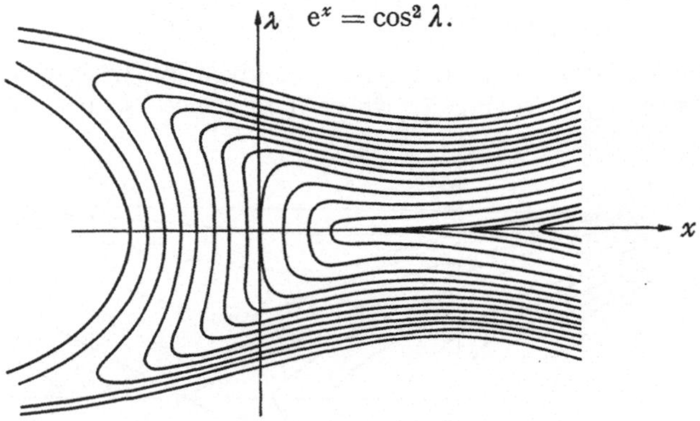

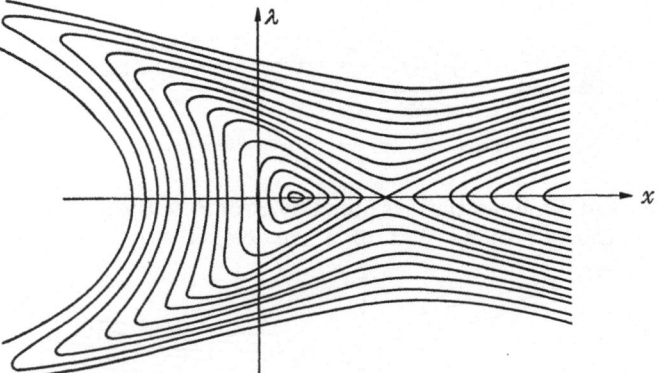

Fig. 9. Level lines $\varrho = $ const for $\gamma_1 = 2/3^{\frac{3}{4}} = 0.877\,383$.

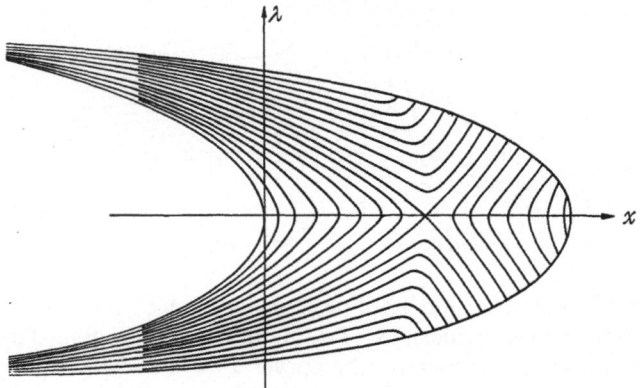

Fig. 10. Level lines $\varrho = $ const for $2^{\frac{3}{4}} < \gamma_1 < 1$.

Fig. 11. Lines of zero acceleration.

The loci (6.1) for different values of γ_1 are shown in Fig. 11. From (6.2) it is seen that the line along which $d^2\lambda/dr^2 = 0$ is the same as the line $\theta = 0$, Eq. (5.5), as is obvious from physical considerations.

7. The problem of the theory of the allowed cone is to ascertain, for every point (x, λ) and every direction (η) the plane orbit in the meridian plane and to discover whether this orbit may, or may not, go to infinity. For it is clear that if a trajectory goes to infinity in the meridian plane, so does the space trajectory derived from it by considering the motion of the meridian plane; and if it does not, neither does the space orbit. The only available method of ascertaining whether at a given point there are directions along which particles coming from infinity may arrive (i.e., allowed directions) is to integrate the pair of differential equations (5.6), (5.7) and to discover, if possible, what kind of trajectories form the boundary between allowed and forbidden directions. The main difficulty here already pointed out before, is that these differential equations are not integrable and therefore one is forced to resort to numerical or mechanical methods of integration. The only instance where these equations may be integrated is on the equatorial plane $(\lambda = 0)$. Since this case already exhibits some of the features of the general case we proceed to examine it briefly.

The equations of motion are now (3.13) and (3.14) with $\lambda = 0$ but it is more convenient to use (3.13) and (3.16). We have [10]

$$r^2 \varphi' = 2\gamma_1 - \frac{1}{r}, \tag{7.1}$$

$$r'' = \frac{4\gamma_1^2}{r^3} - \frac{6\gamma_1}{r^4} + \frac{2}{r^5}, \tag{7.2}$$

$$r^2 r' = (r^4 - 4\gamma_1^2 r^2 + 4\gamma_1 r - 1)^{\frac{1}{2}}. \tag{7.3}$$

A solution of these equations is easily seen to be $r = 1$, $\gamma_1 = 1$, $\varphi' = 1$. This is a circular (periodic) orbit in which the particle moves with unit angular velocity and unit linear velocity. It is the simplest of all the periodic solutions of our differential equations. It is unstable, as may be readily seen qualitatively from a consideration of the potential field $P(r, 0, 1)$, or quantitatively from general theorems of POINCARÉ on characteristic exponents, for its characteristic exponent is positive and has the value $1/\sqrt{2}$ [11]. There are an infinity of other periodic orbits in the equatorial plane, all of which are closer to the earth and have now been found [12].

Our problem now is stated as follows: suppose we have a point P on the earth at $\lambda = 0$ and suppose we consider all the trajectories arriving at P with different θ but lying in the equatorial plane $(\lambda = 0)$. What are the allowed and the forbidden directions? More explicitly: what are the directions along which particles coming from infinity may arrive at P?

To begin with, it is clear that since our differential equation (7.2) is of the second order (without terms in r') a possible trajectory coming from infinity will also be a possible trajectory starting at P and going to infinity. In other words, the time direction is reversible. Thus instead of considering all the trajectories starting at infinity and coming to P we may study all the trajectories starting at P and find out which among them go to infinity.

For a given value of θ the corresponding γ_1 is fixed by

$$\gamma_1 = \frac{r^2 \sin \theta + 1}{2r} \tag{7.4}$$

for positive particles, from (3.18). Now, from (7.3),

$$s = \int \frac{r^2 \, dr}{\sqrt{r^4 - 4\gamma_1^2 r^2 + 4\gamma_1 r - 1}} + C \tag{7.5}$$

which can be easily reduced to an elliptic integral.

The roots of the polynomial in the denominator are important since they determine the boundaries of the allowed regions of motion in the equatorial plane. Let $f(r)$ be this polynomial and let us write

$$\frac{f(r)}{r} = p(r) - q(r),$$

where

$$p(r) = r^3 - \frac{1}{r}$$

is an odd function and $q(r) = 4\gamma_1(\gamma_1 r - 1)$ is a family of straight lines. For $f(r) = 0$, $p = q$, therefore, the intersection of $p(r)$ and $q(r)$ fixes the roots of $f(r)$. The result is shown in Fig. 12 where, remember-ing the meaning of r, only the positive half-plane, for positive r, is shown. For negative values of γ_1 the root r_2 continues to rise to infinity. The root r_1 is not shown because it is always negative. It is seen that for values of $r < 1$, γ_1 must be between the limits 0 and 1. For $\gamma_1 < 0$ the earth is in a forbidden region (except for very high energies) and for $\gamma_1 > 1$ the earth is again either in a forbidden region or in an allowed region separated from in-finity by an impenetrable barrier. In neither case may the earth be reached by particles coming from infinity. When $0 < \gamma_1 < 1$, a corresponding range of θ may be calculated by using (7.4) within which all directions are allowed. The value $\gamma_1 = 1$ and the corresponding value of θ

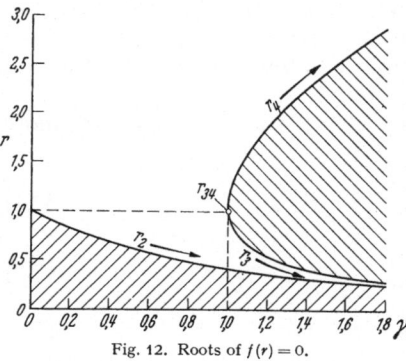
Fig. 12. Roots of $f(r) = 0$.

$$\theta = \arcsin{(r^2 - r)}/r^3 \qquad (r < 1) \tag{7.6}$$

is thus seen to be the boundary between allowed and forbidden directions. For $r > 1$, on the other hand, all directions belonging to the allowed range of γ_1 are allowed directions, and there is no boundary between allowed and forbidden directions. The two limits of γ_1 corresponding to r_2 and r_4 give simply $\theta = -\pi/2$ and $\theta = \pi/2$ respectively, i.e., grazing incidence along the horizontal plane.

The question now arises as to what characterizes the orbit which leaves the earth at $\lambda = 0$ along the east-west plane at an angle θ given by (7.6). Now by eliminating ds between (7.1) and (7.3), we obtain

$$\varphi = \pm \int_{r_0}^{\infty} \frac{\left(\dfrac{2\gamma_1}{r^2} - \dfrac{1}{r^3}\right) dr}{\sqrt{1 - \dfrac{4\gamma_1^2}{r^2} + \dfrac{4\gamma_1}{r^3} - \dfrac{1}{r^4}}} \tag{7.7}$$

which may be expressed in elliptic integrals as follows:

$$\varphi = \arcsin{(K \sin \psi)} \mp \frac{\gamma_1}{\sqrt{2}} F(\alpha, \psi) + C \tag{7.8}$$

where

$$K = \sqrt{\frac{1 + \gamma_1^2}{2}} = \sin \alpha \tag{7.9}$$

and

$$\psi = \arccos \frac{r \gamma_1 - 1}{r \sqrt{1 + \gamma_1^2}} \tag{7.10}$$

and $F(\alpha, \psi)$ is the elliptic integral of the first kind. This holds for $\gamma_1 < 1$. For $\gamma_1 = 1$ we have

$$\varphi = 1\nu - \frac{\gamma_1/\sqrt{2}}{\sqrt{1 + \gamma_1^2}} \log \cot \nu \qquad (7.11)$$

where

$$\nu = \frac{1}{2} \arcsin \frac{r-1}{r\sqrt{2}}. \qquad (7.12)$$

Finally for $\gamma_1 > 1$

$$\varphi = \psi \mp \frac{\gamma_1}{\sqrt{1 + \gamma_1^2}} F(\alpha, \psi) + \varphi_0 \qquad (7.13)$$

where

$$K = \sin \alpha = \sqrt{\frac{2}{1 + \gamma_1^2}} \qquad (7.14)$$

and

$$\psi = \arcsin \sqrt{(2r\gamma_1 + r^2 - 1)/2r^2}. \qquad (7.15)$$

From (7.11) and (7.5) it can be shown that, when $\gamma_1 = 1$, and θ is given by (7.6), as $r \to 1$, $s \to \infty$ and $\varphi \to \infty$. This characterizes an orbit asymptotic to the periodic circular orbit at $r = 1$. At the point of the earth in question all orbits with $\gamma_1 < 1$ go to infinity, all those with $\gamma_1 > 1$ return to the earth, and the orbit $\gamma_1 = 1$ is an asymptotic orbit to the circular periodic orbit. The asymptotic orbit is therefore a boundary between allowed and forbidden directions. The situation is illustrated in Fig. 13.

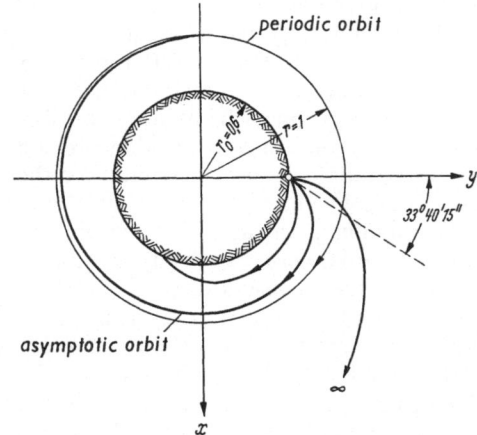

Fig. 13. Asymptotic orbit in equatorial plane.

8. The foregoing example throws considerable light on the importance of periodic and asymptotic orbits for our problem. The circular periodic orbit in the equatorial plane is the first member of a family of unstable periodic orbits in the meridian plane, first discovered by STÖRMER [13], which oscillate symmetrically on both sides of the equatorial plane in the neighbourhood of the pass to infinity ($\gamma_1 < 1$) (Fig. 7). These orbits cut the x-axis and the line $P = 0$ orthogonally and have the property of being the farthest periodic orbits with respect to the earth. STÖRMER was able to show that for certain values of γ_1 there is, in addition to the orbit discussed above, henceforth to be referred to as the "outer periodic orbit", an inner orbit, also symmetrical with respect to the equator and oscillating on either side of it inside the bowl (Fig. 10). This orbit will be called the "inner periodic orbit" [14]. The two orbits together will be alluded to as the "principal periodic orbits". STÖRMER early showed that, besides the pair of principal periodic orbits, there are an infinity of other periodic orbits. It has since been shown [15] that all of them cross the equator between the two principal orbits.

STÖRMER's method for the determination of periodic orbits, in particular the pair of principal periodic orbits, was a method of successive approximations. The differential equations of motion in the meridian plane, (5.6) and (5.7) (or alternately the corresponding equations in the r, λ-plane) were integrated numerically by a step-by-step method. A trajectory was started at right angles to the equator from an arbitrarily chosen point x. If x is suitably chosen, say $x = x_0$, then the trajectory reaches the lines $P = 0$ with zero velocity at a point S_0 and returns back on itself, oscillating back and forth between S_0 and its symmetrical counterpart S_0', and is therefore a periodic orbit. If the value of x is chosen too small, then near the outer periodic orbit the trajectory turns towards the dipole, near the inner orbit away from it. If x is too large near the outer orbit, the trajectory turns towards infinity, while near the inner orbit it turns towards the dipole. As a result of a considerable amount of work STÖRMER was able to calculate the pair of principal periodic orbits for two values of γ_1 ($\gamma_1 = 0.80$ and $\gamma_1 = 0.97$).

The method employed by STÖRMER is too cumbersome and, for practical reasons related to the choice of the interval of integration and the amount of labour involved, incapable of yielding results of high precision. To obviate these difficulties and to obtain at once the whole family of periodic orbits (outer and inner) LEMAÎTRE devised an ingenious analytical method which embodies in principle the same feature as STÖRMER's method. We now turn our attention to giving a brief summary of LEMAÎTRE's method, leaving out the details of calculation, for which the reader should refer to the original paper.

By eliminating the independent variable σ from the equations of motion (5.6) and (5.7), a complicated differential equation between λ and x is obtained, a class of integrals of which are possible dynamical trajectories. In particular there are solutions which meet the line $P = 0$ orthogonally at a point S_0 and continue beyond into a forbidden region. The extension of a trajectory beyond S_0 is devoid of physical meaning, for there σ becomes imaginary, but it is nevertheless a solution of the differential equation. The point S_0 in fact is a hyperbolic singular point at which two integrals of the differential equation (a regular and a singular) cross each other: the periodic orbit and the line $P = 0$. On the basis of these considerations LEMAÎTRE sought to obtain from the differential equations x as a power series in $\sin^2 \lambda$. The expansions corresponding to the inner and outer periodic orbits then converge up to the points R_0 and S_0 respectively and beyond, but if the initial value of x on $\lambda = 0$ has not been suitably chosen the expansion may still converge, but its derivative decreases without limit, as already evident from STÖRMER's results. This reasoning led LEMAÎTRE to lay down the hypothesis, later confirmed by actual calculation, that the power series giving x in terms of $\sin^2 \lambda$ is an alternating series the coefficients of which decrease regularly in absolute value. For $x > x_0$ the positive terms increase with respect to the negative, and conversely for $x < x_0$. A diagram of the absolute value of the successive terms appears as a zigzag curve for $x > x_0$; for $x < x_0$ the zigzag is reversed; hence from such a diagram it is possible to obtain by interpolation the periodic orbit. The method is well suited for the determination of the whole family of periodic orbits.

To see how this may be done it is necessary to enter into some details. In place of x and λ in (5.6) and (5.7) we introduce the new variables

$$z = \sin^2 \lambda, \quad e^{-2x} = \frac{dy}{dz} = y' \tag{8.1}$$

from which and (5.8)

$$\left(\frac{dz}{d\sigma}\right)^2 = 4z \left[y(1 - z) - 1\right] \tag{8.2}$$

with the help of which σ may be eliminated from (5.6), (5.7). The resultant differential equation with y as dependent and z as independent variable is nonlinear and of the third order, of extremely complicated structure. LEMAÎTRE, however, successfully used it to obtain recurrence formulas among the coefficients of the expansion

$$y = y_0 + y_1 z \left[1 + u_2 z + v_3 (u_2 z)^2 + v_4 (u_2 z)^3 + \cdots \right]. \tag{8.3}$$

By actual calculation LEMAÎTRE found $v_3, v_4, \ldots$ to differ slightly from their limiting values when $\gamma_1 = 1$, so he placed

$$v_K = w_K v_K^*$$

where $v_K^* = \lim\limits_{\gamma_1 \to 1} v_K$ and the coefficients w_K differ slightly from unity.

To devise a method of interpolation for the computation of the whole family of periodic orbits, based on the fact that $w_3, w_4, \ldots$ differ slightly from unity for all values of γ_1, LEMAÎTRE contrived the transformation

$$1 - \gamma_1 r_0 = 5\xi + 6\eta, \tag{8.4}$$

$$1 - r_0^4 = 10\xi + 10\eta \tag{8.5}$$

which defines two small quantities ξ and η in terms of γ_1 and $r_0 = e^{x_0}/2\gamma_1$, where x_0 is the initial value of x on $\lambda = 0$. By introducing the initial conditions y_1, u_0, and u_2 are expressed in terms of ξ and η by

$$y_1 = \frac{1}{4 \gamma_1^2 r_0^2}, \tag{8.6}$$

$$u_0 = 4 - 30\xi - 34\eta, \tag{8.7}$$

$$-\frac{1}{u_2} = 14 - 150\xi - 180\eta + 18\eta v_3. \tag{8.8}$$

Table 2.

ξ	η	r_0	γ_1
0.000	0.00000	1.00000	1.00000
0.001	0.03613	0.89045	0.87396
0.002	0.04678	0.84598	0.83846
0.003	0.05346	0.81231	0.81772
0.004	0.05819	0.78415	0.80451
0.005	0.06170	0.75965	0.79616
0.006	0.06437	0.73778	0.79127
0.007	0.06642	0.71802	0.78895
0.008	0.06801	0.69985	0.78865
0.009	0.06922	0.68315	0.78999
0.010	0.07014	0.66757	0.79266
0.011	0.07081	0.65307	0.79645
0.012	0.07128	0.63945	0.80119

Any pair of consistent values of ξ and η represents in general an arbitrary orbit. By following the reasoning outlined above, LEMAÎTRE proved that the family of periodic orbits is characterized by pairs of ξ and η which satisfy the equation:

$$10\xi^2 + 9\xi\eta - \xi + 1.8\eta^2(v_3 - 2) = 0 \tag{8.9}$$

obtained from the initial conditions and the recurrence formulas.

The relation between γ_1, r_0 and LEMAÎTRE's parameters ξ, η is given in Table 2.

From this table or from a plot of either ξ or η as a function of γ_1 it is abundantly clear that *there is a limiting value of γ_1 for which the inner and outer periodic orbits collapse and then vanish together.* This value, whose existence was first pointed out by LEMAÎTRE and the writer [16] and estimated to be approximately $\gamma_1 = 0.783$, was subsequently calculated by LEMAÎTRE on the basis of the results we are now discussing, to be $\gamma_1 = 0.78856$ and subsequently refined by GODART [17] to 0.788541. It plays a fundamental role in the present theory.

To understand the meaning of this limiting value, suppose we consider the family of trajectories leaving the line $P = 0$ orthogonally, for different values of γ_1, and continue them until the point where the trajectory has an infinite slope. At this point the x-component of the velocity vanishes. We now connect all the points where $dx/d\sigma = 0$ and obtain the "loci of zero x-velocity" shown in Fig. 14. These loci are not given analytically but were found by mechanical integration, as described later [18]. The intersection between these loci and the equator for each value of γ_1 gives the two points (one to the left, one to the right) as which the two principal (inner and outer) periodic orbits intersect the equator. It will be observed that for a value of γ_1 between 0.78 and 0.82 the locus in question becomes tangent to the equator and the two orbits collapse together. The exact value of γ_1 for which this takes place is the value $\gamma_1 = 0.788541$ calculated above.

It will also be seen that for $\gamma_1 = 1$ there is an outer orbit (the circular orbit treated previously) and an inner companion orbit which intersects the equator close to $x_i = -0.1$. For $\gamma_1 > 1$ there is still an inner orbit, but no outer orbit, which has now apparently collapsed into the equatorial plane. It has been proved [19] that for a certain denumerable set of values of $\gamma_1 > 1$ there exist in the equatorial plane *closed* periodic orbits having a finite sumber of loops and turns. This happens whenever the ratio of the period of the oscillation in

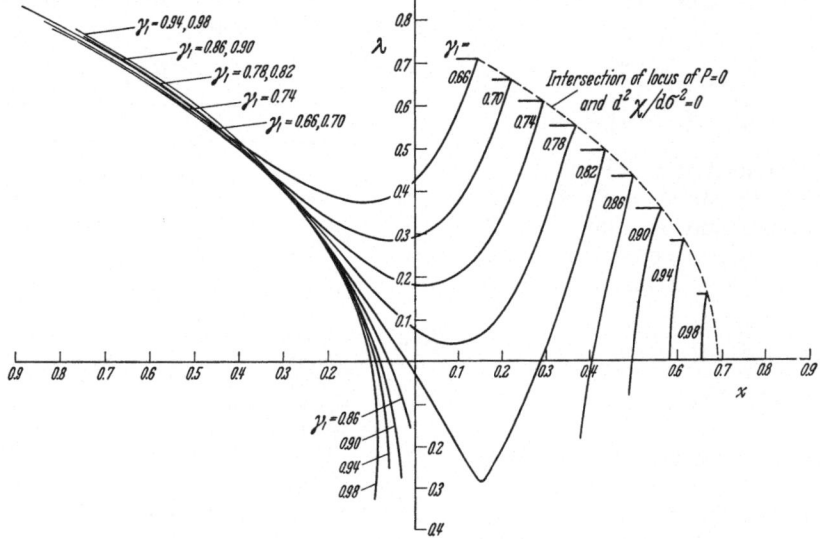

Fig. 14. Lines of zero velocity.

the r-direction to the period in the φ-direction is a commensurable ratio. For all other values of γ_1 the ratio is incommensurable and the orbit, like a conditioned periodic motion, fills up a ring coming as nearly as desired to any point within this ring. The boundaries of the ring in the equatorial plane are shown in Fig. 12. The analogy with the LISSAJOUS figures is apparent. A similar situation exists in space [20].

Returning now to LEMAÎTRE's method for the calculation of the principal periodic orbits, it may be shown by somewhat elaborate manipulation that $\sin \lambda$ is given very approximately by KEPLER's equation

$$\sin \lambda = \sqrt{z_1} \sin(\omega \sigma + \psi_1 \sin 2\omega \sigma) \tag{8.10}$$

where ψ_1 is, a constant characteristic of the orbit and $2\pi/\omega$ is the fundamental period. The orbit may then be analysed into a Fourier series and refined to any desired precision by a variational method. λ and x are expressed by:

$$\lambda = \mu_1 \sin \omega \sigma + \mu_3 \sin 3\omega \sigma + \mu_5 \sin 5\omega \sigma + \cdots, \tag{8.11}$$

$$x = z_0 + z_2 \cos 2\omega \sigma + z_4 \cos 4\omega \sigma + \cdots \tag{8.12}$$

as is readily seen from the symmetry of the principal periodic orbits.

The family of outer periodic orbits which, as will be seen, are most important for our purpose, calculated in this way, is exhibited in Fig. 15. Several of these orbits have been calculated with very high precision: the limiting orbit $\gamma_1 = 0.78856$ to five significant figures by LEMAÎTRE [14]; the outer orbit $\gamma_1 = 0.929898$ to six by LEMAÎTRE and the writer [21] and that for $\gamma_1 = 0.8500008$ by BAÑOS [22].

A very important question has to do with the *stability* of periodic orbits. It has been shown [23] by POINCARÉ's method of characteristic exponents that all the outer orbits are unstable, whereas some of the inner orbits are stable. The values of the characteristic exponents are shown in Fig. 16 and in Table 3 where ω is the fundamental frequency.

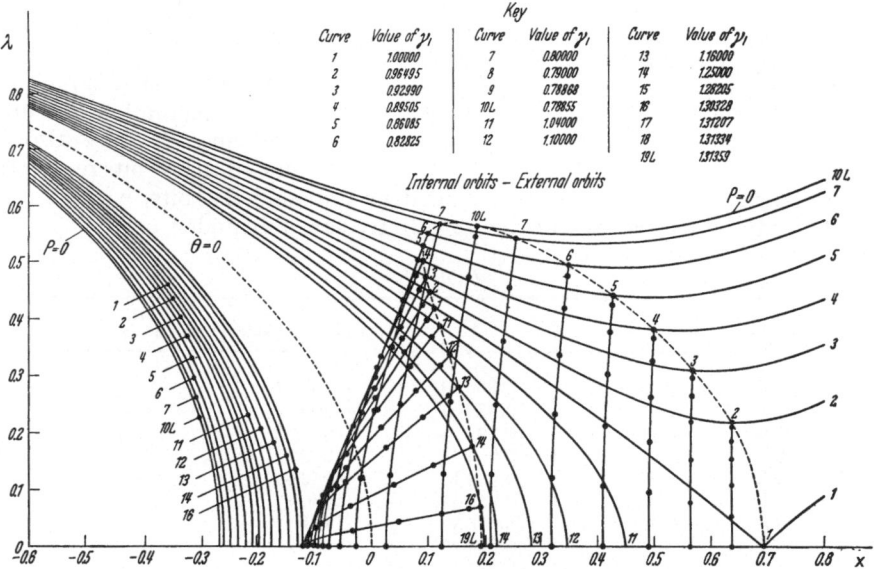

Fig. 15. Family of principal periodic orbits.

The foregoing considerations (Sect. 7) have made clear the importance of asymptotic orbits for the problem of the allowed cone. Before proceeding with the discussion of the orbits which delimit the allowed cone it is necessary to define certain terms for the description and characterization of various orbits [15]. One may define a *section* of an orbit as the segment of it joining two consecutive points where $P(x, \lambda, \gamma_1)$ passes through two consecutive relative maxima. From a study of several thousand trajectories obtained by mechanical integration, as further discussed in a later section, it is known that there exist, between the two principal periodic orbits, self-reversing orbits reaching $P = 0$ which delimit a class of orbital sections, hereafter to be designated as re-entrant toward the dipole, in the following way: each self-reversing orbit has a symmetrical counterpart crossing it at the equator and reversing at a symmetrical point. The resulting symmetrical pair of self-reversing sections delimits a pencil of directions passing through this intersection at the equator. All orbital sections which pass through this equatorial point, within this pencil of directions, are termed re-entrant toward the dipole. Beyond the outer principal periodic orbit there still exist self-reversing orbits. In the same way the self-reversing sections, and those of their symmetrical counterparts, delimit a class of orbital sections termed re-entrant toward infinity. If an orbit contains at least one section re-entrant toward the dipole, it will be named re-entrant towards the dipole; if it contains a section re-entrant toward infinity, it will be termed re-entrant toward infinity.

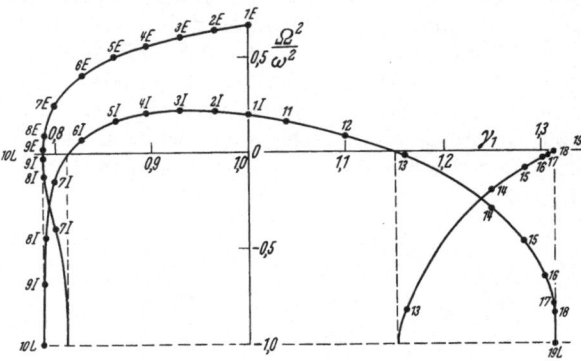

If an orbit contains no section of either kind, it will be termed non-reentrant. Such an orbit necessarily cuts the outer periodic orbit at a non-zero angle. While these three orbital types completely exhaust the totality of orbits, it may be noted that the three kinds of orbital sections that have been mentioned do not exhaust the totality of orbital sections. It may also be noted that orbits re-entrant

Fig. 16. Characteristic exponents of principal periodic orbits.

toward the dipole may be either bounded or unbounded orbits, while non-reentrant orbits and orbits re-entrant to infinity are unbounded orbits.

Table 3.

Orbit No. (Fig. 16)	γ_1	ω	Ω_1 (δx even, $\delta\lambda$ odd)	Ω_2 (δx odd, $\delta\lambda$ even)
1 E	1.000000	0.86603	0.70711	—
2 E	0.964953	0.87964	0.70217	—
3 E	0.929898	0.89278	0.69192	—
4 E	0.895050	0.90473	0.67329	—
5 E	0.860846	0.91418	0.64049	—
6 E	0.828245	0.91842	0.58047	—
7 E	0.800000	0.91052	0.45147	—
8 E	0.790000	0.89420	0.28021	—
9 E	0.788683	0.88558	0.15581	—
10 L	0.7885526	0.88151	0	i 0.88151
9 I	0.788683	0.87722	i 0.15795	i 0.71927
8 I	0.790000	0.86637	i 0.29351	i 0.57286
7 I	0.800000	0.83338	i 0.52462	i 0.30876
6 I	0.828245	0.78044	—	0.21300
5 I	0.860846	0.73617	—	0.29880
4 I	0.895050	0.69832	—	0.31774
3 I	0.929898	0.66546	—	0.31295
2 I	0.964953	0.63661	—	0.29629
1 I	1.000000	0.61104	—	0.27181
11 I	1.040000	0.58506	—	0.23561
12 I	1.100000	0.55121	—	0.15799
13 I	1.160000	0.52221	i 0.46179	i 0.06042
14 I	1.250000	0.48566	i 0.21662	i 0.26904
15 I	1.282050	0.47426	i 0.14821	i 0.32605
16 I	1.303280	0.46712	i 0.09076	i 0.37636
17 I	1.312070	0.46425	i 0.05167	i 0.41258
18 I	1.313340	0.46387	i 0.04060	i 0.42327
19 L	1.3135887	0.46376	0	i 0.46476

From a study of several thousand trajectories and of the potential field in which they arise one may prove the theorem: In the interval $\gamma_1 < 0.7885$ all orbits are unbounded, in the interval $0.7885 \leqq \gamma_1 \leqq 1$ bounded and unbounded orbits may coexist, and in the interval $\gamma_1 > 1$ all orbits are bounded or unbounded according as to whether they lie within the bounded or unbounded regions of motion (Sect. 4). In the *absence of the earth*, therefore, the only limit for allowed directions is $\gamma_1 = 1$ which, by STÖRMER's relation (3.18), fixes, for each latitude and energy, a value of θ. This value of θ fixes in turn the aperture of a circular cone with vertex at the observer; this cone we have called the *Störmer cone*. In this case, almost all directions within this cone are allowed directions for particles coming from infinity. It still remains an unsettled problem whether the exceptional directions corresponding to the bounded motions form a set of zero measure.

In the preceding section the presence of an impenetrable earth was intentionally omitted. It is now required to examine the ways in which orbits, which would otherwise go to infinity, are blocked by the earth. First of all one may immediately distinguish the class of orbits which go directly to infinity and cannot be blocked by the earth. These are the non-reentrant orbits. These orbits constitute the most important region of allowed directions (at any point on the earth and for any energy) and they fill up the inside of the region that we have called the *main cone*. The question now arises as to which orbits make up the boundary of the main cone. These orbits are transition orbits between orbits without re-entrant sections and orbits with sections re-entrant towards the earth, i.e., to orbits re-entrant ine or more times towards the earth. Within the main cone, therefore, all directions are allowed; outside of it, but within the Störmer cone, allowed and forbidden directions are both possible (region of penumbra) depending on whether the corresponding re-entrant trajectory is not, or is, blocked by the earth. Now a transition orbit between re-entrant and non-reentrant orbits must be an orbit (more precisely a half-orbit, i.e., that part of an orbit outside the earth) which has the property of remaining within a bounded region as the time increases without limit. From this we infer that asymptotic orbits form part of the boundary of the main cone, and from the fact that non-reentrant orbits go to infinity only by cutting through the outer principal periodic orbit, but may go to infinity without cutting through the inner orbit, or to the rest of the other periodic orbits (except for very low energies) we deduce that the orbits bounding the main cone must be asymptotic to the outer periodic orbit. Such asymptotic orbits we have called *orbits of the first kind*. It may be proved, by considerations into which we shall not enter here, that the border of the main cone is actually a boundary between allowed and forbidden directions.

It remains to show how trajectories tangent to the earth may form part of the boundary of the main cone. In order that the preceding consideration remain valid, it is necessary that the entire asymptotic orbit remain outside the earth. Hence, if at any point, say x_0, it has a tangent parallel to the λ-axis (infinite slope) it retains its significance only for values of the energy which would place the earth at an $x < x_0$. For energies such that $x > x_0$ the main cone is bounded, not by an asymptotic orbit, but by an orbit which becomes tangent to the earth. Such orbits we have called *orbits of the second kind*, or shadow orbits. They may be classified according to the number of times they are re-entrant towards the earth. More precisely, the main cone is bounded in part by orbits which have one maximum and one minimum in addition to the maxima and minima possessed by the ordinary orbit re-entrant n times. Such orbits have been called simple orbits of the second kind, or *simple shadow orbits*. They form the simple shadow cone. The penumbra bands are partly bounded by orbits of the second kind of a higher order.

As we shall show in successive sections, the study of the main and the simple shadow cones has been thoroughly made. Methods for the study of the penumbra have been devised and its investigation is well advanced.

Finally, it may be mentioned that it has now been proved, as a part of the study of orbits of the second kind, that over a certain range of energy the Störmer cone is never reached by the penumbra. It is known that the main cone is tangent to the Störmer cone only at the equator and along the east-west plane.

9. Once the significance of asymptotic orbits becomes clear, the next problem is to devise means for calculating them. The main difficulty is again the non-integrability of the equations of motion (5.6), (5.7). Leaving for the next section the outline of the method of mechanical integration which has been used for the study of asymptotic and other trajectories, we propose to devote this section to a very brief summary of a method of calculation of asymptotic expansions, and of a new method of numerical integration, which LEMAÎTRE and the writer [21] devised to solve this problem. While this procedure is capable of as high a precision as desired, and yields at once a whole family of asymptotic trajectories, its usefulness is limited by the fact that it takes from one to two hundred computing hours to carry out the necessary work.

The first attempt to calculate asymptotic orbits, and deduce from them the main cone, was made by BOUCKAERT [25] working under LEMAÎTRE's direction. His method, valid for low latitudes and values of γ_1 close to 1, yielded results of very high interest for the understanding of experiments on cosmic radiation (north-south intensity asymmetry). We shall not discuss this method further here.

The calculation of a family of asymptotic orbits, once the periodic orbit to which they are asymptotic has been found by the methods of Sect. 8, naturally falls into two parts. First one must determine an asymptotic expansion valid in the immediate neighbourhood of the periodic orbit and, second, the numerical integration of the whole family of asymptotic orbits is to be carried out by appropriate methods. The first part of the programme involves the integration of the variational equations derived from the equations of motion by using POINCARÉ's classical method of characteristic exponents.

It has been pointed out (in Sect. 8) that any principal periodic orbit, the outer orbit in particular, may be represented by a Fourier series in terms of the phase angle $\omega\sigma + \varphi$ where $2\pi/\omega$ is the fundamental period and φ is an arbitrary constant. This fact coupled with certain results of LEMAÎTRE on asymptotic orbits very close to the equator, led us to postulate, and verify a posteriori by actual calculation, that a family of asymptotic orbits may be represented by a trigonometric series:

$$x = z_0(\sigma) + \sum_p y_p(\sigma) \sin p(\omega\sigma + \varphi) + z_p(\sigma) \cos p(\omega\sigma + \varphi), \qquad (9.1)$$

$$\lambda = \sum_q \mu_q(\sigma) \sin q(\omega\sigma + \varphi) + \nu_q(\sigma) \cos q(\omega\sigma + \varphi), \qquad (9.2)$$

where $p = 2k$, $q = 2k + 1$, and k is an integer (including zero). For actual computation purposes the series is replaced by a finite sum, so that difficult problems of convergence need not arise. The amplitudes z, y, μ, ν must satisfy the condition that $z_p(\sigma)$ and $\mu_q(\sigma)$ tend asymptotically to the Fourier amplitudes z_p, μ_q of the periodic orbit in the limit $\sigma \to -\infty$, while $y_p(\sigma)$ and $\nu_q(\sigma)$ vanish asymptotically, since these coefficients are absent from the Fourier series of a symmetric periodic orbit. The arbitrary constant φ in (9.1) and (9.2) serves to single out and identify individual members of the asymptotic family. Thus, if one succeeds in calculating the amplitudes y, z, etc., then by assigning different values to φ one obtains the entire family of asymptotic trajectories. This is the kernel of the method under discussion, and also the source of its remarkable power.

We now turn our attention to the determination of the asymptotic expansions. Writing

$$E = e^{\Omega \sigma},\qquad (9.3)$$

where Ω is the characteristic exponent of the periodic orbit, the amplitudes may be written as a power series in E:

$$y_p(\sigma) = \qquad y_p' E + y_p'' E^2 + \cdots,\qquad (9.4)$$

$$z_p(\sigma) = z_p + z_p' E + z_p'' E^2 + \cdots,\qquad (9.5)$$

$$\mu_q(\sigma) = \mu_q + \mu_q' E + \mu_q'' E^2 + \cdots,\qquad (9.6)$$

$$v_q(\sigma) = \qquad v_q' E + v_q'' E^2 + \cdots\qquad (9.7)$$

where z_p and μ_q are the Fourier coefficients of the periodic orbit and $E \to 0$ as $\sigma \to -\infty$. Primed terms are first-order terms, double primes indicate second-order terms. The problem is now to find the first- and second-order coefficients and to calculate the characteristic exponent Ω.

Now for a first variation in the coordinates x, λ of the periodic orbit:

$$\delta x = x' E, \quad x' = z_0' + \sum_p (y_p' S_p + z_p' C_p),\qquad (9.8)$$

$$\delta \lambda = \lambda' E, \quad \lambda' = \sum_q (\mu_q' S_q + v_q' C_q),\qquad (9.9)$$

where we write S_p, C_q etc., instead of spin $p(\omega\sigma + \varphi)$, $\cos p(\omega\sigma + \varphi)$, etc. From the equations of motion (5.9) we have

$$\frac{d^2}{d\sigma^2}(\delta x) = X_x \delta x + X_\lambda \delta \lambda\qquad (9.10)$$

$$\frac{d^2}{d\sigma^2}(\delta \lambda) = \Lambda_x \delta x + \Lambda_\lambda \delta \lambda\qquad (9.11)$$

where

$$X = \frac{1}{2}\frac{\partial P}{\partial x}, \quad \Lambda = \frac{1}{2}\frac{\partial P}{\partial \lambda}, \quad X_x = \frac{1}{2}\frac{\partial X}{\partial x}; \text{ etc.}\qquad (9.12)$$

Expressing now the second-order derivatives on the lefthand side of (9.10) and (9.11), as well as δx and $\delta \lambda$ on the righthand side, by the relations (9.8), (9.9), replacing the coefficients $X_x, \Lambda_x = X_\lambda$, and Λ_λ by their Fourier series, and reducing products of sines and cosines, a set of homogeneous equations is obtained by equating to zero the terms independent of σ and the coefficients of the various sine and cosine terms. The determinant of this system must vanish, and this yields the secular equation determining Ω and the coefficients of the first order, y_p', z_p', etc. A very similar calculation yields the second-order terms.

A word of explanation may be useful regarding the Fourier expansions of P, X, Λ and their derivatives. For the periodic orbit, since x and λ are periodic functions, X, Λ as well as P also become periodic functions of σ:

$$X = \sum_K (Y_K S_K + Z_K C_K),\qquad (9.13)$$

$$\Lambda = \sum_K (M_K S_K + N_K C_K),\qquad (9.14)$$

$$P = \sum_K (Q_K S_K + R_K C_K),\qquad (9.15)$$

where again in actual calculation the series are replaced by a finite sum, since beyond a certain order the amplitudes become very small. The problem of the harmonic analysis of the fundamental functions is this: given the Fourier coefficients of x and λ. to find the Fourier coefficients of X, Λ, P, and eventually of X_λ, Λ_x, etc. This problem has been completely solved and convenient computation schedules have been developed. The elaborate details must be left out here.

Once the first and second order terms of the asymptotic expansion are known, the asymptotic orbits in the vicinity of the periodic orbit are also known. To continue them, further recourse must be taken to a method of numerical integration. Instead of integrating for individual trajectories, as in Störmer's method, we have preferred to integrate for the amplitudes $y_p(\sigma)$, $z_p(\sigma)$, etc., in (9.1), (9.2) thus obtaining a whole family of asymptotic orbits. This has necessitated the development of new methods of numerical integration which very likely will be applicable to other problems. The asymptotic expansions are then used solely to construct the tables of successive differences which are indispensable for starting any numerical integration.

The differential equations to be integrated are of the form

$$\frac{d^2 y_K}{d\sigma^2} - 2K\omega \frac{dz_K}{d\sigma} - K\omega^2 y_K = Y_K, \tag{9.16}$$

$$\frac{d^2 z_K}{d\sigma^2} + 2K\omega \frac{dy_K}{d\sigma} - K\omega^2 z_K = Z_K \tag{9.17}$$

and a similar pair for μ_K, ν_K. Introducing the complex variables $x = y + iz$, $X = Y + iZ$ and taking the case $K = 1$ as typical, the equations above may be written

$$\frac{d^2}{d\sigma^2} (x\, e^{i\omega\sigma}) = X\, e^{i\omega\sigma} \tag{9.18}$$

which, upon double integration and elimination of integration constants by taking second differences, yields

$$\Delta^2 x\, e^{i\omega\sigma} = \Delta^2 \iint X\, e^{i\omega\sigma}\, d\sigma^2. \tag{9.19}$$

Let w be the chosen interval of integration in the independent variable σ and let t, defined by

$$t = \frac{\sigma - \sigma_0}{w} \tag{9.20}$$

be the ordinal number of the integration point, σ_0 be the arbitrary origin of σ. Writing

$$a = i w \omega \tag{9.21}$$

(9.19) becomes

$$\Delta^2 x\, e^{at} = w^2 \Delta^2 \iint X\, e^{at}\, dt^2. \tag{9.22}$$

The left-hand side of (9.22) can be readily computed by assigning to t values $0, -1, -2, \ldots$

$$\Delta^2 x\, e^{at} = x_0 - 2x_{-1} e^{-a} + x_{-2} e^{-2a} \tag{9.23}$$

where x_0 is the desired unknown. To evaluate the right-hand side of (9.22), we have, placing $D = \dfrac{\partial}{\partial a}$

$$\Delta^2 \iint X\, e^{at}\, dt^2 = \sum_m \frac{1}{m!} (D+1)(D+2) \ldots (D+m) \frac{1 - 2e^{-a} + e^{-2a}}{a^2} \Delta^m X_{-1} \tag{9.24}$$

obtained by using the symbolic form of the backward Gregory-Newton formula. From (9.23) and (9.24) we obtain x_0 in terms of known quantities. For practical purposes of calculation it becomes necessary to separate real and imaginary components. Writing the right-hand side of (9.24) in the form $\sum a_m = \sum (b_m - i c_m)$ we have

$$y_0 = 2 y_{-1} \cos \omega w - y_{-2} \cos 2\omega w - 2 z_{-1} \sin \omega w +$$
$$+ z_{-2} \sin 2\omega w + w^2 \sum_m b_m \varDelta^m Y_{-1} + w^2 \sum_m c_m \varDelta^m Z_{-1}, \Bigg\} \quad (9.25)$$

$$z_0 = 2 z_{-1} \cos \omega w - z_{-2} \cos 2\omega w + 2 y_{-1} \sin \omega w -$$
$$- y_{-2} \sin 2\omega w + w^2 \sum_m c_m \varDelta\, Y_{-1} - w^2 \sum_m b_m \varDelta^m Z_{-1}. \Bigg\} \quad (9.26)$$

For higher harmonics ω should be replaced by $K\omega$. For $K = 0$ (9.26) reduces to STÖRMER's formula of numerical integration [26]. The coefficients of numerical integration for all harmonics and differences up to the tenth have been calculated and tabulated. Formulas have also been developed for halving and doubling the interval of integration [27].

With the help of the theory outlined here, two asymptotic families (for $\gamma_1 = 0.929898$ and $\gamma_1 = 0.850008$) have been calculated to very high precision (six significant figures over most of the range of integration). Twenty steps were made in one instance and thirty-four in another. Once the integration has been carried out, it is easy to construct

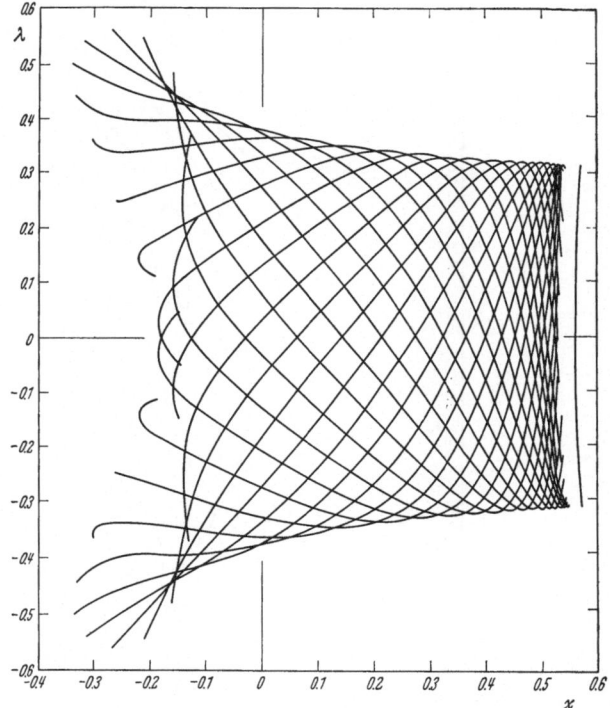

Fig. 17. Example of an asymptotic family of orbits ($\gamma_1 = 0.929898$).

the ephemeris of the family of asymptotic orbits. Two examples of asymptotic families calculated in this way are shown in Figs. 17, 18.

The behaviour of the different harmonics in the expansions (9.1) to (9.4) is worthy of note. They all depart at first slowly from their asymptotic values and then execute oscillations of ever-increasing amplitude and correspondingly sharper maxima and minima. New harmonics of higher order appear as the integration proceeds and the first maximum of higher harmonics occurs at progressively larger values of σ.

The characteristic exponents of the outer orbits decrease as γ_1 goes from $\gamma_1 = 1$ to $\gamma_1 = 0.78856$ and vanish for the latter value, in accordance with well-known theorems of POINCARÉ [28] (see Fig. 16 and Table 3).

By the method outlined in this section, as many asymptotic families may be calculated with as high a precision as desired, and these may be carried as far away from the periodic orbit as may be wished, the only limitation being the amount of labour required. Since the latter is very considerable when several hundred asymptotic orbits are needed, it becomes necessary to use methods which, while yielding results of lower precision, require an incomparably smaller effort and can be carried through in reasonable time. The method of mechanical integration developed by V. BUSH is admirably adapted for finding, not only asymptotic, but other orbits as well. We devote the next section to an outline of this method, which has played a leading role in the solution of our problem.

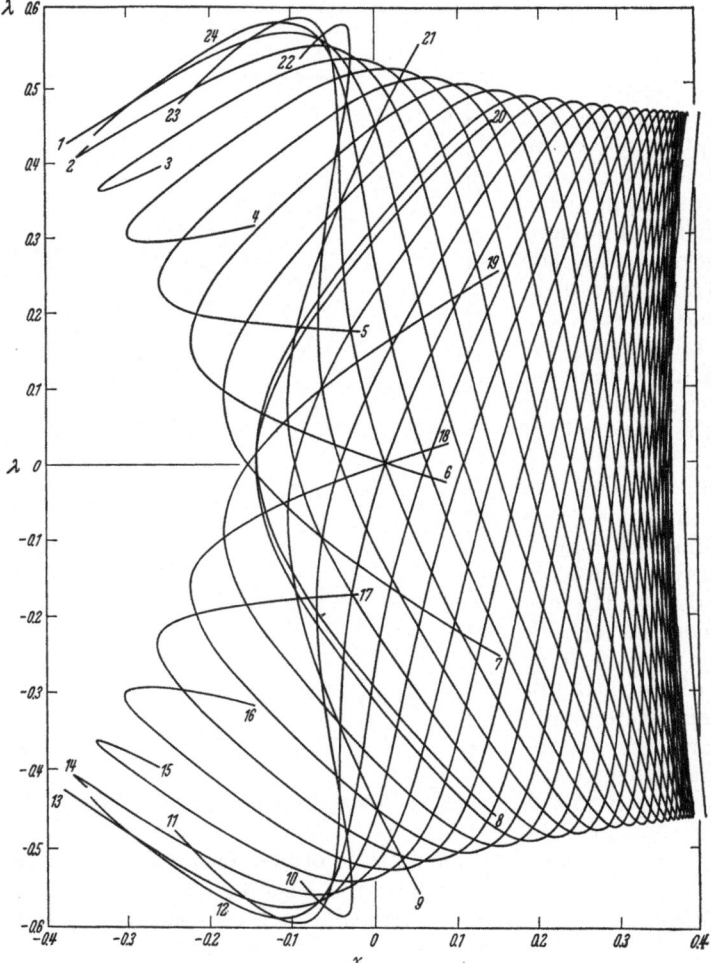

Fig. 18. Example of an asymptotic family of orbits ($\gamma_1 = 0.85$).

10. In 1930, V. BUSH [29] invented and constructed at the Massachusetts Institute of Technology the differential analyser, a machine for the solution of differential equations. The essential organ of this machine is the integrating unit, or integrator. It consists of a horizontal disk rotating about a vertical axis on which rolls without slipping a small wheel rotating about a horizontal axis. The distance between the wheel and the centre of the disk can be adjusted at will. Let w be this distance and let $d\theta$ and $d\varphi$ be the angles rotated by the wheel and the disk, respectively. Then, from the condition of rolling, taking the radius of the wheel as unity,

$$d\theta = w \, d\varphi \qquad (10.1)$$

and hence

$$\theta = \int w \, d\varphi. \qquad (10.2)$$

If w is so controlled as to reproduce a known function of any variable dependent on φ, and the rotation of the disk is made to depend on the value of another variable, the angle turned through by the wheel yields the integral of that function with respect to the chosen variable.

The small torque of the rolling wheel is amplified several thousand times and its rotation is fed into a series of horizontal shafts, so interconnected as to reproduce the operations demanded by the solution of the differential equations. The known functions are plotted on so-called input tables and fed into the machine by manual operation in the following way; suppose it is required to introduce into the machine a known function, let us say $w(x)$. Then the function $w(x)$ is first plotted in Cartesian coordinates and placed on an input table. This contains a pointer which can be displaced in two perpendicular directions. The displacement in one direction is controlled from the machine by coupling with the shaft the rotation of which is proportional to x. By turning a crank the operator displaces the pointer in the perpendicular direction so as to keep it always on $w(x)$. The rotation of the crank is fed back into another shaft, which thus rotates proportionally to $w(x)$. The solution of the equation is obtained on an "output table" in which the displacement of a pen in one direction is controlled by coupling to the shaft the rotation of which is proportional to the independent variable, and the displacement in the perpendicular direction is governed by coupling with the shaft giving the required solution. Further details of the differential analyser and of its operation will be found in Bush's original paper.

A first integration of the differential equations (5.6), (5.7) yields

$$\frac{dx}{d\sigma} = \int (a \, e^{4x} - e^x + \cos^2 \lambda) \, e^{-2x} \, d\sigma \qquad (10.3)$$

$$\frac{d\lambda}{d\sigma} = \int \left(e^{-2x} - \frac{1}{\cos^4 \lambda} \right) \sin \lambda \cos \lambda \, d\sigma, \qquad (10.4)$$

and second integration of the left-hand members with respect to σ yields x and λ. The differential analyser at the Massachusetts Institute of Technology has six integrators and five input tables, and its capacity is just enough to solve the system (5.6), (5.7) through (10.3) and (10.4). The functions placed on these five input tables and the operations performed by the six integrators are given in the following table:

Table 4.

Input tables	1	2	3	4	5	
Functions	e^{-2x}	$a \, e^{4x} - e^x$	$\cos^2 \lambda$	$1/\cos^4 \lambda$	$\sin \lambda \cos \lambda$	
Integrators	1	2	3	4	5	6
Operations	$\int e^{-2x} d\sigma$	$\int \sin \lambda \cos \lambda \, d\sigma$	$dx/d\sigma$	$d\lambda/d\sigma$	x	λ

A sketch of the machine connections is given in Fig. 19. The output table, as seen, is controlled so as to plot x as abscissa and λ as ordinate. Provision is also made to record x, λ, $dx/d\sigma$ and $d\lambda/d\sigma$ at suitable intervals of σ, the operation being performed by an automatic printer.

The scale factors are determined from the range of x and λ which it is desired to explore. For our investigations they were so chosen that x could vary from -1.0 to $+0.7$, λ from -0.7 to 1.0, $dx/d\sigma$ and $d\lambda/d\sigma$ from -0.5 to $+0.5$. These last two limits were deduced from a study of the function P, Eq. (5.8).

The initial conditions are introduce into the machine knowing the initial point (x, λ) and the initial slope $\tan \eta$ at (x, λ). The knowledge of x and λ is used to set the starting points on the input tables. To set the starting points on each of the six integrators it is necessary to know the initial values of the functions

$$a\,e^{4x} - e^{x} + \cos^2 \lambda, \quad e^{-2x}, \quad e^{-2x} - 1/\cos^4 \lambda, \quad \sin \lambda \cos \lambda, \quad dx/d\sigma \quad \text{and} \quad d\lambda/d\sigma.$$

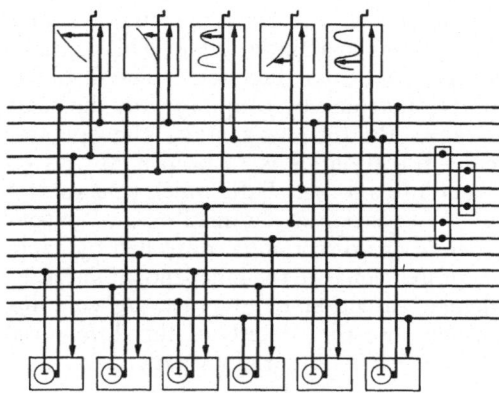

Fig. 19. BUSH's machine connections.

The first four are known from the initial values of x and λ, the last two from the initial slope by the relations

$$\left.\begin{aligned} \frac{d\chi}{d\sigma} &= \sqrt{P}\cos \eta, \\ \frac{d\lambda}{d\sigma} &= \sqrt{P}\sin \eta \end{aligned}\right\} \quad (10.5)$$

where P is given by (5.8). For convenience the trajectories were started in most cases from suitably chosen points, e.g., the equator, the line $\theta = 0$, a line $x = \text{const}$, etc. Once the machine is connected up and the initial conditions are placed on it, which with proper preparations is done in a few minutes, the machine is able to trace a trajectory in the interval of a few minutes. The longest trajectories (of several tens of units of σ) requires about 20 minutes.

A typical method [30] of finding asymptotic trajectories, which may be suitably modified to find other desired trajectories, e.g., orbits of the second kind, is as follows: one chooses a starting point, say on the equator, and starts a trajectory with an arbitrary slope, in the direction towards the periodic orbit; the initial slope is then adjusted until the trajectory neither intersects nor falls short of the periodic orbit. By a suitable method of interpolation it is possible to determine the critical angle with a precision of a few thousandths of a radian by making from five to ten trials, each lasting from four to six minutes. That the critical angle is actually determined with such a precision is confirmed by comparing the results of the calculations outlined in the preceding Sect. 9 with the machine results. A typical comparison is shown in Table 5 for $\gamma_1 = 0.93$.

Table 5.

x	η (machine)	η (calculated)	Diff.
0.150	0.890	0.887	0.003
0.225	0.959	0.953	0.006
0.300	1.053	0.045	0.008

For individual runs the precision depends on the length of the trajectory (in units of σ). For short runs (of a few units of σ) the precision is about one degree. For long runs (of more than ten units) the precision drops to about five degrees.

A comparison between machine and calculated values, for trajectories starting from the line $\theta = 0$, and $\gamma_1 = 0.85$, is shown in Fig. 20.

11. Having described a reliable and fast method of obtaining asymptotic and other trajectories we now turn our attention to a description of asymptotic orbits and some of their outstanding characteristics. The knowledge of them we then use to obtain the main cone.

To begin with, we should mention the non-oscillating asymptotic trajectory existing for a value of γ_1 between 0.9313 and 0.9314. This trajectory, already calculated by Störmer [31] is osculating to the line $\theta = 0$ and proceeds directly to the pole at $r = 0$ or $x = -\infty$. All asymptotic trajectories oscillate about this orbit, i.e., on either side of the line $\theta = 0$ for sufficiently small values of x, and they oscillate outwards in the same manner as inwards. Examples of asymptotic families are given in Figs. 17, 18. Another example of an asymptotic family, belonging to $\gamma_1 = 0.93$

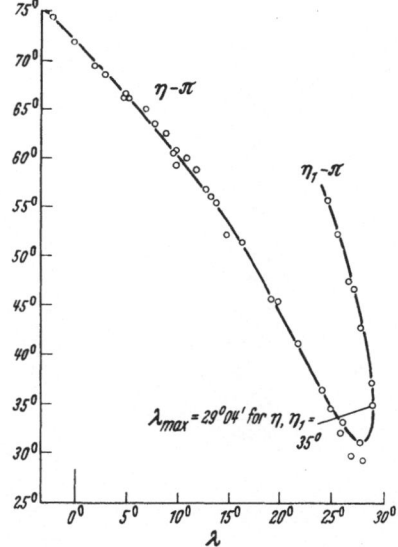

Fig. 20. η vs. λ along line $\theta = 0$ ($\gamma_1 = 0.85$).

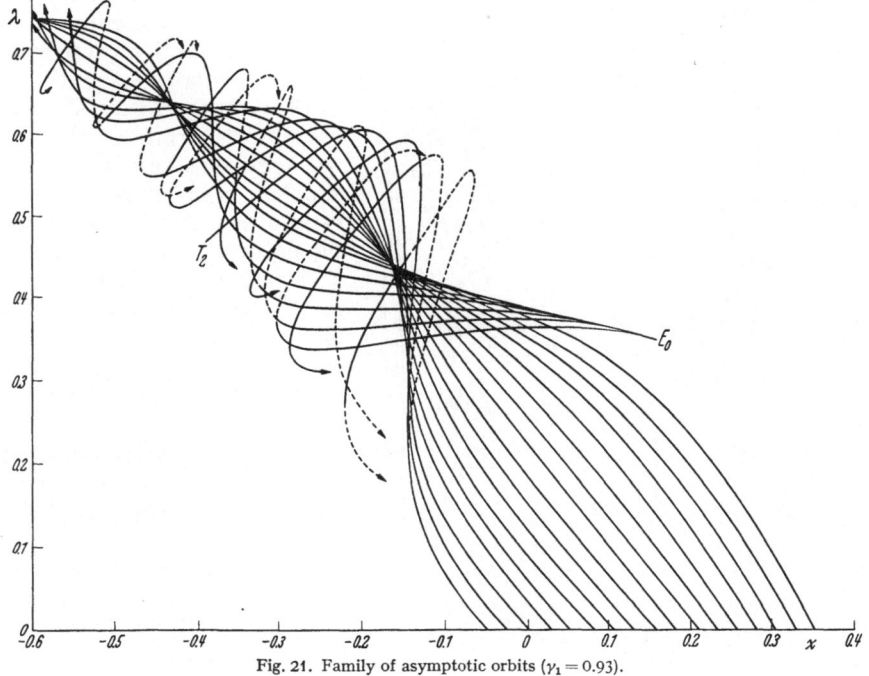

Fig. 21. Family of asymptotic orbits ($\gamma_1 = 0.93$).

and determined by means of Bush's machine, but extending to much higher latitudes and smaller values of x, is shown in Fig. 21.

A family of asymptotic orbits possesses an extensive family of envelopes of the highest importance for our problem, which we are now about to describe.

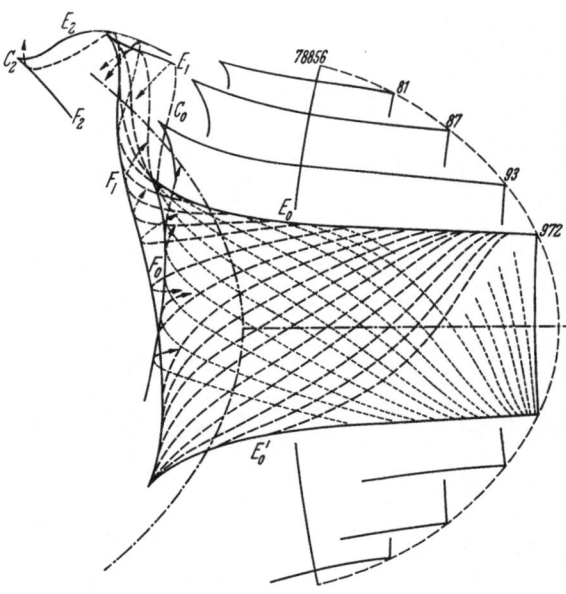

Fig. 22. Envelopes of asymptotic orbits

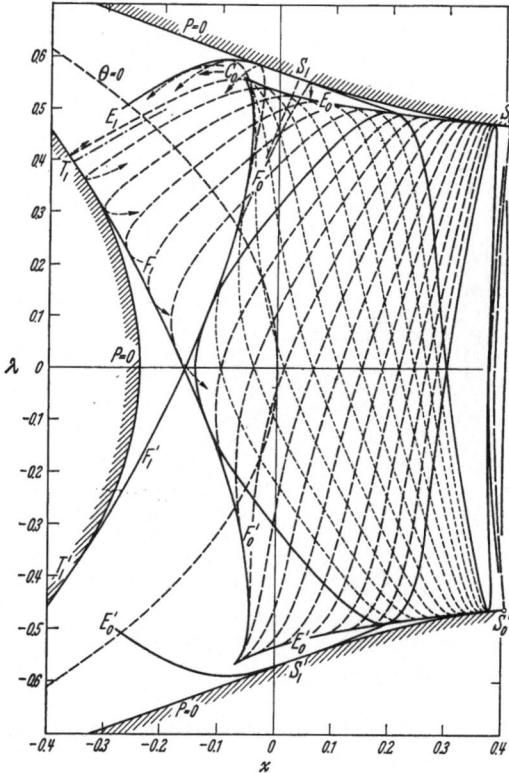

Fig. 23. Envelopes of asymptotic orbits ($\gamma_1 = 0.85$). "Queen" orbit (doubly asymptotic) drawn in full line.

Beginning at the periodic orbit, there is an envelope (Fig. 22) denoted by E_0 which first proceeds in a direction fairly parallel to the x-axis and then rises rapidly towards a cusp C_0 where it meets an envelope F_0 proceeding upwards. The system of envelopes proceeds in a similar manner; corresponding branches are denoted by E_1, E_2, E_3, ... and F_1, F_2, F_3, ... These envelopes are tangent to the line $P = 0$ (Fig. 23) at points denoted by S for the E-envelopes and T for the F-envelopes, with appropriate indices. At any S or T there is a trajectory turning back on itself, i.e., a self-reversing doubly asymptotic orbit. On either side of it the trajectories proceed in opposite directions. The point S_0, in particular, is the top of the periodic orbit. Any envelope and any cusp has a symmetrical counterpart, for the same x and negative λ, which will be denoted by a corresponding primed symbol.

A remarkable class of doubly asymptotic orbits is the class of the symmetric orbits, of which the simplest member is the "queen" orbit of Fig. 23 [22]. The remarkable nature of this orbit appears in a clear light if we recall for a moment the calculations of Sect. 9, for the amplitudes of the harmonic components of displacement at values of σ equidistant from the vertex are entirely different. Remembering that the phase singles out individual members of a family, we may conjecture the following theorem valid in the absence of the earth (cf. Sect. 8): All asymptotic orbits are unbounded (in one time direction) as σ increases without limit, except for a denumerable set of measure zero of values of φ which corresponds to doubly asymptotic orbits.

All orbits are divided into sections by their points of contact with their envelopes, and all asymptotic trajectories tend towards periodic orbits while performing an infinite number of oscillations between E_0 and E_0'. There is a known theorem in the theory of kinetic foci, which is a direct consequence of certain results of KORTEWEG and a theorem of STURM, that states that any trajectory infinitely near an asymptotic orbit intersects each section once only in any finite domain of σ. The trajectory in the infinitely close neighbourhood will therefore intersect each section once only, but finally will get away from it either by intersecting the periodic orbit (non-reentrant orbit) or by returning inwards (reentrant orbit). We thus have to distinguish between two kinds of sections: northern sections (dashed in Figs. 22, 23) are those for which an infinitely close trajectory intersecting the asymptotic orbit at an angle *smaller* than its own slope is a non-reentrant orbit. Southern sections (dotted) are such that if the intersection is at an angle *greater* than the slope of the asymptotic orbit, the neighbouring trajectory is non-reentrant. From a study of the trajectories obtained with the differential analyser we have concluded that sections coming from E_0 are northern, those coming from E_0' are southern. They form the northern and southern boundaries of the main cone. If we denote the slopes of the northern and southern sections by η and ζ respectively, we have the result that all trajectories between η and ζ come directly from infinity. The region between η and ζ lies within the main cone of which asymptotic trajectories, in agreement with the results of a preceding section, are the generators.

Each time that an asymptotic trajectory touches an envelope the sections on either side of the point of contact are of a different kind. It is then readily seen that an envelope is a double point on the main cone, because there two asymptotic directions coincide. At such points the main cone has a tangent plane whose trace on the unit sphere is parallel to the meridian plane. Cusps are triple points and the trace of the cone on the unit sphere has then a point of inflection with a tangent parallel to the meridian. These general principles suffice for the construction of the main cone. This will be done in a following section (Sect. 11).

Directions where the velocity vector tangent to the space trajectory lies in the meridian plane correspond to the line $x = 2 \log \cos \lambda$, in the conformal map, Eq. (5.5), for which $\theta = 0$ by STÖRMER's formula (3.18). Once families of asymptotic trajectories are obtained it is quite feasible to measure for each asymptotic trajectory cutting through this line the slope η and the corresponding value of the latitude λ. η may then be plotted as a function of λ for different values of γ_1. The result is shown in Fig. 24. Particular attention should be paid in this connection to the significance of the envelopes and cusps. The intersection of an envelope with the line $\theta = 0$ is shown in this diagram as a point where the η, λ-curve has a tangent parallel to the η-axis. These are the "loci of maximum latitude". A point where a cusp lies on $\theta = 0$ is shown on the diagram as a double point; its characteristic feature is that there two branches of the η, λ-curve for the same γ_1 intersect, one of which has a tangent parallel to the η-axis. The value of γ_1 for which C_0 lies on the line $\theta = 0$ is somewhere near $\gamma_1 = 0.94$. The following cusps $C_1, C_2, C_3, \ldots$ lie almost on it for $\gamma_1 = 0.93$. STÖRMER's results quoted in the preceding section make it very probable that these values form a progression, the limiting value of which is about 0.9313. For all other values of γ_1 the cusps lie alternatively on either side of the line $\theta = 0$, they are finite in number, and reach up to a certain latitude depending on the value of γ_1. There is always a trajectory, the "cuspidal orbit", which goes through all the cusps in succession. It may be clearly seen in Figs. 21 and 22. For $\gamma_0 = 0.9313$ the cuspidal orbit becomes osculatory to the line $\theta = 0$. This is precisely STÖRMER's orbit referred to in the preceding section.

8*

A *"trajectory of the third kind"* is a trajectory which is at the same time asymptotic and tangent to the earth. Such orbits are important because they indicate where the region of shadow, i.e., the simple shadow cone more fully discussed below, meets the main cone and its boundary becomes a boundary of the main cone.

The analysis embodied in Fig. 24 may be readily used to find out what is the least energy [32] that a particle must have to arrive, in the meridian plane, at a given angle with the zenith and within the main cone. Such knowledge is very important for the interpretation of directional measurements of cosmic ray intensity.

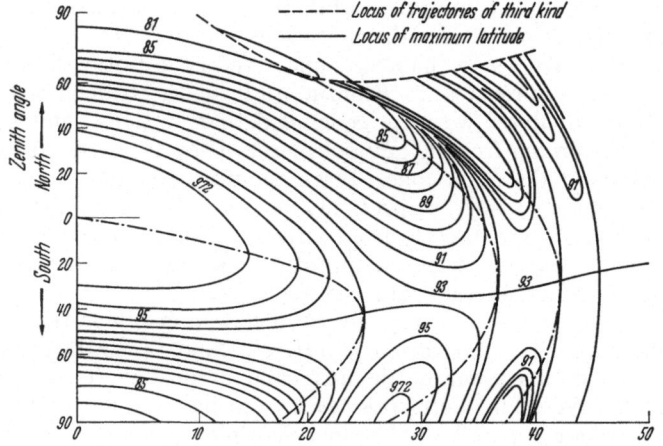

Fig. 24. Analysis of η vs. λ along line $\theta = 0$ for different values of γ_1.

12. We adopt the following representation of the allowed cone (main cone, penumbra, etc.): we imagine the unit sphere centred at the observer and consider the orthogonal projection of the intersections of the generators of the allowed cone (orbits of the first and second kinds) with the unit sphere, on the horizontal plane. If θ and η are the angles fixing the direction of a trajectory at a point x, λ for particles of energy such that the surface of the earth would be at x, the coordinates of the representative point are

$$X = \sin \theta, \tag{12.1}$$

$$Y = \cos \theta \sin \eta. \tag{12.2}$$

For positive particles, θ is counted positive eastwards, η positive northwards. θ is computed for any point (x, λ) from STÖRMER's formula (3.18) or (5.4) and η is given by (5.3). The particle energy r for which the earth would be at x is given by (5.1) together with (3.11).

For the systematic utilization of asymptotic trajectories obtained with the differential analyser, the following line of attack was adopted: the inclination η, along lines $\lambda = \text{const}$ for each of the values of γ_1 studied (i.e., $\gamma_1 = 0.81$ to 0.99 by steps of 0.01), was measured, noting at the same time the value of x (and therefore of r) where the intersection between the asymptotic orbit and the line $\lambda = \text{const}$ took place. Graphs were then prepared giving η as a function of λ (0° to 35° by intervals of 5°) for different values of γ_1. An envelope is characterized in the η, r-diagrams by a maximum or minimum value of r; a cusp by a point of inflection with tangent parallel to the η-axis. These peculiarities are found also in the cone's representation as its projection on the horizontal plane, and have already been discussed.

The main cones for $\lambda = 0°$, $20°$ and $30°$ are reproduced in Figs. 25 to 27. These have been drawn for positive particles, and for the northern hemisphere. The allowed regions are therefore to the west of the boundary. For negative particles the main cones are the mirror images of those shown, on the north-south

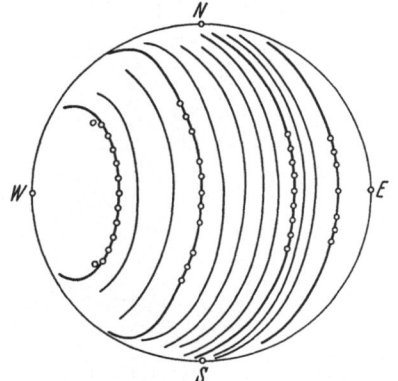

Fig. 25. Main cones at the equator ($\lambda = 0°$). Thick lines: 0.45, 0.5, 0.6 Störmer.

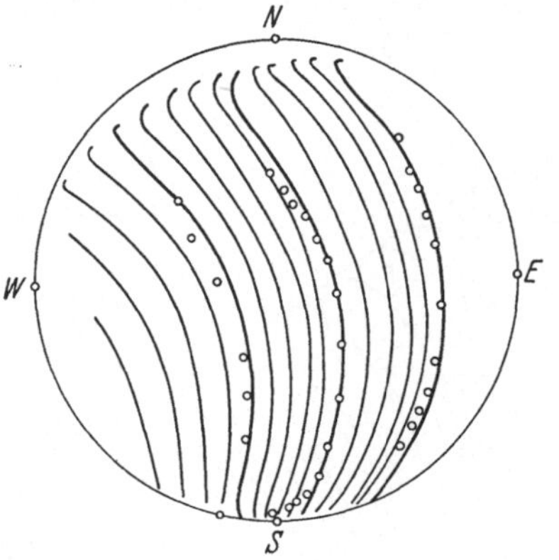

Fig. 26. Main cones at $\lambda = 20°$. Thick lines : 0.45, 0.5, 0.6 Störmer.

plane. For the southern hemisphere, the main cones are the mirror images of those for the northern hemisphere, but on the east-west plane. The small circles are the points calculated by BOUCKAERT [25]. The extreme northern boundary is stopped nearly at the point where the boundary determined by asymptotic orbits connects with the simple shadow cone (next section). A number of important physical consequences can be drawn from the study of the main cone.

13. The problem of ascertaining the simple shadow cone for all latitudes λ and energies r is essentially quite straightforward. It is only necessary to obtain for each λ and r the family of all orbits possessing sections tangent to the earth at r and having no loops, i.e., the family of simple shadow orbits. The angles (η, θ) defined by this family give the simple shadow cone for the point λ and the energy r.

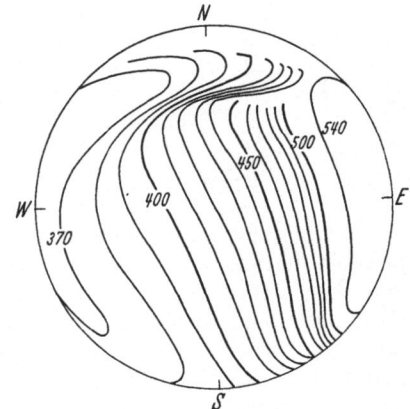

Fig. 27. Main cones at $\lambda = 30°$. Energies in millistörmers.

In the first approximation, where the earth's magnetic field is that of a dipole, such families of simple shadow orbits have been obtained with the differential analyser for all values of r from 0.1 to 0.9 by steps of 0.1 and γ_1 from 0.62 to 0.94 by steps of 0.04; except for the different type of initial conditions the set-up of the machine was exactly the same as that used for the determination of asymptotic orbits. The procedure was to fix γ_1 and the particle energy referred to the earth (i.e., the abscissa of the starting points) and to vary λ. The sequence of orbits has the appearance shown in Fig. 28. The lower limit of λ is characterized by the fact that the orbit through it is a self-reversing orbit, and therefore is a transition orbit between a simple shadow orbit and penumbral orbits, i.e., orbits of the second kind and higher order, with at least one loop. The upper limit of λ is singled out by the fact that the orbit through it no longer has a minimum of x along the line $x = \text{const}$ but has an inflection point and is therefore a transition orbit between simple shadow orbits and orbits which have maxima at $x = \text{const}$ and minima within the earth. All such orbits are surely in shadow and are therefore not generators of the simple shadow cone for the given energy.

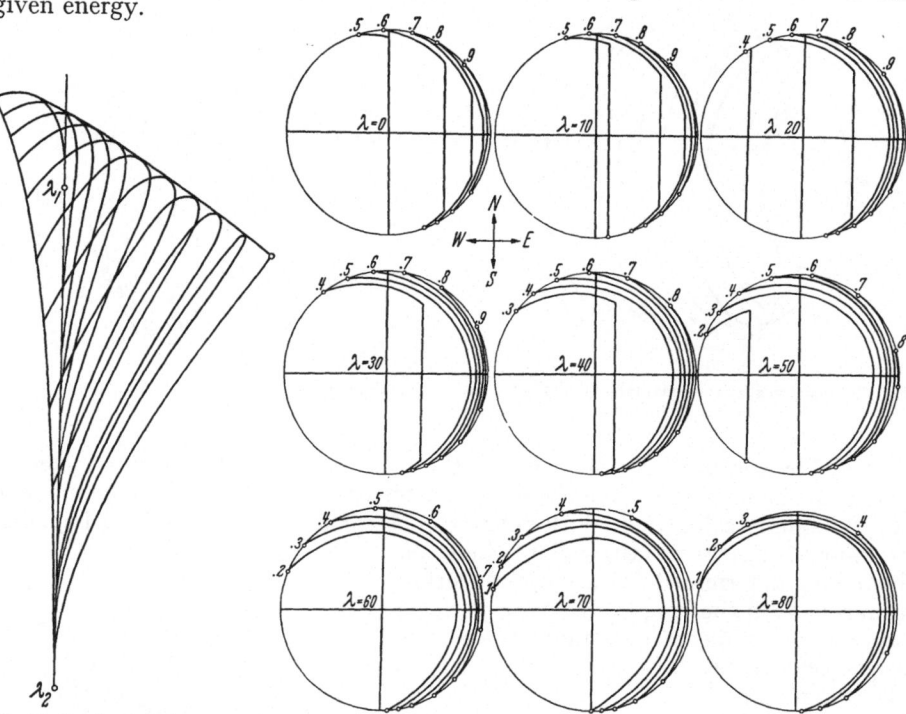

Fig. 28. Family of shadow orbits.   Fig. 29. Simple shadow cones in dipole field.

Further, for each value of γ_1 there is a definite sub-region of the region of possible motion, within which simple shadow orbits may be started. The upper boundary is given by the locus of zero x-acceleration [Eq. (6.1), Fig. 11]. The lower boundary is given by the locus of first minima of all reversing shadow orbits of the type shown in Fig. 28. This is the locus reproduced in Fig. 14, obtained as described in Sect. 8. Once the region of admissible starting points has been mapped out for each value of γ_1 the remaining programme of integration is clearly defined and consists in finding families of simple shadow orbits for sets of values of x and γ_1 covering the range of existence of such orbits, until enough data are accumulated for the determination of the simple shadow cone over its entire range of existence. For each orbit of each sequence of simple shadow orbits the values of λ and η were obtained at the point at which it crosses the earth's surface. For each such sequence of orbits a curve was plotted giving η as a function of λ for the whole interval $-\pi/2 < \eta < \pi/2$. These curves, one for each γ_1 were then assembled into families, one family for each value of r. From these the simple shadow cones may be obtained, much as in the case of the main cone. The result is shown in Fig. 29, where again the cones are drawn for positive particles and for the northern hemisphere.

It will be observed that for certain latitudes and energies the simple shadow cone ends abruptly at STÖRMER's cone ($\gamma_1 = 1$). Actually the limit $\gamma_1 = 1$ is not reached by the penumbra for all values of the energy. It will further be seen that the simple shadow cone of a given energy opens up gradually with increasing latitude, until it covers the complete hemisphere. The latitude at which this happens corresponds to that at which simple shadow orbits disappear, which takes place where the locus of zero x-velocity meets the locus of zero x-acceleration at the boundary $P = 0$. The last two intersect at points satisfying the condition

$$r^2 = \cos \lambda \tag{13.1}$$

or

$$\gamma_1 = r^3. \tag{13.2}$$

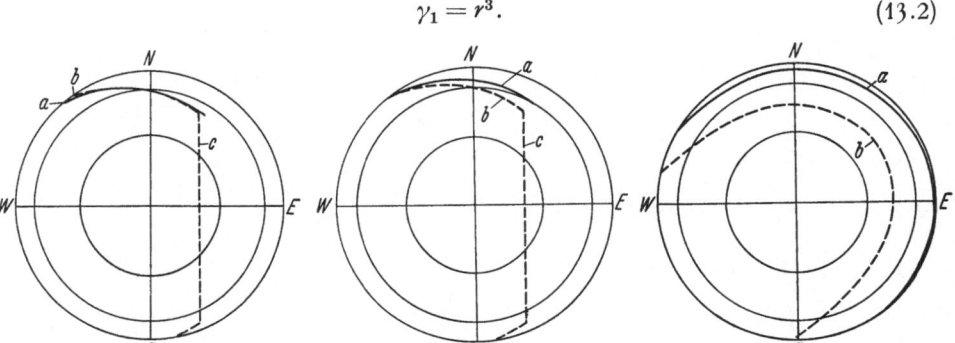

Fig. 30. Simple shadow cones for protons of 8.61 Bev at $\lambda = 30°$ N and $\varphi = 8°\ 46'$ W geographic longitude. (a) Eccentric dipole and quadrupole field, (b) dipole field, (c) STÖRMER's cone.

Fig. 31. Simple shadow cones for protons of 8.61 Bev at $\lambda = 30°$ N and $\varphi = 95°\ 35'$ W geographic longitude. (a) Eccentric dipole and quadrupole field, (b) dipole field, (c) STÖRMER's cone.

Fig. 32. Simple shadow cone for protons of 0.172 Gev at $\lambda = 70°$ N and $\varphi = 107°\ 11'$ W geographic longitude. (a) Eccentric dipole and quadrupole field, (b) dipole field.

14. Simple shadow trajectories, however, have the property that, in the critical section between the point of arrival and the point of tangency, they stay at a distance from the earth of the order of magnitude of the earth's radius. Since the magnetic quadrupole of the earth contributes on its surface as much as 15% of the total field, its influence on these trajectories is not negligible. For this reason, and because of the importance of the shadow cone in the interpretation of experimental results, the integration of the equations of motion taking into account both the eccentric dipole and quadrupole has now been carried, out [34]. The equations of motion are now

$$\ddot{\boldsymbol{R}} = \frac{e}{mc}\,(\boldsymbol{V} \times \boldsymbol{H}) = \frac{e}{mc}\,[\boldsymbol{V} \times \boldsymbol{V}\,(V_D + V_Q)] \tag{14.1}$$

where $\boldsymbol{R}$ is the position vector of the particle, e and m its charge and (relativistic) mass, respectively; $\boldsymbol{V}$ its velocity and V_D, V_Q the potentials of the dipole and quadrupole, respectively. The numerical integrations have been carried out with the help of IBM computing machines [35]. The results are shown in Figs. 30 to 32. It is seen that, independently of the longitude effect due to the eccentricity of the earth's magnetic center, to be dealt with more fully below, the new shadow cone at high latitudes is more open than that shown in Fig. 28. As a consequence the geomagnetic cut-off depends both on latitude and longitude and in many cases is lower than that derived from the cones in Fig. 29, to be more fully discussed in a later section of this paper.

15. All of the directions contained within the main cone represent angles of incidence for trajectories coming directly from outside the earth; i.e. without forming loops (in the meridian plane). These are the non-reentrant orbits mentioned previously.

Penumbra orbits, on the other hand, must always have reentrant sections. Examples of such orbits, with two reentrant sections, are exhibited in Fig. 34. More complicated trajectories have more and

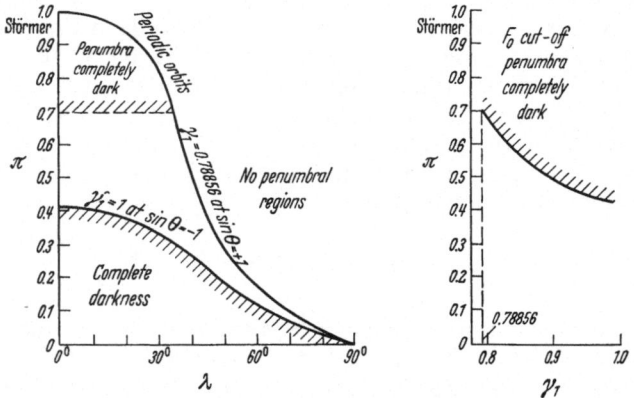

Fig. 33. Penumbra orbits.

Fig. 34. Types of singly and doubly re-entrant orbits.

more loops, their number increasing to infinity. Schremp pointed out [37] that all reentrant sections which characterize penumbral trajectories must cross the equator between the inner and outer periodic orbits. It follows that all penumbral

Fg. 35. Energy limits for the existence of penumbra.

orbits occur only in the range $0.788541 < \gamma_1 < 1$. In addition, the impenetrability of the earth (Fig. 33) is an important factor determining the penumbra bands.

A reentrant trajectory always has its minimum value of x somewhere behind the F_0 envelope of asymptotic orbits. Hence energies below those corresponding to the F_0 envelope must have no penumbra for that and higher values of γ_1.

This value of γ_1, above which the penumbra becomes shadow, is called the F_0 cut-off (Fig. 35).

All reentrant orbits can be generated by starting at points along the equator between the inner and outer principal periodic orbits and varying the initial angle η' within the angle α defined by the pair of self-reversing orbits at that point (Fig. 36). Let $\varphi_0 = 2\eta'/\alpha$. An example of a family of such orbits is shown in Fig. 37. The values of $\alpha/2$ for different y_0 ($y_0 = |x_0|/D + d/D$, see Fig. 36) and γ_1 are shown in Table 6.

For certain values of γ_1 a lower cut-off for penumbra orbits than the F_0 cut-off mentioned previously can be assigned. This is determined by the intersection of the locus $dx/d\sigma = 0$ with the envelope F_0; the abscissa of this intersec-

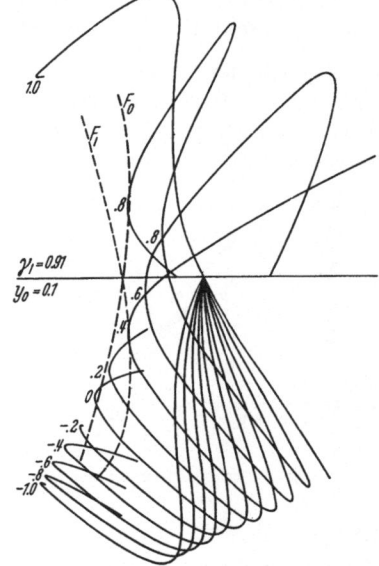

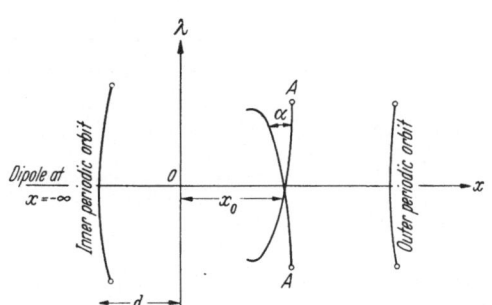

Fig. 36. Angle defining pencil of re-entrant orbits at point on equator.

Fig. 37. Family of penumbra trajectories. Label on each is value of φ_0.

tion will be denoted by x_A. Values of x_A for several values of γ_1 and corresponding values of energy r are given in Table 7 and Fig. 38.

Table 6. *Values of $\alpha/2$ in radius.*

y_0	γ_1						
	0.99	0.97	0.95	0.93	0.91	0.89	0.88
0.05	—	0.285	—	0.173	0.130	—	—
0.10	—	0.430	0.355	0.293	0.230	0.188	—
0.20	0.638	0.550	0.480	0.413	0.321	0.270	—
0.30	0.710	0.622	0.522	0.447	0.370	0.307	0.264
0.40	—	0.638	0.533	0.454	0.375	0.310	0.274
0.50	—	0.634	0.525	0.437	0.337	0.291	—
0.60	—	0.593	0.480	—	0.319	0.259	—
0.70	—	0.525	—	0.340	0.266	—	—
0.80	—	0.405	0.304	0.240	0.190	0.148	—
0.90	—	0.234	0.158	0.123	0.099	0.078	—

The penumbra for $\lambda = 20°$, based on the results of the analysis sketched above, is shown in Figs. 39 to 41. A similar analysis for other latitudes has not yet been carried out. Such an analysis is particularly difficult at high latitudes, where the lighted bands of the penumbra presumably fill up the region between the main cone and the STÖRMER and shadow cones.

To summarize, there are three principal parts of the allowed cone: the main cone, the simple shadow cone, and the penumbra. Within the main cone all directions are allowed, outside the simple shadow cone all are forbidden, in the penumbra there are allowed as will as forbidden directions. At the equator the main cone is the completely allowed cone and there is no prenumbra. At intermediate latitudes the penumbra is not negligible, at high latitudes the penumbra consists almost entirely of bands of allowed directions, and the simple shadow cone is essentially the completely allowed cone.

16. The longitude effect [38] arises because of the asymmetry of the real geomagnetic field about the earth's center. According to Gauss' classical analysis, this field can be represented by an expansion of Legendre polynomials about any arbitrary point chosen as origin, in particular about the earth's center. As shown by Schmidt [39] this expansion takes its simplest form, only one term

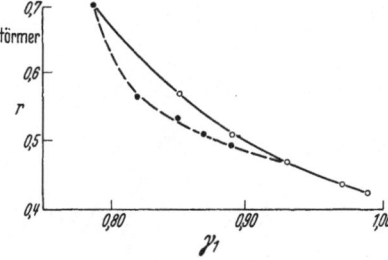

Fig. 38. Energy cut-off for penumbra.

Table 7.

γ_1	x_A	r (Störmers)
0.82	− 0.070	0.568
0.85	− 0.095	0.535
0.87	− 0.118	0.511
0.89	− 0.125	0.496
0.93	− 0.143	0.466
0.97	− 0.170	0.435

of the first order and one of the second remaining, if the origin is chosen at the "magnetic center of the earth". If the dipole and the quadrupole corresponding to the first and second order terms are placed at the magnetic center [40] there are two immediate consequences: first, the distance from a point on the earth at any latitude is a periodic function of its longitude and second, the angle between the line from a point on the earth and the earth's magnetic center, and the plane of the quadrupole, is also a periodic function of longitude. Both of these factors affect the allowed cone and therefore there must be a longitude effect.

The complete theory of the longitude effect thus depends on the integration of the equations of motion in the field of the dipole and the quadrupole, a problem which has not yet been completely solved. In addition the magnetic field due to ionospheric currents probably also perturbs critical trajectories, for example, the trajectories which determine the shadow cone. Calculations using the dipole and quadrupole components of the earth's internal field, and the magnetic field due to ionospheric currents can be attempted when a good model of the latter is available.

Preliminary results on the longitude effect are shown in Figs. 30 and 31. The work is still in progress.

17. Perhaps the most important consequence of the theory developed in these pages, from the experimental point of view, is the existence of the geomagnetic energy cut-off. This is shown in Figs. 42 to 46 [41]. The corrections due to distance from the earth's magnetic center, without taking into account the quadrupole, are shown in Fig. 47 to 49 for three different latitudes. In Figs. 42 to 46 the lower curve, labeled E_1 gives the least energy determined by the simple shadow cone or the Störmer cone, the former for low latitudes, without the

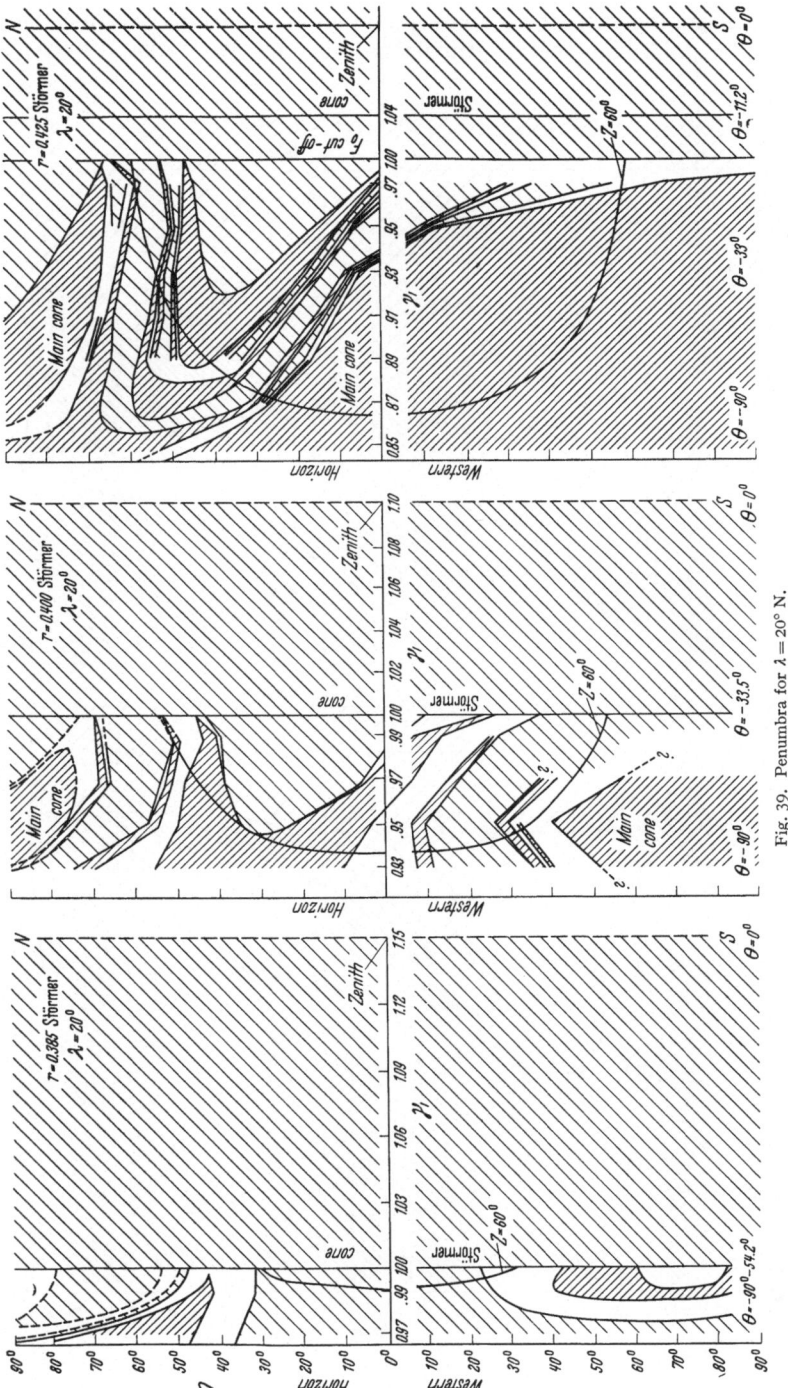

Fig. 39. Penumbra for $\lambda = 20°$ N.

quadrupole field. The upper curve labelled E_2 gives the energy above which all energies are allowed by the earth's magnetic field. This limit is determined by the main cone, again not considering the quadrupole field.

18. Still another aspect of geomagnetic effects concerns the question of the distant albedo. A high-energy primary cosmic particle striking a nucleus in the

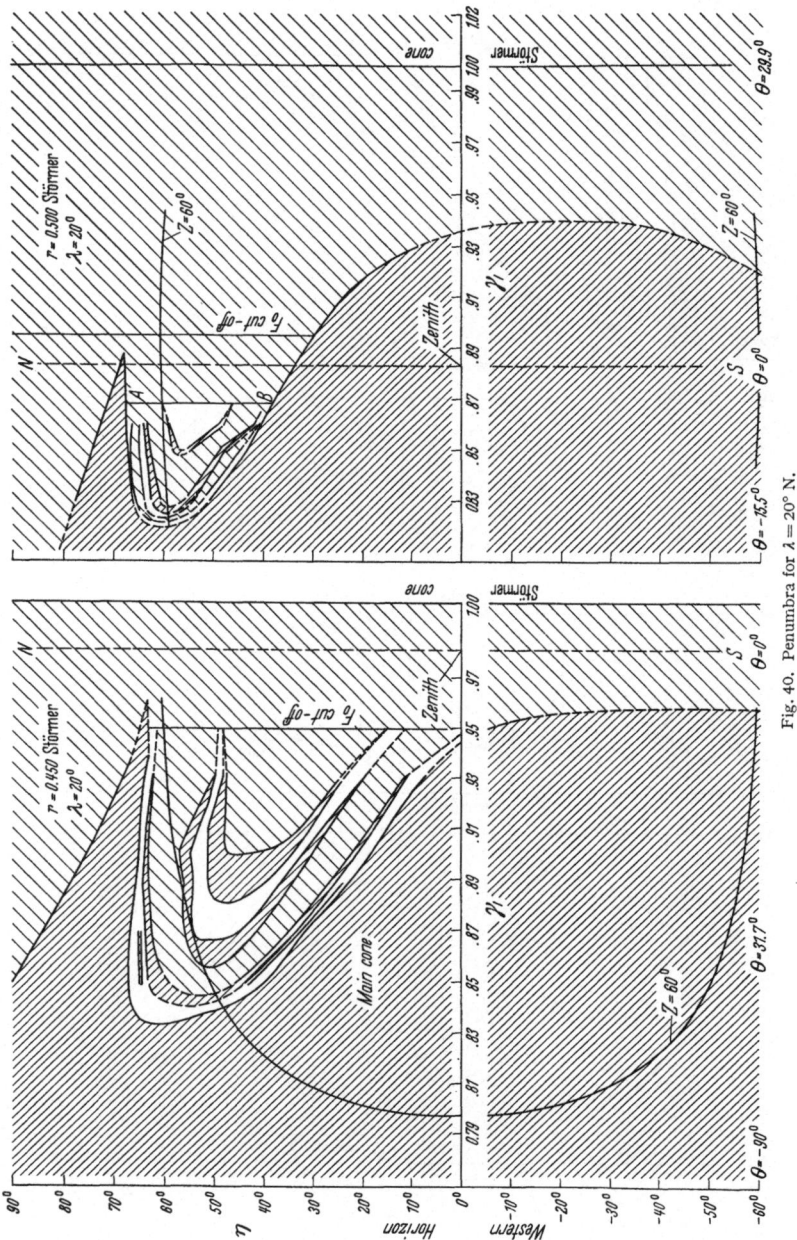

Fig. 40. Penumbra for $\lambda = 20°$ N.

atmosphere produces a "star" and some of the secondary particles in this star are caught by the earth's magnetic field, describe a trajectory and are returned to the earth at some other point [42]. Such *secondary albedo particles* cannot be distinguished from primary particles and will contribute their share to the

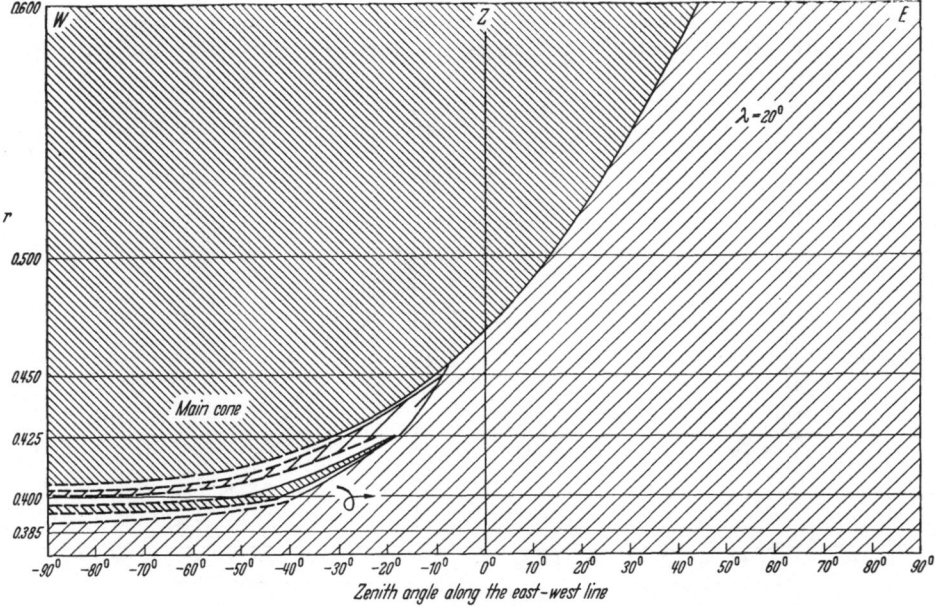

Fig. 41. Penumbra as a function of zenith angle and energy for $\lambda = 20°$ N.

observed intensity. This distant albedo should not be confused with the "*spash albedo*" which is due to secondary particles shot upwards and immediately caught by the measuring instrument.

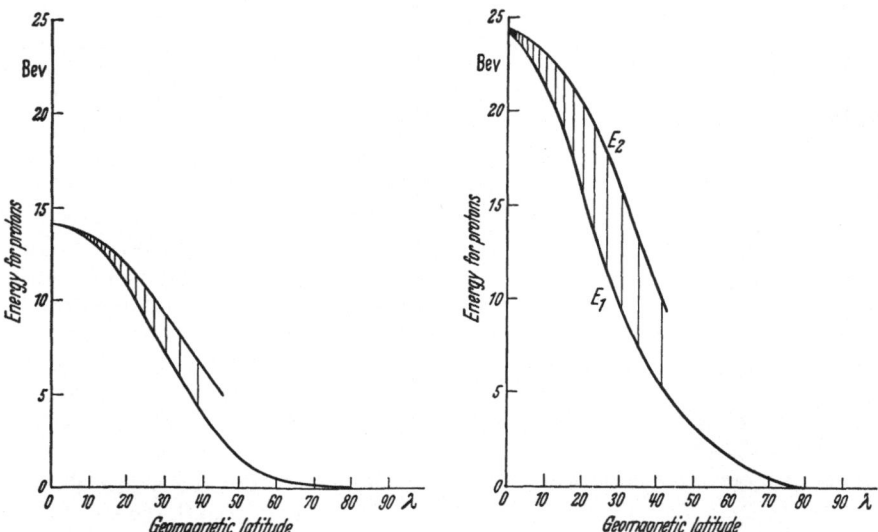

Fig. 42. Cut-off for vertical protons.

Fig. 43. Geomagnetic cut-off for protons at 45° E.

The equations of motion are (13.1) or

$$\frac{d^2 \boldsymbol{R}}{dt^2} = C \frac{d\boldsymbol{R}}{dt} \times \left[\nabla\left(\frac{z}{r^3} + \alpha \frac{xy}{r^5} \right) \right] \tag{18.1}$$

where the speed is chosen equal to 1 and the earth's radius equal to 0.5. $C = -m_1 a q/4 p c$, where $a = 6.37 \times 10^8$ cm is the earth's radius, q is the particle's

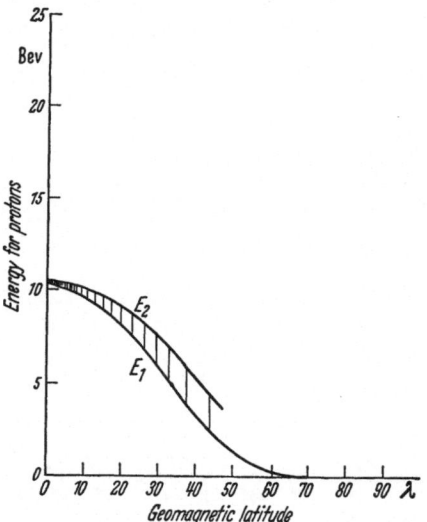

Fig. 44. Geomagnetic cut-off for protons at 45° W.

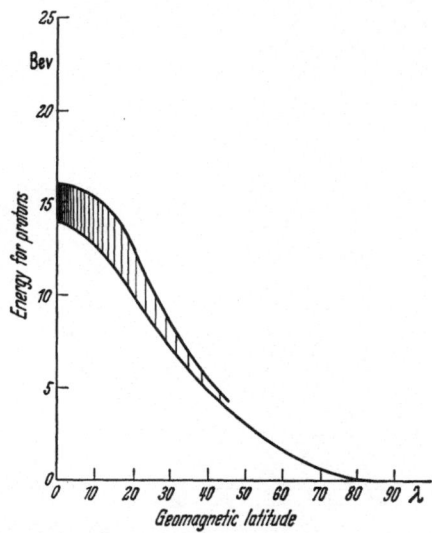

Fig. 45. Geomagnetic cut-off for protons at 45° N.

charge, p its relativistic momentum and c the velocity of light $\alpha = 3 m_2/4 m_1$, $m_1 = M_d/a^3 = 0.3097$ gauss, $m_2 = M_c/a^4 = 0.0224$ gauss and M_d, M_c are the dipole and quadrupole moments, respectively. The magnetic center is taken at the point $(-344, 150, 96)$ km [40].

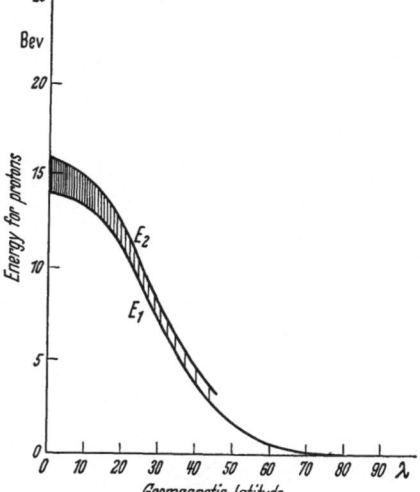

Fig. 46. Geomagnetic cut-off for protons at 45° S.

From a study of about 1000 trajectories of secondary protons computed by MILNE's method [43] with the help of IBM machines, the albedo cones were determined. The simple albedo cone is defined as the solid angle within which particles of a given energy arrive at a point of incidence from other points of the earth following trajectories without intermediate minima (non-reentrant trajectories). Second, third, etc., order albedo cones correspond similarly to trajectories having one, two, etc. intermediate minima (trajectories with one, two, etc. reentrant sections).

Simple albedo cones are shown in Figs. 50 to 52. Within the cones of Fig. 51, northeastern directions predominate. Directions outside these cones belong to higher order albedo cones, if at all, because no proton of these energies can come from infinity. In Fig. 52 the cones are open for large zenith angles only.

As already pointed out, the longitude effect is due both to the eccentricity of the geomagnetic center and to the quadrupole field. The longitude effect for the distant albedo along the geomagnetic equator is exhibited in Fig. 53. The small drawing at the lower right shows the directions and relative magnitudes of

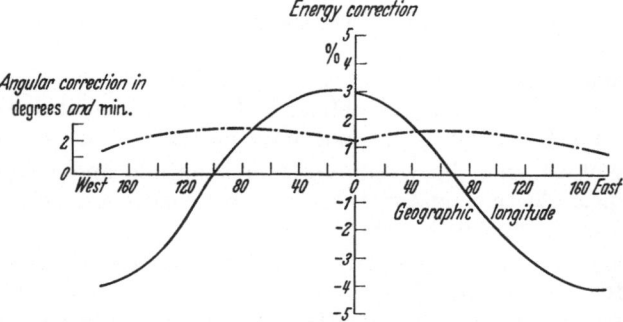

Fig. 47. Corrections due to the eccentric dipole along geographic equator.

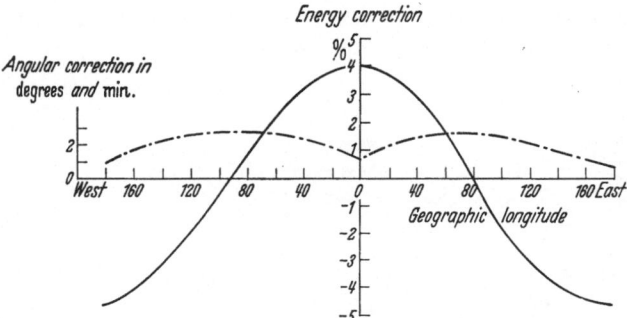

Fig. 48. Corrections due to eccentric dipole along geographic parallel 20° N.

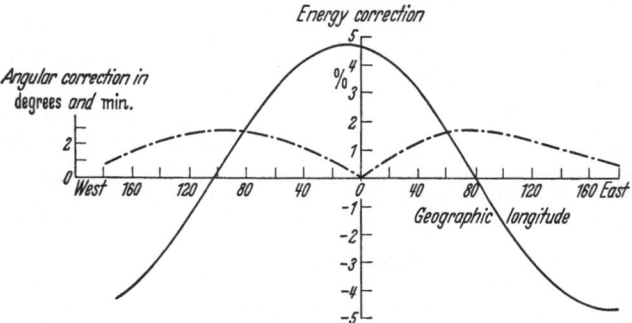

Fig. 49. Corrections due to eccentric dipole along geographic parallel 40° N

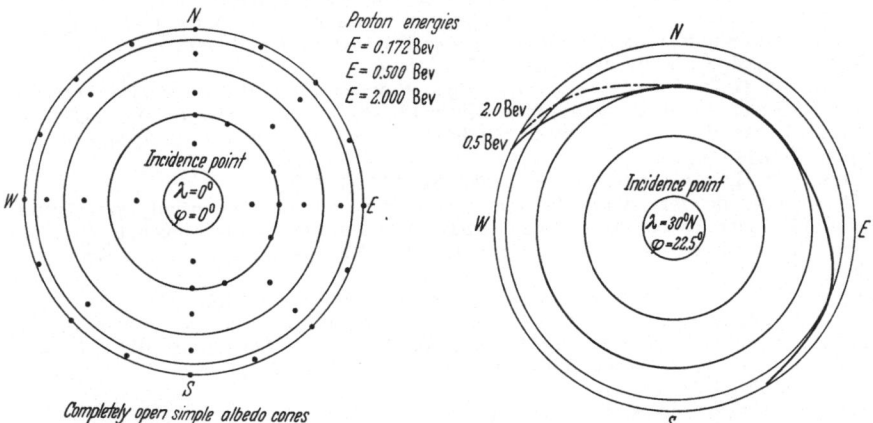

Fig. 50. Completely open simple albedo cones.

Fig. 51. Simple albedo cones $\lambda = 30°$ N, $\varphi = 22.5°$ E. Geographic zenith angle and azimuth.

the field components at the three points. As points 2 and 3 are equidistant from the magnetic center, their longitude effect is due to the quadrupole field only. It is seen that the latter shifts the cone from northest to southwest.

The points of emission lie within $\pm 5°$ in longitude and latitude from the point of incidence. At the equator, they lie to the east of the point of incidence within $\pm 14°$ latitude and $6°$ longitude.

In these pages we have attempted to give a summary of the theory of the geomagnetic effects of cosmic radiation, not including the theory of impact zones for particles emitted by the sun which belong more properly to solar-terrestrial relationships. It is clear that, in spite of the very considerable amount of work which has been devoted to this problem during the last 50 years, there are still some questions which are not completely solved.

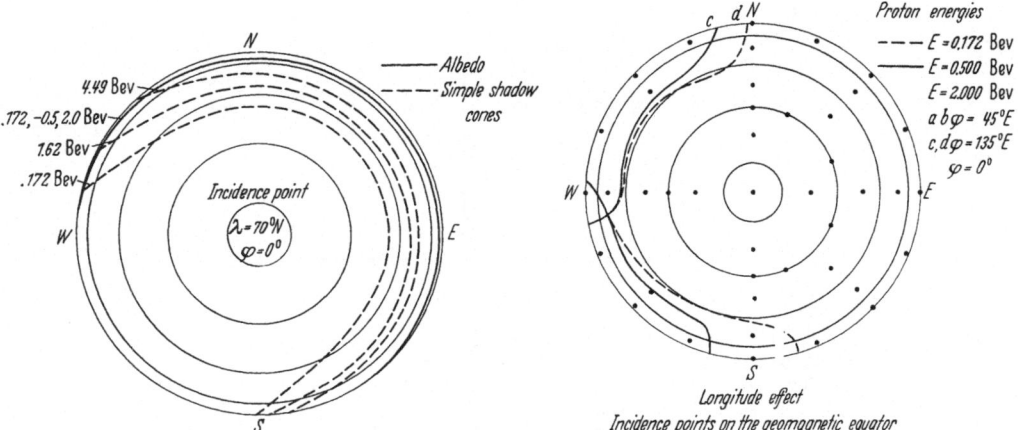

Fig. 52. Simple albedo cones $\lambda = 70°$ N, $\varphi = 0°$. Geographic zenith angle and azimuth.

Fig. 53. Longitude effect along geomagnetic equator. Geographic zenith angle and azimuth.

References.

[1] Clay, J., and H. P. Berlage: Naturwiss. 20, 687 (1932).

[2] Compton, A. H.: Phys. Rev. 41, 111 (1932).

[3] C. Störmer's original work is scattered in a great many publications. For the present problem his most important contributions are contained in Z. Astrophys. 1, 237—274 (1932), where numerous references to his earlier work will be found.

[4] Vallarta, M. S.: J. Franklin Inst. 227, 1 (1939).

[5] Lemaître, G., and M. S. Vallarta: Phys. Rev. 43, 87 (1933). See also Fermi, E., and B. Rossi: Rend. R. Accad. Naz. Lincei 17, 346 (1933). — Swann, W. F. G.: Phys. Rev. 44, 224 (1933). — Störmer, C.: Publ. Univ. Obs. Oslo No. 10, 1934 and Phys. Rev. 45, 835 (1934).

[6] Lemaitre, G.: Ann. Soc. Sci. Bruxelles 54, 162 (1935).

[7] The introductory material in this and other sections goes back essentially to Störmer's work [1]. Spherical coordinates seem to have been first used in this problem by Lemaître and Vallarta [4]. See also Lemaître, G., M. S. Vallarta and L. Bouckaert: Phys. Rev. 47, 434 (1935).

[8] Lemaître, G., and M. S. Vallarta: Phys. Rev. 49, 720 (1936).

[9] For example in connection with the galactic rotation effect, or the problem of space closure of periodic orbits. See Graef, C., and S. Kusaka: J. Math. Phys. 17, 43 (1938). — Vallarta, M. S., C. Graef and S. Kusaka: Phys. Rev. 55, 1 (1939). — Godart, O.: J. Math. Phys. 20, 207 (1941).

[10] Compare Graef and Kusaka: loc. cit. [9].

[11] For the theory of periodic orbits and their characteristic exponents se Lifshitz, Jaime: J. Math. Phys. 21, 284 (1942), and the literature quoted therein.

[12] GRAEF, C., and S. KUSAKA: loc. cit. [9]. — LIFSHITZ, J.: J. Math. Phys. **21**, 94 (1942).
[13] STÖRMER, CARL: Z. Astrophys. **1**, 237 (1930).
[14] LEMAÎTRE, G.: Ann. Soc. Sci. Bruxelles **54**, 194 (1935). — GODART, O.: Ann. Soc. Sci. Bruxelles **58**, 27 (1938). — BAÑOS, ALFREDO, HÉCTOR URIBE and JAIME LIFSHITZ: Ref. Mod. Phys. **11**, 137 (1939). — LIFSHITZ, JAIME: loc. cit. [11].
[15] SCHREMP, E. J.: Phys. Rev. **54**, 158 (1938).
[16] LEMAÎTRE, G., and M. S. VALLARTA: loc. cit. [5].
[17] GODART, O.: Ann. Soc. Sci. Bruxelles **58**, 38 (1938).
[18] SCHREMP, E. J.: loc. cit. [15].
[19] GRAEF and KUSAKA: [9].
[20] GODART, O.: J. Math. Phys. **20**, 207 (1941).
[21] LEMAÎTRE, G., and M. S. VALLARTA: Ann. Soc. Sci. Bruxelles **56**, 102 (1936).
[22] BAÑOS jr., ALFREDO, Phys. Rev. **55**, 621 (1939). — Journ. Math. Phys. **18**, 211 (1939).
[23] GODART, O.: loc. cit. [14]. — LEMAÎTRE, G., and M. S. VALLARTA: loc. cit. [21]. — BAÑOS, A., H. URIBE and J. LIFSHITZ: loc. cit. [14]. — LIFSHITZ, J.: loc. cit. [11]. — An example of a periodic orbit not belonging to the principal family is given by HUTNER, R. ALBAGLI: Phys. Rev. **55**, 109 (1939).
[24] LEMAÎTRE, G., and M. S. VALLARTA: loc. cit. [21].
[25] BOUCKAERT, L.: Ann. Soc. Sci. Bruxelles **54**, 174 (1935); see also TCHANG YONG-LI: Ann. Soc. Sci. Bruxelles **59**, 285 (1939). — LEMAÎTRE, G., M. S. VALLARTA and L. BOUCKAERT: Phys. Rev. **47**, 344 (1935).
[26] STÖRMER, C.: Congrès Internat. des Mathématiciens (Privat, Toulouse) 243, 1921.
[27] LEMAÎTRE, G., and M. S. VALLARTA: loc. cit. [21]. — BAÑOS jr., A.: loc. cit. [22].
[28] LIFSHITZ, J.: loc. cit. [11].
[29] BUSH, V.: J. Franklin Inst. **212**, 437 (1931). — It seems very likely that machine methods of integration were first used on a large scale in connection with this problem.
[30] LEMAÎTRE, G., and M. S. VALLARTA: Phys. Rev. **49**, 719 (1936); **50**, 493 (1936).
[31] STÖRMER, C.: Terr. Mag. Atmos. Elect. **3**, 31 (1931).
[32] VALLARTA, M. S.: Phys. Rev. **74**, 1837 (1948).
[33] SCHREMP, E. J.: loc. cit. [15].
[34] VALLARTA, M. S., R. GALL and J. LIFSHITZ: Phys. Rev. (in press).
[35] For details see GALL, R., and J. LIFSHITZ: Phys. Rev. **101**, 1821 (1956) and Sect. 17, this article.
[36] HUTNER, R. ALBAGI: Phys. Rev. **55**, 15 (1939). — LEMAÎTRE, G.: Ann. Soc. Sci. Bruxelles **54**, 162 (1935). — HUTNER, R. ALBAGI: Phys. Rev. **55**, 614 (1939). — TCHANG YONG-LI: Ann. Soc. Sci. Bruxelles, **59**, 285 (1939).
[37] SCHREMP, E. J.: Phys. Rev. **54**, 153 (1938).
[38] VALLARTA, M. S.: Phys. Rev. **47**, 647 (1935).
[39] SCHMIDT, A.: Z. Geophys. **2**, 38 (1926).
[40] CHAPMAN, S., and J. BARTELS: Geomagnetism, II, p. 651. Oxford: Clarendon Press 1940. — CHARGOY, A.: Rev. Mex. Fis. **2**, 1 (1953).
[41] VALLARTA, M. S.: Phys. Rev. **74**, 1837 (1948).
[42] GALL, R., and J. LIFSHITZ: Phys. Rev. **101**, 1821 (1956). — Proc. of VI. Internat. Cosmic Ray Conference, Varenna, 1957 (in press).
[43] MILNE, W. E.: Numerical Calculus, p. 135. Princeton: Princeton University Press 1949

[Manuscript received January 1958].

Experimental Results of Flights in the Stratosphere.

By

ERNEST C. RAY.

With 15 Figuies.

Introduction.

Balloon flights, and more recently airplane and rocket flights, have played a vital role in the history of cosmic ray research. It was first shown that cosmic radiation was of extraterrestrial origin when KOLHÖRSTER found that the intensity increased as one went up in the atmosphere. Later flights by many different people helped determine the nature of the primary radiation, the variation of its intensity in space and time, the nature of its interactions in the atmosphere, and the properties of the secondary radiations produced by these interactions.

This article is primarily concerned with the interactions of the primary cosmic ray beam with the atoms and nuclei in the atmosphere, and with the nature of the secondary radiations resulting from these interactions. A few effects due to the influence of the earth's magnetic field on the primary radiation will be mentioned. A detailed presentation of geomagnetic effects is given in VALLARTA's contribution to this volume.

In Chap. I of the present article the various intensities of the total charged radiation in the atmosphere are discussed. Chap. II is concerned with measurements which have been made of the nature and rate of occurrence of electromagnetic and nuclear reactions in the atmosphere. Current practice is to use mostly the results of quantum electrodynamics in treating electromagnetic interactions, rather than experimental results. On the other hand, studies of the new heavy mesons and hyperons seem to be progressing most rapidly through the use of accelerators. This will be the case more and more as accelerators produce greater energies. Chap. II reports primarily on nuclear reactions, and the discussion is restricted to those topics of importance for cosmic rays as a geophysical phenomenon rather than as a source of data for nuclear physics. Chap. III presents the available data on intensities of various kinds of particles within the atmosphere. These are usually somewhat fragmentary.

The qualitative nature of the development of the cosmic ray beam as it descends in the atmosphere is well understood. There do not remain any important quantitative discrepancies. It is to be hoped that ultimately more precise information will be acquired on certain aspects of the phenomena, notably the material in Chap. III. There is also some question about the fragmentation probabilities discussed in Sect. 6. It is to be expected that when these numerical questions are answered, we shall have a complete picture of the behavior of cosmic radiation in the atmosphere.

I. Intensities of cosmic radiation in the atmosphere.

1. Directional intensities. The directional intensity, j, is defined such that $j(\vartheta, \varphi)\, d\omega\, dA$ is the number of particles crossing per second the element of area dA and coming from within the element of solid angle $d\omega$. The direction of arrival is normal to the element dA. As indicated, the directional intensity is a function of zenith angle ϑ and azimuth φ. The vertical direction at the point of observation is $\vartheta = 0$. The vertical intensity will be denoted by $j(0)$ with the then meaningless azimuthal angle suppressed in the argument.

This definition is an ideal one. In practice one measures an average over some finite solid angle of the quantity defined above. If this finite solid angle is sufficiently small the distinction is usually unnecessary. In accordance with custom we shall ignore it.

It is inappropriate to discuss here at any length the details of the instrumentation which makes possible a measurement of the directional intensity. The measurement usually depends on the use of a device called a counter telescope. The heart of this apparatus is the so called coincidence circuit, an electronic device with two or more inputs and a single output. If an electrical pulse appears on each of the inputs within a length of time of the order of microseconds to milliseconds, the output will produce a single pulse. If any one of the inputs receives no pulse during this length of time, nothing appears on the output. If a separate counter, for instance a Geiger counter, is used to supply each of the inputs for such a circuit, and the counters are arranged parallel to each other and in one plane, nearly all of the outputs of the coincidence circuit will correspond to the traversal by a high speed particle of all of the counters. From the geometry of the counting arrangement one then knows that the particle arrived at the counters from within a certain solid angle. Often one or two percent of the pulses produced by the coincidence circuit correspond to the simultaneous traversal by one particle of part of the counters and by another of the rest. This effect can easily be corrected for by a semi-theoretical treatment.

When the counting rate of such a counter telescope is known, there remains the problem of calculating from this datum the directional intensity. This can always be done by dividing the counting rate of the telescope by a number, called the geometric factor, which depends on the dimensions of the counters, the placement of them with respect to each other, the dependence of j on ϑ and φ over the solid angle seen by the telescope, and the fraction of the particles traversing a counter which produces a pulse on its output. This last quantity is called the efficiency. For details of the calculation of the geometric factor, see POMERANTZ[1] and GREISEN[2].

The value of the directional intensity depends on altitude, latitude, longitude, zenith angle, azimuth, and time. Time variations are discussed elsewhere in this volume. They are chiefly due to changes in the intensity of radiation arriving at the top of the atmosphere. There are also some marked meteorologically associated time variations, but these also do not concern us here. In addition to these dependences of the directional intensity itself, the measured value is subject to variation which depends on the thickness of the walls of the detector used in the measurement. This effect is caused by the exclusion of particles with less than a certain energy from the counting volume by the absorptive power of the walls. Counters are available which have walls thin enough to admit electrons of about 0.15 Mev and protons of about 4 Mev.

First we discuss the variation of the vertical intensity with altitude, latitude, and longitude. Fig. 1 is a plot of vertical intensity against thickness of air overhead in g/cm². The latitude of observation is the parameter. The curves are

[1] M. A. POMERANTZ: Phys. Rev. **75**, 1721 (1949).
[2] K. GREISEN: Phys. Rev. **61**, 212 (1942).

drawn after BIEHL, NEHER, and ROESCH[1]. These curves, as noted by the authors, are subject to some uncertainty because several corrections, mostly for time variations of the primary radiation, were necessarily applied. Nevertheless, these curves are certainly quite close to the truth, at least for the epoch 1948. The curves for different latitudes represent data taken from balloon flights made on widely different occasions. An airplane flight made at an atmospheric depth of 310 g/cm² during the summer of 1948 supplied data which were used to normalize these curves in an effort to reduce them to a single epoch. The validity of this procedure is open to strong doubt, but the practical consequences of this doubt are not likely to be appreciable.

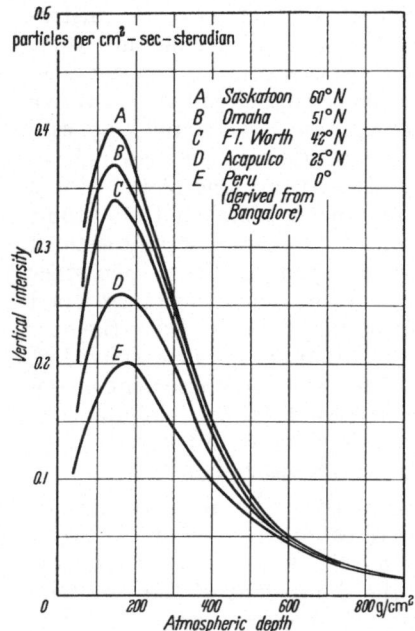

Fig. 1. Total vertical intensity as a function of altitude at several latitudes. (After BIEHL et al.[1].)

The latitudes given in this figure are geomagnetic latitudes. Geomagnetic latitude in this article means the angle between the radius vector from the geographic center of the earth to the point of observation and a plane passing through the geographic center at right angles to the magnetic axis of the erarth as determened by the harmonic analysis of the earth's field.

These curves show that the vertical intensity at all depths is least at the equator and increases monotonically toward higher latitudes. All other intensities show the same behavior in this respect as does the vertical intensity. Results of measurements in the southern hemisphere show a symmetric behavior. That is, the geomagnetic equator is nearly the locus of minima of cosmic ray intensity. There are, however, certain small deviations of considerable theoretical importance[2].

The altitude dependence of the vertical intensity is quite characteristic of the response of almost all detectors throughout the atmosphere. The maximum in such curves was first discovered by PFOTZER[3].

Data taken at higher latitudes than those shown in Fig. 1 for years before about 1948 show no important change from the data for SASKATOON. Recently some long period time changes in the primary intensity have been discovered at high latitudes[4]. This effect is small in terms of total particle intensities both above and within the atmosphere

The longitude variation of the vertical intensity has not been measured. Judging from the effect observed with other intensities there should be a variation of some 10% at the equator. This question is considerably clouded at the present because of the results of such studies as those of SIMPSON et al.[2].

The dependence of the vertical intensity near the top of the atmosphere is rather well explained by the assumptions that the primary particles are positively

[1] A. T. BIEHL, H. V. NEHER and W. C. ROESCH: Phys. Rev. **76**, 914 (1949).
[2] J. A. SIMPSON, K. B. FENTON, J. KATZMAN and D. C. ROSE: Phys. Rev. **102**, 1648 (1956).
[3] G. PFOTZER: Z. Physik **102**, 23, 41 (1936).
[4] H. V. NEHER: Phys. Rev. **103**, 228 (1956).

charged and that they move in the field of the earth's magnetic dipole as found from a harmonic analysis of the field intensities measured at points on the earth's surface. This description will eventually be modified in the light of such work as that of SIMPSON et al.[1].

The altitude dependence of the intensity will be discussed in Chap. II, below.

The dependence of directional intensity on zenith angle and azimuth has also been studied extensively. The zenith angle variation is much the larger of the two. We shall review that first.

Itis customary to fit the zenith angle dependence of the directional intensity in the atmosphere with the function $j = j(0)\cos^n\vartheta$. This rarely produces a fit of high accuracy, but for most practical cases it is quite adequate. As one goes from sea level to the top of the atmosphere, one finds that n decreases from about 2 to zero. At the highest altitudes, the intensity has a minimum in the vertical direction and rises somewhat with increasing zenith angle. This behavior is conveniently represented by the function $j = j(0)(1 + C\sin\vartheta)$. C is a positive number considerably less than unity. A set of results typical of middle latitudes was obtained by WINCKLER, STROUD, and SHANLEY[2]. BIEHL et al. obtained result for a range of latitudes[3].

The azimuthal variation is usually negligible. At the equator it is closely a $\sin\varphi$ variation, with φ measured from north through east. The amplitude is about 10% of the average intensity for the zenith angle at which the azimuthal variation is measured. At middle latitudes the azimuthal variation is less than

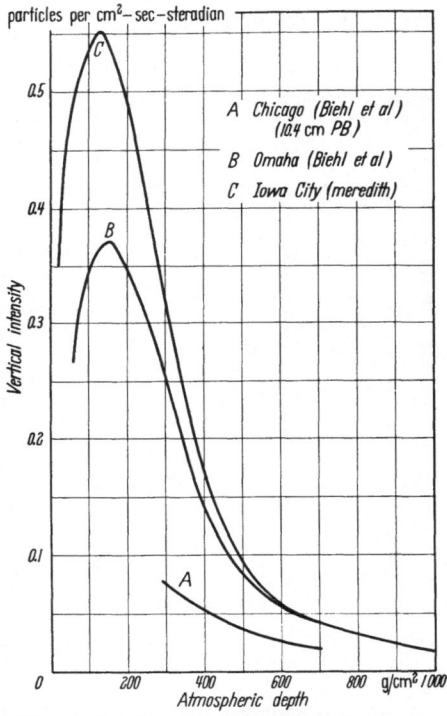

Fig. 2. The dependence of vertical intensity on counter wall thickness as a function of altitude. (After BIEHL et al.[3], and MEREDITH[4].)

10% and within the statistical errors. The primary radiation itself should have a very considerable azimuthal variation, according to the theory of the motion of charged particles in the earth's magnetic field. Measurements made with a ČERENKOV detector telescope which had a high efficiency for the detection only of downward moving particles show that the expected azimuthal variation is indeed present for these. The upward moving secondaries considerably reduce the azimuthal variation of the total intensity within the atmosphere[5].

The above discussed results were obtained using detectors with wall thicknesses of a few g/cm². Thus while they are not restricted in their applicability to particles of only the highest energies, neither do they include the very slow particles. Fig. 2 is a plot of the vertical intensity as a function of atmospheric

[1] See footnote 2, p. 132.
[2] J. R. WINCKLER, W. G. STROUD and T. J. B. SHANLEY: Phys. Rev. **76**, 1012 (1949).
[3] See footnote 1, p. 132.
[4] L. H. MEREDITH: Master's thesis, State University of Iowa, 1952 (unpublished).
[5] J. R. WINCKLER and K. ANDERSON: Phys. Rev. **93**, 596 (1954).

depth which shows the effect of wall thickness. The monotonic altitude dependence of the intensity measured under the lead absorber shows that the Pfotzer maximum is caused by the altitude dependence of the formation of secondary particles by interactions of the primaries with atoms and nuclei in the atmosphere. This point is discussed in Chap. II, below. The curves A and B represent data taken by BIEHL et al.[1]. Curve C is after L. H. MEREDITH[2].

2. Omnidirectional intensities. The omnidirectional intensity, J, is defined as the number of particles crossing per second a sphere with unit cross sectional area. It is evident that

$$J = \int j(\vartheta, \varphi)\, d\omega \tag{2.1}$$

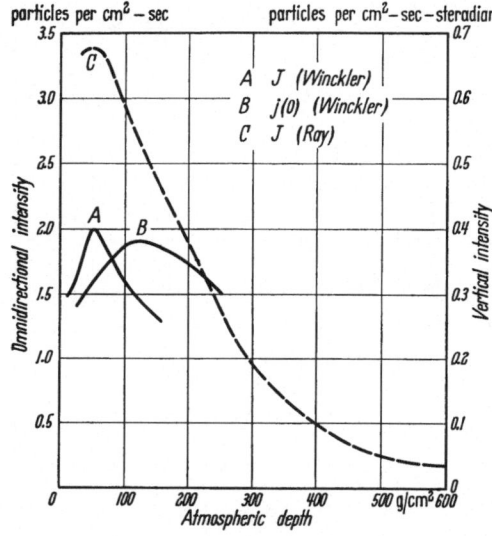

Fig. 3. Omnidirectional intensity as a function of altitude. (After WINCKLER et al.[3], and RAY[4].)

where the integration is carried out over a sphere situated at the point of observation. Most of the determinations of this quantity depend on numerical integrations of directional intensities obtained as discussed in Sect. 1, using Eq. (2.1).

Fig. 3 shows typical results for J as a function of altitude and, for comparison, $j(0)$ (see Sect. 1). These curves represent data taken by WINCKLER, STROUD, and SHANLEY. One notices that the Pfotzer maximum for the omnidirectional intensity occurs at a higher altitude than does that for the vertical intensity. This is quite reasonable since particles arriving at a given depth in the atmosphere from a large zenith angle will have traversed more material than will those arriving vertically. Thus one expects the effects of the atmosphere on an inclined cosmic ray beam to occur more rapidly with increasing atmospheric depth.

Various attempts have been made to measure J directly using a single counting geometry. In this case one has to calculate the geometric factor which relates the counting rate to J. Further, since the dependence of j on ϑ and φ is not measured, one must find a counter arrangement such that the geometric factor does not depend on the angular distribution of the radiation. A spherical counter of course satisfies this condition exactly. Such counters have been built.

A single cylindrical counter, particularly when mounted with its axis horizontal, is a fairly good approximation to a sphere for the angular distributions actually found in the atmosphere. This is especially so if the length of the counter is not very much greater than the diameter. A single counter with a length-to-width ratio of 3.5 has, when mounted with its axis horizontal, a geometric factor which changes by about 8% when the zenith angle dependence of the directional intensity goes from isotropy to $\cos^2 \vartheta$.

[1] See footnote 1, p. 132.
[2] See footnote 4, p. 133.
[3] See footnote 5, p. 133.
[4] E. C. RAY: Thesis, State University of Iowa, 1953 (unpublished).

A noticeable improvement on this technique is achieved by measuring the counting rate of a cylindrical counter with its axis vertical at the same time that one measures the rate of an identical counter with its axis horizontal. If one doubles the rate of the horizontal counter and adds to the result the rate of the vertical counter, the geometric factor for the sum is remarkably independent of n for a zenith angle dependence of the radiation of $\cos^n \vartheta$ where n goes from zero to 2.

A straightforward calculation yields the following formulae for the geometric factors. The notation $G_H(n)$ stands for the geometric factor appropriate to a counter with its axis horizontal in a radiation field such that $j = j(0) \cos^n \vartheta$. [Note that in these considerations one tacitly assigns all of the radiation to the upper hemisphere. That is, $j(\vartheta > \pi/2) = 0$.]

$$G_H(0) = G_V(0) = \frac{\pi}{4} a l \left(1 + \frac{a}{2l}\right) \varepsilon,$$

$$G_H(1) = \frac{8}{3\pi} a l \left(1 + \frac{\pi a}{8 l}\right) \varepsilon,$$

$$G_H(2) = \frac{9\pi}{32} a l \left(1 + \frac{a}{3 l}\right) \varepsilon,$$

$$G_V(1) = \frac{2}{3} a l \left(1 + \frac{\pi a}{4 l}\right) \varepsilon,$$

$$G_V(2) = \frac{3\pi}{16} a l \left(1 + \frac{a}{l}\right) \varepsilon.$$

In these equations, the subscript H refers to a counter with its axis horizontal, V refers to a counter with its axis vertical. The counter length is l, its diameter is a. The counter efficiency is ε. These geometric factors are to be used in the relation

$$N = G(n) J$$

where N is the measured counting rate.

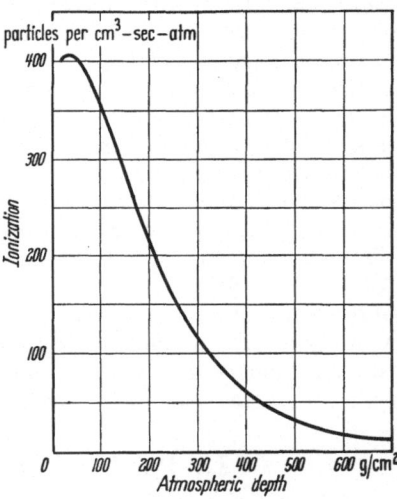

Fig. 4. Ionization as a function of altitude. (After MILLIKAN, NEHER and PICKERING[1].)

A flight at Iowa City, Iowa in the spring of 1953 yielded the result shown in the dashed curve of Fig. 3[2]. The double counter method mentioned above was used. The walls of the counters were 30 mg/cm² of aluminum. The particles measured by WINCKLER et al. traversed 2.6 g/cm². The two results are appropriate to the same latitude. It is believed that the higher counting rate in the 1953 measurements results from the inclusion of particles with energy so low that they are unable to penetrate the telescope of WINCKLER et al.

3. Ionization. A great deal is known about the ionization produced throughout the atmosphere, due mostly to the efforts of H. V. NEHER and a host of collaborators[1]. Fig. 4, after NEHER, shows the dependence of ionization on altitude. I is the number of ion pairs which would be formed per second in a cubic centimeter of air at a pressure of 1 atmosphere and a temperature of 20° C by cosmic radiation. These data were taken with a special ionization chamber. This chamber consists of a known volume of gas which surrounds a sensitive vibration free electroscope. By measuring the rate of discharge of the electroscope, NEHER determined the number of ion pairs formed in the chamber per second. It is assumed that the charge is all collected, that is, that recombination does not

[1] R. A. MILLIKAN, H. V. NEHER and W. H. PICKERING: Phys. Rev. **66**, 295 (1944).
[2] See footnote 4, p. 134.

occur. There has been considerable investigation of the behaviour of an ionization chamber in a radiation field. The most elementary work shows that for small potentials on the collecting electrode recombination does occur. As the collecting potential is increased, however, the number of ion pairs measured per second increases until it reaches a plateau. It is commonly assumed that this plateau occurs because no recombination occurs and all ion pairs are collected. It has been questioned whether in dense columns of ionization such as are left by multiply charged particles in their traversal of material some recombination does not still occur. At any rate, the usual measurements of ionization must be fairly good, and in the case of NEHER's data the consistency is very good since he has taken great pains to make his various detectors as nearly alike as possible and then to intercalibrate them carefully. In addition, he very early took extensive steps designed to correct, to a large extent, for long term time variations in the primary radiation.

One of the possible uses of this sort of data is in the problem of the energy balance. If one knows the average energy required to form an ion pair, commonly taken as 32 eV for air, then the number of ion pairs formed in a column of atmosphere with unit cross section can be used to estimate the energy dissipated in the atmosphere as ionization. If one adds to this the energy expended in non-ionizing processes, for instance as kinetic energy of neutrinos and as binding energy in the formation of nuclear stars, one can obtain a value for the total energy dissipated by cosmic radiation in the atmosphere. A comparison of this with estimates of the energy brought in by the primary radiation then serves as a check on our understanding of the processes which cosmic rays produce in the atmosphere and on our understanding of geomagnetic theory. These two estimates agree moderately well.

II. Particle interactions in the atmosphere.

4. Qualitative review. The primary cosmic radiation undergoes both electromagnetic and nuclear interactions in the atmosphere. The electromagnetic interactions are now largely treated theoretically, using quantum electrodynamics A review of the results of such treatments is given by ROSSI[1]. We shall restrict ourselves here to a discussion of the role played by these interactions and to the results of measurements concerned with nuclear reactions.

For cosmic radiation, the electromagnetic processes of importance are elastic scattering, excitation and ionization, bremsstrahlung, the Compton effect, and pair production. The first three of these processes are interactions in which the incident particle is charged. The other two are interactions of photons with matter.

Elastic scattering is the process in which an incident charged particle is deflected by the Coulomb field of an atomic nucleus. When so deflected, the incident particle of course emits one or more photons. The collision is nevertheless called elastic if the energy carried away by these photons is negligible compared with the energy of the incident particle. This process is of importance since it broadens the angular distribution of the low energy radiation.

Excitation and ionization are processes in which one or more electrons in an atom absorb energy from an incident particle and as a result rise to higher quantum states. If the higher state is a bound state, the process is called excitation. If it is a free state, the process is called ionization. In connection with cosmic ray phenomena, one usually speaks of these processes in connection with incident charged particles. When incident photons ionize atoms, one calls it the *Compton effect*.

---

[1] B. ROSSI: High Energy Particles. New York 1952.

Bremsstrahlung is the event in which an incident charged particle undergoes a close collision with the nucleus of an atom. Just as in the case of elastic scattering, the interaction is by way of the Coulomb field, but now the photon created during the acceleration of the incident particle carries away an important part of the energy originally held by the incident particle.

The processes of excitation, ionization, and bremsstrahlung are important because they continually slow down energetic charged particles travelling through matter. Bremsstrahlung is important in this connection only for electrons. For all other particles, the bremsstrahlung cross section is negligible at the usual energies. The theoretical formula giving the rate of energy loss of charged particles travelling in matter is called the Bethe-Bloch formula.

In general, the Compton effect is a collision between a photon and a free electron. In practice, the target electron is bound in an atom, but the theory is valid if the electron is ejected from the atom with an energy large compared with its original binding energy.

Pair production is a process in which a high energy photon is converted into an electron and a positron. In order that energy and momentum may simultaneously be conserved, it is necessary that this event occur in the Coulomb field of a nucleus. The nucleus then takes up the excess momentum.

The *electromagnetic processes* produce two results of importance in cosmic radiation. In the first place, excitation, ionization, and bremsstrahlung continually take energy from high speed charged particles, eventually bringing them to rest. Since at any height in the atmosphere the charged particles present have a continuous distribution of energies, and since a particle of a given energy can penetrate farther into the remaining atmosphere than can a particle of smaller energy, the electromagnetic processes have a tendency to decrease the number of charged particles as the cosmic ray beam travels downward through the atmosphere. At the same time, these same interactions, as well as the nuclear reactions to be discussed below, continually produce new charged particles, tending to increase the number of such particles present. Whether the tendency to increase or to decrease the number prevails depends on the mean energy of particles in the beam. Only particles with sufficiently great energies produce appreciable numbers of these secondary particles, and it turns out that as one descends from the top of the atmosphere one finds the number of particles to be increasing only for the first 50 g/cm^2 or so of atmospheric depth. Below this the mean energy of the beam is low enough that its attrition outweights the production of new particles. The place of maximum particle intensity is called the *Pfotzer maximum*.

The other important effect of the electro-magnetic reactions is called a *cascade shower*. This process can be initiated by either a highly energetic electron or photon. During the early stages of the shower, the number of electrons and photons increases enormously by bremsstrahlung, pair production, and the Compton effect. At the same time, the average energy per particle decreases. Finally the number of particles with sufficient energy to further multiply begins to decrease and the shower dies out. Ultimately all of the particles are brought to rest by the usual processes.

It is thought that there are no electrons or photons in the primary cosmic radiation. High energy photons come from the decay of π^0 mesons, and electrons come from the decay of μ mesons.

The *nuclear reactions* of importance in cosmic radiation all involve incident particle energies which are very high by laboratory standards. Till recently a considerable amount of work was done on nuclear physics as a subject itself, using the cosmic ray beam as a source of nuclear projectiles. While many useful results have been collected in this way, these developments will not be discussed here. Only those nuclear events will be considered which substantially contribute to our understanding of cosmic radiation as a geophysical phenomenon. The choice is largely justified since recent developments in accelerator work have made feasible the production in the laboratory of beams of particles in the Bev range.

Nuclear reactions induced by particles with energies of several Bev show a very characteristic appearance when observed with nuclear emulsions. There are several reaction products except for the lowest energy events. The low energy products are emitted nearly isotropically in the laboratory system, while the particles which produce minimum ionization are emitted in a narrow cone with an axis roughly an extension of the path of the particle which initiated the reaction. Since many of the ejected particles are neutral and hence produce no visible tracks, the degree of alignment may be poor, particularly for low energy events where few secondary particles are produced. An event of this sort is called a *star* because of its appearance in a photographic plate.

Most of the reaction products of stars are protons, π mesons, and neutrons. Since these particles have large cross sections for interaction with nuclei, and since many of them are emitted with high energies, these secondary particles themselves initiate stars lower in the atmosphere. Thus there develops what is called the nucleonic cascade. The neutral π meson, the so-called π^0, decays into energetic photons in distinct preference to suffering any other fate. These photons initiate electromagnetic showers as discussed above. The charged π mesons decay into μ mesons which decay into electrons. These also lead to showers.

Stars initiated by singly charged and uncharged particles apparently do not produce highly energetic multiply charged secondaries. These are produced in important numbers, however, when the incident particle is multiply charged. In this event the secondary particles always carry smaller charges than does the incident particle. Apparently these multiply charged secondaries are fragments of the initiating particle. The process is called *fragmentation*. The cosmic ray beam incident on the top of the atmosphere contains appreciable numbers of the nuclei of various elements up through iron. As this beam progresses through the atmosphere, it is continually enriched in the lighter elements by fragmentation of heavier ones. This effect is of great importance in interpreting the results of measurements of the intensities of multiply charged particles in balloon flights.

Stars of moderately high energy emit appreciable numbers of heavy mesons and hyperons. This is of little importance except perhaps near the equator where the mean energy of the primary radiation is high.

The nuclear reactions, by two different mechanisms, remove energy from the cosmic radiation and render it indetectable. Since most nuclear reactions in the atmosphere involve nitrogen and oxygen as target nuclei, the disruption of these nuclei requires the expenditure of energy. Secondly, the decay of charged mesons produces energetic neutrinos as well as charged particles. The indetectable neutrinos carry away an appreciable amount of energy.

All of the interactions, particularly near the top of the atmosphere, contribute in some degree to the phenomenon called *albedo*. This is a flux of secondary particles above the atmosphere resulting from upward moving particles formed in interactions below the point of observation and downward moving particles which have been guided to the point of observation by the earth's magnetic field from interactions which occurred elsewhere. These particles must somehow be removed from consideration when one wishes to calculate intensities of primary radiation from data collected in rocket flights above the atmosphere. No definitive treatment of the problem, either experimental or theoretical, has yet been done.

The above considerations are the framework for the material discussed in the following section.

5. Stars. Cloud chambers and nuclear emulsions have furnished most of the data on stars. Nuclear emulsions have been the most popular tools for use in balloons because they weigh much less than does a comparable sensitive volume in the form of a cloud chamber.

Several different measurements of properties of a track of developed grains left by the passage of an energetic charged particle can be made in an effort to learn something about the particle. Some of these properties are delta ray density, gap frequency, grain density, and multiple Coulomb scattering. Delta rays are short crooked tracks left by secondary electrons ejected from atoms by the passage of the primary particle. A gap is a portion of the track of some minimum length in which there are no developed grains. The last two techniques were used in much of the work reported below. Grain density is the number of developed grains per unit path length. In using multiple Coulomb scattering one determines the mean angle of deviation of the track from its initial direction in proceeding a unit distance. The deviation from the original direction is the cumulative effect of many small changes in direction produced by Coulomb scattering. These two effects depend on the particle parameters in different ways. By measuring both quantities one can simultaneously determine the mass and energy of the particle which produced the track. Resolution among kinds of particles is usually good. Protons with energies greater than about 700 Mev are not distinguishable from π mesons with energies greater than about 1 Bev. Deuterons and tritons are not well separated, although they probably could be if considerably smaller statistical errors were obtained.

A review of work up till the middle of 1951 is given by Rossi [1]. A series of papers by the Bristol group is of particular importance [2].

The Bristol group was primarily concerned with stars initiated by particles with energies less than about 10 Bev. They separate the secondary particles formed in nuclear disintegrations into three groups according to the grain density along the tracks they cause in the emulsion.

Thin tracks, which they call tracks of shower particles, are those with a specific ionization of less than 1.4 times minimum. These are chiefly π mesons of energy greater than 80 Mev or protons of energy greater than 500 Mev. These particles are emitted in a cone of rather small half angle with its axis a crude extension of the direction of travel of the initiating particle. The number of these tracks associated with a star is denoted by n_s.

Gray tracks are those produced by protons in the energy range 25 to 500 Mev. Tracks with this range in grain density are also produced by deuterons and tritons, and about 18% of them are due to π mesons with energy less than 80 Mev. The number of gray tracks associated with a star is denoted by N_g.

Black tracks are produced by protons with energy less than 25 Mev, deuterons with energy less than 50 Mev, tritons of energy less than 75 Mev, and α particles with energy less than 800 Mev. The number of black tracks is denoted by N_b.

Gray tracks and black tracks, when considered as a single group, are called *heavy tracks*, and $N_g + N_b$ is denoted by N_h. Stars are classified according to N_h, n_s, and a suffix n, p, α, π, ... to indicate the initiating particle. For example,

Table 1. *Analysis of 15 300 stars observed at 68 000 ft* (CAMERINI et al. [2f]).

n_s	N_h	3	4	5	6	7	8	9	10	11	12	13	14	15	16	17	18	19	20	21	22	23	24	25	26	27	28	29	30	Total
0	n	2499	2132	1499	938	576	367	238	169	124	100	89	67	47	37	29	14	18	11	7	3	4	1	1	3	2	1		1	8977
	p	225	310	356	305	204	137	116	92	75	45	49	53	32	32	22	12	10	11	9	8	3	2	4	1	2	1	1	1	2117
1	n	133	194	195	199	128	97	88	63	51	53	28	34	23	13	25	15	18	3	8	1	2	1	1	2	1	1		1	1378
	p	91	146	151	135	96	79	52	50	44	36	28	25	24	15	15	18	11	6	10	5	1	8	6	3	2	3		3	1063
2	n	33	36	44	51	37	26	26	23	20	17	16	27	17	15	15	9	3	4	2	7			2		1		2	3	437
	p	24	42	28	49	32	24	29	29	22	15	20	23	20	10	13	7	7	7	3	8	1	2	3	2	1		1	2	424
3	n	9	5	8	7	6	10	5	4	4	5	2	2	5	2	1	1	2	5	1	1	1	2	1	1	2			1	83
	p	13	15	25	17	14	14	12	12	18	10	11	16	12	7		7	4		2		6	2	2		1	1			229
4	n	3	6	7	6	7	2	6	2	2	2	1	4	5	2	8	3	4	5	2	1	1	2	1		1	1	1		61
	p	3	11	15	11	11	9	9	10	5	4	6	2		6		1	4		4	2	5	4	2	1	3		1	1	148
5	n	3	2		4	3	1	1	5	1	4	2	4	1	4		4	1	5	1	1	1	3	1	2	1				37
	p	8	5	8	9	7	5	6	6	5	5	2	6	4	1	3	2	1	2	2	3		2	1	1	1	1			97
6	n		2	3	3	4	1	3	1	1	5	5	1	2		3	1		1		2	1	2		1					28
	p		5		7	5	2	5	2	2		2	6	6	1	1	4	1	2	1	3	1	1							68
7	n															2														6
	p			1	2	2		3	4	3	5	1	2	6		1	1	1		1	2	3			2			1	2	39
8	n		2				1	1					1	1	2		2				2		2			1	1	1		4
	p					4	2		3	3						1		1				3				1				34
9	n		1	1	2		1	1											1										1	6
	p	1	1	1		1	1	1										1							3	1				14
10	n		2			2												1						1						4
	p					2			3				2																	13
11	n																													0
	p				1		1	1							1		1				1		1							6
12	n										1																			0
	p										0						1				1		1							6
13	n																						0							0
	p												0		2		1	1					1							3
14	n														2	1	1	1			2	1						2		0
	p			1	1	1	1	1	3								1	1		1	1	1	1					1		9
15	n																	1												3
	p		1	1	1	1	1	1									1				1	1	1					1		11

a star of type $3+4p$ would be a star induced by a proton which had 3 heavy tracks and 4 thin tracks.

The data discussed by the Bristol group [2f] consist of identification as to type of 15300 stars found in 87 cm^3 of emulsion exposed at 68000 ft. Scattering and grain density measurements were made on 2000 tracks associated with these stars. Grain density and angular distribution were measured for 3070 thin tracks and 1508 gray tracks. The energy of the particle producing the star was measured in 200 cases.

Table 1 shows the distribution of star types among the 15300 stars. The notation n or p in the left column of the table specifies the initiating particle. The n type stars include cases where for any reason the initiating particle could not be identified. For example, stars initiated by multiply charged particles would appear in this group since the track of the initiating particle can not be distinguished from the heavy tracks produced by isotropically emitted secondary particles. Most of the n type stars, particularly those of low multiplicity, are produced by neutrons. The p type stars may be produced by π mesons as well as by protons.

It is evident from the table that most of the very small stars are generated by neutrons, while most of the large stars are generated by charged particles. For example, less than 10% of the stars with 3 heavy tracks and no thin tracks are produced by charged particles. On the other hand, of those stars with more than 7 thin tracks or more than 22 heavy tracks, or both, 75% are produced by charged particles.

While it is not exhibited in the table, other considerations show that a small fraction of the charged generating particles are α particles. This reflects the small abundance of α particles compared to singly charged particles. Of all stars created by particles with energies greater than about 1 Bev, about 6% are formed by π mesons. Most of the charged particles which generate stars are protons.

The multiplicity of secondaries produced by a star tends to increase as the energy of the initiating particle increases. Fig. 5 is a plot, after CAMERINI et al., of the values of $\overline{N}_h$ and $\overline{n}_s$, the averages of the corresponding quantities, as a function of the kinetic energy of the initiating particle. Since the thin tracks are due to particles of greater energy than are the heavy tracks, $\overline{n}_s$ varies more rapidly with energy at low energies than does $\overline{N}_h$. In spite of this strong dependence of $\overline{n}_s$ on the energy of the initiating particle, one can not use this as a means of determining a reliable value for the energy of the star. Fig. 6 exhibits the multiplicity distribution for stars of various energies. It is evident from this figure that the multiplicity establishes only a poorly defined lower limit on the initiating energy.

The angular distribution of the secondary tracks with respect to the direction of the incident particle is also a matter of interest. Fig. 7 shows the angular distribution of thin tracks in the laboratory system. It is seen that these particles are emitted in a rather narrow cone about the direction of the incident particle. The width of the angular distribution decreases slowly as n_s increases. This is presumably due to the rough dependence of n_s on energy with the corresponding increase in the velocity of the center of mass as n_s increases.

The particles causing gray tracks are also emitted preferentially in the forward direction, but in a broader cone than are the shower particles. Fig. 8 shows the angular distribution of the gray tracks as a function of the energy of the particles producing the tracks.

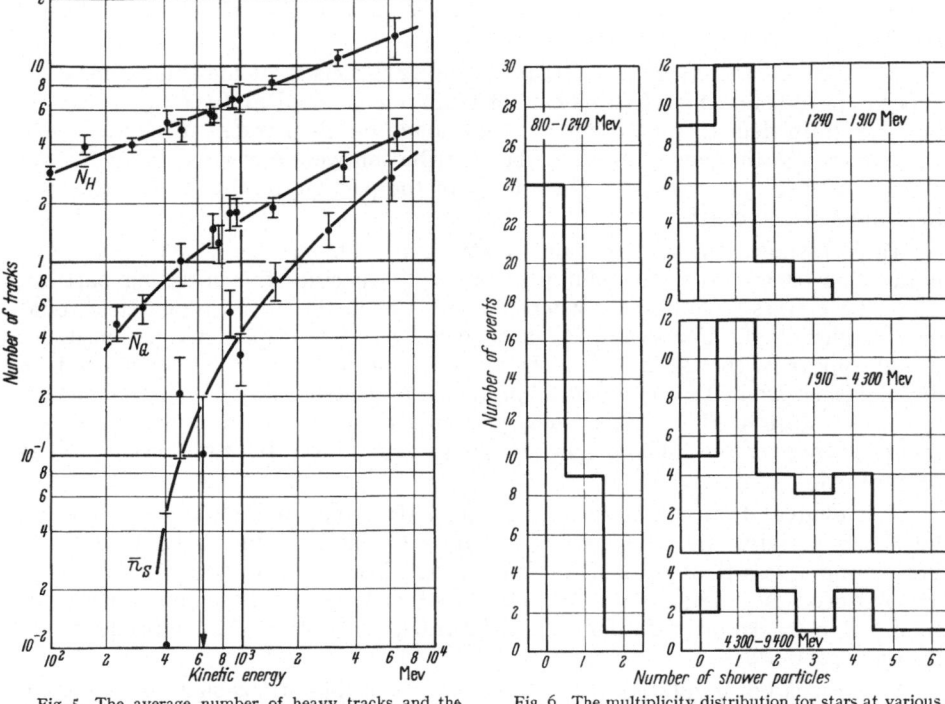

Fig. 5. The average number of heavy tracks and the average number of shower particles in stars as a function of kinetic energy. (After CAMERINI et al. [2f].)

Fig. 6. The multiplicity distribution for stars at various energies. (After CAMERINI et al. [2f].)

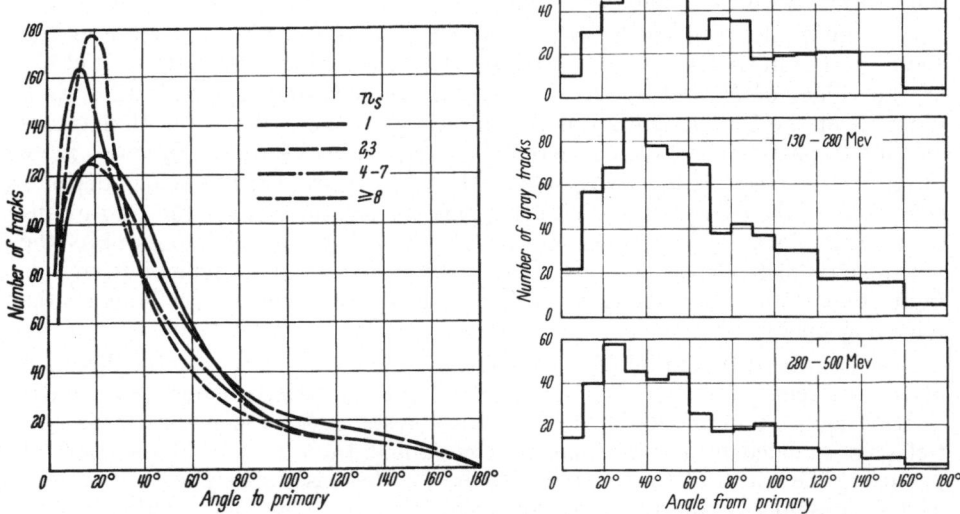

Fig. 7. The angular distribution of thin tracks in stars. (After CAMERINI et al. [2f].)

Fig. 8. The angular distribution of gray tracks in stars. (After CAMERINI et al. [2f].)

The black tracks radiate nearly isotropically.

Fig. 9 shows the energy spectra of π mesons, protons, and deuterons and tritons emitted by stars. To within experimental errors, the differential energy spectrum for π mesons is proportional to $E^{-1.5}\,dE$, where E is the total energy. This form of the spectrum is applicable in the range of total energy 250 Mev $\leq E \leq$ 1400 Mev [2d].

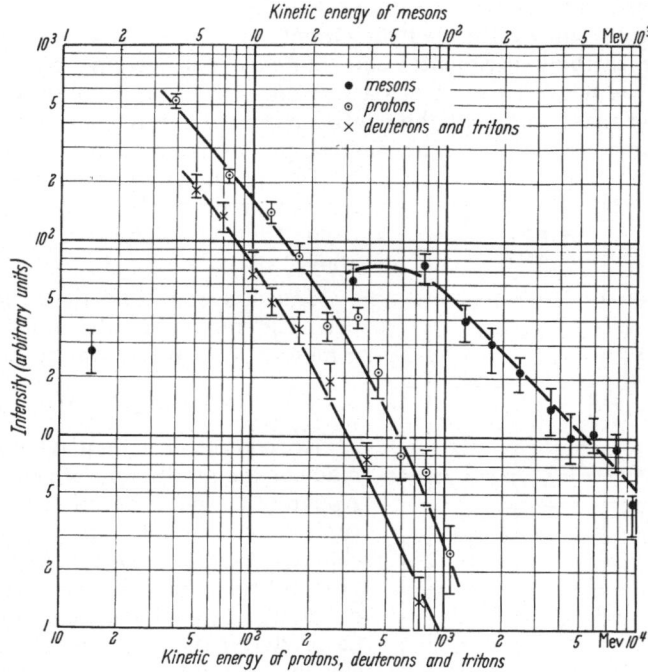

Fig. 9. The energy spectra of secondary π mesons, protons, and deuterons + tritons in stars. (After CAMERINI et al. [2d].)

The mean free path for nuclear interactions of the shower particles is geometric to within experimental errors. Table 2 gives the results of determinations by the Bristol Group [2d].

Table 2. *Mean free path for nuclear interactions of secondary particles.*

Type of particle	Energy Mev	Mean free path in emulsion (g/cm²)
π meson	150–1500	82 ± 35
Unidentified shower particle . .	{ 150–5000 (meson) 600–5000 (proton) }	} 102 ± 27
Proton	40–100	125 ± 63
Proton	100–600	117 ± 39
Deuteron or triton . .	75–250	47 ± 18

The geometric mean free path for the emulsion used is 90 g/cm². Only the deuterons and tritons have a mean free path which differs from this by more than the experimental error.

In addition to the charged particles which produce visible tracks, stars also emit neutral particles. These are chiefly π^0 mesons and neutrons.

The π^0 mesons were studied by CARLSON, HOOPER and KING [2e]. They collected data on the occurrence of electron pairs produced by the two photons into which a π^0 meson decays. They measured the total energy of the pairs of electrons which they found, and plotted an energy spectrum. They calculated a theoretical spectrum by using conservation of energy and momentum to relate the velocity and rest mass of the presumed neutral parent of the photons to the photon energy, and by then assuming that the photons were emitted isotropically in the rest system of the neutral parent. The comparison of the theoretical spectrum with the experimental one is in excellent agreement with the assumptions that the neutral parent has a rest mass $m = (295 \pm 20)\ m_e$ where m_e is the electron rest mass, and that the neutral parents have the same energy spectrum as the charged π mesons. The neutral parent is therefore called a π^0 meson.

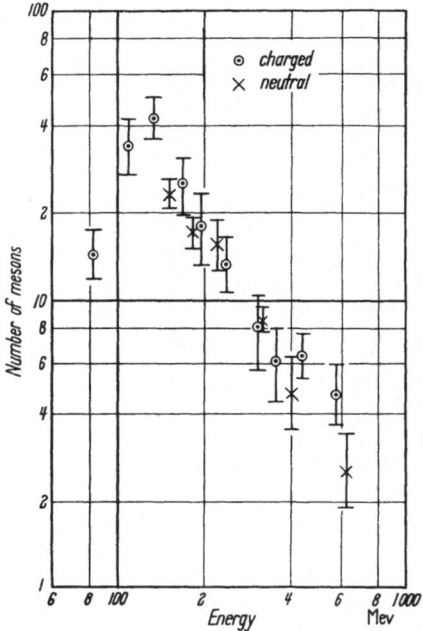

Fig. 10. The energy spectrum of π^0 mesons from stars. (After CARLSON, HOOPER, and KING [2e].)

Fig. 10 shows the energy spectrum of π^0 mesons produced by stars, and, for comparison, that of charged π mesons discussed above.

CARLSON et al. find the ratio of the number of π^0 mesons to the number of charged π mesons formed per star to be

$$N(\pi^0)/N(\pi\pm) = 0.45 \pm 0.10.$$

No completely adequate study has been made of the production of neutrons by stars. Apparently the neutron multiplicity of a typical star is not very different from the charged particle multiplicity[1]. The Bristol group showed that if one assumed roughly equal numbers of neutrons and protons to be produced then the energy brought into a star by the initiating particle is not very different from the accountable energy carried away by the products of the interaction [2f]. Neutron intensities in the atmosphere have been studied. This point is discussed in Sect. 10.

The above data refer to the nuclear interactions which occur in emulsions. The target nuclei are mostly silver and bromine. In applying the results of such studies to the behavior of cosmic radiation in the atmosphere, it is of vital importance that one know how the various cross sections are related to those for targets of nitrogen and oxygen. The simplest thing to do is to assume that the cross section is simply geometric for all nuclei. BROWN[2] finds that the cross section for nuclear interactions at 10600 ft is closely proportional to $A^{\frac{2}{3}}$ where A is the mass number of the target nucleus. He studied the effect for helium, nitrogen, neon, and argon, using a cloud chamber. BARBOUR[3] has studied a certain aspect of stars using a technique in which foils of various metals are sandwiched between sheets of emulsion. He studies star production in platinum, gold, tin, nickel, copper, and aluminum. He excludes shower particles from consideration. He finds that the cross section for production of all stars with four or more gray and black tracks is proportional to $A^{0.7\pm0.12}$. One must be careful in using the measured values of this cross section to notice what cross section is

[1] W. C. G. ORTEL: Phys. Rev. **93**, 561 (1954).
[2] W. W. BROWN: Phys. Rev. **93**, 528 (1954).
[3] I. G. BARBOUR: Phys. Rev. **93**, 535 (1954).

measured. For example, emulsion data often exclude stars in which one or two ionizing prongs are produced. BROWN's work excludes interactions in which less than 8 Mev is dissipated in his detector by products of the interaction. Stars with no ionizing prongs are not detected by the devices used in obtaining the above data.

The latitude and altitude dependence of star formation were studied by LORD[1]. He reports data for stars having more than two prongs. Table 3, taken from LORD's paper, gives the rate of star production in units of stars/cm³/day at two latitudes and a range of altitudes for each latitude. The data are given separately for various sizes of stars.

Table 3. *Rate of star production/cm³/day.* (After LORD.)

Type of star (number of prongs)	Geomagnetic latitude 52 to 56° N Atmospheric depth (g/cm²)						Geomagnetic latitude 27 to 29° N Atmospheric depth (g/cm²)	
	14.9	47.4	63.7	81.3	121	677	14.9	47.4
>2	2390	2030	2150	2040	1610	22.0	575	425
3–5	1030	1059	1162	1210	820	18.2	215	187
6–9	742	619	616	520	422	2.6	127	106
>9	618	352	372	235	290	1.2	238	132
>16	223	122	128	27	116	0.29	105	61

These data are for stars produced in emulsion.

All of the data so far discussed concern stars initiated by particles with energies less than about 10 Bev. Most of the primary cosmic rays have energies less than this. Therefore the collection of information about higher energy stars proceeds slowly.

A review of work up to the spring of 1954 is given in [3].

The high energy stars show the same general characteristics as the low energy ones we have already discussed. There is a narrow cone of lightly ionizing particles together with a number of heavily ionizing particles which are emitted isotropically. The higher energy stars have much greater multiplicities of product particles than do the stars with energies below 10 Bev. A star with $n_s > 200$ was reported by TEUCHER [3]. As a sample of this sort of star, the Teucher star will be analyzed in some detail.

The star was initiated by a particle with charge determined by delta ray count to be $Z = 8 \pm 1$. TEUCHER takes it to be an O^{16} nucleus. From the observed six black prongs, he concluded that the target nucleus was silver or bromine. There were also 23 gray tracks and 221 shower particles. The incident O^{16} nucleus and a silver target could supply 55 protons and 68 neutrons. After deducting the number of nucleons among the black tracks, one finds about 20 protons available as direct contributions to the shower.

Fig. 11 shows the angular distribution of the shower particles with respect to the direction of the incident nucleus. Notice that the vertical scale is logarithmic. About half of the particles lie in a cone with half angle of 10°. TEUCHER estimates the kinetic energy of the incident nucleus to be about 70 Bev/nucleon.

In the same series of papers [3], LAL, PAL and RAMA report work that they have done on the ratio of number of π^0 mesons to n_s, and compare their work

---

[1] J. J. LORD: Phys. Rev. **81**, 901 (1951).

with that of others at various energies. They find that for incident energies between 50 Bev/nucleon and 50000 Bev/nucleon this ratio is $n_{\pi^0}/n_s = 0.40 \pm 0.04$ and independent of energy within the statistics.

DANIEL et al.[1] found that for stars with energies greater than 10 Bev about 10% of the shower particles are protons. They also obtain a value $n_{k^\pm}/n_{\pi^\pm} = 0.6 \pm 0.5$. LAL et al. [3] calculate a value for this ratio of 0.25. They regard this as an upper limit.

LAL et al. also agree with the Bristol group that the nuclear interaction cross section of secondary particles from stars is geometric.

6. Fragmentation of multiply charged primary nuclei.

When a nuclear interaction is initiated by a multiply charged primary nucleus, a new effect occurs in addition to the events characterizing proton induced stars. A closely collimated beam of secondary particles is produced from the breakup of the incident nucleus. We saw in the previous section that stars produced by incident protons and neutrons have, among their relativistic secondaries, mostly protons, neutrons, and π mesons. The relativistic jet from a star induced by a multiply charged particle commonly includes multiply charge fragments. This is apparently the only mechanism for production in appreciable numbers of relativistic multiply charged nuclei.

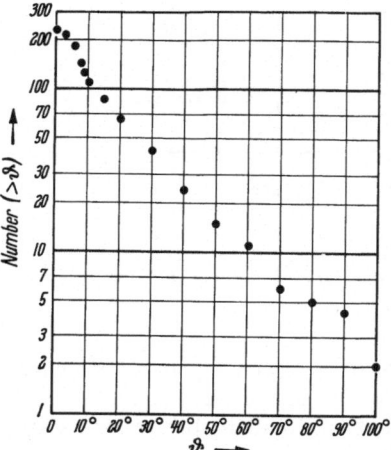

Fig. 11. The angular distribution of shower particles in a very high energy jet. (After TEUCHER [3].)

Fragmentation is of considerable importance in interpreting the measurements of intensities of multiply charged components near the top of the atmosphere. Because of the low intensities of these components in the primary beam adequate statistics are available only from balloon flights. The nuclear interaction mean free paths in air for the various components lie in the range from 20 g/cm² to 50 g/cm². Balloon data are collected at atmospheric depths which are all reasonably large fractions of these mean free paths. Accordingly, in order to obtain values for primary intensities, one must make appreciable corrections to the data for adulteration and depletion of the beam by fragmentation. Secondly, of course, the altitude dependence of the intensity of multiply charged nuclei is of interest for its own sake, and fragmentation must be taken into account in understanding this.

There does not seem yet to be a consensus regarding the values of the so-called fragmentation probabilities. The chief published values are given in Table 4. These were obtained by NOON and KAPLON[2].

The heavy group refers to initiating particles with $Z > 10$. The medium group has $6 \leq Z \leq 10$. The entries "41°" and "55°" refer to the geomagnetic latitudes at which the data were obtained. H events are those for which $N_h \geq 4$; L events are those for which $N_h \leq 4$ (see Sect. 5). This division is made with the view that the events with few black and gray evaporation tracks largely involve targets of small atomic number, while those with many of these tracks involve

[1] R. R. DANIEL, J. H. DAVIES, J. H. MULVEY and D. H. PERKINS: Phil. Mag. **43**, 753 (1952).
[2] J. H. NOON and M. F. KAPLON: Phys. Rev. **97**, 769 (1955).

Table 4. *Fragmentation probabilities.* (Noon and Kaplon[1].)

(a) Medium group

	P_{MM}	P_{ML}	$P_{M\alpha}$	P_{Mp}
H events				
41°	0.03 ± 0.03	0.22 ± 0.08	1.17 ± 0.19	3.75 ± 0.35
55°	0.07 ± 0.05	0.13 ± 0.07	1.0 ± 0.18	4.13 ± 0.38
L events				
41°	0.18 ± 0.08	0.54 ± 0.13	1.0 ± 0.2	1.78 ± 0.25
55°		0.52 ± 0.13	1.9 ± 0.26	1.69 ± 0.24
L + H events				
41°	0.09 ± 0.04	0.36 ± 0.07	1.09 ± 0.13	2.89 ± 0.19
55°	0.03 ± 0.02	0.32 ± 0.07	1.44 ± 0.16	2.94 ± 0.22
Hydrogen events				
41°		0.83 ± 0.37	1.16 ± 0.46	1.0 ± 0.41
55°		0.57 ± 0.28	1.71 ± 0.49	1.0 ± 0.37

(b) Heavy group

	P_{HH}	P_{HM}	P_{HL}	$P_{H\alpha}$	P_{Hp}
H events					
41°	0.15 ± 0.08	0.23 ± 0.09	0.5 ± 0.14	2.04 ± 0.28	5.2 ± 0.4
55°	0.19 ± 0.1	0.14 ± 0.08	0.19 ± 0.1	2.8 ± 0.37	5.58 ± 0.5
L events					
41°	0.3 ± 0.12	0.2 ± 0.1	0.55 ± 0.17	1.8 ± 0.3	3.95 ± 0.44
55°	0.26 ± 0.18	0.37 ± 0.14	0.37 ± 0.14	2.05 ± 0.34	3.05 ± 0.4
L + H events					
41°	0.22 ± 0.05	0.22 ± 0.05	0.52 ± 0.1	1.93 ± 0.21	4.65 ± 0.31
55°	0.23 ± 0.08	0.25 ± 0.08	0.28 ± 0.08	2.45 ± 0.25	4.37 ± 0.33
Hydrogen events					
41°	0.2 ± 0.2	0.4 ± 0.28	0.8 ± 0.4	1.8 ± 0.6	5.0 ± 1.0
55°		1.0 ± 0.71		1.0 ± 0.71	3.0 ± 1.2

targets of large atomic number. Noon and Kaplon conclude that about 20% of the events involving heavy target nuclei lie in the class $N_h \leq 4$. *Hydrogen events* are events in which a proton track is observed at a large angle to the jet of fragments. They assume that in these cases the target was a proton from hydrogen in the emulsion and that the observed track is the recoil proton. The quantity P_{IJ} is the average number of J type nuclei produced by an incident I type nucleus where at the same time N_h has a value in the range indicated in the left hand column of the table. The nuclear types are denoted by H, M, and L (for heavy, medium, and light). Here, L corresponds to $3 \leq Z \leq 5$, M corresponds to $6 \leq Z \leq 10$, and H corresponds to $Z > 10$.

A significantly different set of values has recently been obtained by P. Fowler[2]. He and Noon and Kaplon agree on the probabilities of fragmentation to carbon and above. Their disagreement is strongest for fragmentation to Li, Be, B. Fowler's results for fragmentation to elements lower than carbon are all smaller than the fragmentation probabilities obtained by Noon and Kaplon. The two sets of data agree concerning interaction cross sections. Fowler's data were taken from an emulsion stack flown in Italy. Measurements of α particle intensities in Italy and Texas show these two places to be equivalent from the point of view of the latitude effect in the primary intensity. Fowler attributes the disagree-

---

[1] See footnote 1, p. 146.
[2] P. Fowler: Phil. Mag., March **1957**.

ment to the methods of charge identification used. FOWLER used tracks only when they had lengths of 3 mm or more in one emulsion. NOON and KAPLON accepted tracks down to 0.75 mm in two of their stacks and 0.45 mm in the third. FOWLER believes that these short tracks accepted by NOON and KAPLON caused them to make serious errors in charge determination for the light elements. On the other hand, the discrepancy could also be resolved if one assumed that FOWLER had missed tracks of light particles actually present.

HÄNNI[1] has also obtained some fragmentation probabilities at a latitude of about 40°. For the most part, his results agree both with NOON and KAPLON and with FOWLER, since his statistical errors are rather large. For fragmentation to light elements, his results agree statistically with NOON and KAPLON and not with FOWLER. In all cases of fragmentation to α particles, his results agree with either.

No data exist which are adequate to resolve the discrepancy. For the present paper we adopt the data of NOON and KAPLON since these are the ones used by authors the work of whom we discuss in later sections. The resolution of the present discrepancy must await further work.

Table 5. *Fragmentation probabilities in air and light absorbers.*

	Air	Emulsion-L	Light absorber
P_{HH}	0.25	0.28	0.16 ± 0.16
P_{HM}	0.27	0.24	0
P_{HL}	0.48	0.45	0.83 ± 0.37
$P_{H\alpha}$	2.07	1.9	1.66 ± 0.55
P_{Hp}	4.29	3.45	5.5 ± 2.2
P_{MM}	0.13	0.18	0.06 ± 0.06
P_{ML}	0.42	0.53	0.35 ± 0.14
$P_{M\alpha}$	1.42	1.40	1.59 ± 0.39
P_{Mp}	1.70	1.73	2.7 ± 0.66

In addition to the data used in obtaining Table 4, NOON and KAPLON report some interesting curiosities. In six cases they observed a heavy incident nucleus to fragment without other visible tracks being produced, that is, with $N_h = n_s = 0$. Their calculations show that about one of these can have been due to Coulomb interaction. They attribute the rest to interaction with an extended neutron of either the target or incident nucleus.

Whether or not the values given in Table 4 are correct for interactions in emulsion, it is evidently necessary to make corrections before trying to apply them to interactions in air. NOON and KAPLON estimate this correction from a crude geometric picture of the nucleus together with the assumption that glancing collisions, by exciting the incident nucleus only weakly, break it into a few large fragments while nearly head-on collisions, by exciting the incident nucleus a great deal, produce many small fragments. A sketch of the calculation is given in their paper. The column labelled "Air" in Table 5 gives the results of this calculation. The values are to be thought of as averages for the latitude range 41 to 55°. The column labelled "Emulsion-L" gives the average of the results for L events given in Table 4. The column labelled "Light absorber" gives the results of measurements of these probabilities made by NOON and KAPLON on nuclear interactions in sheets of gelatin and cellulose acetate which were interposed between the sheets of emulsion on some of their flights.

The statistical errors are large for events in the light absorber. Nevertheless, it is evident that there is general agreement among the various sets of numbers.

NOON and KAPLON do not give data regarding the angular distribution of fragments. The fragments are closely collimated about the extension of the motion of the incident nucleus. It is common practice in correcting heavy particle intensities to the top of the atmosphere to assume that the collimation is perfect.

[1] F. HÄNNI: Helv. phys. Act a **29**, 281 (1956).

7. The absorption of primary particles. In addition to their role in producing secondary particles, nuclear interactions also attenuate the incident beam by absorbing primary particles. Rossi discusses this matter in some detail [1].

As one comes down from near the top of the atmosphere, one finds that the intensity of protons, neutrons, and π mesons decreases roughly exponentially with an absorption length of about 120 g/cm². Since, in addition to the absorption of these particles by nuclear events, new particles of the same type are created, the mean free path for nuclear collision is evidently less than this. If one assumes that the cross section is not greater than geometric (the evidence indicates that is roughly geometric, see Sect. 5) then the mean free path is not less than 68 g/cm².

At the same time that nuclear interactions are taking place, electromagnetic interactions slow down all of the charged particles. Near the top of the atmosphere, where the mean energy of the particles is high, these are of little importance. Rossi states that ionization loss decreases the effect of reproduction somewhat and brings the absorption length slightly nearer to the collision mean free path than it otherwise would be. We shall neglect this effect.

Table 6. *Collision mean free paths* (Noon and Kaplon [1]).

Latitude	Emulsion	Air (calculated)
41°	$\lambda_M = (59.4 \pm 7.8)\,\text{g/cm}^2$ $\lambda_H = (34.7 \pm 5.6)\,\text{g/cm}^2$	
55°	$\lambda_M = (60.4 \pm 10.3)\,\text{g/cm}^2$ $\lambda_H = (45.5 \pm 10)\,\text{g/cm}^2$	
Average	$\lambda_L = (61.7 \pm 19.4)\,\text{g/cm}^2$ $\lambda_M = (59.6 \pm 6)\,\text{g/cm}^2$ $\lambda_H = (36.5 \pm 4.8)\,\text{g/cm}^2$	$\lambda_L = 31.5\,\text{g/cm}^2$ $\lambda_M = 26.5\,\text{g/cm}^2$ $\lambda_H = 18.0\,\text{g/cm}^2$

Noon and Kaplon [1] report the results of some fairly recent measurements of mean free paths for collision in various materials and for various kinds of particles. Table 6 summarizes their results. The values for air are the result of a correction applied to the emulsion data. The same crude model for a nuclear reaction was used as for the corresponding correction which they applied to the fragmentation probabilities (Sect. 6). In the table, λ_X is the collision mean free path of the X component. The subscripts specify the range in atomic number. For L, $3 \leq Z \leq 5$; for M, $6 \leq Z \leq 10$; for H, $Z > 10$.

Webber has obtained a value for λ_α, the collision mean free path for α particles, using a Čerenkov detector [2]. His work was done at San Angelo, Texas. The geomagnetic latitude is 41.5°. The balloon rose slowly to its peak altitude of 18.5 g/cm², so that good statistics were available for the vertical intensity of α particles as a function of atmospheric depth. Noon and Kaplon give a theoretical treatment which relates the vertical intensity of a given component to the collision mean free paths and fragmentation probabilities. Upon using the values given by Noon and Kaplon, Webber had two undetermined parameters, one of which was the path length λ_α. A least squares fit to the altitude dependence of his data gave $\lambda_\alpha = (46.5 \pm 3.0)\,\text{g/cm}^2$. The error is that due to the errors in the other parameters. The absorption mean free path for α particles as measured in this experiment was 58 g/cm².

Noon and Kaplon give the following simple treatment of the absorption of nuclear radiation in the atmosphere. They assume the radiation to be a parallel beam. They also assume that at each fragmentation reaction the products travel in the same direction as the incident nucleus and have the same energy per nucleon as does the incident particle. They neglect ionization. They obtain the

---

[1] See footnote 1, p. 146.
[2] W. R. Webber: Nuovo Cim. **4**, 1285 (1956).

equation

$$\frac{dj_I(x)}{dx} = -\frac{1}{\lambda_I} j_I(x) + \sum_{I' \geq I} j_{I'}(x) P_{I'I}/\lambda_I \Bigg\}$$

with

$$I, I' = \alpha, L, M, H.$$

(7.1)

The solutions for this equation for $j_H(x)$ and $j_M(x)$ are

$$j_H(x) = j_H^0 \exp(-x/\lambda_H'),$$
$$j_M(x) = j_M^0 \exp(-x/\lambda_M') + $$
$$+ (\alpha_{HM} P_{HM}/\lambda_H)[j_H^0 \exp(-x/\lambda_M') - j_H(x)].$$

(7.2)

The solutions for $j_L(x)$ and $j_\alpha(x)$ are quite complex. (See NOON and KAPLON, and WEBBER, respectively.) In the above equations, j is the directional intensity, $x = l \sec \vartheta$ is the thickness of material traversed. The atmospheric depth is l and ϑ is the zenith angle at which the directional intensity is measured. The other quantities are

$$\alpha_{IJ} = \lambda_I' \lambda_J' (\lambda_J' - \lambda_I')^{-1} > 0,$$

and

$$\lambda_I' = \lambda_I (1 - P_{II})^{-1}.$$

The heavy group is simply absorbed exponentially. The medium group is depleted exponentially by the first term on the right and enhanced by fragmentation of the heavy group according to the second term on the right.

III. The intensities of the components.

8. Multiply charged nuclei. Recently good progress has been made in measuring the vertical intensity of α particles. This has come about largely through the use of Čerenkov counter telescopes. Most of the effort has been expended in order to learn the values of this intensity above the atmosphere at various latitudes. The α particle intensity is of particular interest in connection with geomagnetic effects since, as we saw in Sects. 5 and 6, secondary α particles are formed essentially only in fragmentation processes, and a successful correction for these secondaries has been rather easy because the fragments are emitted in such a narrow cone in the forward direction. At the same time, the α particle intensities are sufficiently high that it is not difficult to get good statistical accuracy.

In the course of learning to correct their balloon flight data to the top of the atmosphere, the workers in this area have also achieved a good understanding of the intensity of α particles as a function of atmospheric depth. Sects. 6 and 7 review the material which contributes to such an understanding.

Fig. 12 shows the altitude dependence of the vertical intensity of α particles at San Angelo, Texas, according to WEBBER[1]. The two curves shown exhibit his quantitative interpretation of the data. These curves are based on the fragmentation theory of NOON and KAPLON, together with their values of the fragmentation parameters.

The first step in constructing these curves was to fit the solution for α particles of Eq. (7.1) to the data. In doing this, values more or less well known from other sources were used for all of the parameters except j_α^0 and λ_α'. The fit was used to determine these. An estimate was made of $P_{\alpha\alpha}$, and this made possible the determination of an estimated value of λ_α'. The curve which goes through the experimental points is the fit of the fragmentation theory to the data. The

---

[1] See footnote 2, p. 149.

straight line is the exponential absorption which would presumably occur if their were no heavier nuclei in the primary cosmic radiation. The difference between the curves represents the production of α particles by the fragmentation of heavier components. The α intensity decreases rather rapidly with altitude, and becomes a quite insignificant contribution to the total intensity below 100 g/cm³ or so.

The vertical intensities of the other multiply charged components are less well known. The intensities are so low that no statistically significant intensity-altitude curve has yet been measured. WEBBER obtained a value for the vertical intensity of particles with charge of 3, 4 or 5. His value of $j(0) = (3.11 \pm 0.33)$ m⁻² sec⁻¹ steradian⁻¹ was obtained in a constant altitude Skyhook balloon flight at San Angelo, Texas. The atmospheric depth was 18.5 g/cm². By using values from other sources for all of the fragmentation parameters except j_L^0 he

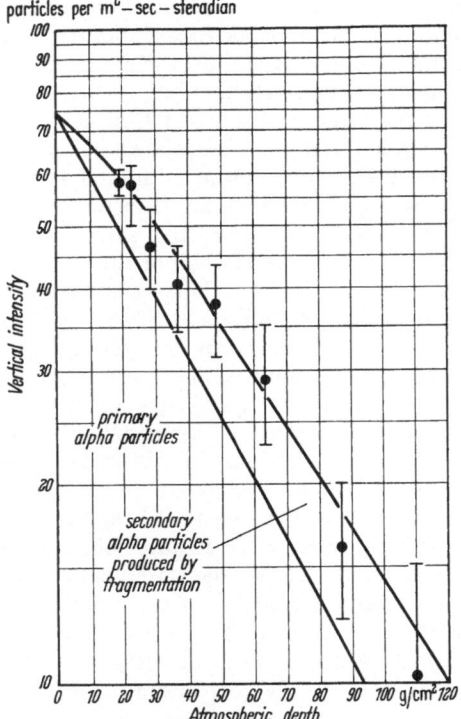

Fig. 12. Vertical intensity of alpha particles as a function of atmospheric depth. (After WEBBER [1].)

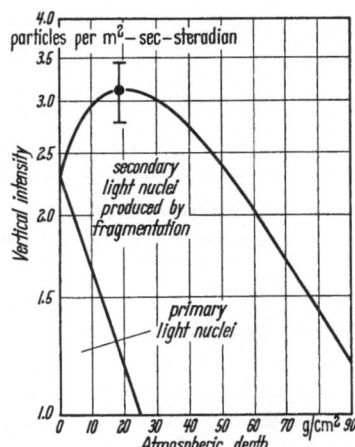

Fig. 13. Vertical intensity of light nuclei $(3 \leq Z \leq 5)$ as a function of atmospheric depth. (After WEBBER [1].)

obtained $j_L^0 = (2.28 \pm 0.55)$ m⁻² sec⁻¹ steradian⁻¹ for the value above the atmosphere. Fig. 13 shows the result for L particles in the same fashion that Fig. 12 does for α particles.

In the same fashion, WEBBER obtained at 18.5 g/cm² the value $j(Z \geq 6) = (5.12 \pm 0.60)$ m⁻² sec⁻¹ steradian⁻¹ and for the estimated value of $j_{Z \geq 6}^0$, (9.2 ± 1.2) m⁻² sec⁻¹ steradian⁻¹. Using the sum of the two Eqs. (7.2), one can easily construct the altitude curve according to the fragmentation theory of NOON and KAPLON.

9. Protons and π mesons. The separate measurement of the intensities of the various singly charged components within the atmosphere presents difficulties which have not yet been completely overcome. Our current knowledge of the altitude dependence of these intensities depends largely on theoretical considerations. The available data are the intensity of the hard and soft components

---

[1] See footnote 2, p. 149.

as a function of altitude, the rate of star formation as a function of altitude, reasonably good values for the cross sections for nuclear interaction of the various constituents of the atmospheric beam, the lifetimes of the unstable particles, and reasonably good values of proton intensity near the top of the atmosphere and of the intensities of μ mesons, protons, and electrons at sea level. In addition there is a fairly good value of the intensity of fast μ mesons as a function of altitude. The next few sections use these data in an effort to determine, as nearly as possible, the intensities of the various components within the atmosphere.

The general approach to this problem was first given by ROSSI [4]. Several treatments have been carried out in connection with the problem of the energy balance [5]. All of these treatments amount, with more or less elaboration, to assuming that the proton intensity for all protons with energies greater than some large fraction of 1 Bev decreases exponentially as atmospheric depth increases. It is usually assumed that these protons have an absorption path of about 120 g/cm², the same as the absorption path for all of the star producing radiation. This assumption is apparently good for sufficiently high energy protons. For low energy protons, where ionization loss is important, and where star production ceases to become important, one expects a different result. MESSEL has obtained some solutions of the problem of the nucleonic cascade which essentially agree with this picture [5, III]. In addition, CLARK[1] has made some measurements from which he obtained an analysis of the high energy charged particle intensity in the atmosphere. He finds agreement with his results if he assumes that the protons with energy above 1 Bev have an absorption length of 110 g/cm². His data were obtained at Lexington, Massachusetts (55° N). If one assumes that the intensity of the hard component at the top of the atmosphere consists only of protons and α particles, one obtains then the intensity of fast protons as a function of atmospheric depth. If this result is subtracted from the charged star producing radiation, the intensity of protons with energy less than 1 Bev is obtained. In this process, one neglects the contribution from π mesons, since their lifetime is too short for them to be able to contribute appreciably to the measured intensities. An appreciable correction must be made for α particles at small atmospheric depths. Fig. 14 gives the result for counting rate due to protons as found by CLARK. It is difficult to reduce this to an absolute intensity with any great confidence, because while he reports the active dimensions of his counters (length 13.7 cm, diameter 2.41 cm) as well as their separation (22.2 cm) many of his events involve nuclear interactions in a 20 cm thick block of lead which he placed between the counters. The change in direction which particles suffer at nuclear interactions then presumably materially alters the geometric factor. If one assumes that the geometric factor is independent of altitude, one can estimate absolute intensities. In typical cases this amounts to neglecting

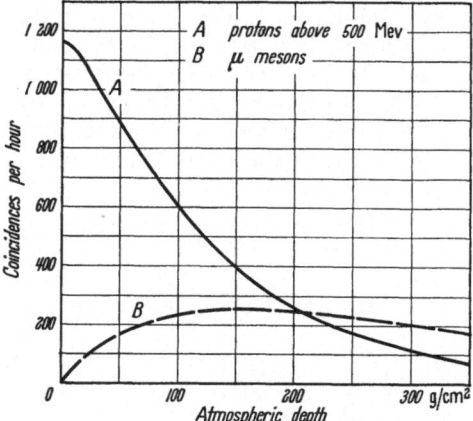

Fig. 14. The dependence of relative vertical intensities of protons and of fast μ mesons on atmospheric depth. (After CLARK[1].)

[1] M. A. CLARK: Phys. Rev. 87, 87 (1952).

variations of the order of 10%. CLARK uses a value of $j(0) = 0.19\ \mathrm{cm^{-2}\,sec^{-1}\,sterad^{-1}}$ for the directional intensity of protons at the top of the atmosphere at Lexington, Massachusetts.

These results are in essential agreement with those obtained by other considerations [4], [5]. The errors are of course large and difficult to estimate. The errors are greater than 20%, but the intensities are probably incorrect by a factor considerably less than 2.

10. μ mesons. Except very near the top of the atmosphere, μ mesons are the only charged particles of appreciable intensity, besides protons, which can penetrate several centimeters of lead. Once one has determined the α particle intensity near the top of the atmosphere, the remaining penetrating particles which in addition do not make nuclear reactions must then be μ mesons. CLARK[1] has made a determination of fast μ meson intensities using this technique. His results are shown in Fig. 14. The contribution of slow mesons is apparently much less than the errors which must be assigned to the values for fast mesons. We shall neglect this contribution.

The values obtained by CLARK are in fair agreement with those obtained from somewhat different considerations by DYMOND [5, II]. The agreement is excellent between atmospheric depths of about 225 g/cm³ and 175 g/cm². At smaller depths, DYMOND's value falls below CLARK's, so that at a depth of 50 g/cm², CLARK's value is about one and one half times as large as DYMOND's. The discrepancy then decreases again as one continues toward the top of the atmosphere. DYMOND obtained his estimates by subtracting a theoretical intensity-depth curve for protons due to MESSEL [5, II] from the measured hard component. The disagreement seems simply to reflect the general uncertainties in our picture of cosmic rays in the atmosphere, but it is encouraging that the discrepancy is no larger.

The energy spectrum of μ mesons has been very well measured at sea level and about 50° N latitude. By using a one-dimensional diffusion theory, together with the vertical intensity as a function of atmospheric depth, one can then construct a μ meson generation function as a function of atmospheric depth. This work is discussed by PUPPI [5, III].

11. Neutrons. Properties of neutrons within the atmosphere below an atmospheric depth of about 200 g/cm² have been studied extensively by SIMPSON[2]. His data were obtained with boron trifluoride proportional counters enriched in B^{10}. For the measurement of slow neutrons he used two such counters, one surrounded by cadmium, the other surrounded by an equivalent amount of brass. Fast neutrons are counted in the same way by both counters. The brass surrounded counter also is sensitive to slow neutrons, but the counter with cadmium shielding does not detect neutrons with energies below about 0.4 ev. He measured the fast neutron counting rate using the same equipment except that the two counters were surrounded in addition with paraffin blocks. The paraffin moderated the fast neutrons so that they could be counted with reasonable efficiency. The cadmium surrounded counter served to monitor the background. This equipment was flown to atmospheric depths of about 220 g/cm² in an airplane. The results of the fast neutron intensity measurements do not yield absolute intensities, since the geometric factor is unknown.

Table 7 gives the measured absorption mean free paths for fast neutrons in air. These values fit data in the range of atmospheric depths 200 g/cm³ to

---

[1] See footnote 1, p. 152.
[2] J. A. SIMPSON: Phys. Rev. **83**, 1175 (1951).

600 g/cm². At greater altitudes, one expects that the neutron intensity must show a maximum. YUAN[1] finds such a maximum at an atmospheric depth of about 100 g/cm² at about 52° N. These neutrons are produced in stars. Since most of the stars are themselves produced by protons or neutrons, these neutrons then are part of the nucleonic cascade in the atmosphere.

SIMPSON has also studied the relative neutron intensity as a function of latitude and longitude. At an atmospheric depth of 306 g/cm² he found the counting rate of his fast neutron detector above the latitude knee to be about 4.4 times the counting rate at the minimum near the geomagnetic equator.

Table 7. *Absorption mean free paths for neutrons* (SIMPSON[2]).

Geomagnetic latitude (°)	Mean free path (g/cm²)
0	212 ± 4
19	206 ± 4
40	181 ± 3
53	157 ± 2
65	157 ± 3

SIMPSON extrapolated his data to the top of the atmosphere, using the data of YUAN as a guide. He then assumed that the neutrons produced a given region of the atmosphere migrate a distance before absorption which is negligible compared with the absorption mean free path of the star-producing radiation. From this he concluded that the difference between the areas under curves at two different latitudes was proportional to the number of neutrons produced per cm³ per Bev energy interval by incoming protons with energies in the range between the cut-off energies for arrival at the two different latitudes. These neutrons of course include those produced at advanced stages in the nucleonic cascade as well as the immediate secondary neutrons. The results of these calculations are shown in Table 8.

Table 8. *Neutron production in an atmospheric column* (SIMPSON[2]).

Primary energy range (Bev)	Relative neutron production per cm²-atm per primary energy range
> 14	10
10.5 — 14	18
4.3 — 10.5	50
1.3 — 4.3	84
0.4 — 1.3	small

The latitude variation of the production of ionization is not nearly as great as this. While the relative ionization production per unit energy range of primaries shows a peak in the region 1.6 to 4.9 Bev of primary energy, this peak is only about 1.8 times the value in the energy range above 16 Bev. The corresponding ratio for neutrons, taken from the table above, is about 8.

During the course of the above work, SIMPSON found a large fractional time dependence of the neutron intensity which occurred while relatively small or no changes were occurring in charged particle detectors within the atmosphere. This effect was particularly marked at high latitudes. From this result together with consideration of Table 8, he concluded that neutron counters are more sensitive to changes in primary intensities occurring mostly for low energy primaries. He has since exploited this property of the nucleonic radiation in the atmosphere in studying time variations of the low energy primary radiation. The emphasis of this work has been on the use of low energy cosmic radiation as a tool for probing the electromagnetic environment of the earth and the sun.

12. Electrons and photons. Electrons and photons are the only remaining particles present with appreciable intensity in the atmosphere. Bremsstrahlung and pair production rapidly degrade the energy of this radiation as it passes through matter. As a result, several centimeters of lead are able to filter it out

---

[1] L. C. L. YUAN: Phys. Rev. **81**, 175 (1951).
[2] See footnote 2, p. 153.

of the cosmic ray beam. The hard component, then, has no contribution from electrons and photons. On the other hand, it has been shown that slow mesons and protons comprise only 10 or 20% of the soft component. No other radiation is present in this component to an appreciable extent. Consequently, one can subtract even relatively poorly known values of slow meson and slow proton intensities from the measured soft component, and obtain a reasonably accurate value for the sum of the intensities of electrons and photons. The discussions in Sects. 1 and 2 concerning the intensities measured with very thin walled counters must be taken as something of a qualification in the present connection. It seems most reasonable to ascribe this very soft radiation to electrons and photons, although there is no direct experimental evidence for this. If one does make this assumption, one is forced to conclude that the intensities of very soft electrons and photons are quite large and that the measured values depend strongly on the thickness of the walls of the counters with which the measurements are made.

The separation of the soft component into electron intensity and photon intensity apparently cannot be made satisfactorily. There is little evidence about the intensity of photons within the atmosphere. The efficiency of a Geiger tube for counting photons is 2% or less. A single counter isolated from other matter will then count mostly only electrons unless the true photon intensity is several times as large as the true electron intensity. A Geiger counter telescope will count essentially no instances of a photon traversing the telescope and being detected separately by each

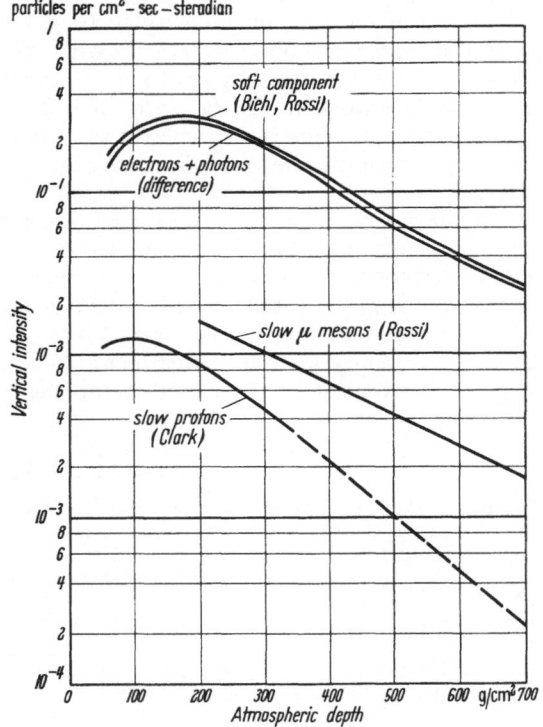

Fig. 15. The analysis of the soft component.

counter. However, any photon, regardless of its direction of travel, may produce a Compton electron which traverses all of the counters in the telescope. As a result, even a telescope may be able to detect photons with appreciable efficiency.

If one assumes that true photon intensities in the atmosphere are not more than two or three times the electron intensities, one can reasonably neglect the contributions of photons to the soft component. This is the course usually adopted.

Fig. 15 gives an analysis of the soft component according to the above considerations. The values for the intensity of the soft component as given by BIEHL et al. were used (see Sect. 1). The values given for slow proton and slow meson intensities are not very accurate. Obviously this makes only a little difference in the accuracy of the intensity of electrons and photons.

Summary and conclusion.

We now have reviewed the available data which bear on the conversion of primaries into mesons and electrons. From a look at the altitude dependence of the various particle intensities as given in Chap. III, it is evident that most of the conversion occurs at atmospheric depths less than about 300 g/cm². Star formation furnishes the link between the primary nucleonic radiation and the μ mesons and electrons which comprise most of the radiation at great depths. In order to completely understand the total intensity at small depths, one must also take account of fragmentation which rapidly degrades the charge spectrum of incoming primaries. We have not devoted any attention to the heavy mesons. These are of too low an intensity to be detected by the usual charged particle detectors. They are observed chiefly with the emulsion technique. Nevertheless, because in their decay they emit neutrinos which carry away appreciable energy in an undetectable form, they are, in principle at least, of importance for the energy balance. This is particularly so near the equator where the mean energy of cosmic ray primaries is high. This point has been discussed by Komori[1] and Benzi[2].

While we have a good qualitative understanding of the processes occurring in the high atmosphere and of their effects on the relationships among the various particle intensities as a function of atmospheric depth, our quantitative information is in some respects rather crude. A better knowledge of the vertical intensities of protons and either photons or electrons would go far toward completing this picture. A really complete treatment would require also a knowledge of the angular distribution of the various components.

General references.

[1] Rossi, B.: High Energy Particles. New York 1952. — This excellent book covers, up to the middle of 1951, the tools, both experimental and theoretical, of use in investigating the properties of cosmic radiation in the atmosphere. It also reports the important results of measurements in the atmosphere and gathers them into a coherent picture.

[2a] Brown, Camerini, Fowler, Heitler, King and Powell: Phil. Mag. 40, 862 (1949).

[2b] Camerini, Coor, Davies, Fowler, Lock, Muirhead and Tobin: Phil. Mag. 40, 1073 (1949).

[2c] Fowler, P. H.: Phil. Mag. 41, 169 (1950).

[2d] Camerini, Fowler, Lock and Muirhead: Phil. Mag. 41, 413 (1950).

[2e] Carlson, Hooper and King: Phil. Mag. 41, 701 (1950).

[2f] Camerini, Davies, Fowler, Franzinetti, Muirhead, Lock, Perkins and Yekutieli: Phil. Mag. 42, 1241 (1951).

[2g] Camerini, Davies, Franzinetti, Lock, Perkins and Yekutieli: Phil. Mag. 42, 1261 (1951). — These papers, published by the Bristol group, are the foundation of our understanding of nuclear events in the energy range from 1 Bev to about 10 Bev.

[3] Nuovo Cim. 12, Suppl. 2, 338—363 (1954). — This supplement is devoted to high energy events. A great deal of the material discusses heavy mesons and hyperons. The pages noted are concerned with jets.

[4] Rossi, B.: Rev. Mod. Phys. 20, 537 (1948). — This is a general review article on the status of the understanding of cosmic ray phenomena in the atmosphere as of 1948. While much of it is supplanted by more recent reviews, it is still a good short outline of atmospheric phenomena.

[5] Wilson, J. G.: Progress in Cosmic Ray Physics. Vol. I, II, III. Amsterdam and New York 1952, 1954, 1956. — These excellent review volumes cover almost all aspects of cosmic ray physics. Each volume consists of a number of chapters each contributed by an expert on the subject of his chapter.

---

[1] H. Komori: Progr. Theor. Phys. 13, 205 (1955).
[2] V. Benzi: Nuovo Cim. 11, 686 (1954).

Penetrating Showers.

By

KURT SITTE.

With 22 Figures.

I. Introduction.

1. Discovery and classification. The occurrence of groups of associated penetrating particles, with a frequency well in excess of that expected for random incidence, has been known for a long time and has repeatedly been commented upon. Thus, for instance, the "explosive showers" seen in early work with multiple-plate cloud chambers (e.g. FUSSELL[1], POWELL[2]) represented a phenomenon evidently different from the already well-known electronic cascades, and caught the eager interest of the theoreticians (EULER and HEISENBERG[3]). Underground cloud chamber studies (BRADDICK and HENSBY[4]) gave additional proof for the existence of showers of hard particles, possibly originating from a single interaction. Typical photographs, taken from the work of FUSSELL, and of BRADDICK and HENSBY, are reproduced in Fig. 1 and 2. But a systematic investigation of the phenomenon started only after WATAGHIN's pioneer experiments[5] had demonstrated the efficacy of counter techniques in this field, and the analysis of AUGER[6] and of JÁNOSSY[7] had established beyond doubt that these "penetrating showers" are indeed a new cosmic-ray phenomenon distinct from the cascades of electrons and photons.

However, even these systematic efforts were not immediately rewarded by success in our understanding of the effect, and the results of the early experiments were full of contradictions and disagreements. The most urgent problem to be solved was that of identification of the primaries of the penetrating showers, and for this purpose, the investigation of the absorption, or of the transition effect, of the shower-producing radiation was the obvious procedure. But even in successive studies of the effect by the same authors with slightly different apparatus, nothing better than qualitative agreement could be achieved. Thus JÁNOSSY and INGLEBY[8], and JÁNOSSY[7], found an approximate saturation of the shower rate under 5 cm Pb, and from a comparison of the transition effect in Pb and Al concluded that the cross section for shower production varies approximately as Z^2. But in the later work of JÁNOSSY and ROCHESTER[9], saturation

---

[1] L. FUSSELL: Phys. Rev. **51**, 1005 (1936); see also J. C. STREET: J. Franklin. Inst. **227**, 765 (1939).

[2] W. M. POWELL: Phys. Rev. **60**, 413 (1941).

[3] H. EULER and W. HEISENBERG: Ergebn. exakt. Naturw. **17**, 1 (1938).

[4] H. J. J. BRADDICK and G. S. HENSBY: Nature, London **144**, 1012 (1939).

[5] G. WATAGHIN, M. DAMY de SOUZA SANTOS and P. A. POMPEIA: Phys. Rev. **57**, 61, 339 (1940).

[6] P. AUGER: Unpublished (quoted by L. JANOSSY: Cosmic Radiation, p. 350. 1948).

[7] L. JÁNOSSY: Proc. Roy. Soc. Lond., Ser. A **179**, 361 (1942).

[8] L. JÁNOSSY and P. INGLEBY: Nature, Lond. **145**, 511 (1940).

[9] L. JÁNOSSY and G. D. ROCHESTER: Proc. Roy. Soc. Lond., Ser. A **183**, 181 (1944).

was found only at a much larger thickness. A clue for the reason of the discrepancy is already contained in the empirical formula by which the latter

Fig. 1. An "explosive shower" observed in a cloud chamber (experiment by L. Fussell, reported by J. C. Street).

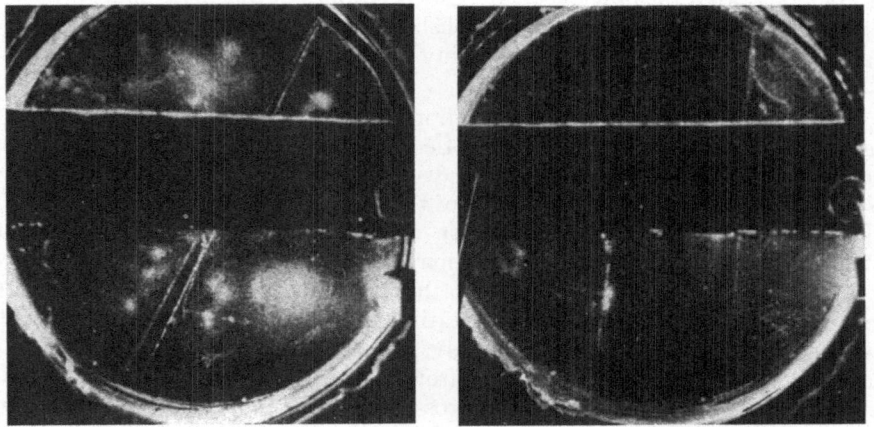

Fig. 2. A group of penetrating particles observed in an underground cloud chamber (experiment by H. J. J. Braddick).

authors represent their results for the rate R of penetrating showers as a function of the absorber thickness x:

$$R = 0.19 + 0.29 \, (1 - e^{-\mu x}). \tag{1.1}$$

The apparatus of Jánossy and Rochester, schematically shown in Fig. 3, can be triggered both by groups of penetrating particles incident from the atmosphere, and by showers of penetrating particles locally produced in the absorber T.

It is, therefore, tempting to interpret the two terms of Eq. (1.1) as the contributions from two components: the constant representing the effect of groups of penetrating particles incident on T, and the second expression due to events originated in T by single primaries with an interaction mean free path of $1/\mu$. The experiment gave a value for $1/\mu \approx 50$ g/cm² Pb.

Following up this idea, JÁNOSSY and BROADBENT[1] added to the experimental arrangement an "extension tray" (tray E in Fig. 3). This tray could be placed in distances from 0.5 to 9m from the "P-set", and in parts of the experiment was likewise covered by a lead absorber. Coincidences $(P+E)$ between tray E and the penetrating showers in the P-set, and anticoincidences $(P-E)$ were both recorded: the first evidently due to extensive showers which contained penetrating particles, the second to single particles capable of originating showers in the P-set. It was found that the transition effect of the two components was

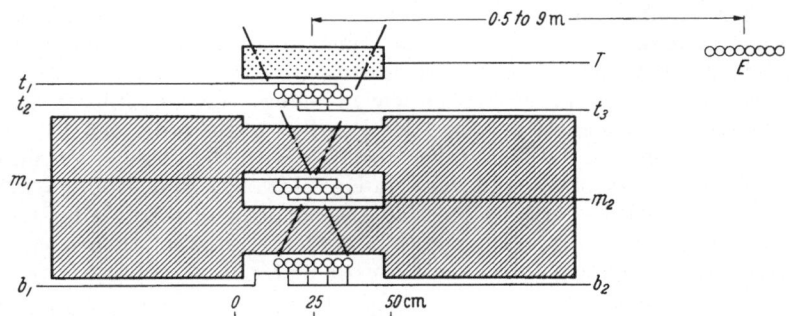

Fig. 3. The experimental arrangement used by JÁNOSSY and ROCHESTER (1944). For triggering, each of the 7 groups $A_1, A_2, A_3, B_1, B_2, C_1, C_2$ had to be struck by at least one particle.

entirely different. That of the coincidences $(P+E)$ was essentially identical with the typical "Rossi curves" of extensive air showers, rising sharply to a maximum at small thicknesses of lead, while the transition effect of local showers $(P-E)$ showed that the rate of these showers was still far from saturation under 10 cm Pb. The events hitherto called "penetrating showers" were thus demonstrated to consist of two distinct phenomena with different physical properties, denoted by the authors as "local penetrating showers" [the anticoincidences $(P-E)$] and "extensive penetrating showers" [the coincidences $(P+E)$]. From then on, studies of penetrating showers would generally be concerned with one of the components only, and following the lead of JÁNOSSY and BROADBENT, experimental arrangements could be designed which would isolate the desired effect.

We know today that the two apparently distinct phenomena are only two phases of the same effect. "Extensive" showers begin as "local" interactions, and it is only because we observe them far away from their point of origin that they have acquired their special features. In fact, we may well say that all the variety of sea-level cosmic ray components descends from "local penetrating showers" produced near the top of the atmosphere. Nevertheless, the classification introduced by JÁNOSSY and BROADBENT proved most helpful—indeed indispensable—in the further study of penetrating showers.

In the following, we shall be concerned almost exclusively with local interactions. Unless otherwise specified, the term "penetrating shower" will therefore refer to the local penetrating events of the JÁNOSSY-BROADBENT classification.

[1] D. BROADBENT and L. JÁNOSSY: Proc. Roy. Soc. Lcnd., Ser. A **190**, 497 (1947); **191**, 517 (1947); **192**, 364 (1948).

2. The nature of the primaries of penetrating showers. The apparatus of Fig. 3 requires the incidence of charged particles on the top counter tray A, and since penetrating showers are observed even with a zero absorber thickness Σ above it, it can be concluded that in this case the showers must be initiated by charged primaries. Using a variation of the same arrangement in which the top counter tray, doubled for better geometry and extended to cover the sides of T, is connected in anticoincidence with the trays B, C and D, JÁNOSSY and ROCHESTER[1] proved that penetrating showers of very similar features could also be produced by neutral primaries. In particular, it was again noted that while an absorber thickness Σ of 5 cm Pb affected the shower rate only slightly, its increase to 35 cm reduced the measured anticoincidence rate quite considerably. Both for charged and neutral primaries, the observation appeared to be consistent with a primary mean free path of the order of 5 cm Pb.

This result clearly rules out high-energy electrons and photons as the shower-originating radiation. In that case, one would expect the interaction to be of electromagnetic nature, and a mean free path of the order of 10 radiation lengths is surely incompatible with such a view. Moreover, an estimate of the absolute intensity of the shower primaries led to a flux of the order of 1/10000 of the total cosmic ray intensity. This, taken together with the observation of a barometer coefficient of about -12% per cm Hg (JÁNOSSY and ROCHESTER[2]), is quite consistent with the assumption that the particles responsible for the production of penetrating showers are nucleons, and the showers themselves are nuclear interactions taking place in the absorber material of the apparatus.

A few years later, the occurrence of showers of penetrating particles deep underground—observed already in the early work of BRADDICK and HENSBY—was repeatedly confirmed[3,4]. All these experiments were carried out at depths where the nucleonic component is for all practical purposes completely absorbed. Hence those showers must be due to μ-mesons, and the question was whether, in order to give rise to such events, these particles have to be endowed with some sort of nuclear force. It was estimated by GEORGE and EVANS[3] that a cross section of the order of some 10^{-30} cm^2/nucleon could be understood on the basis of electromagnetic interactions only: it results from the photonuclear reaction of the virtual photons into which the field of a fast-moving μ-meson can be analysed. The experimental result of $(5.3 \pm 1.7) \times 10^{-30}$ cm^2/nucleon seemed to be in good agreement with this view. For some time, however, there was considerable uncertainty whether all interactions underground which result in the ejection of penetrating particles, could be interpreted in this way. In particular, BRADDICK, NASH and WOLFENDALE[5] reported a cross section of about 5×10^{-29} cm^2/nucleon for the production of pairs of associated penetrating particles in lead. But later careful investigations[6-9] yielded consistently smaller cross sections, and better agreement with the photonuclear model of GEORGE and EVANS.

[1] L. JÁNOSSY and G. D. ROCHESTER: Proc. Roy. Soc. Lond., Ser. A **182**, 180 (1943).
[2] L. JÁNOSSY and G. D. ROCHESTER: Proc. Roy. Soc. Lond., Ser. A **183**, 186 (1944).
[3] E. P. GEORGE and J. EVANS: Proc. Phys. Soc. Lond. **63**, 1248 (1950).
[4] K. GREISEN, G. COCCONI and L. M. BOLLINGER: Phys. Rev. **82**, 294 (1950).
[5] H. J. J. BRADDICK, W. F. NASH and A. W. WOLFENDALE: Phil. Mag. (7) **13**, 1277 (1951).
[6] E. AMALDI, C. CASTAGNOLI, A. GIGLI and S. SCIUTI: Nuovo Cim. **9**, 969 (1952).
[7] V. APPEPILLAI, A. W. MAILGAVANAM and A. W. WOLFENDALE: Phil. Mag. (7) **45**, 1059 (1954).
[8] H. J. J. BRADDICK and B. LEONTIC: Phil. Mag. (7) **45**, 1287 (1954).
[9] G. CASTAGNOLI, A. GIGLI and S. SCUITI: Nuovo Cim. **10**, 893 (1953). — P. E. ARGEN, A. GIGLI and S. SCUITI: Nuovo Cim. **11**, 530 (1954).

More recently, FOWLER[1] has cast doubt on the validity of the way in which the Weizsäcker-Williams method is used in the photonuclear model. But a discussion of this argument, or of further details concerning interactions initiated by μ-mesons, does not fall into the scope of this report. We shall confine ourselves to a survey of showers produced by particles of the nucleonic component.

3. The nature of the shower secondaries. It was of course suspected from the very beginning of the work on penetrating showers, that the penetrating particles emitted in these interactions were mesons. But real progress in the identification of the shower secondaries was only made when cloud chambers controlled by shower-selecting counter arrays were applied to the problem (ROCHESTER[2]). Probably the first unambiguous example of meson production in a penetrating shower was reported by ROCHESTER, BUT-LER and RUNCORN[3] in 1947, in a photograph reproduced in Fig. 4. The meson track is marked *"m"*: its curvature, range and ionisation are consistent with those of a particle of a mass around 200 m_e. Besides, a knock-on electron visible just below the lead permits an additional check on the mass of the colliding particle, and leads also to a mass of 290 $m_e \pm 50\%$.

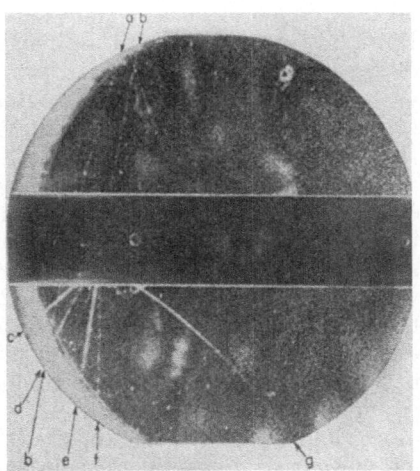

Fig. 4. An example of meson production in a penetrating shower (track "g") observed by ROCHESTER, BUTTER and RUNCORN.

A problem that seriously puzzled the investigators at the time of these early cloud chamber studies, was the comparatively frequent occurrence of large electron cascades associated with the penetrating showers, and apparently originating in the same collision. This phenomenon could only be understood when, a few years later, the existence of the π^0-meson and its decay mode were discovered.

Another drawback in those experiments was that the accuracy of the techniques of mass determination proved too low to decide whether the mesonic shower secondaries were π- or μ-mesons. This important point was cleared up by an ingenious experiment by PICCIONI[4] which made use of the qualitative difference in the behavior of π^-- and μ^--mesons stopping in light and heavy elements. π^--particles coming to rest, will in both cases interact with nuclei, while μ^--particles slowing down in light elements will decay with the emission of an electron, and only in a heavy absorber nuclear interactions will prevail.

PICCIONI's apparatus recorded delayed coincidences of a counter tray C surrounded by either carbon or sulphur, with two other counter trays A and B separated from each other by lead shields sufficient to eliminate all but penetrating particles. Coincidences were registered whenever C was discharged with a delay of the order of a microsecond after A and B. Thus, a good fraction of the μ-mesons stopping in the absorber near C would be counted. It could then be shown that for single penetrating particles, the ratio of the delay events in carbon and sulphur, per unit time and unit mass, was about 1.8—as expected

[1] G. N. FOWLER: Proc. Phys. Soc. Lond. A **68**, 482 (1955).
[2] G. D. ROCHESTER: Proc. Roy. Soc. Lond., Ser. A **178**, 464 (1946).
[3] G. D. ROCHESTER, C. C. BUTLER and S. K. RUNCORN: Nature, Lond. **159**, 227 (1947).
[4] O. PICCIONI: Phys. Rev. **77**, 1 (1950).

for μ-mesons in view of the positive excess—, while for particles originating in locally produced penetrating showers, no significant difference between the two rates was found. This is consistent with the asumption that the shower secondaries are π-mesons, since in this case, both in light and heavy absorbers, the negative particles will be absorbed in the nucleus and only the positive particles will decay into μ-mesons whose subsequent disintegration, with the well-known delay of the order of microseconds, is then registered. Thus it was shown that not more than a small fraction of the shower secondaries can be μ-mesons.

Whilst these experiments had given proof for the *presence* of π-mesons in penetrating showers, they had not yet permitted to estimate the *abundance* of π-mesons among the shower particles. This became possible when, with the development of photographic emulsions sensitive to minimum-ionising particles, the new technique of nuclear emulsions was applied to the study of showers. C. F. POWELL's group in Bristol did most of the pioneer work in this field, and their data will take up a substantial part of this report. At this stage, it will suffice us to quote one of the results obtained by FOWLER[1], and by CAMERINI et al.[2] concerning the composition of the shower particles. Their measurments prove that about 80% of all the (relativistic) "shower" tracks are due to π-mesons.

Two further short remarks may be added to this historical introduction. First, it was thought for some time (AUGER et al.[3], JÁNOSSY[4]) that the penetrating particles observed in extensive air showers could not be "ordinary" mesons, since in traversing an absorber, they multiply much stronger than a μ-meson would, and yet penetrate much further than an electron could. It was therefore thought that they might possess properties intermediate between those of an electron and those of a μ-meson, and indeed be light, "λ"-mesons of a mass between 3 and 10 m_e. However, in deriving this conclusion, the penetrating power of low-energy photons in a cascade was underestimated (COCCONI, TONGIORGI and GREISEN[5]), and the possibility of nuclear interactions, of local penetrating showers occurring in extensive showers, was neglected (SITTE[6]). The λ-meson hypothesis was therefore abandoned.

Secondly, particles other than "ordinary" π- or μ-mesons were found to emerge from local penetrating showers. This was shown by ROCHESTER and BUTLER[7] in 1947 when they reported the first convincing evidence for the existence of heavy unstable particles, and opened up a wide new field of investigation into the structure of matter. But this entire subject is treated elsewhere; in our discussion of penetrating showers, we can restrict ourselves to emphasizing the historical connection, and to a few comments below.

II. Attenuation and interaction mean free path of the shower particles.

4. The techniques of experimentation. Already in the first studies of penetrating showers with multiple-plate cloud chambers (FRETTER[8], GREGORY and TINLOT[9]), numerous photographs showed quite clearly that the secondaries emitted

[1] P. II. FOWLER: Phil. Mag. **41**, 169 (1950).
[2] CAMERINI, J. H. DAVIES, P. H. FOWLER, C. FRANZINETTI, W. O. LOCK, D. H. PERKINS and G. YEKUTIELI: Phil. Mag. **42**, 1241 (1951).
[3] P. AUGER, J. DAUDIN, A. FREON and R. MAZE: C. r. Acad. Sci., Paris **226**, 169, 569 (1948).
[4] L. JÁNOSSY and C. B. A. McCUSKER: Nature, Lond. **163**, 181 (1949). — L. JÁNOSSY: Cosmic Radiation, p. 103. London: Butterworth Ltd. 1949.
[5] G. COCCONI, V. COCCONI TONGIORGI and K. GREISEN: Phys. Rev. **75**, 1068 (1949). — K. GREISEN: Phys. Rev. **75**, 1071 (1949).
[6] K. SITTE: Phys. Rev. **75**, 340 (1949).
[7] G. D. ROCHESTER and C. C. BUTLER: Nature, Lond. **160**, 855 (1947).
[8] W. B. FRETTER and W. E. HAZEN: Phys. Rev. **70**, 230 (1946). — W. B. FRETTER' Phys. Rev. **73**, 41 (1948).
[9] B. P. GREGORY and J. H. TINLOT: Phys. Rev. **81**, 667, 674 (1951).

in the collisions are themselves capable of further nuclear interactions. Thus it became evident that the primary collision is not "catastrophic" in the sense that it eliminates, once and for all, further multiplication in subsequent nuclear interactions. Therefore a distinction must be made between the *interaction* mean free path—the average distance the particle will travel before colliding with a nucleus—, and the *attenuation* (or absorption) mean free path of the primaries—the average path length after which multiplicative nuclear processes cease. If these processes should still have the character of penetrating showers, i.e. consist in the ejection of associated hard particles, the experimenter must take good care to record such nuclear interactions only, and to exclude others, such as electronic cascades initiated by μ-mesons. Indeed, several early experiments were falsified by those undesirable admixtures. Analyzing these measurements, SALVINI[1] and SITTE[2] emphasized the necessity of using more than one tray of shielded counters for the selection of local penetrating showers.

Translated into terms of experimental techniques, the requirements for proper design of the apparatus are now clear. They are demonstrated in the schematic diagram of a model experiment shown in Fig. 5: Our "*P*-set" will consist of three (or more) trays of detectors A, B, C, separated from each other by a sufficient amount of heavy absorber, the thickness of which may be varied according to the experimenter's demands on the penetration of the

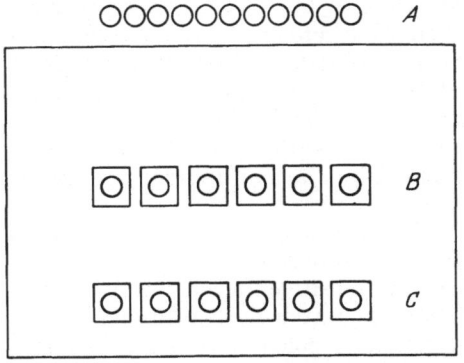

Fig. 5. A model "*P*-set" for a penetrating-shower experiment. According to the choice of primaries, either none or exactly one counter of A should be struck, together with at least m B-counters and at least n C-counters ("events $A_0 B^m C^m$ or $A_1 B^m C^m$").

secondaries. If Geiger counters are used, it is advantageous to insert layers of about $\frac{1}{2}$ to 1 in. of heavy absorber also between the individual counters of B and C, so as to reduce the number of knock-on events registered. Only events in which counters in both shielded trays B and C are struck, will be accepted as penetrating showers. Let us say the triggering conditions are arranged so that *at least m* counters in A must be discharged, in coincidence with *at least n* counters in B and *at least p* counters in C: we call this a "shower $A^m B^n C^p$". If instead events with *exactly m* counters in A, n in B, and p in C are selected, we speak of a "shower $A_m B_n C_p$".

An absorber T of the material to be investigated can be placed above tray A. If, now, the attenuation mean free path is to be measured, the rates of events $A^m B^n C^p$—with at least some of the m, n, p not less than 2—are taken as a function of the thickness of absorber T. If, on the other hand, the purpose of the experiment is a measurement of the interaction mean free path, a "multiplicity" m of 0 or 1 in the top tray A must be strictly enforced: zero for neutral, one for charged primaries. In other words, we record now showers $A_0 B^m C^p$ or $A_1 B C^p$. With suitable variations, the method can be equally well used if the shower detector is an ionisation chamber or a scintillator, instead of a tray of Geiger counters.

Measurements of the interaction mean free path can of course also be carried out in a multiple-plate cloud chamber, either absolutely by a comparison of the rates of showers originating in the individual plates, or relatively by comparing

[1] G. SALVINI: Nuovo Cim. (9) **7**, 786 (1950).
[2] K. SITTE: Phys. Rev. **78**, 714 (1950).

11*

the rates of events initiated in plates of different materials. If counter control is to be used, the absolute measurement is somewhat complicated by the necessity of applying corrections for the variation of the triggering probability with the location of the shower origin. In this respect, the technique of straight counter experiments is therefore easier and less subject to errors. On the other hand, both multiple-plate cloud chambers and photographic emulsions are well suited for studies of the interaction mean free paths of shower secondaries.

5. Attenuation of the shower-producing radiation in air. Even without entering into details of the theory of nucleonic cascades, one recognizes immediately that the attenuation of the shower-producing radiation in air and in dense materials present entirely different problems. In the first case, evidently only the nucleons are capable of carrying on the cascade process, while the π-mesons, owing to their short life time, will participate only if their energies are extremely high. In the second, nucleons and π-mesons can both contribute to the propagation of the cascade. Thus, the attenuation mean free path in air is essentially a measure of the degree of inelasticity of the nucleon-nucleon interactions (a point to be discussed in more detail in Sect. 21), and that in dense materials is deter-minded by the multiplicity and the interaction mean free path of the shower secondaries.

Measurements of the attenuation mean free path Λ of the primaries of pene-trating showers have been carried out with counter arrangements[1-4], and with electron-sensitive nuclear emulsions[5,6]. In some of these experiments, data have been taken at balloon altitudes, i.e. at 70000 feet and above, and in this case a simple exponential attenuation law of the form $I = I_0 \exp(-x/\Lambda)$ can no longer be postulated. The shower primaries, which at sea level and even at mountain altitudes are strongly collimated in vertical direction, become almost isotropically distributed at those extreme altitudes, and the variation of their intensity with the atmospheric depth x is given by the Gross transformation

$$I = I_0 \left\{ \exp(-x/\Lambda) - \frac{x}{\Lambda} \int_{x/\Lambda}^{\infty} \frac{e^{-y}}{y} \, dy \right\}. \tag{5.1}$$

If this is taken in account, satisfactory agreement is found between the various methods. For all high-energy events—that is, for primary energies of the order of 10^9 eV and above—, TINLOT's value of (118 ± 2) g/cm² can be used; for stars of lower energy, a slightly higher attenuation mean free path—up to about 140 g/cm²—is observed.—Measurements taken by ALEKSEEVA and VERNOV[7] at 70000 feet and above, indicate an atmospheric transition effect near the top of the atmosphere.

It will be demonstrated in the following sections that for the bulk of the penetrating showers registered in the standard equipment, the primary energies lie in the region of 10^9 to 10^{10} eV. Only more recently, a few measurements in energy regions well above 10^{10} eV have been reported, and in view of the impor-

[1] J. H. TINLOT: Phys. Rev. **73**, 1476 (1948); **74**, 1197 (1948).

[2] E. P. GEORGE and A. C. JASON: Nature, Lond. **161**, 218 (1949). — Proc. Phys. Soc. Lond. A **62**, 243 (1949); **63**, 1081 (1950).

[3] S. N. VERNOV and T. N. CHARAKHCHYAN: Dokl. Akad. Nauk SSSR. **69**, 629 (1949.)

[4] H. SCHULTZ: Z. Naturforsch. 9a, 419 (1954).

[5] R. H. BROWN, U. CAMERINI, P. H. FOWLER, H. HEITLER, D. T. KING and C. F. POWELL: Phil. Mag. **40**, 862 (1949). — U. CAMERINI, P. H. FOWLER, W. O. LOCK and H. MUIRHEAD: Phil. Mag. **41**, 413 (1950).

[6] J. J. LORD and M. SCHEIN: Phys. Rev. **77**, 19 (1950).

[7] K. L. ALEKSEEVA and S. N. VERNOV: Dokl. Akad. Nauk SSSR. **69**, 317 (1949).

tance of the question of a possible dependence on the primary energy of the degree of inelasticity in a nuclear collision, two of these experiments will be mentioned here.

Working with a counter. hodoscope in connection with a liquid scintillator, SITTE et al.[1] recorded the frequency of showers of high multiplicity of penetrating secondaries, and of electrons in penetrating showers, at altitudes of 3260 and 4300 m. From the absolute frequency of the events, and from the multiplicity distribution of the secondaries, they conclude that these showers were initiated by primaries of energies between 10^{10} and 10^{11} eV. Their results, reproduced

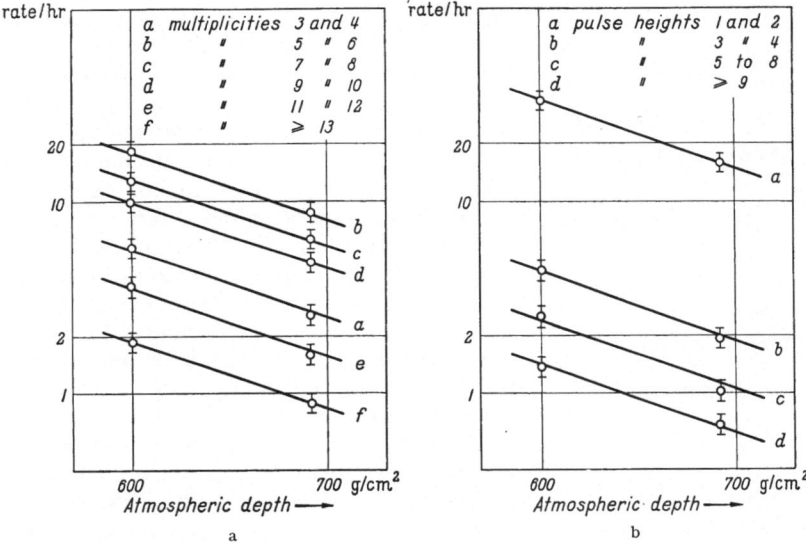

Fig. 6. (a) Rates of showers grouped according to their multiplicity recorded at Echo Lake and at Mt. Evans. (b) Rates of showers grouped according to scintillation pulse height, recorded at Echo Lake and at Mt. Evans. — The solid lines represent attenuation with a mean free path of 117 g/cm² (SITTE, FROEHLICH and NADELHAFT, 1955).

in Fig. 6a and b, indicate no change in the attenuation over the entire range of primary energies; they are all quite compatible with a mean free path $\Lambda = 120$ g/cm².

Even higher energies were studied by KAPLON et al.[2] with nuclear emulsions flown in balloons, and exposed at mountain altitudes. These authors selected showers of primary energies in the order of 10^{12} eV—already bordering on the region of extensive showers—and again found an attenuation mean free path of (129 ± 15) g/cm². It can, therefore, be concluded that at least up to 10^{12} eV, the energy ransfer to the mesonic secondaries remains reasonably constant, and the collisions remain partially elastic. The same conclusion is also reached from an analysis of air shower data (see COCCONI's article in this volume), and is of considerable importance for the interpretation of high-energy nuclear collisions.

6. Attenuation in dense materials. A determination of the attenuation mean free path of shower primaries can, in principle, provide an answer to two important questions: First, a comparison of the mean free paths in different materials may tell us whether a simple geometrical model can adequately describe the phenomenon of shower production, and second, a comparison of the attenuation mean

[1] K. SITTE, F. E. FROEHLICH and I. NADELHAFT: Phys. Rev. **97**, 166 (1955).
[2] M. F. KAPLON, J. Z. KLOSE, D. M. RITSON and W. D. WALKER: Phys. Rev. **91** (1953).

free paths in dense materials with that in air may give evidence about the participation in the nuclear cascade of the shower secondaries.

As to the second point, it is clear that the composition of the cosmic-ray component capable of initiating high-energy interactions (the "nucleo-active" or *N*-component, will change with the transition from a light material like air, to a dense absorber—if the π-mesons can be considered as *N*-particles. As it has been pointed out above, the atmospheric *N*-component will even at sea level contain not more than a few percent of π-mesons, while in a dense material, the number of π-mesons produced in every collision will exceed that of the nucleons by a considerable factor, and their decay probability in one interaction length is no longer overwhelmingly large. Hence the new equilibrium composition established after a few collision lengths, will include many more π-mesons than nucleons, and their interactions will play an important part in the propagation of the cascade.

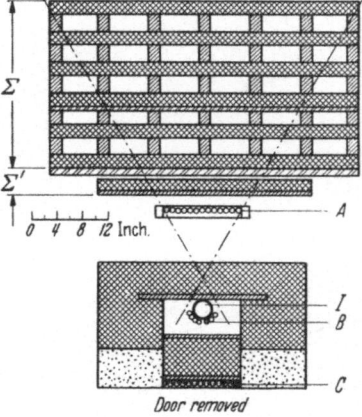

Fig. 7. The experimental arrangement of BRIDGE and REDIKER (1952) to measure the absorption of the *N*-component in condensed matter. Three trays of Geiger counters *A*, *B*, *C* and the ionization chamber *I* are used to register the events.

At the same time, this consideration illustrates a certain difficulty in the experimental technique of absorption measurements. Because of the "transition effect" resulting from this change of the equilibrium compositions, the intensity of the shower-producing radiation will not decrease with increasing absorber thickness according to a simple exponential law, but may at the beginning even increase. Measurements taken at only a few layer thicknesses, and in particular under not very thick absorbers, can therefore easily be misinterpreted, and will systematically yield too high estimates of the apparent attenuation mean free path. Evidently, unambiguous results can only be obtained from experiments where either the observations were carried out over a very large range of absorber thicknesses, or else the apparatus is provided with a thick permanent shield to which in parts of the measurements further layers of the absorber material are added.

A second experimental difficulty stems from the fact that in a sense material, the nuclear cascade will not spread out very much and the recording equipment, instead of registering a single shower, may well respond to the superposition of several events which are all part of the general nuclear cascade propagated in the absorber. Consequently, one may surmise the occurrence of nuclear interactions when in fact shower production has already ceased, and the range of the cascade may be overestimated.

Let us now discuss the adequacy of the various experimental methods for the purpose of measuring attenuation lengths. Three techniques have been widely used: Photographic emulsions, ionization chambers (frequently in combination with counters), and counter arrangements. Of those, the emulsions have the advantage that the recognition of the interactions, and hence the elimination of spurious events, is most easily assured. Their disadvantage is that the rate of showers recorded is usually low, and hence the statistical accuracy insufficient for all but low-energy events and under small absorber thicknesses. Ionization chambers permit registration over a longer period of time, and consequently better statistics, but the analysis and classification of the showers is a much more complex, and never quite unambiguously solved, problem. An excellent example of such an analysis is contained in a paper by BRIDGE and REDIKER[1] to which the reader is referred for details. The apparatus of BRIDGE and REDIKER, consisting of three trays of Geiger counters in coincidence with a high-pressure ionization chamber, is sketched in Fig. 7.—Counter experiments finally, suffer from the same difficulty in the elimination of spurious events, and offer the same advantages of good statistics, as ion chamber work. If hodoscope techniques are uses, as in the experiments of AZIMOV et al.[2], a reliability at least equal to that of a good ion chamber experiment can be claimed for their results.

Table 1. *The attenuation mean free path, in g/cm², for carbon, aluminium, iron and lead.*

Author	Material				Remarks
	C	Al	Fe	Pb	
BERNARDINI et al.[3] . .		200		300	Stars in emulsions. Transition effect in Pb
GEORGE et al.[4]	166 ± 7			310 ± 20	Stars in emulsions
ROSSER and SWIFT[5] .				380 ± 65	Penetrating showers in emulsions
SCHEIN and FAHY[6] . .				434	Ionization chamber
STINCHOMB[7]			240 ± 20	350 ± 70	Ionization chamber
BRIDGE and ROSSI[8] .			320 ± 70	430 ± 90	Ionization chamber. Transition effect
BRIDGE and REDIKER[9]				a) 420 ± 100 b) 440 ± 50	Ionization chamber. Transition effect a) larger pulses b) smaller pulses
REDIKER	a) 310* b) 350**				Ionization chamber, a) low penetration b) high penetration
AZIMOV et al.	216 ± 15		344 ± 25	482 ± 31	Counter hodoscope
COCCONI[10]			310 ± 60	380 ± 60	Counters

* "Showers $AI_{03}BC_0$" in REDIKER's notation.
** "Showers $A I_{03}BC_1$" in REDIKER's notation.

Data obtained in a number of typical experiments[3-10] are summarized in Table 1. It must be emphasized that this list is representative but not complete.— In the last column, the technique used is indicated, and where deviations from a simple exponential attenuation was observed, the remark "transition effect" is added. At least for the two heavy materials, lead and iron, the agreement reached with different methods appears satisfactory—the one low value found in the emulsions may perhaps be ascribed to an appreciably lower primary energy—; but in the case of carbon, the situation is less clear. However, part

[1] H. S. BRIDGE and R. H. REDIKER: Phys. Rev. **88**, 206 (1952).
[2] S. A. AZIMOV, N. A. DOBROTIN, A. L. LYUKIMOV and K. P. RYZHKOVA: Dokl. Akad. Nauk SSSR. **90**, 51 (1953).
[3] G. BERNARDINI, G. CORTINI and A. MANFREDINI: Phys. Rev. **74**, 845, 1878 (1948).
[4] E. P. GEORGE and A. C. JASON: Proc. Phys. Soc. Lond. A **62**, 243 (1949). — J. C. BARTON, E. P. GEORGE and A. C. JASON: Proc. Phys. Soc. Lond. **64**, 175 (1951).
[5] W. G. V. ROSSER and M. W. SWIFT: Phil. Mag. (7) **42**, 856 (1951).
[6] E. F. FAHY and M. SCHEIN: Phys. Rev. **75**, 207 (1949).
[7] G. T. STINCHCOMB: Phys. Rev. **83**, 422 (1951).
[8] H. S. BRIDGE and B. ROSSI: Phys. Rev. **75**, 810 (1949).
[9] R. H. REDIKER: Phys. Rev. **95**, 526 (1954).
[10] G. COCCONI: Phys. Rev. **75**, 1074 (1949).

of the discrepancies may be due to the technical difficulty of placing sufficiently thick carbon absorbers on top of the detector set. Generally, not much more than about 100 g/cm² could be used, and that is, as we shall see, only about one *collision* mean free path, and hence not enough to ensure equilibrium among the N-particles. Under this condition, the figures quoted represent upper limits rather than true values of the attenuation length.

Nevertheless, certain qualitative conclusions can be derived from these data, and in view of the complexity of the phenomenon, more ambitious efforts of quantitative analysis should perhaps better be postponed. Generally speaking, it will be noted that the long mean free paths found cannot be explained without the assumption that π-mesons as well as nucleons participate in the propagation of the nuclear cascade, and this with a collision cross section comparable to that of the primary N-component. This is seen if one tries to compute, from the well-established value $\Lambda_{air} = 120$ g/cm², the attenuation mean free paths in heavy materials according to a model of a semi-transparent nucleus (cf. ROSSI[1]). The calculation gives in all cases much too small values, and the role of the π-mesons as "nucleo-active" particles is, therefore, established beyond doubt.

7. The interaction mean free path of shower primaries. Experimentally, the methods of determining the interaction mean free path λ of shower primaries have less pitfalls than those for the measurement of the attenuation length Λ; theoretically, λ is the fundamentally simpler and more directly significant quantity. Thus, it is not surprising that of the considerable amount of work that has gone into experiments to study the interaction mean free paths of shower primaries in various substances, a rather consistent picture has begun to emerge, and that the quantitative agreement between the numerous investigators has generally been good, once the technical pre-requisites outlined in Sect. 4 had been recognized

In comparing the results for different materials, it is convenient to define a "geometrical" mean free path λ_g, as corresponding to the geometrical nuclear cross section $\sigma_g = \pi R_0^2 A^{\frac{2}{3}}$:

$$\lambda_g = \frac{A^{\frac{1}{3}}}{\pi R_0^2 N} \tag{7.1}$$

where N is AVOGADRO's number, and $R_0 = 1.37 \times 10^{-13}$ cm, the older value for the nucleon radius. Some values for λ_g are listed in Table 2 which shows a numbers of experimental results obtained in carbon, aluminium, sulphur, iron and lead, with the techniques discussed above.

As it is seen, all investigators are in essential agreement on two main points: (1) that the interaction mean free path is shorter for harder or larger showers (i.e., presumably for showers of higher primary energies) and longer for showers of less penetration and/or less multiplicity; and (2) that for heavy materials, the mean free paths of the more complex and harder showers is very near the geometric value, while in light materials, even showers of large penetration and multiplicity still give a λ larger than the geometrical λ_g. In this second point, the only exceptions are the data of BROWN[2]; but since in his experiment only showers of exceptionally high complexity were recorded, this discrepancy must not be taken too seriously.

In spite of the satisfactory agreement, a word of warning should be added. In all the experiments which fall in the scope of our analysis, it is not really the *interaction* mean free path that was measured, but the mean free path for *shower production*, or more particularly, the mean free path for the production of pene-

[1] B. ROSSI: High Energy Particels, p. 490 ff. New York: Prentice-Hall 1952.
[2] R. R. BROWN: Phys. Rev. **87**, 949 (1952).

trating showers of a pre-arranged structure. If it is correct to assume that in every collision of a high-energy nucleon, a large fraction of its energy will be transferred to a group of penetrating secondaries, then those two mean free paths are identical, and the experiments are rightly described as measurements of the interaction length. But if small energy transfers are not too unusual, or a transfer to non-penetrating secondaries not too exceptional—for instance—

Table 2. *The interaction mean free paths λ, and the "geometrical" interaction mean free paths λ_g, in g/cm^2, for carbon, aluminium (or sulphur), iron and lead.*

Author	Materials				Remarks
	C $\lambda_g = 64$	Al (S) $\lambda_g = 86$	Fe $\lambda_g = 105$	Pb $\lambda_g = 164$	
Cocconi[1] . . .	100 ± 15		113 ± 15	160 ± 15	Counters. Charged primaries
Walker[2] . . .				a) 180 ± 10 b) 150 ± 8 c) 140 ± 10	Counters. Charged primaries a) 4 or 5 counters discharged b) 5 counters discharged c) 7 counters discharged
Walker et al.[3]	a) 81 ± 5 b) 80 ± 7				Counters a) charged primaries b) neutral primaries
Sitte[4]				a) 196 ± 13 b) 162 ± 10	Counters. Charged primaries a) sec. pen. 100 g/cm^2 b) sec. pen. 200 g/cm^2
Boehmer and Bridge[5] . . .	a) 106 ± 6 b) 103 ± 10 c) 85 ± 12			a) 311 ± 34 b) 220 ± 35 c) 143 ± 34	Counters. Neutral primaries a) Events $A_0 B^2 C^1$ b) Events $A_0 B^2 C^2$ c) Events $A_0 B^3 C^2$
Brown[6] . . .	a) 65 ± 5 b) (89 ± 12)	76 ± 7 (sulphur)	115 ± 12		Counters. Charged primaries a) at 2765 m alt. b) at 130 m alt.
Rediker[7] . . .	a) 136 ± 12 b) 111 ± 11				Ion chamber a) low multipl. b) high multipl.
Tinlot and Gregory[8] . .		140		200	Cloud chamber Charged primaries
Walker et al.[9]	76 ± 8				Cloud chamber
Froehlich et al.[10]				a) 190 ± 15 b) 158 ± 12	Cloud chamber Charged primaries a) $E\ 2 \times 10^9$ ev b) $E\ 2 \times 10^{10}$ ev

as a result of fluctuations in the charge distribution to π^0-mesons, then the techniques applied will systematically overestimate the interaction mean free path.

A case of this sort is schematically shown in Fig. 8: A primary entering the top absorber T interacts in it, emitting only low-energy ionizing secondaries and perhaps π^0-mesons whose decay cascade is entirely absorbed in T. The primary

---

[1] G. Cocconi: Phys. Rev. **75**, 1074 (1949).
[2] W. D. Walker: Phys. Rev. **77**, 686 (1950).
[3] S. P. Walker, W. D. Walker and K. Greisen: Phys. Rev. **80**, 546 (1950).
[4] K. Sitte: Phys. Rev. **78**, 714 (1950).
[5] H. W. Boehmer and H. S. Bridge: Phys. Rev. **85**, 863 (1952).
[6] R. R. Brown: Phys. Rev. **87**, 949 (1952).
[7] R. H. Rediker: Phys. Rev. **95**, 526 (1954).
[8] J. H. Tinlot and B. P. Gregory: Phys. Rev. **81**, 667 (1951).
[9] W. D. Walker, N. W. Duller and J. D. Sorrel: Phys. Rev. **86**, 865 (1952).
[10] F. E. Froehlich, E. M. Harth and K. Sitte: Phys. Rev. **87**, 504 (1952).

will then continue with only a small energy loss, and be capable of initiating another penetrating shower in the P-set. Writing X for the thickness of the top absorber, we see that the flux $s(E)$ of unaccompanied primaries of energy E incident on A is not $s(E) = s_0(E) \cdot e^{-X/\lambda}$ with $s_0(E)$ the differential primary spectrum above T, but

$$s(E) = s_0(E) \cdot \left\{ 1 - \int_0^X e^{-x/\lambda} \, w(E, X - x) \, dx/\lambda \right\} \qquad (7.2)$$

where $w(E, X - x)$ stands for the probability that an interaction produced at depth x would not be missed. If the energy distribution of the showers recorded in the P-set were known, the expression (7.2) could be evaluated since w can be estimated from the data of cloud chamber and emulsion experiments. But to deter-

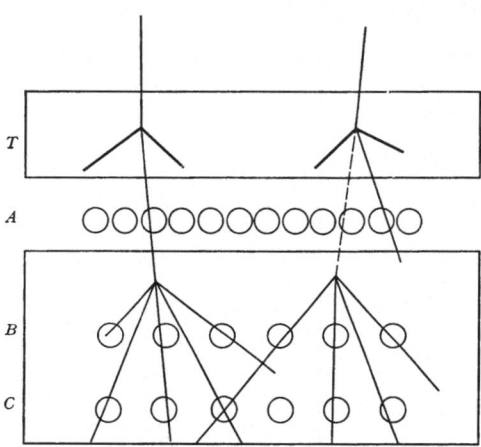

mine it is in general a difficult task, though in a few cases it has been done with apparent success. Basically, the pattern for the procedure is the following: Write $P(E)$ for the probability that a shower-initiating collision occurs between trays A and B, $\pi(E, n)$ for the multiplicity distribution of the shower secondaries in this collision, and $p(n)$ for the triggering probability of an event with multiplicity n. Then the rate R of recorded showers is

$$R = \int_\varepsilon^\infty s(E) \, P(E) \, \pi(E, n) \, p(n) \, dE \qquad (7.3)$$

where ε stands for the minimum energy required to trigger the set. — $s(E)$ is known, $P(E)$ is a constant for the arrangement if λ does not depend on the energy, and $\pi(E, n)$ can be taken

Fig. 8. Schematical picture of "missed" interaction: Primaries interacting in the top absorber with small energy dissipation, with or without charge exchange, and initiating a second shower in the P-set.

from experimental data; only $p(n)$ has to be computed in each case. This is difficult if the absorber layers between A, B and C are large, and hence transition effects significant.

The experiment of Boehmer and Bridge[1] listed in Table 2 provides one example where these uncertainties were comparatively small. These authors quote for the interaction mean free paths of their (neutral) primaries, values ranging from (311 ± 34) g/cm² for events $A_0 B^2 C^1$, to (143 ± 30) g/cm² for events $A_0 B^3 C^2$. But a detailed analysis[2] reveals that indeed their data are entirely compatible with the assumption that the *interaction* mean free path of the shower primaries of all energies above about 1 GeV has the geometrical value, 164 g/cm². Thus, though most other experiments could not be analysed with equal confidence, we cannot dismiss the possibility that in heavy elements, the *interaction* cross section of energetic nucleons is for all primary energies practically geometrical, but that not all these interactions result in observable penetrating showers. We shall return to this point later on.

8. Nuclear interactions in light elements: "Transparency". It has already been mentioned that the results summarized in Table 2 indicate for *light* elements, even at high primary energies, a cross section for shower production of appreciably

[1] H. W. Boehmer and H. S. Bridge: Phys. Rev. **85**, 863 (1952).
[2] K. Sitte: Acta Phys. Austriaca **6**, 167 (1952).

less than the geometrical value, in view of the criticism put forward above, it is of particular importance to know whether these deviations are real, or ar perhaps again due to flaws in the experimental techniques.

Theoretically, it is easy to understand that if high-energy nuclear collisions can be viewed as an interaction between the incident shower primary and the individual nucleons of the target nucleus, the collision probability is not determined by the geometrical cross section of the *nucleus* alone. For a small nucleus, the incident particle might have a certain chance to traverse the target without actually coming to interact with any one of its constituents; for a large nucleus, this probability is much smaller. The picture of an only partially opaque nucleus, introduced by SERBER et al.[1] to describe experiments on neutron scattering, has been used by COCCONI[2] to compute, from the observed "transparency" of light nuclei, the "mean free path in nuclear matter" for high-energy interactions. The outline of their arguments given below follows ROSSI's presentation[3].

Assume that the nucleons are distributed at random in a sphere of radius $R_n = R_0 \cdot A^{\frac{1}{3}}$. A nucleon passing at a distance b from the center of the nucleus (see Fig. 9) will then, in its traversal of the nucleus, sweep out the roughly cylindrical volume V:

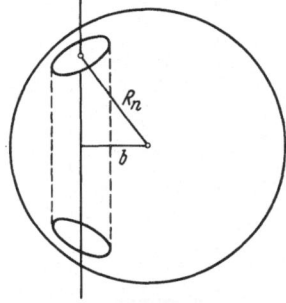

$$V = 2\pi r_i^2 \sqrt{R_n^2 - b^2} \qquad (r_i \ll R_n) \qquad (8.1)$$

where r_i is the "interaction radius" of the incident particle. Since the average number of nucleons in V is

$$\frac{VA}{V_n} = \frac{3}{4\pi} \frac{V}{R_0^3} \qquad (8.2)$$

Fig. 9. Collision of a nucleon with a nucleus.

$\left(V_n = \dfrac{4\pi}{3} R_0^3 \cdot A \text{ is the nuclear volume}\right)$, the probability that the primary traverses the cylinder without interaction is $\exp(-VA/V_n)$, and the total interaction cross section σ_i is given by

$$\sigma_i = 2\pi \int_0^{R_n + r_i} [1 - \exp(-VA/V_n)]\, b\, db. \qquad (8.3)$$

Introducing the "interaction mean free path in nuclear matter" λ_n, according to

$$\lambda_n = \frac{V_n}{\pi r_i^2 \cdot A}, \qquad (8.4)$$

(8.5) becomes in the limiting case $r_i \ll R_n$

$$\sigma_i/\sigma_g = 1 - t(\lambda_n/R_n) \qquad (8.6)$$

where

$$t(\lambda_n/R_n) = \frac{\lambda_n^2}{2\pi R_n^2}\left[1 - e^{-2R_n/\lambda_n} - 2\frac{R_n}{\lambda_n} e^{-2R_n/\lambda_n}\right] \qquad (8.7)$$

is the "average transparency" of the nucleus, that is, the average probability for an incident nucleon to pass through the nucleus without interaction. The relation between the interaction mean free path λ_i and the geometrical mean

[1] R. SERBER: Phys. Rev. 72, 1114 (1947). — S. FERNBACH, R. SERBER and T. B. TAYLOR: Phys. Rev. 75, 1352 (1949).

[2] G. COCCONI: Phys. Rev. 75, 1075 (1949).

[3] B. ROSSI: High Energy Particles, p. 359ff. New York: Prentice-Hall 1952.

free path λ_g is, accordingly,

$$\lambda_i = \lambda_g \cdot \sigma_g/\sigma_i = \frac{\lambda_g}{1 - t(\lambda_n/R_n)} . \tag{8.8}$$

However, the assumption $r_i \ll R_n$ is not good enough for light nuclei, and to obtain satisfactory results, a numerical integration of (8.3) has to be carried out. Curves obtained in this way by R. W. SAFFORD, and taken again from ROSSI's

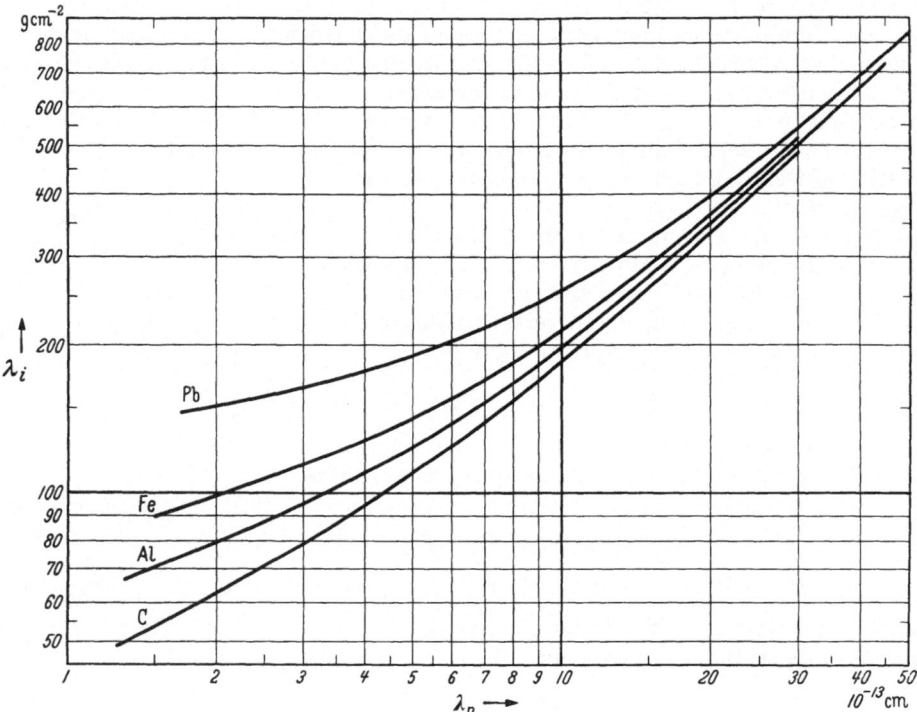

Fig. 10. The mean free paths for shower productions, λ_i, in various substances, plotted as functions of the mean free path in nuclear matter, λ_n (from ROSSI, "High-Energy Particles", p. 364).

book, are shown in Fig. 10. From the ratios σ_i/σ_g shown, the interaction mean free paths λ_i are then immediately computed according to (8.8).

We may then use the data of Table 2 (p. 169) to determine the quantity λ_n, the mean free path in nuclear matter. Taking as the most reliable values for carbon $\lambda_i = 80$ to 85 g/cm², and for iron $\lambda_i = 115$ g/cm², we compute from Fig. 10 for both elements, in good agreement, $\lambda_n \approx 3.2 \times 10^{-13}$ cm. As a check, we find that with this value, the interaction mean free path in lead still retains a value very close to the geometrical $\lambda_g = 164$ g/cm².

An attempt to express the mean free paths in various substances as a function of one or two simple parameters has also been made by FROMAN[1]. The value he gives for λ_n is larger than the one quoted here(4.4×10^{-13} cm), as is that of COCCONI's first paper. The difference, however, is mostly due to the choice of r_i, and should not be given too much significance.

9. Critical discussion of the experiments. More recent results. In spite of the apparent consistency of these results, serious doubt can be cast on the reliability

---

[1] D. FROMAN: Phys. Rev. **88**, 172 (1952).

of the conclusions, not only because of the crudeness of the model used, but also because of deficiencies of the experimental techniques applied. Let us begin with the latter.

As it has been pointed out above, there are basically two types of experiments in which the interaction mean free path is measured: the first of the kind sketched in Fig. 5 (or with equivalent equipment), in which the rate of showers initiated by unaccompanied primaries is recorded as a function of the top absorber thickness; and the second a cloud chamber study, usually based on a comparison of the production rate in the light elements with that in a standard heavy element like lead. Thus, the first method will give reliable results only if all interactions in the top absorber are efficiently eliminated: a condition which is satisfied only if, with overwhelming probability, the energy transfer in this collision will be large enough to cause the ejection of at least one penetrating secondary in forward direction. But if, as it has been shown above, it is far from safe to accept this premise even for collisions in heavy absorbers and with moderate primary energies, one may well suspect even more serious errors resulting from its application to experiments in light materials where peripheral, "glancing" collisions will be relatively more frequent, and plural interactions within the target nucleus correspondingly rare. It is, therefore, quite possible that experiments of this type will give an erroneously large mean free path.

Work with counter-controlled cloud chambers, on the other hand, suffers from the fact that the shower selection is usually based on the demand for a certain minimum secondary multiplicity striking the detection trays. This introduces a bias in favour of showers originating in heavy elements, since for equal primary energy, the multiplicity of showers initiated in heavy nuclei is larger than that of light-nuclei showers. This latter fact has been confirmed by a number of investigators (e.g. Lovati et al.[1]), and will be discussed in more detail later on; for the cloud chamber experiments to determine of interaction mean free paths it has the consequence that the average primary energy of heavy-nuclei showers recorded with the equipment will be lower than that of the light-element showers, and if no correction for this bias is made, the mean free path in the light element will again be overestimated.

Once these dangers are realized, it is easy to see what remedies to apply. In counter experiments, it suffices to split the top absorber T into several thinner layers, and place under each of them a counter try which must be struck by an unaccompanied primary if a shower originating in the P-set is to be accepted. Experiments of this kind have been carried out by Camerini and Thom at the Chacaltaya mountain laboratory, and by Cicchini, Cardoso and Ghielmetti at Mina Aguilar, Argentina; but so far, no final results have been reported[2].

A cloud chamber experiment in which correction for the differences in primary energies was made, and besides a comparison between various light-element showers was used instead of a comparison with lead, was carried out by Asko-with and Sitte[3]. They report considerably smaller mean free paths and transparencies, as summarized in Table 3. In all cases, the average primary energy was of the order 25 GeV. If from a comparison of these values of λ_i, a mean free path λ_n in nuclear matter is computed, the best fit is with a $\lambda_n = (2.55 \pm 0.25) \times$

---

[1] A. Lovati, A. Mura, G. Salvini and G. Tagliaferri: Nuovo Cim. **7**, 943 (1950).

[2] *Note added in proof.* The Mina Aguilar experiment has been completed. Its results do not bear out the above predictions. The observed mean free paths remain essentially unchanged when the absorbers are split. — I am obliged to Dr. Ghielmetti for communicating his results in manuscript.

[3] J. G. Askowith and K. Sitte: Phys. Rev. **97**, 159 (1955).

10^{-13} cm: significantly shorter than the value derived above. As an illustrative side result, it may be mentioned that with this short mean free path, the probability that even in a nucleus as light as carbon the primary undergoes more than one interaction, is about 50%. In aluminium, it increases to around 60%, and is in lead almost 90%.

An entirely different way to eliminate the errors due to absorption of the secondaries has been chosen by CERVASI et al.[1] in a hodoscope-counter experiment. Instead of the permanent production layer of the P-set in Fig. 5, they let their showers originate in a variable thin production plate of various materials, and record the shower rate as a function of the thickness of that plate. Of course only thin targets can be used ($\ll \lambda_i$), and hence the total rates are low; but with the good angular resolution of their equipment, and through the use of a large hodoscope of Geiger counters, the authors are able to make all necessary corrections much more safely than in most similar experiments. From the absolute rates it can be estimated that the average primary energies were again of the order of 20 to 30 GeV. The surprising result of this very thorough investigation is that for the three pure elements studied: Pb, Fe and C, the cross sections follow very precisely an $A^{\frac{2}{3}}$-law, without any indication of a transparency even of the carbon nucleus; only for hydrogen, extrapolated from measurements in paraffin, a deviation is observed. Although it is possible that in this kind of experiment a certain residual error favouring the detection of light-nuclei showers might be made, it is hard to see how such an effect could mask a transparency of some 20%. One is, therefore, undoubtedly led to the conclusion that this experiment, like the one mentioned before, compels us to revise the older estimates of the degree of transparancy in light nuclei, and of the values of the mean free path in nuclear matter derived from them.

Table 3. *Interaction mean free paths λ_i, and transparencies t, in aluminium, carbon, and lithium* (ASKOWITH and SITTE).

Material	Al	C	Li
λ_i (g/cm²) .	88 ± 3.5	72 ± 3.5	62 ± 5
t (%) . . .	6 ± 3	12 ± 4	15 ± 7

It is, of course, quite plausible that part of the difficulties one encounters in the interpretation of these experimental results, derives from the crudeness of the model of the nucleus which has been used. We have all reasons to believe that the nucleus does not conform precisely to the picture of a semi-opaque sphere of constant density; in particular, the experiments of HOFSTADTER and his co-workers[3] have given ample proof that the charge distribution in nuclei is much better represented by a two-parameter model characterised by a "half-density distance" c and a "skin thickness" t. Moreover, in all cases the nuclear dimensions determined from electron scattering experiments turn out much smaller than the "geometrical" size of our spherical model.

Noting that also cosmotron experiments with 1.4 GeV neutrons yield interaction cross sections much lower than those computed from a uniform-sphere model with $R_0 = 1.4 \times 10^{-13}$ cm, WILLIAMS[1] attempted to reconcile these data with the large cross sections at cosmic ray-energies by assuming that the nucleon-nucleon cross section, πr_i^2 in the notation used above, is itself energy-dependent. Apart from uniform density distribution, he investigated several other shapes, among them the Gaussian model

$$\varrho_g = \varrho_0 \exp\left(-r^2/R^2\right) \tag{9.1}$$

[1] M. GERVASI, G. FIDECARO and L. MEZZETTI: Nuovo Cim. (10) **1**, 300 (1955).
[2] Cf. R. HOSTADTER: Rev. Mod. Phys. **28**, 214 (1956).
[3] R. W. WILLIAMS: Phys. Rev. **98**, 1387, 1393 (1956).

and the "tapered" model

$$\left.\begin{array}{l} \varrho_t = \varrho_0, \quad r \leqq R \\ \varrho_t = \varrho_0(2r^3/R^3 - 9r^2/R^2 + 12r/R - 4), \quad R \leqq r \leqq 2R \\ \varrho_t = 0, \quad r \geqq 2R. \end{array}\right\} \qquad (9.2)$$

The latter gives a reasonable fit with the cosmotron data if $R = 0.736 \times 10^{-13}$ cm and $\varrho_0 = 0.166 \times 10^{-39}$ nucleons/cm³ are chosen. This corresponds to a square-well radius of $1.19 A^{\frac{1}{3}} \times 10^{-13}$ cm, in good agreement with HOFSTADTER's electron-scattering experiments.

With the nuclear parameters thus determined, the cosmic ray cross sections can then serve to compute the value of the elementary nucleon-nucleon cross section σ_i. WILLIAMS' calculations lead to a cross section $\sigma_i = \left(120^{+30}_{-20}\right)$ mb at cosmic ray energies — around 30 GeV —, against the 43 mb obtained at 1.4 GeV.

WILLIAMS' conclusions will need some revision in details, since the experimental data used in his analysis are taken from the older measurements of interaction lengths described in Sect. 7 which, as it has been pointed out, are almost certainly too high in the case of the light elements. To illustrate this, we have inserted in Fig. 11, taken from WILLIAMS' paper, the data of ASKOWITH and

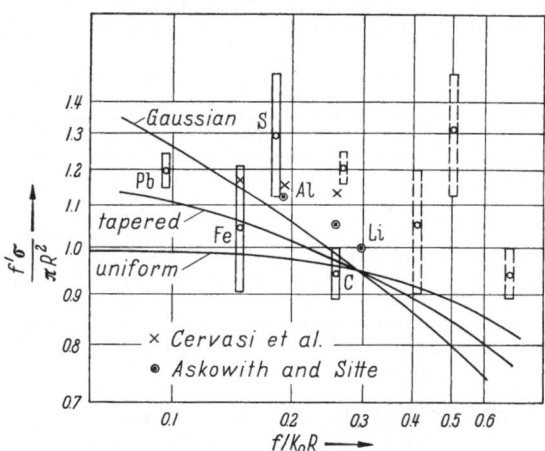

Fig. 11. Opacity, $\sigma_i/\pi R^2$, as a function of the mean free path in nuclear matter divided by the nuclear radius R for the uniform-density model, and for the tapered and the Gaussian model [WILLIAMS (1955)]. Some further points taken from more recent measurements are added.

SITTE (black points) and of CERVASI et al. (crosses). The experimental uncertainties for the new points are not shown; they are of the same order as those of the other points. — It is seen that while this additional evidence is qualitatively compatible with WILLIAMS' arguments, the quantitative agreement is so far unsatisfactory. Moreover, recent experiments of BRENNER and WILLIAMS[1] failed to confirm the assumption of a cross section increasing with energy: Their measurements in iron with primary energies of the order of 20 to 100 GeV yield the very low elementary cross section of ~ 20 mb. An alternative viewpoint, derived from the analysis of interactions of extremely high energies, will be discussed in the following chapters. But most of all, more experimental work both with very light and with medium-heavy target nuclei is urgently needed.

In this connection, it may be appropriate to add that a critical reexamination of another conclusion which has been widely, but perhaps too quickly, accepted may well be worth while. We refer to the statement reported in Sect. 7 that in heavy elements, the cross section for shower production remains geometrical even for the highest primary energies. This appears indeed to follow from a cursory inspection of the results of practically all experiments in that field; but a more careful survey reveals that in the majority of the cases, the results obtained at the highest energies, while usually still, or nearly, consistent with the geometrical value, indicate an increasing cross section. A glance at Table 2 will demonstrate this fact; more examples either covering clearly defined range of average of primary energies, or corresponding to very high primary energies, could be added. In the first category, we mention the initial slopes in the absorption curves of Froehlich and Sitte[1]: at primary energies estimated to lie just under 10 GeV the authors find a mean free path of (160 ± 8) and (167 ± 9) g/cm^2; at around 20 GeV, they report (135 ± 12) g/cm^2. In the second category, Gottlieb's results in gold[2] — (145 ± 15) g/cm^2 — may be quoted. In view of differences in techniques, evaluation, and average primary energy it is difficult to summarize these results without prejudice and arbitrariness, but if we take the papers quoted here and in Table 2 as a representative sample, we arrive at the picture shown in Table 4: the values observed at primary energies around or just below 10 GeV scatter around the geometrical mean free path, with an average close to that length. At primary energies well in excess of 10 GeV, the experimental uncertainties would in no single experiment permit to exclude the geometrical

Table 4. *Ratio* λ_i/λ_g *of the observed interaction mean free paths* λ_i *in heavy elements, to its geometrical value* λ_g, *in two ranges of the primary energy* E_p.

Author	$E_p \lesssim 10\,\text{GeV}$	$E_p > 10\,\text{GeV}$
Cocconi	0.98 ± 0.10	—
Walker	0.92 ± 0.06	0.85 ± 0.07
Sitte	0.99 ± 0.06	—
Gottlieb	—	0.90 ± 0.10
Boehmer and Bridge	1.33 ± 0.21	0.87 ± 0.20
Froehlich et al.	0.97 ± 0.08	—
Froehlich and Sitte	1.02 ± 0.06	0.83 ± 0.07
Average	1.01 ± 0.04	0.86 ± 0.05

mean path, but the weighted average over four experiments makes such a large value appear very unlikely. Again, further and more accurate experiments are called for in order to ascertain the existence and to affirm the right interpretation of the phenomenon.

Note added in proof. Remarkable indications that this suspected variation of the elementary interaction cross section does in fact exist, and may even be quite considerable, have recently been observed in the study of high energy jets in nuclear emulsions carried out by the Bristol group[3]. A careful examination of the distribution of origins of the jets in the emulsion stacks shows that the fraction of events starting near the top of the stack is much too high to be compatible with a cross section of the order of the traditional "geometrical" value. Indeed, an elementary cross section well above 60 mb may be needed to explain the experimental distribution. It is significant — as will be discussed in more detail later — that at the same time, the average inelasticity of the interactions tends to decrease. Thus, the total energy transferred by the primary to shower secondaries *per unit path in* g/cm^2 does not change appreciably.

10. The interaction mean free path in hydrogen. One may well ask why nothing has been said, so far, about what must appear the most obvious method to decide this question: a direct determination of the elementary collision cross section σ_i or of the mean free path λ_n in nuclear matter, through the measurement of the interaction mean free path in hydrogen. The answer is that, unfortunately, numerous attempts to do this have failed to elucidate the situation, and that the experimental results reported by various authors are by no means in full agreement.

In part, the reason can be found in purely technical difficulties. Evidently, the use of liquid hydrogen as attenuation layer in an arrangement of the type

---

[1] F. E. Froehlich and K. Sitte: Phys. Rev. **97**, 151 (1955).

[2] M. B. Gottlieb: Phys. Rev. **82**, 349 (1951).

[3] I am indebted to Dr. D. H. Perkins for communicating these results in discussions, and permitting their use before publication.

of Fig. 5 introduces a host of tricky problems. Consequently, few cosmic-ray experiments of this sort have been reported, and those few remained rather inconclusive. Thus, VIDALE and SCHEIN[1], in a counter experiment operated at 90 000 feet altitude, derived from their data an interaction cross section between the geometrical value (61.5 mb) and twice that area. On the other hand, WEAVER[2], using a counter-controlled cloud chamber in a magnetic field and comparing the rates of showers produced in targets of Pb, Al, C and liquid hydrogen, found that in the 626 hours of his operation with an average target thickness of 2.28 g/cm², not a single shower was observed which originated in the hydrogen, although the equipment should have recorded at least 6 such events if the cross section for shower production in hydrogen had the geometrical value. He concludes, therefore, that the penetrating showers originating in other nuclei are mostly due to plural interaction in the target nuclei, and that in the region of primary energies observed in his experiment—about 6 GeV—a real multiple production of mesons in a nucleon-nucleon collision does not occur in more than 15% of all collisions.

This straightforward experiment serves well to illustrate the difficulties encountered in any attempt to determine the interaction length in hydrogen from a comparison of the shower rates. Even more than for other light elements, the objections already discussed in the preceding section most be raised: since counter control, or any similar selection system, is tied to a certain fixed *multiplicity*, the average *energy* of the shower primaries will vary with the target element, and will be higher in hydrogen than in a complex nucleus. Let us recall that even the uniform nuclear model demands already in a nucleus as light as C, a probability of the order of 50% for the occurrence of second-generation collisions in the target nucleus; any non-uniform model in general agreement with HOF-STADTER'S conclusions would, with the large elementary cross section then required, lead to an even larger probability of plural effects. Before more and better experimental evidence becomes available, it is, therefore, practically impossible to derive from such measurements safe quantitative conclusions about the mean free path in hydrogen. The following discussion of some further experiments will make this point even more abundantly clear.

In several investigations, an attempt has been made to measure the cross section for high-energy interactions in hydrogen by comparing the rates of shower production in carbon and in hydrocarbons such as paraffin or oil. Some of these[3-5] were satisfied with the qualitative conclusion that in nucleon-nucleon collisions at high energies multiple production of mesons does occur. Of the others, most authors agree that the mean free path in hydrogen is probably longer than that corresponding to the geometrical cross section of 61.5 mb. Thus, WALKER et al.[6] found with an apparatus selecting showers of primary energies around 20 to 30 GeV if originating in Pb or C, a mean free path of (31^{+35}_{-7})g/cm² (or a cross section between 25 and 70 mb), still just consistent with the geometrical value. They also concluded from their observations that a considerable fraction of the interactions originating in hydrogen may have been missed by their triggering arrangement.—A larger mean free path was reported by FROMAN et al.[7]

[1] M. L. VIDALE and M. SCHEIN: Nuovo Cim. **8**, 1 (1951).

[2] A. B. WEAVER: Phys. Rev. **90**, 86 (1953).

[3] G. BERTOLINI, M. CINI, P. COLOMBINO and G. WATAGHIN: Nuovo Cim. **9**, 407 (1952).

[4] O. HAXEL and M. SCHULTZ: Z. Naturforsch. **9a**, 178 (1954).

[5] I. MIURA, T. MATANO, Y. TOYODA and T. MURAYAMA: Progr. Theor. Phys. **13**, 115 (1955).

[6] W. D. WALKER, N. M. DULLER and J. D. SORRELS: Phys. Rev. **86**, 865 (1952).

[7] D. FROMAN, J. KENNEY and V. H. REGENER: Phys. Rev. **91**, 707 (1953).

who, from a comparison of the collision lengths of neutral primaries in H_2O [(113 ± 12) g/cm^2] and in D_2O [(123 ± 13) g/cm^2] derive a value $\sigma_D - \sigma_p = (13 \pm 15)$ mb, and from the absolute values of the two mean free paths, conclude that also the cross sections σ_n and σ_p of interactions with neutrons and with protons must be much smaller than the geometrical value. In their experiments the primary energies were probably considerably smaller. Rediker[1], measuring collision lengths in graphite and oil, calculated a cross section between 0 and 70 mb—again including the geometrical value but favouring a smaller one.— Finally, also Cervasi et al.[2] reported a cross section somewhat less than the geometrical area: the average between the data obtained with charged and neutral primaries is about 47 mb. This result is perhaps particularly significant since these authors, working with a much more sensitive shower selection, found essentially no transparency even in C and Al.

The question must now be asked whether it is possible to reconcile the results indicating a small cross section for shower production in hydrogen, with Williams' calculations demanding a large total cross section for nucleon-nucleon collisions. Here again, data are lacking for a satisfactory quantitative analysis, but from semi-qualitative arguments it would seem that this should be possible, provided only that the degree of inelasticity in a nucleon-nucleon collision is rather small. In other words, at not too high energies most interactions in hydrogen, failing to meet the usual multiplicity conditions, will not be counted as "penetrating showers". — In this case, the difference in the multiplicities of the interactions in hydrogen and even a light nucleus would be appreciable—thus accounting for the long mean free path for shower production measured in hydrogen experiments—while the difference in the multiplicities of interactions in, say, C and Pb showers might be small since already in a carbon nucleus, the primary would dissipate so much energy in its successive collisions that further interactions, as they might occur in the larger Pb nucleus, cannot contribute in a large measure to the meson production. This point will be taken up again in the following sections.

11. The interaction mean free path of the shower secondaries. It is of considerable interest for our understanding of the elementary interactions and of the propagation of the nuclear cascades, to know whether all shower secondaries. are themselves nuclear-active and contribute equally to successive radiative collisions, or whether this effect is restricted to the nucleons. Consequently, this question has been the subject of numerous investigations with cloud chambers, counter arrangements, and emulsions. In view of the quite appreciable differences in the types of events which are efficiently selected with these various techniques, it is not surprising that the earlier results were frequently contradictory. Thus, for instance, the first cloud chamber experiments were carried out with comparatively thick producer plates in which, owing to complete absorption of all charged secondaries, some interactions remained unrecorded. Furthermore, large-angle scattering was then usually not included among the events of interest. Consequently, the quantity measured in these experiments was a mean free path for the production of very energetic secondary showers rather than that for nuclear interactions in general; but it took some time before this difference was duly recognized. Similar, and in some cases even more serious, qualifications must be made in the case of some earlier counter experiments. Evidently, the superior technique for a determination of the interaction length of

[1] R. H. Rediker: Phys. Rev. **95**, 526 (1954).
[2] M. Cervasi, G. Fidecaro and L. Mezzetti: Nuovo Cim. (10) **1**, 300 (1955).

shower secondaries, is that of the nuclear emulsions. With this method, the recognition of all collisions is easier, and their classification safer than with the others. A certain difficulty which could, in principle, lead to discrepancies, might arise from the fact that the distances involved are always very small,

Table 5. *Ratio of the observed interaction mean free paths, λ_i, to the geometrical values λ_g, for charged shower secondaries.*

Author	λ_i/λ_g	Remarks
FRETTER[1]	4.6	Cloud chamber. Pb. Uncorrected
BUTLER et al.[2]	2.4 ± 0.8	Cloud chamber. Pb. Uncorrected
BARKER and BUTLER[3] . .	1.24 ± 0.20	Cloud chamber. Pb. Corrected
LOVATI et al.[4]	1.85 ± 0.60	Cloud chamber. Pb. Corrected
LOVATI et al.[5]	1.22 ± 0.25	Cloud chamber. Pb. Corrected
BROWN and McKAY[6] . .	1.95 ± 0.50	Cloud chamber. Pb. Corrected
HARTZLER[7]	1.45 ± 0.40	Cloud chamber. Au. Corrected.
DULLER and WALKER[8] . .	$1.10 {}^{+\,0.5}_{-\,0.3}$	Cloud chamber. Pb. $E_s \gtrsim 4$ GeV
	$4.0 {}^{+\,3.0}_{-\,2.2}$	$E_s \sim 2$ GeV
	$10 {}^{+\,8}_{-\,3}$	$E_s \sim 1$ GeV
	$1.33 {}^{+\,1.0}_{-\,0.4}$	C. $E_s \gtrsim 4$ GeV
	$6.2 {}^{+\,15}_{-\,1.8}$	$E_s \sim 2$ GeV
	$23 {}^{+\,12}_{-\,7}$	$E_s \sim 1$ GeV
ASKOWITH and SITTE[9] . .	a) 2.2 ± 0.5 b) 1.15 ± 0.15	Cloud chamber. Pb. (a) Shower production (b) All interactions
PICCIONI[10]	>11	Counters. Fe. No corrections
COOL and PICCIONI[11] . .	1.33	Counters. C and Pb
FROEHLICH and SITTE[12] .	2.6 ± 0.6	Counters. Pb. $E_\pi \sim 8$ GeV
	$2.6 {}^{+\,2.9}_{-\,0.6}$	$E_\pi \sim 20$ GeV
HARDING and PERKINS[13] .	$\gtrsim 1.33$	Emulsions. All secondaries
CAMERINI et al.[14]	0.9 ± 0.4 1.1 ± 0.3	Emulsions. All interactions. $E_\pi \lesssim 1.5$ GeV "Shower particles"
LAL, PAL and RAMA[15] . .	1.05 ± 0.2	Emulsions. Primary energies 50 — 1000 GeV/nucleon

[1] W. B. FRETTER: Phys. Rev. **76**, 511 (1949).

[2] C. C. BUTLER, W. G. V. ROSSER and K. H. BARKER: Proc. Phys. Soc. Lond. A **63**, 145 (1950).

[3] K. H. BARKER and C. C. BUTLER: Proc. Phys. Soc. Lond. A **64**, 4 (1951).

[4] A. LOVATI, A. MURA, G. SALVINI and G. TAGLIAFERRI: Phys. Rev. **77**, 284 (1950).

[5] A. LOVATI, A. MURA, C. SUCCI and G. TAGLIAFERRI: Nuovo Cim. **8**, 271 (1951).

[6] W. W. BROWN and A. S. McKAY: Phys. Rev. **77**, 342 (1950).

[7] A. J. HARTZLER: Phys. Rev. **82**, 359 (1951).

[8] N. M. DULLER and W. D. WALKER: Phys. Rev. **93**, 215 (1954).

[9] J. G. ASKOWITH and K. SITTE: Phys. Rev. **97**, 159 (1955).

[10] O. PICCIONI: Phys. Rev. **77**, 1 (1950).

[11] R. COOL and O. PICCIONI: Phys. Rev. **87**, 531 (1952).

[12] F. E. FROEHLICH and K. SITTE: Phys. Rev. **97**, 151 (1955).

[13] J. HARDING and D. PERKINS: Nature, Lond. **164**, 285 (1949).

[14] U. CAMERINI, P. H. FOWLER, W. P. LOCK and H. MUIRHEAD: Phil. Mag. **41**, 413 (1950).

[15] D. LAL, YASH PAL and RAMA: Nuovo Cim. Suppl. **12**, 347 (1954).

and that the survey thus includes short-lived secondaries which are missed in other experiments.

A list, representative rather than complete, of the results of a number of experiments with cloud chambers[1], counters[2], and emulsions[3], is given in Table 5. In order to illustrate the importance of the corrections for scattering and "lost" interactions mentioned above, some of the older data are included, and the ratio λ_i/λ_g of the secondary interaction mean free path to the geometrical collision length is listed so that the values measured in the various—though mostly heavy or medium-heavy—target materials can be compared. The general picture is evident: with improved technique, the observed cross sections tend to approach the geometrical value, at least for sufficiently high particle energy. This energy dependence is particularly well demonstrated in the results of DULLER and WALKER[4] who succeeded in determining the energy of the shower secondaries from the geometry of the showers they produce; and only the measurements of FROEHLICH and SITTE[5] appear, at first, to disagree with that tendency. But even this contradiction may not be as serious as it seems. The latter authors, intent on studying a possible difference in the mean free paths for shower production by π-mesons and by nucleons, had to impose rather stringent conditions for shower selection which might lead to an erroneous interpretation of the data if the properties of the meson-induced showers—like multiplicity, geometry, and so on—differ from those of showers of nucleonic origin. Such a difference might well exist, particularly in the energy region involved in those experiments where hyperon production may play a significant part. But this question is already outside of the scope of this chapter; suffice it to say as a summary that, apparently, the *interaction* mean free path of shower secondaries reaches its geometrical value already at energies of the order of a few GeV and remains reasonably constant also at higher energies. The mean free path for *shower production*, however, is longer at least up to about 4 GeV and perhaps even around 10 GeV: a feature which we shall have to remember in the discussion of the intranuclear cascade processes.

III. Meson production in penetrating showers.

12. Survey of the theories of meson production in nucleon-nucleon collisions. When research on penetrating showers entered its quantitative phase, the hopes of the theorists rose high that from its results, clues could be derived to resolve the fundamental difficulties in which meson theory found itself at that time. The production of mesons in high-energy nucleon-nucleon collision appeared to be ideally suited to serve as a proofstone for the various models and concepts out of which meson theories had been constructed.

Unfortunately it soon became clear that penetrating shower data would not necessarily provide evidence on the finer points of meson theory. It may well be that the production of mesons is very insensitive to the details of the mechanism of the collision, and that a picture rather in analogy to the liquid-drop model and the compound nucleus idea of low-energy nuclear physics may prove adequate. Or, to use the definitions of LEWIS' survey of the theories of meson production[6], the possible theories can be divided into two groups, of which only one, the "pre-existing field" theories, involves consideration of the exact features

---

[1] See footnote 1—9, p. 179.
[2] See footnote 10—12, p. 179.
[3] See footnote 13—15, p. 179.
[4] See footnote 8, p. 179.
[5] See footnote 12, p. 179.
[6] H. W. LEWIS: Rev. Mod. Phys. **24**, 241 (1952).

of nuclear forces. The theories of this group can best be described in the language of the Weizsäcker-Williams method developed for problems of electromagnetic radiation: Here as there, the collision is treated as the interaction of two particles surrounded by a "pre-existing" field, and the outcome is primarily determined by the interaction of this field with the incident particle. We may say that part of the field is "shaken loose" in the collision, and transformed into free particles "emitted" in the collision. Against this view, the group of "excited field" theories assumes that essentially, the collision has only the effect of transferring a large amount of energy and momentum into the "collision volume", and that subsequently, this "hot" region of strong interaction will release energy in the form of quanta of its field.

From the point of view of the experimenter, another kind of classification has been most useful: that according to the multiplicity of the elementary collision. If, as in the case of electromagnetic interactions, as a rule only one quantum is emitted per collision, we speak of "plural theories", since in this case a shower containing n mesonic secondaries will be due to a series of n successive collisions inside a target nucleus. This idea was used by HEITLER and his co-workers[1], following his theory of radiation damping. With some modifications, it has been upheld until quite recently[2], although the experimental results clearly contradicted one of the most evident conclusions of any plural theory: the condition that a shower of n mesons must also contain $(n+1)$ nucleons. HEITLER and TERREAUX[3] have found a way to get around this difficulty by postulating that since after an energetic collision, primary and secondary nucleon will move essentially parallel and with high energy, both—or in a later generation, even more—particles may collide with the same further nucleons in the target nucleus, thus releasing two or several mesons in a composite collision. We shall return to this picture later.

The alternative is the production of several mesons in a single nucleon-nucleon collision, envisaged in all "multiple" theories. Successive collisions in one nucleus, each resulting in the creation of secondaries and thus multiplying the total of shower particles in an intranuclear cascade, are of course not ruled out; but they are not needed in order to explain the predominance of mesons among the shower secondaries.

In recent accelerator experiments, multiple production of π-mesons in proton-proton collisions has been clearly established as a quite common process[4]. The following discussion will, therefore, be restricted to the most important types of "multiple" theories only. Furthermore, the ambiguities in "pre-existing" field theories are such that only "excited-field" theories will be considered. Also, the presentation will include only those points which are considered essential for the following interpretation of the experimental data, and no attempt will be made to judge the merits of the theories.

Phenomenologically the simplest model is that used by FERMI[5]. It is based on the assumption that the entire energy of the colliding energetic nucleons—in

---

[1] J. HAMILTON, W. HEITLER and H. W. PENG: Phys. Rev. **64**, 78 (1942). — W. HEITLER: Rev. Mod. Phys. **21**, 113 (1949). — W. HEITLER and L. JANOSSY: Proc. Phys. Soc. Lond. A **62**, 374 (1949); **62**, 669 (1949).

[2] For literature, cf. H. MESSEL: Progress in Cosmic Ray Physics, edit. by J. G. WILSON, Vol. II, p. 135. 1954.

[3] W. HEITLER and C. TERREAUX: Proc. Phys. Soc. Lond. A **66**, 929 (1953). — C. TERREAUX: Helv. phys. Acta **24**, 551 (1951).

[4] E.g. M. M. BLOCK, E. M. HARTH, V. T. COCCONI, E. HART, W. B. FOWLER, R. P. SHUTT, A. W. THORNDYKE and W. L. WHITTEMORE: Phys. Rev. **103**, 1483 (1956).

[5] E. FERMI: Progr. Theor. Phys. **5**, 570 (1950). — Phys. Rev. **81**, 683 (1951).

the centre-of-mass system pictured as flat discs—is dissipated in the collision volume, and that in this region a genuine thermodynamical equilibrium is established. Mesons, nucleons and antinucleons are then evaporated from this volume, and since its linear dimensions are of the order $1/\varkappa = \hbar/mc$, the energy density in it is about $\varkappa^3 E_0$ (E_0 is the primary energy) and the temperature $(\varkappa^3 E_0)^{\frac{1}{4}}$. This leads directly to three important conclusions which in principle can be subjected to experimental checks:

1. The average multiplicity $\langle n \rangle$ of the shower varies with the primary energy E_0 according to $\langle n \rangle \propto E_0^{\frac{1}{4}}$.

2. The energy distribution of the shower secondaries in the center-of-mass system resembles a Maxwell distribution. Also, all kinds of particles will, on the average, be produced in equilibrium.

3. The angular distribution of the secondaries is isotropic in the center-of-mass system for head-on collisions though more collimated in forward and backward direction for collisions with a larger impact parameter.

Thus, high-energy meson production is, in the Fermi theory, a purely thermo-dynamical process not affected by the special features of meson fields. However, serious objections can be raised against this radical picture: In order to produce thermodynamical equilibrium in the extremely short time available, the coupling between the particles in the "hot" interaction volume must be very strong. But if that is so, we have no right to treat them as entirely independent free particles, as it is done in the thermodynamical theory. In other words: the time available is too short, and the volume too small, in order to permit achievement of a genuine equilibrium. This state can be reached only after the wave packets representing the particles present after the collisions, have expanded to a larger volume; until then, the interaction must be considered as continuing. But by then, the temperature of the collision will be lower than that computed according to the Fermi theory.

Instead of restricting themselves to purely thermodynamical considerations, LANDAU[1] and HEISENBERG[2] have proposed theories that might be termed semi-classical. LANDAU'S approach resembles that of Fermi, but inclusion into his calculations of interactions during the expansion phase makes his results differ in two points strongly from those of Fermi: the energy spectrum of the second-aries is very much less uniform, and the degree of inelasticity can be much lower than unity. HEISENBERG, on the other hand, builds his theory on the similarity of the propagation of the strongly-interacting meson waves emerging from the collision, with the emission of shock waves in an explosion. In the first order, the theory is again a classical one since we are dealing with a phenomenon involv-ing high quantum numbers, and quantum effects can be taken into account with the help of the correspondence principle. The main differences between the results of HEISENBERG'S theory and that of Fermi are in the following points:

1. The average multiplicity n varies with the primary energy E_0 according to

$$\langle n \rangle = \langle \varepsilon \rangle / m_\pi c^2 \cdot \log (1 - \beta^2)^{-\frac{1}{2}},$$

where $\langle \varepsilon \rangle$ stands for the mean value of the maximum fraction of energy that is available in the center-of-mass system. The multiplicity of the interaction is subject to very large fluctuations, in conformity with the impact parameter.

[1] L. LANDAU: Izv. Akad. Nauk. SSSR. **17**, 51 (1953). — S. Z. BEDENKU and L. LANDAU, Suppl. Nuovo Cim. **3**, 15 (1956).

[2] W. HEISENBERG: Z. Physik **113**, 61 (1939); **126**, 569 (1949). — Nuovo Cim., Suppl. **6**, 493 (1949).

2. The energy distribution of the shower secondaries is of the type of a spectrum, $dn \propto dE/E^{\alpha+1}$ ($\alpha > 0$), rather than similar to a Maxwell distribution.

3. Though a large part of the energy is carried off by mesons of comparatively low energy in all directions, the most energetic mesons will travel close to the direction of the primary. The angular distribution will in general not change much with the impact parameter—the energetic mesons always taking up the angular momentum—but it will depend strongly on the primary energy.

An interesting model which can be adjusted to all similar theories has been discussed by ZATSEPIN[1], and, independently, by KRAUSHAAR and MARKS[2]. They assume that in the primary collision, two highly excited nucleons are formed, and that their life-time is long enough in order to ensure separation of the particles before their decay. The disintegrations of the two excited particles can therefore be treated as independent events, for instance according to the thermodynamical model (KRAUSHAAR and MARKS). ZATSEPIN has pointed out that this model predicts, or can be made to explain, several disturbing experimental results observed in high-energy collisions: the surprisingly high elasticity of those interactions, the possibility of asymmetry of forward and backward cone (in general, the angular distribution will of course again show strong but symmetrical forward and backward collimation), and the differences in the jets originated by nucleons, and by the energetic secondaries of a high-energy collision.

Note added in proof. Since also in the following, reference to the "two-center model" of this kind will be made repeatedly, two further short remarks may be in order:

(i) Any theory which makes use of the concept of meson emission from an "expanding cloud"—such as HEISENBERG's or LANDAU's—should necessarily come to adopt the two-center model. At very high energies, the relativistic contraction of the colliding particles will not permit the ejection of mesons of the known average secondary energies during the short period of actual overlapping. The bulk of the secondaries will thus be emitted only after the collision partners have already separated.

(ii) One may even use this picture to describe the phenomenon quantitatively, i.e. to derive from it a secondary spectrum, and thus to test in an empirical way the basic ideas involved. Considering the expanding field as a rather turbulent "hot" gaseous cloud, parts of which may be lost by ejection, it appears plausible to impose two restrictions on the emission of secondaries: (a) particles of a de Broglie wavelength exceeding the cloud dimension cannot be formed, and hence are not emitted at that stage of expansion; and (b) particles of de Broglie wavelengths much less than the dimensions of the cloud, formed in its interior, will in general not be emitted because of the short life of such turbulences. *It then follows that at any given stage of expansion, practically only such mesons can escape whose de Broglie wavelength just coincides with the size of the cloud.* In consequence, the energy spectrum of the secondaries is determined by the rate of expansion of the meson cloud. If a suitable equation of state is chosen for the "hot gas", the entire process can be treated quantitatively. At any rate, it is easily seen that—in particular if also emission of one or a few high-energy secondaries from the original collision volume is permitted, i.e. if a "three-center model" is used—the general features of the experimental data, including the occasional transfer of a very large fraction of the primary energy to a single secondary, are well understood in this way. The results of the calculations, so far encouraging, will be presented elsewhere.

The significance of a success of an empirical theory of this kind would, of course, not lie in a confirmation of such details as have to be introduced, but in a check concerning the validity of basic concepts on dimensions and identities up to extremely high energies.

The experimental evidence will be discussed in the following sections. As it has been said before, no attempt will be made to derive conclusions about the correctness of the theories; the material at our disposal does not permit to argue for more than plausibility or comprehensiveness.

13. Methods of determination of the primary energy. Every quantitative study of meson production in showers must evidently begin with a determination of

[1] V. I. ZATSEPIN: International Conference on Cosmic Radiation, Varenna 1957 (report on previous work).

[2] W. L. KRAUSHAAR and L. J. MARKS: Phys. Rev. **93**, 326 (1954).

the nature and energy of the shower primary. While the first is generally a fairly simple task, the second is as a rule much more difficult, and sometimes impossible to solve in an unambiguous way.

In principle, measurements can be carried out directly on the primary, or on the secondaries with the purpose of deducing the energy of the shower-initiating particle from the effects it creates. For a direct measurement, the technique of a magnetic cloud chamber would appear the obvious choice, if its usefulness were not seriously impaired by the very small range of energies to which it can be applied with satisfactory accuracy. Even momenta up to 10 GeV/c can be measured only if very strong fields extending over a sufficiently large path are supplied. The shower itself would have to be initiated in, or above, a second— perhaps a multiple-plate—cloud chamber. This results in a drastic reduction of the solid angle of collection, and makes the experiment unattractive.

Scattering measurements in a multiple-plate cloud chamber give useful information only for primary energies well below 10 GeV, and thus are not suited for this special problem. However, somewhat better results can be achieved with the help of scattering measurements in nuclear emulsions. This technique, as described in more detail in another part of this Encyclopedia[1], did permit to push the limit of measurable primary energies well beyond 10 GeV. It has been used repeatedly since the early investigations of CAMERINI et al.[2] demonstrated its advantages. More recently, though, indirect methods have been widely applied, and have made it possible to extend the observations to energies several orders of magnitude higher than any direct method could give.

Two methods of indirect measurements have been worked out. The first, less accurate but more straightforward, aims at determination of the primary energy from a study of the energy dissipation of the nuclear cascade originated by the primary. This can be done either by measuring the integral path length of the cascade, in the manner used by FROEHLICH et al.[3], or by determining the total ionization produced under various layers of absorber, as in the experiments of VERNOV and his collaborators[4] on the soft shower component, and more recently by NIKHOLSKY et al.[5, 6] on primaries up to very high energies—a truly "calorimetric" method. Statistically, this method yields significant results; in individual cases, the errors may be large and the evidence obtained scanty.

The second approach to indirect energy measurements is based on simple kinematic considerations. In a nucleon-nucleon collision, naturally no distinction can be made between the incident particle and the "target" nucleon. Viewed in the center-of-mass system (the "C-system" in the following, with "L-system" for that in which the laboratory is at rest), the two particles are identical, and hence both must be considered in an equal role as the progenitors of a meson shower. Consequently, this shower will exhibit a twofold symmetry: axially, with respect to the line connecting the two colliding nucleons, and—even more important— specularly, with respect to the plane of the collision. After the interaction, one-half of all the mesons produced will on the average move in

[1] M. M. SHAPIRO: Vol. XLV.

[2] U. CAMERINI, J. H. DAVIES, P. H. FOWLER, C. FRANZINETTI, W. O. LOCK, D. H. PERKINS and G. YEKUTIELI: Phil. Mag. **41**, 1291 (1950).

[3] F. E. FROEHLICH, E. M. HARTH and K. SITTE: Phys. Rev. **87**, 504 (1952).

[4] E.g. S. N. NERNOV, N. L. GRIGOROV, G. T. ZATSEPIN and A. E. CHUDAKOV: Izv. Akad. Nauk SSSR. **19**, 493 (1955).

[5] NIKHOLSKY and coll., report by N. I. DOBROTIN: International Conference on Cosmic Radiation, Varenna 1957.

[6] GRIGOROV, MURZIN and RAPAPORT: J. Exp. Theor. Phys. **34**, 506 (1958). — GRIGOROV et al.: Proc. Moscow Cosmic Ray Conf. I, p. 130 (1959).

forward, and one-half in backward, direction in the C-system. Seen in the L-system, the mesons travelling in the direction of the primary will all be found within the narrow cone (or "jet" if the primary energy is sufficiently high) into which all "forward" directions are compressed by the Lorentz transformation, while the other half will spread out over the remaining large solid angle[1]. The situation is illustrated by the schematic drawings of Fig. 12a and b, showing the angular distribution of a typical shower in the C-system and in the L-system.

The observation of the opening half-angle $\vartheta_{\frac{1}{2}}$ of the cone within which onehalf of all particles are confined in the L-system permits, therefore, the determination of the velocity β_c of the C-system with respect to the laboratory. Writing ϑ_i', p_i', U_i', β_i' for the angle of emission, the momentum, the total energy and velocity of the i-th particle of a shower in C, and $\vartheta_i, p_i, U_i, \beta_i$ for the same quantities in L, we have the general relation

$$\tan \vartheta_i$$
$$= \frac{\beta_i' \sin \vartheta_i'}{\beta_i' \cos \vartheta_i' + \beta_c} \times$$
$$\times \sqrt{1 - \beta_c^2} \qquad \left.\right\} (13.1)$$

which in the extremely relativistic case, $\beta_i' \approx \beta_c \approx 1$, reduces to

$$\tan \vartheta_i = \tan (\vartheta_i'/2) \times$$
$$\times \sqrt{1 - \beta_c^2}. \qquad \left.\right\} (13.2)$$

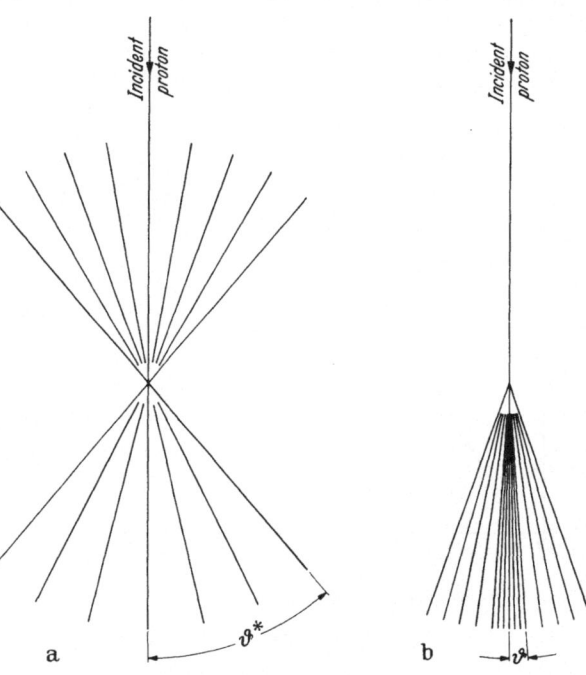

Fig. 12a and b. Schematic representation of a high-energy interaction (a) in the center of mass system; (b) in the laboratory system. —— Instead of ϑ^* read ϑ'.

Hence, if $\vartheta_{\frac{1}{2}}'$ is the half-angle of the cone in C into which a certain fraction of all shower secondaries is emitted, the same fraction will be found in L within $\vartheta_{\frac{1}{2}}$ according to

$$\tan \vartheta_{\frac{1}{2}} = \tan (\vartheta_{\frac{1}{2}}'/2) \cdot \sqrt{2MC^2/U} = \tan (\vartheta_{\frac{1}{2}}'/2) \cdot \gamma_c^{-\frac{1}{2}}. \qquad (13.3)$$

$M c^2$ is the rest energy, U the total energy in L of the primary, and γ_c its total energy in units of its rest mass in C.—In particular, one-half of all shower particles will be confined to a half-angle $\tan \vartheta_{\frac{1}{2}} = \gamma_c^{-\frac{1}{2}}$: a relation which can easily serve to measure the primary energy γ_c.

WALKER et al.[2] introduced into this method another valuable criterion: realizing that it can claim correctness only for a genuine nucleon-nucleon interaction, they noted that under the assumption of isotropic distribution of the secondaries in the C-system, the fraction $F(\vartheta')$ of all particles lying in C inside of the cone of opening half-angle ϑ' is

$$F(\vartheta') = \sin^2 (\vartheta'/2). \qquad (13.4)$$

---

[1] Exceptions are possible, for more precisely, only the total forward *momentum* must equal that ejected backward. Theories of the kind proposed by ZATSEPIN, or by KRAUSHAAR and MARKS (cf. preceding Sect. 12) permit even strong asymmetry in the particle numbers.

[2] W. D. WALKER and N. M. DULLER: Phys. Rev. **93**, 215 (1954).

Consequently, (13.3) gives for the fraction $F(\vartheta)$ of all particles observed in L inside of ϑ, the relation

$$\tan^2\vartheta = \frac{1}{\gamma c^2} \cdot \frac{F(\vartheta)}{1 - F(\vartheta)}. \tag{13.5}$$

In order to check the reliability of this relation, WALKER and DULLER drew up the "F-plots" according to (13.5) for a number of stars studied by the Bristol group in nuclear emulsions. Their result is reproduced in Fig. 13. It shows a surprisingly good fit between the experimental points and the computed curve, and a satisfactory agreement between the value of the primary energy determined from the plot, and directly from scattering measurements. Thus, it appears that at least for energies of this order of magnitude the method of the "F-plots" is well applicable, and hence the angular distribution of the secondaries in the C-system not far from isotropic.

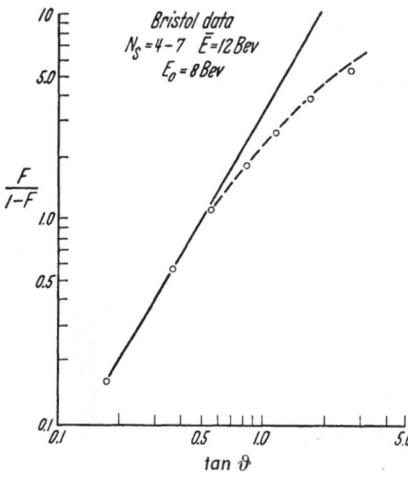

Fig. 13. "F-plot" of the angular distribution of shower secondaries published by CAMERINI et al. for events of 4-7 particles from DULLER and WALKER (1954).

But for still higher energies, neither this nor the other basic assumption—that the shower originated in a single nucleon-nucleon encounter—can be correct, and the method has to be modified. A suitable procedure has been developed by CASTAGNOLI[1], and by v. LINDERN[2]: First, note that for high primary energies, the *integral* distribution F_1^+

$$F_1^+ = \frac{\gamma_i^2 \sin^2\vartheta_L}{1 + (\gamma_i^2 - 1)\sin^2\vartheta_L} \tag{13.6}$$

can be written in terms of the variable $u = \gamma_i \vartheta_L$

$$F_1^+ = u^2/(1 + u^2). \tag{13.7}$$

This relation has proved very useful for checks on the accuracy of the basic assumptions made. — If instead of an equal average energy γ_i for all secondaries, a Heisenberg spectrum is introduced, one finds a similar relation (cf. v. LINDERN[2]) for the integral distribution F_2

$$1 - F_2 = 1/(1 + u^2)^{\frac{3}{2}}. \tag{13.8}$$

Note that owing to the preponderance of slow secondaries, in C, the primary energy calculated by the half-angle method of (13.3) or its equivalent,

$$\gamma_L = 2/\tan^2\vartheta_{\frac{1}{2}} \tag{13.9}$$

and identification of $\vartheta_{\frac{1}{2}}$ with the angle of emission of the "forward cone", the primary energy will in the average be overestimated, and should be reduced by a factor 0.77. The deviations are less marked for a Fermi or a Landau secondary spectrum.

A similar expression can be derived for anisotropic emission. It has been shown by v. LINDERN[2] that if the angular distribution (in C) follows a $\cos^n\vartheta$-law, the integral distribution in L satisfies the relation

$$1 - F_3 = 1/(1 + 2u). \tag{13.10}$$

---

[1] C. CASTAGNOLI, G. CORTINI, C. FRANZINETTI, A. MANFREDINI and D. MORENO: Nuovo Cim. **10**, 1539 (1953).

[2] L. v. LINDERN: Z. Naturforsch. **11**a, 561 (1956).

For further details, the reader is referred to v. LINDERN' spaper, where also additional literature is listed.

A further generalisation, applicable also where the condition $\beta_i = \beta_c = 1$ is no longer met—a case of importance for very high primary energies—but where symmetry of forward and backward cone in C can be assumed, has been discussed by EDWARDS et al.[1]. It is evident that when the average secondary energy $\gamma_\pi < \gamma_c$, particles traveling at large backward angles in C will appear in the forward cone of L. The angles of emission in L have a maximum possible value $\vartheta_{\max}$ given by

$$\sin \vartheta_{\max} = \frac{\beta_\pi}{\beta_c} \cdot \frac{\gamma_\pi}{\gamma_c} \tag{13.11}$$

and mere counting of the tracks in the forward cone will lead to an erroneously small value of the half-angle $\vartheta_{\frac{1}{2}}$.

However, for any two particles emitted symmetrically in C (i.e., at angles ϑ and $\pi - \vartheta$), the L-system angles ϑ_f and ϑ_b satisfy

$$\gamma_c^2 \tan \vartheta_f \tan \vartheta_b = \frac{\sin^2 \vartheta}{(\beta_c/\beta_\pi)^2 - \cos^2 \vartheta} \tag{13.12}$$

leading to forward-backward symmetry if $\beta_\pi \geqq \beta_c \approx 1$, and ϑ not too small, with

$$- \log \gamma_c = \frac{1}{n_s} \sum_{k=1}^{n_s} \log \tan \vartheta_k \tag{13.13}$$

—a relation already obtained by CASTAGNOLI.

If $\beta_c/\beta_\pi = -1$, the median angle $\vartheta_{\frac{1}{2}}$ satisfies

$$\gamma_c \tan \vartheta_f = \tan \vartheta_f / \tan \vartheta_{\frac{1}{2}} = (\tan \vartheta)/2 \tag{13.14}$$

and corresponds to the center of gravity of the distribution of $\tan \vartheta$ in a logarithmical scale. Symmetry of emission in C is demonstrated by symmetry of the distribution of those values relative to the point $\tan \vartheta_{\frac{1}{2}}$.

Furthermore, the Bristol authors show in a thorough analysis of the method that the errors of the procedure resulting from its various simplifications are not too serious. For nucleon-nucleon collisions described by a single symmetrical C-system, it is probably the most adequate method of energy determination. Its limitations become critical only in two cases: firstly, if or when the course of the interaction requires description in a two-center model (cf. the discussion in the preceding Sect. 12); and secondly, if the mass of the two colliding systems is not equal—as in the "composite collisions" discussed below. In this latter case, the symmetry considerations in the (single) C-system remain valid, but the relation between the energy γ_C and the rpimary energy γ in L becomes for a simultaneous collision with n nucleons

$$\gamma_C = (\gamma/2n)^{\frac{1}{2}}. \tag{13.15}$$

Therefore, the energy γ_N determined in the usual way under the assumption of a nucleon-nucleon encounter, is related to the true primary energy γ by

$$\gamma = n \gamma_N. \tag{13.16}$$

The discrepancies can be very large, and the question as to the single or composite nature of the interaction is of crucial importance.

---

[1] B. EDWARDS, J. LUSTIG, D.H. PERKINS, K. PINKAU and J. REYNOLDS: Phil. Mag. 3, 237 (1958).

14. Multiplicity and angular distribution of the shower secondaries. The number of shower particles produced in a nuclear interaction has been studied as a function both of the primary energy, and of the nature of the target nucleus. Unfortunately, it was not always possible to separate those two factors; thus, in most showers oberved in emulsions the nature of the interacting nucleus remains unknown, while in many counter and even cloud chamber experiments the primary energy is undetermined.

The problem of the multiplicity and multiplicity distribution of π-meson production in showers up to a few 10^{10} eV was attacked already in the earlier emulsion work of the Bristol group, in particular in the comprehensive study of CAMERINI et al.[1] to which reference was already made repeatedly. Primary energies were determined from multiple scattering and the relative frequency of π-mesons among the shower secondaries was assumed to equal $\frac{2}{3}$, as inferred from scattering measurements in the energy region where a distinction between π-mesons and protons still is possible. The multiplicity was found at first to rise rather fast with the primary energy, and then to level off gradually. Large fluctuations in the numbers of mesons originated by primaries of the same energy were also already noted.

Very similar data were reported from a cloud chamber study by FROEHLICH et al.[2]. In this case the target material was lead, and one of the aims was to compare the multiplicity of π-meson production in this heavy nucleus with that found in emulsions. An indirect method of energy measurement by determination of the integral path length of the nucleonic cascade was used, and the fraction of π-mesons among the relativistic shower particles was obtained from a comparison of the number of neutron-induced secondary interactions with that of showers originated by charged secondaries. In a recent re-analysis of this experiment, the average primary energies in the showers of each of the four energy groups chosen in this experiment were measured from the angular distribution of the secondaries of all showers in the group. Although this test may involve, in the range of energies concerned (1 to 10 GeV), quite appreciable uncertainties and

Table 6. *Average primary energies E_0, determined from the angular distribution of the shower secondaries ("F-plot method") for the particles in the four energy groups of Froehlich et al.*

E_0 (GeV) from integral path length	1—2	2—4	4—6	6—10
E_0 (GeV) from F-plot method	1.5 ± 0.5	2.8 ± 0.6	4.7 ± 0.6	7.3 ± 0.6

even systematic errors, the agreement between the values thus determined and the earlier estimate can nevertheless be considered as proof that, within the accuracy of 10 to 20% originally claimed, the energy data are reliable. Since the results have not been published elsewhere, they are reproduced in Table 6.

Very few other systematic investigations of the energy dependence of shower multiplicities have been reported; usually, only one or a few values are given. Thus, a slow variation with the primary energy was reported by GRIGOROV[3] in his summary on a number of experiments of USSR-workers, in which the shower detectors were sent aloft in balloons at various geomagnetic latitudes, and the primary energy interval defined by the differences in the cut-offs. In the range of around 20 to 30 GeV, cloud chamber data using the F-plot method

---

[1] U. CAMERINI, J. H. DAVIES, C. FRANZINETTI, W. O. LOCK, D. H. PERKINS, and G. YEKUTIELI: Phil. Mag. **42**, 1261 (1951).

[2] F. E. FROEHLICH, F. M. HARTH and K. SITTE: Phys. Rev. **87**, 504 (1952).

[3] N. L. GRIGOROV: Uspekhi Fiz. Nauk **58**, 599 (1956).

of energy estimates were obtained by WALKER and DULLER[1] and by ASKOWITH and SITTE[2]. At even higher energies, emulsion studies have provided further evidence, but the nature of the target nucleus is then no longer known with certainty. The most serious difficulty in evaluating high-energy data, however, stems from the fact that at energies above a few 10^{10} ev, it is certainly no longer possible to assume that the bulk of the shower secondaries derives from the first nucleon-nucleon interaction. On the other hand, these energies are not yet high enough to justify the use of the simplifying assumptions about the intranuclear cascade which become applicable at primary energies above a few 10^{11} eV,

as we shall see in Sect. 17. Hence only very general evidence can be derived from these investigations, among which in particular the work of the Bristol group (DANIELS et al.[3], BRISBOUT et al.[4]) and of KAPLON and RITSON[5] in Rochester contain a large amount of useful data. The most striking feature in these investigations is the very large range of fluctuations in the multiplicities: for practically identical primary energies, the number of secondaries frequently varies by factor of 10 or more. This is a wide spread even if one keeps in mind that, in these interactions, deviations from the mean can be ascribed to no less than four individual causes: the impact

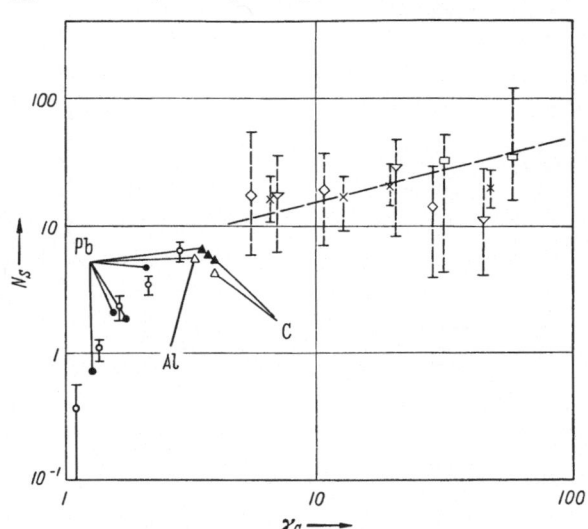

Fig. 14. Multiplicity of meson production as a function of the energy in the C-system. ○ CAMERINI et al. (emulsions), ● FROEHLICH, HARTH and SITTE, △ DULLER and WALKER, ▲ ASKOWITH and SITTE, ◇ DANIELS et al. (emulsions), × KAPLAN and RITSON (emulsions), □ BRISBOUT et al. (emulsions).

parameter of the incident particle with respect to the center of the target nucleus (which defines the length of the path of the primary inside the nucleus), the fluctuation in the number of nucleonic collisions on this path and in the impact parameters of these encounters, the uncertainties in the number of secondaries ejected in each collision, and the neutral-to-charged ratio among the secondaries thus created.

Fig. 14 summarizes the data of the papers mentioned above. In the energy region above 50 GeV, the (dashed) error lines indicate, not statistical standard errors but the range between maximum and minimum multiplicity observed. Evidently, the data are by far insufficient to permit a decision between rival theories, or a quantitative interpretation as a guide to the theories. Just as an illustration, the heavy dashed line in the graph was drawn to indicate the slope of the multiplicity—vs.—energy curve as predicted by the Fermi theory. It is

[1] W. D. WALKER and N. M. DULLER: Phys. Rev. **93**, 215 (1954). — W. D. WALKER, N. M. DULLER and J. D. SORRELS: Phys. Rev. **86**, 865 (1952).

[2] J. G. ASKOWITH and K. SITTE: Phys. Rev. **97**, 159 (1955).

[3] R. R. DANIELS, J. H. DAVIES, J. H. MULVEY and D. H. PERKINS: Phil. Mag. **43**, 753 (1952).

[4] F. A. BRISBOUT, C. DAHANAYAKE, A. ENGLER, Y. FUJIMOTO and D. H. PERKINS: Phil. Mag. (8) **1**, 605 (1956).

[5] M. F. KAPLON and D. M. RITSON: Phys. Rev. **88**, 386 (1952).

not in contradiction to the experimental findings: this is the extent of the conclusion that can be asserted.

The variation of the number of shower particles, for a given primary energy, with the nature of the target nucleus, is in parts already shown by the data of Fig. 14. At low and medium energies, the difference is small between heavy and medium-heavy elements (Ag, Br in emulsions, and Pb), but conspicuous between light elements and lead. This latter fact has been established already in counter and cloud chamber experiments (e.g. Lovati et al.[1]), though it must be kept in mind that in most experiments with those techniques, a strong bias for a set multiplicity and hence for a varying primary energy is imposed. Emulsion studies at somewhat higher energies, differentiating according to the number of heavy prongs between interactions in Ag and Br on the one hand, and the light elements on the other (Parsons et al.[2]), gave multiplicities of 7.7 and 6.4 respectively. The classification used is, of course, not foolproof and does not prevent the inclusion of glancing collision in heavy nuclei among the light-element group. The "emulsion chamber" technique of the Rochester group[3] showed at still higher energies average multiplicities of about 24 in lead, 18 in brass, 14 in Ag, Br, and likewise 14 in the light elements.

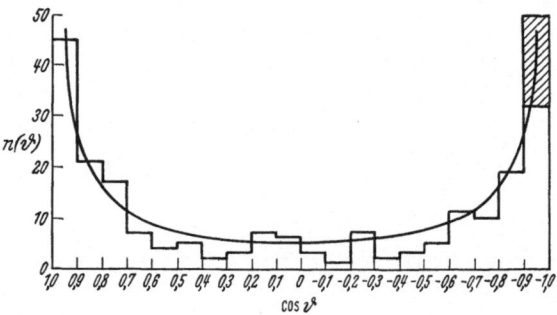

Fig. 15. Histogram of the angular distribution of shower particles in the C-system for the very energetic star reported by Lal, Pal, Peters and Swami. The smooth curve corresponds to the distribution calculated according to the Fermi theory with a medium impact parameter.

Turning our attention now to an analysis of the angular distribution, we meet a situation which is only slightly more satisfactory. Apart from the low-energy region where, as we have seen, an isotropic or near-isotropic emission in the C-system seems to prevail—but for which we have no really adequate theory—, the overall picture obtained from the detailed study by numerous investigators of the emission angles in jets[4-9] shows two important features which were stressed in almost all reports: a strong forward and backward collimation of the most energetic shower secondaries, and a practically isotropic distribution of the slower ones. A typical example is reproduced in Fig. 15, taken from a paper by Lal et al.[4]. In this case, only the angles have been measured; Schein et al.[6] showed that valuable additional evidence can be obtained if the energies of those shower particles is determined which belong to the diffuse cone of the L-system— the backward cone of the C-system—, and hence are slow enough to permit accurate scattering measurements. In this way it can be shown in the C-system a large fraction of the momentum is carried off by a few very energetic secondaries in directions not far from the axis of the collision.

[1] A. Lovati, A. Mura, G. Salvini and G. Taglliaferri: Nuovo Cim. 7, 943 (1950). — Phys. Rev. 77, 284 (1950).

[2] R. W. Parsons, F. A. Brisbout and V. D. Hopper: Phys. Rev. 95, 193 (1954).

[3] M. F. Kaplon, D. M. Ritson and W. D. Walker: Phys. Rev. 90, 716 (1953).

[4] D. Lal, Yash Pal, B. Peters and M. S. Swami: Proc. Ind. Acad. Sci. 36, 751 (1952).

[5] G. Bertolini, D. Pescetti and L. Tallone: Nuovo Cim., Suppl. 12, 338 (1954).

[6] G. R. Glasser, D. M. Haskin and M. Schein: Phys. Rev. 99, 1555 (1955).

[7] L. V. Lingern: Z. Naturforsch. 11a, 340 (1956).

[8] E. Lohrmann: Z. Naturforsch. 11a, 516 (1956).

[9] International Conference on Cosmic Radiation, Varenna 1957: Reports by M. Schein, A. E. Chudakov, D. H. Perkins, and M. Oda.

Note added in proof. The analysis of the angular distribution of jet secondaries has made great progress in recent times. In a brief summary, the following essential points are noted: (1) While in the range of lower primary energies (i.e. approximately for $E_0 \lesssim 10^{11}$ eV), the emission appears to be in general at least very nearly isotropic in the C-system, significant departures from this isotropy are found already at moderately high energies (about 10^{11} eV $\lesssim$ $E_0 \lesssim 10^{12}$ eV). This had already been noted by v. LINDERN [Z. Naturforsch. **11**a, 340 (1956)], LOHRMANN [Z. Naturforsch. **11**a, 561 (1956)] and others, and was first interpreted as an increasing tendency to preferential forward- and backward-emission. (2) At even higher primary energies, the distribution curves show a quite marked bias against emission at right angles to the line of flight in the C-system. The angular distribution becomes "double-humped". As it was first shown by the Polish-Czech emulsion group [CIOK, DANYSZ, GIERULA, JURAK, MIESOWICZ, PERNEGR, VRANA and WOLTER, Nuovo Cim. **6**, 1409 (1957)], and later by COCCONI [Phys. Rev. **111**, 1699 (1958)] and others, this feature is readily understood in terms of a two-center model, be it of the "excited nucleon" type (TAKAGI; KRAUSHAAR and MARKS; see also FARLEY, Proc. Moscow Cosmic Ray Conf. I, p. 254), or of the "fireball" type (COCCONI). In general, it appears that within each of the two independent emission centers, the ejection of secondaries again occurs isotropically. — Modifications may be needed to account for the unequal masses of the two centers in cases of nucleon-nucleus collisions (SITTE and KATZ, Proc. Moscow Cosmic Ray Conf. I, p. 249), and for possible emission from a third center: the original collision volume before separation, as suggested on p. 183 (Note added in proof). But all the important characteristic features of the angular distribution of jet secondaries seem to be well described by a model of independent isotropic ejection of secondaries from separated emission centers.

15. The ratio of neutral to charged secondaries. Electron production in penetrating showers, though observed already in early cloud chamber investigations, remained an enigma for many years in spite of the remarkable analysis of ROCHESTER[1] which demonstrated with a surprising degree of correctness some of the basic features of the hypothetical source of the electron cascades. Some years later, the shower studies of CARLSON et al.[2], and of GREGORY and TINLOT[3], share with accelerator experiments the honour of confirming the existence of the neutral π-meson. Soon afterwards, the relative frequency of the production of neutral and charged secondaries, a quantity of decisive importance for the foundations of meson physics, could be accurately measured in a number of experiments, both with cloud chambers and with nuclear emulsions.

Among the work with the former technique, the experiment of SALVINI and KIM[4] is an outstanding example. In order to eliminate the danger of a bias in favour of interactions containing many *charged* particles—otherwise inherent in work with counter-controlled cloud chambers—these authors mounted a NaI-crystal scintillator in the top shelf of their multiple-plate cloud chamber, and triggered their equipment whenever a star was formed within the crystal. The development of the electron cascade in the thin lead plates of the chamber was then followed, so that the numbers, the angular distribution, and the energy of the decay photons could be studied. The result, increasing a ratio of $1:2$ for the frequencies of π^0's and charged π-mesons, was perhaps the first reliable experimental proof for the validity of the hypothesis of charge independence in meson production at high energies.

Most of the later precision work was done with nuclear emulsions. The technique then consists in counting the number of electron pairs that are associated with a high-energy shower or jet, in particular near the origin of the shower. A number of simplifying assumptions are usually made:

(i) that the detection efficiency for such pairs is 100%;

[1] G. D. ROCHESTER: Proc. Roy. Soc. Lond., Ser. A **178**, 464 (1946).
[2] A. G. CARLSON, J. E. HOOPER and D. T. KING: Phil. Mag. **41**, 701 (1950).
[3] B. P. GREGORY and J. H. TINLOT: Phys. Rev. **81**, 667, 675 (1951). — B. P. GREGORY, B. ROSSI and J. H. TINLOT: Phys. Rev. **77**, 299 (1950).
[4] G. SALVINI and Y. KIM: Phys. Rev. **85**, 921 (1952); **88**, 40 (1952).

(ii) that the angular distribution of the neutral shower secondaries is the same as that of the charged particles, and the angular distribution of the photons extends that of their parents;

(iii) that the bulk of the photons converted into electron pairs, derives from the decay of π^0's, and that practically all π^0's decay in this manner;

(iv) that the lifetime of the π°-meson is known, and hence is mean range before decay can be calculated.

The validity of the first three assumptions has been frequently checked and found satisfactory; the errors introduced may be of the order of 1%. The fourth has been the subject of extensive investigations which do not fall into the scope of this article; suffice it to quote as safely established that the lifetime of the neutral π-meson is of the order of 10^{-15} sec. Under these conditions, the number of *first* electron pairs, i.e. of the pairs created directly by the conversion of the decay photons and not in subsequent stages of the cascade development, can easily be written down. More electron pairs may be expected as the result of conversion of "bremsstrahlung", either of the energetic electrons among the "first" pairs or emitted by nucleons or mesons directly in the nuclear interaction. But calculations show (SCHIFF[1], OEHME[2]) that the latter effect becomes significant only at extremely high energies, or for particles of higher spin. Nevertheless, it remains a serious problem to distinguish between those associated

Table 7. *Ratio of the number of π^0-mesons to that of all charged shower secondaries, $R = N_{\pi^0}/N_s$, in showers of various primary energies E_p. Asterisks indicate that the value has been obtained from measurements on a single star.*

E_p (GeV)	$R = N_{\pi^0}/N_s$	Observer
10—20	0.50 ± 0.10	SALVINI and KIM[3]
∼25	0.48 ± 0.11	DANIEL et al.[4]
∼100	0.40 ± 0.08	LAL, PAL and RAMA[5]
∼500	0.33 ± 0.07	DANIEL et al.[4]
∼1000	0.46 ± 0.09	KAPLON, WALKER and RITSON[6]
∼5000	*0.44 ± 0.13	NAUGLE and FREIER[7]
∼6000	0.383 ± 0.044	BRISBOUT et al.[8]
∼20000	*0.25 ± 0.10	MULVEY[9]
∼30000	*0.44 ± 0.13	KOSHIBA and KAPLON[10]
Average	0.416 ± 0.032	

decay pairs, and the bremsstrahlung pairs. Its solution has been attempted in two ways: either by restricting the counting to a very small interval, beginning with a distance large enough to assure that all π^0-mesons have already decayed and ending it at a distance which is still only a small fraction of one conversion unit (3.75 cm); or else, by grouping the electron pairs according to their energy and their location with respect to the shower axis, thus eliminating by inspection

[1] L. SCHIFF: Phys. Rev. **76**, 89 (1949).

[2] R. OEHME: Z. Physik **129**, 573 (1951).

[3] G. SALVINI and Y. KIM: Phys. Rev. **85**, 921 (1952); **88**, 40 (1952).

[4] R. R. DANIEL, J. H. DAVIES, J. H. MULVEY and D. H. PERKINS: Phil. Mag. **43**, 753 (1958).

[5] D. LAL, YASH PAL and RAMA: Nuovo Cim., Suppl. **12**, 743 (1954). — Proc. Ind. Acad. Sci. **39**, 127 (1954).

[6] M. F. KAPLON, W. D. WALKER and M. KOSHIBA: Phys. Rev. **93**, 1424 (1954).

[7] J. E. NAUGLE and P. S. FREIER: Phys. Rev. **92**, 1086 (1953).

[8] F. A. BRISBOUT, C. DAHANAYAKE, A. ENGLER, Y. FUJIMOTO and D. H. PERKINS: Phil. Mag. (8) **1**, 605 (1956).

[9] J. H. MULVEY: Proc. Roy. Soc. Lond., Ser. A **221**, 367 (1954).

[10] M. KOSHIBA and M. F. KAPLON: Phys. Rev. **97**, 193 (1955).

the likely bremsstrahlung pairs. For details of the procedure, the reader is referred to the original papers; the good agreement between the results obtained with all the variations used is proof that several techniques can be applied successfully.

The results of a number of studies are summarized in Table 7 which contains also the average energies of the showers investigated. It is seen that over the entire wide range of the observations, there is no significant variation of the neutral-to-charged ratio. The implications of this finding will be discussed below.

16. The frequency of "strange" particles. In the preceding sections, no attention has been given to the circumstance that the shower secondaries may in fact not be uniform in their nature but consist of a variety of different particles. Actually, the presence among them not only of nucleons, but also of heavy unstable fundamental particles has, of course, long been known. Indeed, quantitative evidence on the frequency of these "strange" particles is one of the most urgent and important problems in shower research, for it could provide clues not only concerning the basic character of the nucleon-nucleon interactions, but also concerning the very nature of these particles. An equilibrium theory of the Fermi type, for instance, demands equipartition of the energy among the components; other theories lead to different predictions. Similarly, certain general rules about the relative frequency of the "strange" particles could be formulated if these particles can be treated as excited states of nucleons and of π-mesons; but these rules would not apply if the "strange" particles are as fundamental as the others. A good measurement of the energy dependence of the frequency of the heavy shower secondaries could, therefore, provide much-needed evidence.

In the low-energy region of 5 to 10 GeV in which most of the cosmic ray studies on the heavy unstable particles have been carried out, and which at present is also already in the range of the new accelerators, the results of this work are undisputed but inconclusive. For details, we must refer to the reports on strange-particle research in other parts of this Encyclopedia (Vol. XLIII); in the context of this article, it will suffice to quote that both accelerator and cosmic ray data in this energy region show a relative abundance of 1 to 2% of charged strange particles among all shower secondaries. These same investigations also prove that there is no decay process other than π^0-disintegration known which transfers efficiently the energy of the penetrating shower component to the electron component.

But it is only in the region of higher primary energies that we might expect more conclusive evidence, and in particular, information on the energy dependence of the frequency of heavy secondaries is of importance. We shall, therefore, discuss some of the resulty reported in the last few years.

At energies only slightly above these mentioned—about 20 to 30 GeV— WALKER and DULLER[1] deduce from their cloud chamber experiment an upper limit of about 15% for all heavy unstable particles. The correct value may well lie appreciably lower. This is not inconsistent with the conclusions derived by DANIEL et al.[2] from a study of stars in emulsions, identifying the heavy particles by scattering measurements. Their results give a fraction of 16 to 18% at primary energies between 10 and 50 GeV. However, at higher primary energies there is no longer such agreement among the various experimenters.

DANIEL et al. interpret their data obtained with the high-energy group— events of 50 to about 2000 GeV, with a mean value at $\sim$550 GeV—as a proof that in this region, the fraction of heavy shower particles has increased to about

[1] W. D. WALKER and N. M. DULLER: Phys. Rev. **93**, 215 (1954).

[2] R. R. DANIEL, J. H. DAVIES, J. H. MULVEY and D. H. PERKINS: Phil. Mag. **43**, 753 (1952).

36 to 40%. This is obtained from an analysis of the neutral-to-charged ratio, $R = 0.33 \pm 0.07$—against $R = 0.56 \pm 0.12$ at low energies—with the assumption that protons do not amount to more than 10% of the shower particles. A similar value was found by Mulvey[1] in the analysis of data obtained from a very energetic star, and more recently by Brisbout et al.[2], who find $R = 0.383 \pm 0.044$ and from this, together with a ratio $Q = 0.25$ for the frequencies of secondary interactions initiated by neutral and by charged shower particles, calculate a relative abundance of about 25% of heavy unstable particles at primary energies of the order 10^{12} to 10^{13} eV.

The correctness of these conclusions has repeatedly been challenged. Other emulsion studies have yielded significantly lower results: Thus, for instance, Kaplon, Walker and Koshiba[3] find in their emulsion chamber experiment an upper limit of $\lesssim 9\%$, while Lal, Pal and Rama[4], again from the neutral-to-charged ratio computed for a large number of events with an average primary energy of about 100 GeV/nucleon, arrive at a fraction $\lesssim 18\%$.

The problem has also been attacked in two entirely different manners. In an experiment based on conventional counting techniques, Sitte et al.[5] attempted to measure the fractional energy transfer to the electron component in nuclear interactions between 10 and 100 GeV primary energy. The penetrating showers under investigation originated in a carbon absorber; a row of narrow counters, their output fed into a hodoscope circuit, served to determine the multiplicities of the charged shower secondaries. The electron component was developed only below this hodoscope tray, in a lead absorber dimensioned so as to bring to maximum development the cascade typical for a 30 GeV primary, and its density was measured in a large liquid scintillator placed below this lead absorber. A further tray of Geiger counters under a 6-in. lead shield ensured that only hard showers were recorded.

Since the position of the cascade maximum increases only logarithmically with the primary energy, and the number of electrons in the neighbourhood of the maximum is a slowly varying function of the absorber thickness, the size of the electron cascade under this fixed absorber is a good measure of the primary energy within the entire range under study. An absolute calibration of the scintillator pulse height as a function of the number of electrons traversing the liquid can be carried out, using as a standard the well-known density distribution of air showers.

The primary energy was determined in two independent ways: from the multiplicity of the initial shower, and from the absolute frequency of the events. None of the two methods can claim an accuracy of better than 10 to 20%, and within this uncertainty, the results agreed. However, for the purpose of this experiment a precise knowledge of the absolute value was not of great importance; it was the determination of a possible variation of the fractional energy transfer to the electron component that counted. If, as the Bristol data indicate, at energies around and above 50 GeV about half of the energy in the shower secondaries is carried off by the heavy particles, this fraction should decrease by a factor of the order 2 between primary energies around 10 GeV—where still practically only π-mesons are produced—and 100 GeV. It is much easier to

[1] J. H. Mulvey: Proc. Roy. Soc. Lond., Ser. A **221**, 367 (1954).
[2] F. A. Brisbout, C. Dahanayake, A. Engler, Y. Fujiomto and D. H. Perkins: Phil. Mag. (8) **1**, 605 (1956).
[3] M. F. Kaplon, W. D. Walker and M. Koshiba: Phys. Rev. **93**, 1424 (1954).
[4] D. Lal, Yash Pal and Rama: Nuovo Cim., Suppl. **12**, 347 (1954).
[5] K. Sitte, F. E. Froehlich and I. Nadelhaft: Phys. Rev. **97**, 166 (1955).

detect such a variation than to measure with precision all the energies involved. The results obtained definitely ruled out a conspicuous decrease, so that it must be concluded either that the heavy secondaries are much less numerous and energetic, or else that a hitherto unknown decay process transfers a substantial part of the strange-particle energy to the electron component, just as π^0-decay does for the light mesons. Since no such process has so far been observed, the second alternative is, of course, much less likely.

Note added in proof. Working along similar lines, GRIGOROV and his collaborators have recently extended the measurements of the fractional energy transfer to the electron component up to primary energies of the order 10^{12} to 10^{13} eV [GRIGOROV, GUSEVA, DOBROTIN, KOTELNIKOV, MURGIN, RYABIKOV and SLAVATINSKY, Proc. Moscow Cosmic Ray Conf. I, p. 143 (1959)]. In their ingenious experiment, the primary energy is determined by the "calorimetric" method described in Sect. 13. The number of electrons emerging from the production layer was counted from their tracks in photographic emulsions. By pin-pointing the location of the electron showers with the help of a hodoscope of crossed ionization chambers, only those of the small emulsion strips have to be developed and replaced which had been struck by a shower. From his data, GRIGOROV finds no significant change in the fractional transfer (i.e., also of the inelasticity k_π) up to at least 10^{13} eV. The implications of this rather remarkable result will be discussed in some detail later on.

More directly, SCHEIN et.al.[1], in their detailed analysis of the "S-star" already mentioned, studied for this individual event of 15 shower secondaries the angular and momentum distributions. Direct momentum measurements on all particles of the backward cone, together with the assumption of equal constitution of forward and backward cone, yield a satisfactory, symmetrical structure of the star in the C-system, consistent with either LANDAU'S or HEISENBERG'S theory (Sect. 12), if all the secondaries are π-mesons. However, if some of them had a higher mass, their energies would have to be much higher—well above that of the π-mesons—and it would be difficult to reconcile the evidence with the demand of specular symmetry. In particular, the presence of nucleon pairs or of hyperons in the backward cone can practically be ruled out since the total energy of such heavy particles would already add up to the maximum energy available in the C-system, and leave nothing for the light secondaries. It was, therefore, concluded that not more than a small fraction of the shower secondaries could be particles other than π-mesons.

Although this argument does not exclude the possible presence of heavy "strange" particles in the *forward* cone, and although in general, owing to the large fluctuations that occur in the production of shower particles, one must always be wary about generalizing evidence from a single event, this analysis is nevertheless the most convincing proof hitherto offered about the nature of the shower secondaries, and it is definitely not in favour of an equipartition theory.

Evidence against the high abundance of "strange" secondaries in energetic interactions has also come forward from yet another side: In the study of the atmospheric effects of very hard μ-mesons, the results on the temperature effect underground have been found consistent with the assumption that their parents are exclusively or predominantly π-mesons. This matter is discussed in the article by EHMERT in Part 2 of this volume.

In summing up the evaluation of the published material with the statement that at all primary energies, π-mesons seem to be by far the most frequent shower secondaries, one may well accentuate two more points. The first is that the discrepancy between the earlier Bristol results and those obtained elsewhere stems

---

[1] R. G. GLASSER, D. M. HASKIN and M. SCHEIN: Phys. Rev. **99**, 1555 (1955).

from their interpretation of the data rather than from disagreement in the experimental results. In fact, all the values of $R = N_{\pi^0}/N_s$ reported from BRISTOL, including the latest and most accurate one of BRISBOUT et al., are consistent with the average listed in Table 7 (p. 192). Yet it is from this R, together with the ratio Q of secondary interactions initiated by neutral shower particles to those produced by charged secondaries, that these authors calculate their abundance ratio $N_{K\pm}/N_{\pi\pm} = 0.33 \pm 0.16$. The reason is that in their analysis, the assumption is made that nucleons are present only in negligible numbers. The neutral particles which initiate secondary showers, for instance, are identified with K^0-particles whose frequency is thought to be of the same order as that of the charged K-mesons. Most other authors, however, believe that even at high primary energies, protons still form a significant component, perhaps 10% of all secondaries. This is indeed the crux of the matter. To put it quantitatively: Writing $f_K = N_K/N_\pi$, $f_p = N_p/N_\pi$, and assuming $N_{\pi^0} = \frac{1}{2}N_{\pi\pm}$, $N_K = N_{K^0}$, $N_p = N_n$ (number of neutrons), we easily verify the relation

$$f_p + f_K = Q/2R.$$

But the situation does not seem to be entirely remedied by this addition. The S-star, apparently without proton secondaries, does also not contain many K-particles. Unless we accept it as a case of large fluctuations, we have to look for further explanation.

This may be found in the distinct possibility that a considerable variation in the number of strange secondaries could result from differences in the nature of the interaction which produces the shower. Even if in a pure nucleon-nucleon collision K-mesons will be created only rarely, one must admit that in a nucleon-nucleus collision, the interactions of π-mesons originated in earlier generations inside the target nucleus may contribute more strongly to the ejection of heavy secondaries. At low energies, this is a well-known characteristic of π-meson collisions, and it is quite plausible that it persists even in high-energy interactions. In this case, showers arising from nucleon-nucleon encounters would differ in their constitution from those of nucleon-nucleus events; and among the latter, interactions originating in a light nucleus might contain fewer heavy particles than those initiated in heavy nuclei. The S-star is almost certainly an example of a nucleon-nucleon collision; the counter-scintillator experiment discussed above recorded showers produced in carbon; and the atmospheric μ-mesons come from interactions with the light nitrogen and oxygen nuclei: in all these cases, K-particles and hyperons should be rare. In emulsions, on the other hand, a large fraction of the stars originate in the comparatively heavy silver and bromine nuclei where intranuclear cascade processes play a more important part. This possible explanation of the discrepancies and inconsistencies about the frequency of strange particles could be examined by a systematic analysis of the neutral-to-charge ratio of stars with different numbers of heavy prongs, but so far, no such investigation has been reported.

17. Nucleon-nucleus interactions at very high energies. The "tunnel" and the "funnel" model. As soon as it had become evident that nucleon-nucleon interactions are not always "catastrophic" in the sense of complete dissipation of the primary energy which would leave none of the shower particles with an energy sufficient to undergo further radiative collision, the necessity of treating penetrating showers originating in the impact of a nucleon on a complex nucleus as an event different in kind from those interactions has become obvious. A remarkable step forward was achieved in the "plural-collision" picture of HEIT-

LER and TERREAUX[1] discussed in Sect. 12. In the region of extremely high energies, it has been generalized and more precisely formulated in the "tunnel model" of McCUSKER and ROESSLER[2].

This theory carries to the logical conclusion the cascade model with small scattering. As it has already been pointed out before (HEITLER and JÁNOSSY[3], HABER-SCHAIM[4]), in very energetic processes where the scattering angles in the collisions are very small, the particles emerging from one interaction will, within the intranuclear mean free path of the order of nucleon dimensions, not separate sufficiently to undergo individual, distinct collisions of the next generation. The result will be a composite collision of all the incident cascade particles with the nucleons along their path, all of them occurring in a narrow, slowly widening channel punched through the target nucleus. If the primary energy is sufficiently high, scattering can be entirely neglected, and the channel becomes a cylindrical tunnel. In general, however, the shape of the collision volume may be anything between the widening, trumpet-like hole of Fig. 16, and the constant-width tunnel.

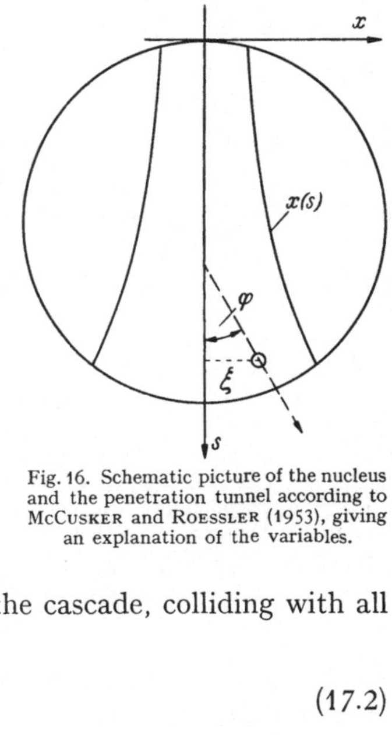

With this tunnel model, McCUSKER and ROESS-LER developed a cascade theory of shower formation into which any model of meson production can be fitted. Briefly, the treatment is built on the following basic ideas:

(i) Collisions at depth s (see Fig. 16) can be considered as interactions of the entire "jet" mass $m(s)$ arriving at this depth, with the target mass of density ϱ;

(ii) The variation of the scattering angle φ with the energy in the C-system (expressed in units of the rest mass) follows

Fig. 16. Schematic picture of the nucleus and the penetration tunnel according to McCUSKER and ROESSLER (1953), giving an explanation of the variables.

$$\Delta \overline{\varphi^2} = \alpha (1/\gamma_c)^2, \qquad (17.1)$$

with a constant α of the order 1.

(iii) Moving down a distance Δs in the tunnel, the cascade, colliding with all nucleons in the volume, "diffuses" according to

$$\Delta \overline{\varphi^2}/2\Delta s = 2D \qquad (17.2)$$

and D, "diffusion coefficient", can be related to the mass of the tunnel and to the primary energy E_0 by

$$D = \frac{\pi}{8} \alpha \varrho c^2/E_0. \qquad (17.3)$$

(iv) The increase in the mass of the jet, resulting from the mass of the target nucleus added to it and from that of the particles created in the interactions, is written in the form

$$dm/ds = \frac{\pi}{4} f \varrho x^2 = \frac{\pi}{4} f \varrho \alpha \overline{\xi^2}. \qquad (17.4)$$

This general representation, using the unspecified factor f to account for all details of the collisions, makes it possible to apply the formalism of the McCusker-

[1] W. HEITLER and CH. TERREAUX: Proc. Phys. Soc. Lond. A **66**, 929 (1953). — CH. TERREAUX: Helv. phys. Acta **24**, 551 (1951).

[2] C. B. A. McCUSKER and F. C. ROESSLER: Nuovo Cim. **10**, 127 (1953).

[3] W. HEITLER and L. JÁNOSSY: Proc. Phys. Soc. Lond. A **62**, 364, 669 (1950).

[4] U. HABER-SCHAIM: Phys. Rev. **84**, 1199 (1951). — Nuovo Cim., Suppl. **12**, 344 (1957).

Roessler theory in conjunction with all theories of meson production, plural or multiple.

On the basis of these assumptions, the authors then proceed to solve the two-dimensional diffusion equation of the Fokker-Planck type for the probability density ψ in a phase space introducing the projections of φ on the $s - x$-plane and on the $s - y$-plane, φ_x and φ_y:

$$\frac{\partial \psi}{\partial s} = -\varphi_x \frac{\partial \psi}{\partial \xi} - \varphi_y \frac{\partial \psi}{\partial \eta} + D\left(\frac{\partial^2 \psi}{\partial \varphi_x^2} + \frac{\partial^2 \psi}{\partial \varphi_y^2}\right) \tag{17.5}$$

under the initial conditions $\overline{\xi^2}(0) = \xi_0^2$ and $\overline{\varphi^2}(0) = 0$. At present the special case most interesting to us is the limit for negligible increase in $\overline{\xi^2}$. If it is assumed that f does not vary with the depth s—probably not too bad an assumption since the decrease in the multiplicity per collision in the tunnel may be compensated by the increase in the number of collisions—this leads to

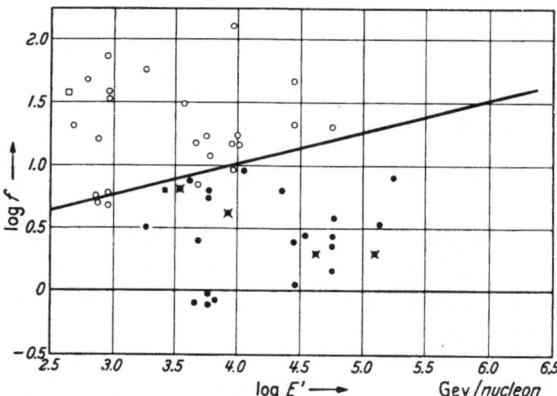

$$m = m_0 + \frac{\pi}{4} f \varrho \alpha \xi_0^2 \cdot s \tag{17.6}$$

or, since $\frac{\pi}{4} \alpha \xi_0^2 = \sigma_0$, the primary interaction cross section,

$$m = m_0 + \sigma_0 f \varrho s. \tag{17.6a}$$

It is then shown that, if an error for ξ^2 not exceeding 5% is admitted, the validity of this solution is restricted to interactions of primary energies E_0 satisfying the relation

$$E_0 \gtrsim 16 A \tag{17.7}$$

(E_0 in GeV) for a *central* hit, and only a little less for an average impact parameter.

Fig. 17. The multiplicity function f and the estimated primary energy for jets obtained since 1953 assuming (a) that the event was a nuclear-nucleon collision (open circles), and (b) that it was a central collision with silver (solid circles). A superimposed cross implies an α-particle primary. The line represents the law (17.9) [from McCUSKER and ROESSLER (1957)].

In order to compare the theoretical results with the experimental data on the multiplicity N_s of shower particles, McCUSKER and ROESSLER assume furthermore that the average mass of the secondaries—π-mesons and heavy particles—is about $\frac{2}{3} m_p$, and that one-third of all secondaries is uncharged. Then one has

$$m = N_s m_p. \tag{17.8}$$

Upper and lower limits for the function $f(E_0)$ can be derived by analyzing the data of all jets under the two extreme conditions of maximum and minimum path length: $s = 2R$, and $s = s_1 = \frac{4}{3} R_0$. In each case, an $E_0 - f$ combination is obtained, and unless very large fluctuations in the multiplicity per nucleon-nucleon collision prohibit an analysis based on average numbers, these two sets of points should form separate groups in a graph. That this is true will be seen from Fig. 17 which is taken from a more recent paper of McCUSKER and ROESSLER[1]. Since the lower boundary of the minimum-path set should give the variation of f with the primary energy in a true glancing collision, the straight line

$$f = (1.0 \pm 0.2)(E_0/m_p c^2)^{0.25 \pm 0.5} \tag{17.9}$$

[1] C. B. A. McCUSKER and F. C. ROESSLER: Nuovo Cim. (10) **5**, 1136 (1957).

may be used to represent the empirical law for the energy dependence of the multiplicity in nucleon-nucleon interaction, at primary energies around and above 10^{12} eV.

The detailed calculations of McCusker and Roessler are based on the plural theory of meson production. Cocconi[1], on the other hand, made use of the Fermi theory applied to the "tunnel model". He considers a nucleon-nucleus interaction as a "composite" collision involving n nucleons in the target nucleus, all participating in equal manner since their distance from each other is only of the order of the range of nuclear forces. In fact, the Fermi picture is probably more correct with the extended "collision volume" of the tunnel than with one of the order $(\hbar/mc)^3$: one of the criticisms listed in Sect. 12 is then invalidated.

The multiplicity function $f(n, \gamma)$ is obtained in the usual way, from the energy W_c available in the C-system for which an elementary calculation gives

$$W_c = (2n\gamma + n^2 + 1)^{\frac{1}{2}} \tag{17.10}$$

or if $\gamma \gg n$,

$$W_c = (2n\gamma)^{\frac{1}{2}}. \tag{17.10a}$$

Since according to Fermi, the multiplicity increases with $\gamma^{\frac{1}{4}}$, one has

$$N_s = K\gamma^{\frac{1}{4}} f(n) \tag{17.11}$$

and the effect of the composite collision described by $f(n)$ can be computed from the energy density ε_c and the temperature T in the collision volume $V \propto n$:

$$\varepsilon_c = \frac{W_C}{V_c} = \gamma_C \frac{W_C}{V_L} \propto \gamma_C \frac{W_C}{n} \tag{17.12}$$

(the suffices L, C again denote laboratory and center-of-mass system), and

$$T \propto \gamma_C^{\frac{1}{4}}. \tag{17.13}$$

But $\gamma_C = (\gamma/2n)^{\frac{1}{2}}$, so that

$$\varepsilon_c \propto \gamma n^{-1} \tag{17.14}$$

and finally, since the number N_s of shower particles is proportional to W_C/T:

$$N_s = K\gamma^{\frac{1}{4}} n^{\frac{3}{4}}. \tag{17.15}$$

Cocconi showed also that, if the half-angle ϑ_s is introduced instead of ϑ_s' [Eq. (13.3)], a new relation is obtained which can easily be tested experimentally:

$$N \cdot Q_s^{\frac{1}{2}} = 1.2 K n. \tag{17.16}$$

For a given nucleus, n is limited by the maximum length of the tunnel $2R$, and thus (17.16) puts an upper limit to the quantity $N \cdot Q_s^{\frac{1}{2}}$.

The objection can be raised against the tunnel model that, unless the nucleus is pictured as a fluid rather than as a compound of particles, a cylindrical collision volume can result only from the extremely rare case of a succession of pure head-on collisions. For any non-vanishing impact parameter, the target nucleon will travel on a path slightly apart from that of the incident primary, so that even in the extreme case of negligible scattering the trajectories of the cascade particles will be parallel but not identical. Each successive collision widens the cross section of the avalanche of shower particles and thus increases the chance for further encounters inside the target nucleus: the "tunnel" becomes a "funnel" even at energies where the spread due to scattering is negligible. This effect leads to a stronger variation of the number of participating nucleons with the atomic number A than it would follow from the "tunnel" model.

[1] G. Cocconi: Phys. Rev. **93**, 1107 (1954).

A simple computation[1] shows that at least for the first 5 to 6 collisions, the increase in the width r_i of the "funnel" can be represented, with an error not exceeding about 5%, by

$$r_i = r_0 \left(1 + \alpha \frac{s}{\lambda}\right) \qquad (17.17)$$

where r_0 is the effective radius related to the elementary cross section σ_0 by $\sigma_0 = \pi r_0^2$, s the depth of the funnel in the target nucleus, λ the "mean free path in nuclear matter" related to r_0 by $r_0 = R_0 (4R_0/3\lambda)^{\frac{1}{2}}$, and $\alpha = 0.38$ a constant. With this, the number n of nucleons taking part in a collision at impact parameter, obtained from the ratio of the "collision volume" V of the cone-shaped funnel to the nucleon volume $V_0 = \frac{4}{3}\pi R_0^3$, is in first approximation given by

$$n = V/V_0 = \frac{1}{3}\alpha \cdot \{[1 + 2\alpha (R_0/\lambda) \sqrt{A^{\frac{2}{3}} - (b/R_0)^2}]^3 - 1\}. \qquad (17.18)$$

Taking as the limits of the impact parameter b:

$$(R_0 A^{\frac{1}{3}} + r_0) = R_0 (A^{\frac{1}{3}} + \sqrt{4R_0/3\lambda})$$

and 0, and using $\langle b \rangle = \frac{2}{3} b_{\max}$, one has for the *average* number $\langle n \rangle$ of the nucleons involved in a collision of a nucleus A

$$\langle n \rangle = \frac{1}{3}\alpha \{[1 + 2\alpha (R_0/\lambda) \sqrt{\frac{1}{3}(A^{\frac{1}{3}} + (4R_0/3)^{\frac{1}{2}})}]^3 - 1\}. \qquad (17.19)$$

Now it can be argued that in view of the empirical validity of the $\gamma^{\frac{1}{4}}$-law for the shower multiplicity, not only the Fermi theory but any theory which does not demand a strong energy dependence of the degree of inelasticity in the collisions, will yield a relation of the form (17.15) and hence a dependence of N_s on $n = V/V_0$ according to

$$N_s \propto (V/V_0)^{\frac{3}{4}} \gamma^{\frac{1}{4}} \qquad (17.20)$$

which can be computed, for the "funnel model", according to (17.19). On the other hand, the "tunnel" model, both in the McCusker and in the Cocconi version, leads to a simple law.

$$N_s \propto \frac{4}{3}(A^{\frac{1}{3}} + r_0/R_0). \qquad (17.21)$$

The values of V/V_0 according to (17.18) and (17.21) are shown in Fig. 18 for a few elements, and the values of $n^{\frac{3}{4}}$ for the tunnel and for the funnel model in Fig. 19.

In comparing experimental jet data with calculations made on the basis of a composite-collision model, attention must be paid to the fact that the primary energies assigned to the jets are practically always determined by methods assuming a nucleon-nucleon collision. As it has been pointed out in Sect. 13, the energy γ_N thus obtained is too small by a factor n, the number of participating nucleons. Consequently, the relation between N_s and γ_N which replaces (17.20) will be

$$N_s \propto (V/V_0) \gamma_N^{\frac{1}{4}}. \qquad (17.22)$$

Unfortunately, the experimental material suitable for comparison with these expressions is scanty and inaccurate. Apart from a few emulsion chamber data of KAPLON, RITSON and WALKER[2]—the statistical insufficiency of which is evidenced by the fact that these authors find equal multiplicities (14) in light-element and in heavy-element collisions in emulsions—both lower than those observed

[1] A number of the calculations on the "funnel" model reported here have been carried out by T. GOZANI in his (hitherto unpublished) Masters Thesis, at The Technion, Haifa.
[2] M. F. KAPLON, D. M. RITSON and W. D. WALKER: Phys. Rev. 90, 716 (1953).

for Cu (18)—there are available only the older measurements of KAPLON and RITSON[1] on interactions in Cu, for which the energies are known and which,

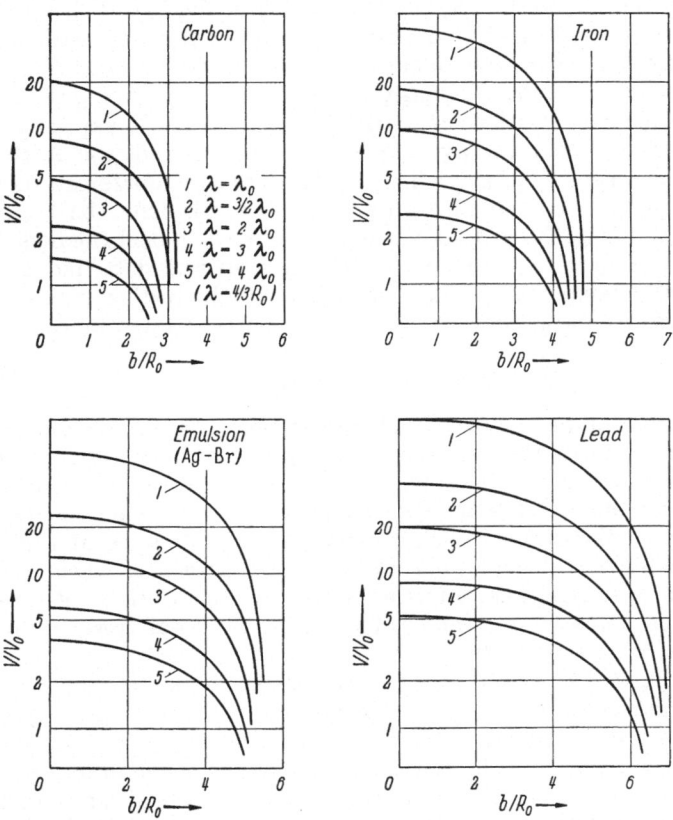

Fig. 18. Collision volumes according to the "funnel" model as a function of the impact parameter b for various values of the mean free path in nuclear matter, in carbon, iron, emulsion (heavy elements), and lead.

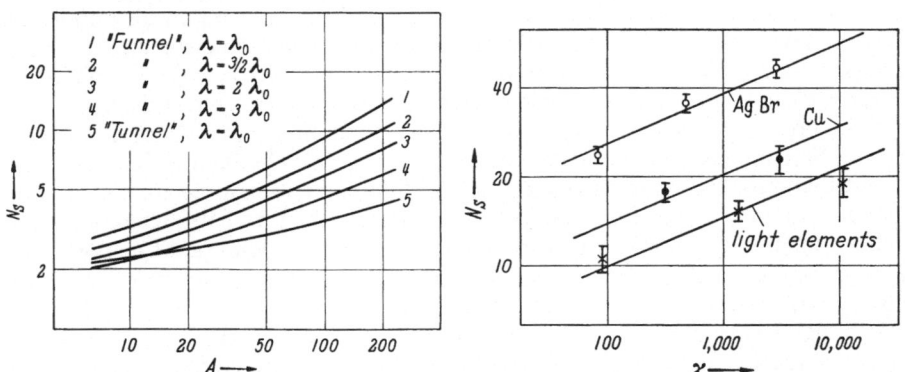

Fig. 19. The quantity $n^{\frac{2}{3}}$ [$n = (V/V_0)$ is the number of nucleons participating in the nucleon-nucleus interaction] as a function of A, for the funnel and the tunnel model.

Fig. 20. Multiplicities N_s in heavy emulsion nuclei (open circles), copper (solid circles), and light emulsion elements (crosses) as a function of the primary energy γ.

therefore, can be compared with the jet data of the Bristol group. In an attempt to carry out this comparison, we have again used the papers of DANIEL et

[1] M. F. KAPLON and D. M. RITSON: Phys. Rev. **88**, 386 (1952).

al.[1], and of BRISBOUT et al.[2], which cover a similar energy range, and have grouped the emulsion jets according to the number of heavy prongs into "light-element" and "AgBr" interactions. The result is shown in Fig. 20: it demonstrates that the $\gamma^{\frac{1}{2}}$-law is well satisfied, and that the multiplicities vary with A in a qualitatively correct manner. For quantitative interpretation, however, one is faced with a difficult problem: Obviously, the grouping does not give an unbiassed selection. Glancing collisions with the heavy nuclei will appear in the light-element group, so that the average impact parameter of the AgBr is smaller than the true mean, $\frac{4}{3}(R_0 + r_0)$. Thus, only a crude estimate can be made, best perhaps by the restriction $0 < b \lesssim \frac{2}{3}(R_0 + r_0)$. If this is done, the "tunnel model" leads to a ratio $N_{AgBr}/N_{Cu} \sim 1.40$, while the "funnel" model predicts $N_{AgBr}/N_{Cu} \sim$ 1.7 to 2.2, according to the choice of λ. The empirical data, indicating a ratio of 1.8, hence favour the "funnel" model, but only further emulsion chamber work could really decide how far this correction is of importance.

IV. Discussion and interpretation.

18. Structure of the nucleus. Total cross sections. In the following, an attempt will be made to bring together the various pieces of evidence presented above, into a consistent picture containing the essential features of nuclear structure and of nuclear interactions at high energies. The very wide energy range and the diversity of the events which provide the clues enforce, at this stage, still rather preliminary and semi-quantitative answers, some of which will be given here only as a stimulation to further work. We shall begin with re-discussing the mean free paths for shower production, or the total cross sections at high energies, for various elements.

In Chap. II we had seen that the "transparency" picture, applied to the *light* elements, demanded a rather large elementary cross section for nucleon-nucleon collision (or a short "mean free path in nuclear matter"), possibly varying with the primary energy. But the measurements in the medium-heavy iron nuclei had yielded conflicting results indicating a much too small cross section if the same model is applied.

A certain clue for the removal of these inconsistencies may be derived from the analysis of the "jets" presented in Chap. IV. We have seen how all-important it was, at very high energies, to distinguish between nucleon-nucleon and nucleon-nucleus interactions. In other words, the effects of the propagation of the intra-nuclear cascade must not be neglected or underestimated: but this is exactly what the ordinary "transparency" model does, including into the considerations only the first interaction. Even at energies of a few 10^{10} eV, the incident primary will not dissipate enough of its energy in the first encounter to become incapable of further radiative collisions, so that the cascade effect and its dependence on the size of the target nucleus will be significant. It then follows, of course, that the true average multiplicity per nucleon-nucleon collision is considerably lower than the values shown, for instance, in Fig. 14.

Although the theory of intranuclear cascades has been worked out thoroughly and in great detail (e.g. MESSEL[3]), we shall here choose a much more elementary, and admittedly much less precise, formulation to indicate why the inconsistencies

[1] R. R. DANIEL, J. H. DAVIES, H. H. MULVEY and D. H. PERKINS: Phil. Mag. **43**, 753 (1952).

[2] F. A. BRISBOUT, C. DAHANAYAKE, A. ENGLER, Y. FUJIMOTO and D. H. PERKINS: Phil. Mag. **1**, 605 (1956).

[3] H. MESSEL: The Development of a Nuclear Cascade, in: Progress of Cosmic Ray Physics, Vol. II, edit. by H. J. G. WILSON. 1954.

in the cross section data are perhaps not as disturbing as they appear at first. We shall make use of the picture of the spreading cascade—the "funnel" model— for this purpose. Qualitatively, one notices immediately that in this model— even if the "funnel" of interacting nucleons does not pierce the entire target nucleus—the number of particles participating in the composite collision, and hence the number of shower particles, in an average collision increases rapidly with A, and that this fact enhances the deviation of the observed cross sections from the geometrical $A^{\frac{2}{3}}$-law, beyond the "transparency" effect. Or to put it differently: It follows immediately that if we use the cross sections measured, say, in lead and in carbon to determine a value of the elementary nucleon-nucleon cross section σ_0, and then compute with this σ_0 the appropriate cross section for iron, the "funnel" model will always lead to a smaller value than the simple transparency picture.

Obviously, the crude funnel model in the formulation of the preceding section cannot be expected to give relevant quantitative results if applied to the low-energy region of interactions below 10^{11} ev. Nevertheless, just in order to put more forcefully the qualitative argument stated above, we shall show in a rough computation that indeed a better agreement with the empirical results can be obtained if the transparency idea is combined with the model of the conical inter-action volume. The calculation will proceed along the following lines:

(i) An event is called a penetrating shower if at least a certain number N_s of charged secondaries is produced. According to (17.20), this requires a certain minimum collision volume which can be computed for each energy, *and is independent of the nature of the target nucleus.*—In the following, we have arbitrarily chosen $N_s = 4$.

(ii) Knowing the "funnel" volume $V(l, \lambda)$ as a function on the intranuclear path length l and of the mean free path in nuclear matter λ, we can determine for each impact parameter b the corresponding average collision volume

$$V(b, A) = \int_0^L V(l - y, \lambda) \exp(-y/\lambda)\, dy/\lambda \qquad (18.1)$$

where y stands for the coordinate, along the primary's path, of the first encounter, and $L = 2(R_0^{2/3} - b^2)^{\frac{1}{2}}$ is the total penetration length.

(iii) From this, and from the minimum volume defined in (i), we can now define a maximum impact parameter $b_{\max}$ up to which a collision will, on the average, comprise a sufficiently large number of nucleons to give N_s shower particles:

$$V(b_{\max}, A) = V_{\min}(N_s, \gamma). \qquad (18.2)$$

(iv) The total cross section for shower production is then obtained from

$$\sigma = \frac{\pi \int_{\gamma_{\min}}^{\infty} b_{\max}^2 (A, \gamma)\, s(\gamma)\, d\gamma}{\int_{\gamma_{\min}}^{\infty} s(\gamma)\, d\gamma} \qquad (18.3)$$

where $\gamma_{\min}$ is the threshold energy for a shower with N_s particles, obtained, for instance, from the condition that the available energy must exceed $\frac{3}{2} N_s \cdot m_\pi c^2$. This leads to

$$\gamma_{\min} = \frac{1 + 3(m_\pi/m_p) N_s (V/V_0)}{2(V/V_0)} \approx \frac{3}{2}(m_\pi/m_p) N_s. \qquad (18.4)$$

In view of the semi-quantitative nature of the argument, the calculations were based on the empirical multiplicity-energy relation observed for high-energy interactions, (17.20). The calibration constant K,

$$K = N_s (V/V_0)^{-\frac{3}{4}} \gamma^{-\frac{1}{4}}, \tag{18.5}$$

Table 8. *The values of the calibration constant K of the empirical multiplicity-energy relation,* $N_s = K (V/V_0)^{\frac{3}{4}} \gamma^{\frac{1}{4}}$, *as a function of* $\lambda/\lambda_0 (\lambda_0 = \frac{2}{3} R_0)$.

	Tunnel model $(\lambda = \lambda_0)$	Funnel model, $\lambda/\lambda_0 =$				
		1	$\frac{3}{2}$	2	3	4
K	0.460	0.127	0.278	0.464	0.758	1.01

was not taken from the Fermi theory (as in COCCONI's paper), but from the experimental data in a pure target material, copper (see Fig. 14 on p. 189). This gives $K(V/V_0)^{\frac{3}{4}} = 3.64$, and for K itself the values listed in Table 8 as a function of the mean free path λ in nuclear matter. However, the absolute calibration will be of use only later; for the moment only the ratio of the cross sections in various substances is of interest. The values obtained according to the procedure outlined above were, therefore, re-calibrated so that the cross section for lead, at a mean free path of $\frac{3}{2} \lambda_0$ or an elementary cross section of about 40 mb, has the empirical value of 2.1 barn. With the rigid-sphere model of the nucleus which

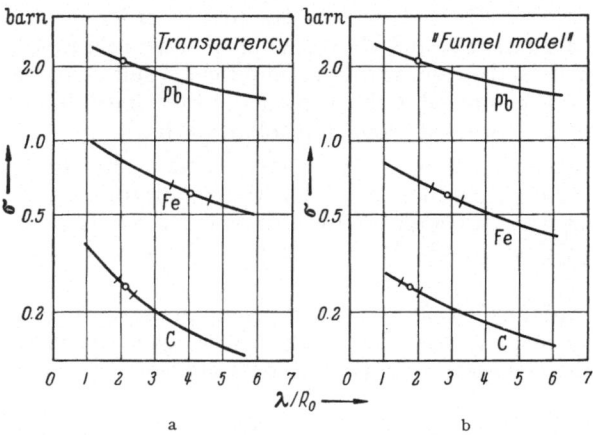

Fig. 21a and b. Total cross section for the production of a penetrating shower with at least 4 charged secondaries, in lead, iron and carbon, as a function of the mean free path in nuclear matter (a) for transparency model of Sect. 8; (b) for the "funnel" model.

has been used for simplicity, the cross sections of iron and of carbon according to the funnel model are then at all λ's both smaller than the corresponding values for the transparency picture. This is shown in Fig. 21, in which also the experimental point of BRENNER and WILLIAMS[1], $\sigma_{Fe} = (0.61 \pm 0.04)$ barn, is indicated, together with the most plausible average values for Pb (2.1 barn) and C [(0.25 ± 0.02) barn]. It is seen that even in this elementary calculation, the funnel model, while not entirely removing the discrepancies, leads to a somewhat better representation. The basic features of the explanation for the disagreement suggested above in a qualitative way are, therefore, very likely correct: Low average multiplicity per nucleon-nucleon interactions, slow energy dissipation in the collisions, and an elementary cross section of the order of 40 mb, together with the model of a spreading cascade of interacting particles.

19. Structure of the nucleus: The ratio of positive to negative secondaries. In the preceding section, the old value for the nucleon radius $R_0 \approx 1.37 \times 10^{-13}$ cm has again been used, in spite of the contrary evidence of HOFSTADTER's experiments on charge distribution mentioned in Sect. 8. Apart from the support

[1] A. E. BRENNER and R. W. WILLIAMS: Phys. Rev. **106**, 1020 (1957).

given to the assumption of the larger value by the cross section measurements in heavy elements, there is only one further reason to accept it. This is the suggestion made on theoretical grounds by JOHNSON and TELLER[1] that the distribution of protons and neutrons in the nucleus should not be the same, with the neutron orbits extending beyond those of the protons. In simple language one might picture the nucleus as a sphere containing in its interior a mixture of protons and neutrons in about equal parts, and on the outsides a shell of neutrons. The Hofstadter experiments would then explore the structure of the inner part, while penetrating-shower experiments would always involve interactions taking place anywhere in the volume.

Up to now, this point can be brought up only in a speculative manner; experimental evidence for or against it appears to be entirely lacking. Just because of that, it might be worth while to point out a few possibilities of experimental checks, and of the possible use of penetrating-shower data for investigation into the structure of nuclei.

The most suggestive line of approach is a precise study of the positive/negative ratio among the shower secondaries. It is easy to write down the value expected for this ratio under the assumption of homogeneous distribution of both kinds of nucleons, for neutral and for charged primaries. It depends, of course, on the multiplicity of the event, varying between a maximum of 2.5 in light and about 2 in heavy elements for a proton primary with one secondary only, to unity at very large multiplicities. For a neutron primary, the negative/positive ratio will exceed 3 in a heavy nucleus at the lowest multiplicity. But even a comparatively thin outer shell of neutrons will change this ratio radically, since the cross section offered is even then an appreciable fraction of the total. In fact, a simple calculation—assuming, for simplicity, an exactly spherical shape of the inner volume containing the mixture, and an untapered outer shell of homogeneous neutron density—shows that a shell of thickness R_0 on a nucleus $Z = 47$, $A = 108$ will for an incident neutron primary reduce the negative/positive ratio, and for an incident proton increase the positive/negative ratio, both to about the same value ~ 3 if a single charged secondary is emitted.

Unfortunately, the experimental material for a systematic investigation is again lacking. The only extensive body of data available are those on the positive excess of atmospheric μ-mesons which have occasionally been used for a study of the multiplicity of meson production (cf. YEIVIN[2]). But already the fact that they result from interactions in light nuclei where the inhomogeneities are probably the smallest, makes their analysis not too promising. It would be much more hopeful if a comparison of the ratios in various elements, and for charged and neutral primaries, could be carried out. The appropriate data could, in principle, be obtained both from emulsion work and from experiments in magnet cloud chambers. In the first case, the frequency of stopped positive and negative mesons in a given momentum interval must be studied; if the momentum at production exceeds about 100 MeV, Coulomb effects in the target nucleus would not interfere and the positive/negative ratio observed in any interval would be characteristic for the whole spectrum. In the second, charge determinations could be carried out on all mesons emitted in forward direction. However, the work is tedious, and in the case of the emulsions, further complicated by the uncertainties of the impact parameter in "heavy-element" collisions selected according to the customary criterion, the number of heavy prongs. The cloud chamber analysis is more recommendable, and could be the by-product of another

[1] M. H. JOHNSON and E. TELLER: Phys. Rev. **93**, 357 (1954).
[2] Y. YEIVIN: Nuovo Cim. (10) **2**, 658 (1955).

shower experiment. At any rate, the possibilities of this simple-minded approach to a study of the relation between charge and mass distribution in nuclei should be kept in mind.

20. Multiplicity distribution of shower secondaries. Some experimental data on the multiplicity distribution of charged shower secondaries have been quoted in Sect. 14, and we repeat here the two main conclusions:

(i) At all primary energies, the number of secondary particles ejected has been found to vary widely;

(ii) at energies around and below 10^{10} eV, the multiplicity and the multiplicity distribution of showers initiated in the heavy emulsion nuclei Ag and Br, and of those originating in lead, are not significantly different.

Penetrating showers, it has been emphasized, are not usually the result of a single nucleon-nucleon collision. The question arises, therefore, if and how far the observed variations may be due to causes other than fluctuations in the number of secondaries emitted in the elementary interaction.

Let us first consider the low-energy results of Camerini et al., and of Froeh-lich et al., reported in Sect. 14. The striking similarity between the data found for interactions in the medium-heavy and in the heavy nuclei has already been interpreted as a sign that for these energies, the primaries have already dissipated most of their energy after traversing a nucleus of the size of Ag or Br at an average impact parameter. From the considerations of Sects. 9 and 17 we have seen that the elementary cross section certainly lies between the "geometrical" value $\sigma_0 = \pi R_0^2$, and about $\frac{1}{3}$ of this value, but is most likely $\frac{2}{3}\sigma_0$. Consequently, the average number of interactions in an emulsion nucleus is of the order 2 to 5, and it follows that the expected statistical fluctuations in the number of collisions together with those in the neutral/charged ratio, can account for a multiplicity spread of the order of that observed. In fact, a simple calculation shows that the propability distribution $P(N_s)$ computed under the assumption of a Poisson law for the frequencies $p(n, b)$ of exactly n collisions at impact parameter b, and of a similar law for the empirical neutral/charged ratio, becomes almost embarrassingly large even if the multiplicity per collision (for the energy dependence of which a power law of the order $\gamma^{\frac{1}{4}}$ may be assumed) is kept rigidly constant. It would appear that this calculation "permits too many collisions".

A reason for limiting the number of "permissible" collisions is seen in a comparison of the data in lead and in the emulsion. It may be assumed that in traversing the silver nuclei, the primary has on the average spent about $\frac{4}{5}$ of its energy in the 2 to 5 collisions suffered. Hence the degree of inelasticity must lie roughly between $\frac{1}{4}$ and $\frac{1}{2}$, according to the value of λ postulated; and the maximum number of collisions the primary can undergo before being slowed down below the threshold energy can thus be estimated for all primary energies. Repeating the calculation of the multiplicity spread with this restriction, one still finds for all but the longest mean free paths a range of multiplicities of sufficient width.

Turning now to the high-energy region where the "jet" theory can be applied, it is again of interest to compare the relative significance of the various factors which determine the spread of multiplicities reported: variation of the path length as a function of the impact parameter b; fluctuations in the number n of nucleons encountered upon traversing the nucleus; deviations from the average neutral/charged ratio: and differences in the number of secondaries per collision. To do so, we shall first revise the Cocconi argument (17.16), in order to determine the number n from empirical material only: Assuming the validity of a relation of the form $N_s = K n^{\frac{1}{3}} \gamma_N^{\frac{1}{4}}$, we have in Table 8 obtained numerical values for K from the Kaplon-Ritson data on Cu-stars. With those, we now compute the numbers n for all the events listed by these authors: except for charge fluctuations, they should not exceed the numbers computed according to the various models at $b = 0$. The results are shown in Table 9. The two last rows give the average numbers $\bar{n}_{max}$ of particles at maximum path length, $b = 0$, and the fictitious "apparent" maximum number n_{max} which would, with the same constants, produce $\frac{3}{2} N_s$—the largest number that can be explained as a result of charge fluctuation only. Clearly, $n_{max} = \frac{3}{2} \bar{n}_{max}$. The result of the survey is

Table 9. *Number n of nucleons participating in the composite nucleon-nucleus interaction in Cu-nuclei reported by Kaplon and Ritson, for the "tunnel" and the "funnel" model.*

N_s	$\gamma \times 10^{-3}$	Tunnel model, $\lambda = \lambda_0$	Funnel model			
			$\lambda = \lambda_0$	$\lambda = \frac{3}{2}\lambda_0$	$\lambda = 2\lambda_0$	$\lambda = 3\lambda_0$
24	4	6.55	23.7	10.9	6.5	4.0
20	1.3	7.2	26.2	12.0	7.15	4.4
9	4.5	2.4	8.7	4.0	2.4	1.45
26	19.5	4.8	17.3	7.9	4.7	2.9
16	1.3	5.8	21.0	9.6	5.7	3.5
24	1.8	8.0	29.0	13.2	7.9	4.85
19	6.1	4.7	16.9	7.7	4.6	2.8
18	1.29	6.6	23.9	10.9	6.5	4.0
24	5.1	6.15	22.3	10.2	6.1	3.75
17	0.5	7.8	28.2	12.9	7.7	4.7
24	8.9	5.35	19.4	8.9	5.3	3.25
16	1.0	6.15	22.3	10.2	6.1	3.75
11	0.64	4.75	17.2	7.9	4.7	2.9
36	19.4	6.7	24.2	11.1	6.6	4.1
15	30.0	2.5	9.0	4.15	2.5	1.5
24	0.8	9.8	35.4	16.2	9.7	5.95
15	0.8	6.1	22.2	10.2	6.0	3.7
15	50.0	2.2	7.9	3.6	2.15	1.3
10	5.0	2.6	9.35	4.3	2.6	1.6
14	7.2	3.3	11.9	5.5	3.25	2.0
$\bar{n}_{max}$		6.0	24.8	12.3	7.2	3.7
n_{max}		9.0	37.2	18.45	10.8	5.55

inconclusive. It appears that all these observations in Cu can be explained on the basis of charge fluctuations alone; and this applies in equal way to "tunnel" and "funnel" model. It would be tempting to carry out a more complete statistical analysis of the material of Table 9: Computing the probability $P(N_s)$ to find exactly N_s charged secondaries, from the probability $p(N)$ to produce $N \geq N_s$ shower particles, and $\pi(N, N_s)$ that N_s of them are charged. The calculation is straighforward, but the empirical material not extensive enough to permit a satisfactory analysis.

However, there are a few cases where the observed multiplicities are definitely to high to be explained as due to charge fluctuations, even at maximum path length. Thus, the event reported by Brisbout et al., with $N_s = 126$ at $\gamma = 10\,000$, or Mulvey's star of 76 secondaries at $\gamma = 1000$[1], lead to excessive values of n. In these cases, the explanation is more likely to be found in a genuine fluctuation of the multiplicity per collision than in density fluctuations in the—already large— collision volume. From rough computations it would appear that a Gaussian

---

[1] A list of numerous jets recently observed has been compiled by McCusker and Roessler. Nuovo Cim. (10) **5**, 1136 (1957).

with a half-width of about 2 for the probability of such deviations would suffice
to cover the observed variations.

21. The degree of inelasticity in high-energy collisions. Apart from multiplicity and multiplicity distribution, the experimental material can probably
offer no clue more promising for an analysis on purely empirical grounds than
the measurements of the degree of inelasticity. On the one hand, its connection
with the theory of nuclear forces is straightforward and intimate; on the other,
it is one of the few quantities which can be obtained from the experiment without
a priori assumptions on the characteristics of the interaction. In fact, there is
no dearth of methods to approach this matter, neither in the high-energy region
of the "jets" nor in that accessible to counter and cloud chamber techniques,
and neither by direct nor by indirect measurements.

To begin with a discussion of the results obtained from indirect studies of
the degree of inelasticity, we refer briefly to the subject discussed at length in
Sects. 5 to 8: the difference between absorption and interaction mean free paths.
It has already been stressed in these sections that especially the well-established
large value of the atmospheric absorption length $\Lambda_{air} = 120$ g/cm², indicates a
"transparency" of the light nuclei not only with respect to first collisions, but
also with respect to energy transfer. The energy dissipation of the incident
primary in a light nucleus is not large enough to eliminate the particle from
the N-component responsible for the further penetrating showers. For detailed
presentation, references may be made to the thorough analysis given by ROSSI[1],
or to the more elementary presentation in some of the older papers (GREISEN
and WALKER[2], VERNOV[3], SITTE[4]). If, as it is now most plausible, a value of
about 85 g/cm² is assumed for the interaction mean free path in air—a quantity
for which no direct measurement has as yet been carried out—, it follows that
the primary must retain after the collision about 50% of its energy. Moreover,
from the fact that the absorption mean free path is practically independent of
the primary energy, at least in the range of about 10^{10} to 10^{13} eV, it can be concluded that also the degree of inelasticity does not vary appreciably in this same
interval—provided, of course, that the elementary cross section σ_0 remains constant.

Another indirect method makes use of the measurement of the ratio of neutral
to charged N-particles of high energy. This quantity has been determined by
a number of investigators[5-10], most of whom agree that in an atmospheric depth
of 600 to 700 g/cm², neutrons are present in an amount of about 0.75 to 0.8 of
the proton component. From this, again an inelasticity of the order of 60%
can be estimated[6,7].

Direct measurements at low energies can be carried out most precisely if
the energy of the meson component is determined by using as a monitor the
electron cascades associated with the penetrating shower, and computing the
total energy transferred to the secondaries with the help of the now well-established
neutral-to-charged ratio of 0.5. This can be done either by counting the total

[1] B. ROSSI: High-energy Particles, p. 490 ff. New York: Prentice-Hall 1952.
[2] K. I. GREISEN and W. D. WALKER: Phys. Rev. **90**, 915 (1953).
[3] S. N. VENOV:International Conference on Cosmic Radiation, Guanajuato 1955 (Summary of earlier work).
[4] K. SITTE: Phys. Rev. **78**, 714 (1950).
[5] W. D. WALKER: Phys. Rev. **77**, 686 (1950).
[6] K. SITTE: Phys. Rev. **78**, 714 (1950).
[7] K. GREISEN, S. P. WALKER and W. D. WALKER: Phys. Rev. **80**, 535, 546 (1950).
[8] M. B. GOTTLIEB: Phys. Rev. **82**, 943 (1951).
[9] A. LOVATI, A. MURA, G. TAGLIAFERRI and TERRANI: Nuovo Cim. **9**, 946 (1952).
[10] H. L. KASNITZ and K. SITTE: Phys. Rev. **94**, 977 (1954).

number of electrons at the maximum of the composite cascade, as has been pointed out in Sect. 17, or by study of the individual cascades near their origin. A thorough analysis of this method has been made by BENDER[1], and results obtained with it have for instance been reported by CRESTI et al.[2], applied to pion-nucleus collisions, and by SITTE[3]. Unfortunately, the results of the latter paper are in error due to insufficient corrections for the bias introduced by the counter control, which leads to slight but systematic deviations if showers originating in different materials are compared. Therefore, the results of a more recent and hitherto unpublished re-analysis of all suitable cloud chamber records of the Syracuse group 1950 to 1954 will be reproduced. Primary energies have been measured either by the "F-plot" method (Sect. 13), or in part of the experiments, from the hodoscope data on the integral path length. BENDER'S expressions were used for the determination of the energy of the neutral meson component, and wherever possible, both methods were applied: in general with consistent

Table 10. *The degree of inelasticity, k, of high-energy interactions as a function of the primary energy E_0, from cloud chamber observations on showers originating in lead, aluminium, and light elements (carbon and lithium).*

Element	E_0(GeV)					
	$2-4$ $\langle E \rangle_0 = 2.8$	$4-6$ $\langle E \rangle_0 = 4.7$	$6-10$ $\langle E \rangle_0 = 7.3$	$10-20$ $\langle E \rangle_0 = 15.5$	$20-30$ $\langle E \rangle_0 = 26.5$	30 $\langle E_0 \rangle = 47$
Pb	0.30 ± 0.15	0.60 ± 0.25	0.60 ± 0.20	0.76 ± 0.10	0.65 ± 0.12	0.70 ± 0.15
Al	—	—	(0.80 ± 0.50)	0.55 ± 0.10	0.60 ± 0.09	0.55 ± 0.12
Li—C	—	—	(0.60 ± 0.40)	0.50 ± 0.08	0.55 ± 0.10	0.40 ± 0.12

results. A summary of the data thus obtained is shown in Table 10 where the average degrees of inelasticity k are listed as a function of the primary energy for lead, aluminium, and light elements (carbon and lithium data combined). The errors of k indicated in the Table include the estimated uncertainties of the primary energy. It is seen that above 10 GeV, the values of k remain essentially constant, and that the degree of inelasticity seems to increase, in this energy interval, with increasing primary energy. Noting the decrease with A of the degree of inelasticity k, it is illustrative to go one step further and to ask whether the value of k corresponding to a nucleon-nucleon collision would not be smaller still than that found for Li and C. Indeed, if one combines the light-element observations on k with the results obtained by various methods for the nucleon-nucleon interaction cross section—all pointing to a value $\sigma_0 \approx 40$ mb—the answer is readily found: The average inelasticity in a nucleon-nucleon interaction should not exceed a fraction of about 0.25 to 0.30. A k of this order would be quite consistent with all results hitherto reported; but it must be emphasized that then the average multiplicity per nucleon-nucleon collision is quite appreciably smaller than the multiplicities found in Fig. 14.

At jet energies, a determination of the degree of inelasticity was carried out in a number of cases where energy measurements on the secondaries became possible[4-9],

[1] P. A. BENDER: Nuovo Cim. (10) **2**, 980 (1955).

[2] M. CRESTI, W. D. B. GREENING, L. GUERRIERO, A. LORIA, G. ZAGO add M. DEUTSCH-MANN: Nuovo Cim. (10) **4**, 747 (1956).

[3] K. SITTE: Nuovo Cim. (10) **3**, 1467 (1956).

[4] C. CASTAGNOLI, G. CORTINI, C. FRANZINETTI and L. SCARSI: Nuovo Cim. **10**, 1539 (1953).

[5] C. C. DILWORTH, S. J. GOLDSACK, T. F. HOANG and L. SCARSI: Nuovo Cim. **10**, 1261 (1953).

[6] T. E. HOANG: J. de Phys. **15**, 337 (1954).

[7] G. BERTOLINO: Nuovo Cim. (10) **2**, 1130 (1955); **3**, 141 (1956).

[8] G. WATAGHIN: Nuovo Cim. (10) **4**, 154 (1956).

[9] J. PERNEGR: Cs. Casopis pro fysiku **6**, 2 (1956).

and a very large spread of k was observed, ranging from under 0.1 to complete inelasticity. Some authors present their data by dividing their jets into two groups according to the multiplicity of the event, while others instead attempt to establish a smooth multiplicity-inelasticity relation. As more data became available, the first procedure appears rather artificial, and the second increasingly plausible. Before summarizing the experimental results, we shall however briefly discuss their interpretation in the frame of the model of composite nucleon-nucleus collisions.

Apart from genuine fluctuations in the energy transfer, we have then as a variable parameter the number n of nucleons participating in the interaction which is a function of the impact parameter b, and determines the mean energy available in the C-system [Eq. (17.10)]. This fact alone would, therefore, lead to a variation in the degree of inelasticity of the collision the magnitude of which is readily calculated. Although in this way obviously only an incomplete description of the phenomenon is obtained, the considerable difference between the results based on the "tunnel" and on the "funnel" model make it appear worth while to compare the predictions of the two models with a summary of the jet data hitherto published. This is shown in Fig. 22. Rather arbitrarily, the calculated inelasticities were re-calibrated to give the coefficients obtained for $b=0$ the value 1.

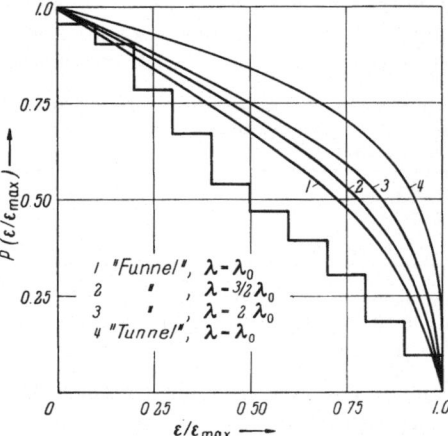

Fig. 22. Inelasticities $k=\varepsilon/\varepsilon_{max}$ in jets: Distribution calculated according to 'tunnel' and 'funnel' models (smooth curves), and experimental values published before 1957 (step curve).

Evidently, the histogram of the experimental values resembles much more a straight line than any one of the calculated curves, but the high probabilities predicted by the "tunnel" model for small numbers of particles does seem to fit even less than the somewhat flatter "funnel" values. However, it is difficult to decide to what degree the neglection of individual fluctuations, and the arbitrary re-calibration, may effect the results, and not too much weight should be given to the computed curves. Only a far more thorough study including multiplicity *and* inelasticity fluctuations would permit reliable conclusions, but for this work, more systematic data have still to be collected.

Note added in proof. More recent investigations in the "jet" region have provided ample material for a re-evaluation of the problem complex. In all systematic studies of high-energy events [e.g. Schein, Haskin, Lohrmann and Teucher, Proc. Moscow Cosmic Ray Conf. I, p. 6 and p. 26 (1959); Duthie, Fisher, Fowler, Kaddoura, Perkins and Pinkau, ibid. p. 30 and 35; Fujimoto, Hasegawa, Kazuno, Nishimura, Niu and Ogita, ibid. p. 41; Vinitsky, Golyak, Takibayev and Chasnikov, ibid. p. 67; Boos and Takibayev, ibid. p. 82) the conclusion was reached that with increasing primary energy, the average degree of inelasticity k_π—or more precisely, the fraction of energy transferred to the π-mesonic secondaries—decreases quite considerably. Events are frequently found where a secondary interaction displays practically the same energy as did the primary collision, but in the region above $\sim 10^{13}$ eV, the *average* value of k_π is certainly not more than 10 to 20%, and may well be even smaller.

This low inelasticity, together with the established fact that the absorption mean free path does not appear to vary sensitively with the primary energy in the energy range concerned, is frequently quoted as an indication that processes other than π-meson production (ejection of heavy unstable secondaries and nucleon-antinucleon pairs) play a significant part in collisions at very high energies—although this claim is not borne out by direct observation. But it is easily seen that the known facts can also be explained by the assumption of an increase of the collision cross section with increasing primary energy γ. A variation of the number of shower secondaries $\bar{n}_s$ with γ according to $\bar{n}_s \propto \gamma^m (m \lesssim \frac{1}{4})$, together with the approximate constancy of the mean secondary energy, requires $\bar{k} \propto \gamma^{m-\frac{1}{2}}$. Since $\bar{k} = \int\limits_0^{b_{max}} k(b) \times 2\pi b \, db/\sigma$ $(\sigma = \pi b_{max}^2)$, it follows that if for all energies $k(0) = 1$, and $k(b) \to 0$ with increasing b, the asymptotic value for the average inelasticity is $\bar{k} \propto 1/\sigma$, and hence the data are compatible with $\sigma \propto \gamma^{\frac{1}{2}-m}$, or approximately $\sigma \propto \gamma^{\frac{1}{4}}$.

While a qualitative interpretation of these results comes easily—for instance along the lines of HEISENBERG's arguments ["Cosmic Radiation", DOVER (1953)] describing the energy available for secondary emission by the integrated overlapping of the meson fields, or else by the overlap volume of colliding "tapered" nucleons—quantitative agreement is not so easily achieved. However, two interesting features of this approach may be stressed: Whatever minimum overlap we choose to define a "shower", it is evident that higher primary energies will permit shower production in increasingly distant collisions, and with a correspondingly lower *minimum* inelasticity. Thus, apart from explaining the trend towards reduced *average* inelasticities, this picture necessitates the conclusion that at the same time, the total cross section for shower production rises with the primary energy, and similarly the anisotropies connected with the increased average impact parameter. On the former point, some supporting experimental evidence has been quoted in Sect. 9. On the latter, some confirmation is found for instance in the work of PERNEGR [Proc. Moscow Cosmic Ray Conf. I, p. 121 (1959)]. It remains here to emphasize that without accepting as creditable this simultaneous decrease in inelasticity and increase in cross section, it would be difficult to reconcile a number of conflicting experimental results. Thus, GRIGOROV's experiments (Sect. 16) showing a *constant* energy transfer to the electron component can be understood if we assume that the reduced yield per collision is about compensated by an increased number of collisions in his top absorbers. In other words, this experiment does not determine the transfer per interaction as intended, but the transfer in a certain fixed amount of material. In the same way, air shower data must be re-interpreted. The usual conclusion from the balance of nuclear and electron component, that even up to air shower energies the inelasticity remains reasonably constant, should be understood to mean "inelasticity per unit mass", but not, as customary, per interaction. Thus, in spite of the new results on inelasticities and cross sections, there would be—on the average—no appreciable change in the longitudinal development of air showers. Significant differences would be found only in the degree of fluctuations in the shower development, and in an increased height of "first" interactions (permitting, for instance, the arrival of groups of μ-mesons originated in the upper 10 to 25 g/cm² of the atmosphere).

22. Conclusions relating to high-energy nuclear interactions in air showers.
A full account on the extensive studies of the nuclear component in large air showers, and of their interpretation, is given elsewhere in this volume. In this section, only a very brief summary of pertinent points will be given. After all, air showers originate as "penetrating showers" like the ones discussed here; only due to the fact that the energy transfer to the electron component is a practically irreversible process, the mixed cascade will in its descent through the atmosphere apportion more and more of its initial energy to this component, and thus resemble a pure electronic cascade.

Two important features of high-energy interactions which have been emphasized above, were clearly confirmed in air shower work:

(i) The comparatively slow dissipation of the energy in the nucleonic component, i.e. the high degree of elasticity of the collisions; and

(ii) the strong collimation of the angular distribution among the shower secondaries ejected in the *first* interaction.

The first of these statements is evident from the conspicuous proportionality between nucleonic and electronic component almost throughout the entire longitudinal development of the air shower, and from the absence of a strong altitude variation of the lateral distribution of the showers which indicates a continuous "rejuvenation" of the electronic cascade from the nucleonic component[1-3]. More directly, it has been borne out by the investigations of the Russian air shower workers[4] who were able to measure the energies of the nucleons still present after a large number of collision lengths, and thus to determine limits for the inelasticities which are quite consistent with the values derived in the previous sections. Again, attention is drawn to the possible increase of the interaction cross sections, and decrease of the inelasticities, with increasing primary energy. If this effect exists, the approximate constancy of the energy transfer to the electron component *per unit mass traversed*, as observed in experiments, must extend to air shower energies.

The second conclusion mentioned above follows, for instance, from the absence of well-defined "multiple cores" in distances of the order of 1 m or more, which must be expected if the secondaries of the first collision were ejected uniformly in the C-system[5-7]. In particular, again the Russian studies of the lateral energy distribution in air shower cores demonstrate quite convincingly that the average transverse momentum of the shower secondaries does not exceed about 10^9 eV even at primary energies of 10^{16} and more. It is remarkably independent of this energy, and very nearly equal in the inner cones of jets at 10^{11} or 10^{12} eV as in extensive air showers. This characteristic feature is in contradiction to the results of FERMI's theory, but is equally well predicted by the theories of LANDAU and of HEISENBERG (Sect. 12).

On the whole, air shower investigations have, therefore, not added much further knowledge about the properties of high-energy interactions but extended quite appreciably the region over which we can now consider as established the most important features of nuclear collisions. In the scope of this article, a more extensive discussion is, therefore, not necessary; for details, the reader is referred to COCCONI's article in this volume.

V. Summary.

23. The results of penetrating shower research. The following brief review will once more summarize those results which by now may be considered as safely established:

(i) Penetrating showers are the result of interactions of energetic nucleons with other nucleons or with nuclei, resulting in the ejection of fast secondaries.

(ii) π-mesons form the bulk of these shower secondaries. Even at primary energies $E_0 \gg m_p c^2$, heavy mesons and/or nucleon-antinucleon pairs do not account for more than at most 15% of all shower particles.

(iii) The shower secondaries are themselves "nucleo-active", i.e. contribute to the propagation of the nucleonic shower, especially in dense materials.

(iv) The distribution of the secondaries is in general, but not always, symmetrical in the center-of-mass system.

[1] K. GREISEN and W. D. WALKER: Phys. Rev. **80**, 535 (1950).

[2] H. L. KASNITZ and K. SITTE: Phys. Rev. **94**, 977 (1954). — K. SITTE, D. L. STIERWALT and I. L. KOFSKY: Phys. Rev. **94**, 988 (1954).

[3] U. V. ANISHENKOV, G. T. ZATSEPIN, I. L. ROSENTAL and L. I. SARITCHEVA: J. Eksp. Teor. Fyz. **22**, 143 (1952).

[4] GRIGOROV N.L.: Nuovo Cim. Suppl. **8**, 730 (1958). International Conference on Cosmic Radiation, Varenna 1957.

[5] W. E. HAZEN: Phys. Rev. **85**, 1552 (1952). — R. E. HEINEMAN and W. E. HAZEN: Phys. Rev. **90**, 496 (1953). — W. E. HAZEN, R. W. WILLIAMS and C. A. RANDALL jr.: Phys. Rev. **93**, 578 (1954).

[6] S. N. VERNOV: International Conference on Cosmic Radiation, Varenna 1957. — NIKHOLSKIJ et al. (presented by N. I. DOBROTIN): Nuovo Cim. Suppl. **8**, 737 (1958).

[7] K. GREISEN and W. D. WALKER: Phys. Rev. **80**, 535 (1950).

(v) In the case of symmetrical showers, the angular distribution in the center-of-mass system is as a rule not isotropical, but shows a distinct collimation in the direction of the collision axis. The average transverse momentum of the secondaries does not exceed 10^9 eV even at very large primary energies.

(vi) The degree of inelasticity in a nucleon-nucleon collision is on the average about 0.3. In nucleon-nucleus interactions, its value may fluctuate widely and may in individual cases lie anywhere between less than 0.1 and unity. It may decrease slowly with increasing primary energy.

(vii) The total cross sections for the production of penetrating showers are for heavy elements, and for interactions with primary energies exceeding a few 10^9 eV, of the order $\pi R_0^2 A^{\frac{2}{3}}$, with $R_0 \approx 1.37 \times 10^{-13}$ cm. This value is largely independent of the energy of the shower-initiating particles, though it may perhaps increase very slightly with increasing primary energy.

(viii) For light elements, the cross sections are reduced as the result of an appreciable "transparency" effect: the primary has a not negligible chance to traverse the nucleus without colliding with a nucleon of the target.

(ix) At very high primary energies, the collision must no longer be pictured as a succession of individual nucleon-nucleon interactions, but is better understood as a composite reaction of an incident particle colliding with a whole group of nucleons in the target nucleus. One may consider as the "collision volume" either the cylindrical "tunnel" punched through the target nucleus by the energetic primary, or if the nucleus is viewed as a compound of individual particles, a widening "funnel".

(x) Quantitatively, the features described above are in general consistent with the assumption of an elementary cross section for nucleon-nucleon interactions at very high energies of the order of 30 to 40 millibarns.

Within the field of those ten points, there are some open questions but no major inconsistencies.

24. The program of penetrating shower research. At this time when artificially accelerated nucleons have become available up to energies of 10 GeV, and may soon reach 25 GeV, there are certainly good reasons to question the value of future research in cosmic-ray initiated penetrating showers. But it needs no undue partisanship to come to the conclusion that there is still a good deal of work to be done in this field, and that it is urgently needed.

In elaborating this statement—albeit briefly—we shall as before deal separately with the two main goals of shower research: the study of the structure of the target nuclei, and the study of the characteristics of high-energy nuclear interactions. Turning our attention to the first problem, it is obvious that much of this work will in future be taken over by the machines—in the region where they operate. To compete with them in accuracy would be futile; but to complement them by extending the results to the energy region still beyond their reach: the very important range between 10 and 100 GeV, would be most desirable. That in order to do so, new techniques will have to be tried out from which later on the faster machine work will richly benefit, is also evident and is no new phenomenon. But perhaps it should be emphasized that this extension of the observations is not just a fad; most likely it will be from a systematic investigation of the energy dependence of the cross sections only that the finer points of nuclear structure can be studied. Thus, it will certainly be of interest to determine not only the total cross sections for shower production in various elements with a better accuracy—much too little is known in the region of medium-heavy and of light elements—, but also the differential cross sections, at high energies, for the production of events of a given multiplicity. Together with studies of the positive/negative ratio, such experiments could yield much-needed contributions to our knowledge about the structure of nuclei.

A special role belongs to the work on very energetic interactions, i.e. in the range above 10^{12} eV. The small de Broglie wavelength of the particles of these energies permits a direct approach to problems of the structure of the elementary particles, as well as of the characteristics of their interactions. In this way, it connects the two fields mentioned above. As to the second subject, it will suffice to recall how frequently even the rough interpretation attempted in this report, remained inconclusive because of lack of sufficient and systematically collected experimental material To mention just one point in particular: It would be highly advisable to continue the work with the technique of emulsion chambers and to extend it in order to obtain data on showers initiated in a variety of target nuclei, in an unbiassed selection. Further suggestions could easily be added, but to enumerate and to outline them would involve a very personal presentation of the problems for which there is no place in this summary. Its aim was to prove that research in cosmic-ray initiated penetrating showers is as much a living subject as ever; if this was successfully done, each research worker will find his own program without difficulty.

Extensive Air Showers.

By

GIUSEPPE COCCONI.

With 19 Figures.

A. A qualitative description.

1. The importance of Extensive Air Showers. The name of Extensive Air Showers (AS) is used to identify the chain of events that is initiated by a Cosmic Ray particle of ultra-relativistic energy interacting with air nuclei in the upper atmosphere.

The products of this interaction, moving practically in the same direction as the primary Cosmic Ray, give rise to a cascade of other interactions, until, in the lower atmosphere, a number of secondaries is created that can be as high as many millions.

All these secondaries arrive practically at the same time over a plane perpendicular to the direction of the original particle, the "axis" of the shower, but, during their traversal of kilometers of air are scattered around the axis up to distances of several hundred meters. The shower thus covers a circular area of may thousands square meters with maximum density in the central region, the "core" of the shower.

The primary Cosmic Rays that give rise to the largest AS in the lower atmosphere are the most energetic particles known to exist in nature.

A knowledge of their properties and of their behavior in interacting with matter obviously is of the greatest interest. Whatever has been learned so far about the physics of energies above 10^{14} ev has been inferred from the study of the Air Showers.

It is unfortunate that no more direct methods of observation of these ultra-relativistic cosmic ray particles are available. However, it must be remembered that if their multiplication in air had not taken place, as would be the case if one tried to make measurements at the top of the atmosphere, there would be no practical chance of observing these particles at all. In fact, the frequency of arrival, e.g., of a primary Cosmic Ray of $\geq 10^{18}$ ev over a surface of one square meter is 1 in 3000 years, hence a detector flow outside the earth has no chance of being hit.

The atmosphere, by multiplying the original particle by millions and by spreading the secondaries over a surface of one square kilometer makes it possible for a small detector located on the ground to be hit once every few weeks by the secondaries of such a particle. This is an extreme case, but even the primaries of 10^{16} ev would be too rare for direct observation, while their AS are easily detected at sea level with rates of many per hour.

One must remember that the energies involved here are still several orders of magnitudes larger than the maximum energies within the reach of the boldest plans proposed thus far for the artificial acceleration of particles. The interactions between protons belonging to two beams, each of 30 Gev, brought to collide one against the other[1], are equivalent to those produced by Cosmic Ray protons of $\sim 10^{12}$ ev.

The knowledge of the properties of the extremely energetic primaries that produce AS is of great importance also in cosmology. These particles certainly come from very far away; no known magnetic field is able to confine protons of 10^{18} ev within the neighborhood of a star, or even of our own galaxy. If, as seems plausible, all or some of the largest showers are produced by single protons, the study of their size, direction of arrival and time variation can give information about properties of the universe that one cannot hope to obtain otherwise.

The emphasis put on the highest energies must not give the impression that AS are a very rare phenomenon; rare are only the very large showers. Actually, the production of a shower constitutes the normal consequence of the interaction with the atmosphere of all primary cosmic rays. The only difference is the scale of the phenomenon. The low energy Cosmic Rays start small showers that completely exhaust themselves before reaching the lower atmosphere. As the primary energy increases, showers of only a few particles will arrive on the earth, and eventually showers of more and more particles. All of these showers are "extensive", since the air scatters the secondaries far from the axis, but unless a good number of secondaries reaches a certain level, it is difficult to show experimentally that the shower exists. Most of the "single" cosmic ray particles detected by a small apparatus near the ground belong to these extensive, but poor, showers. When the number of secondaries reaches some thousands and the core lands within some ten meters from the detector, then it becomes quite probable that two or more particles of the shower arrive simultaneously over a reasonable area, providing an experimental clue for the existence of the AS.

Of the particles present in AS about 95% on the average are electrons and photons, some percent are μ-mesons and only about one percent consists of particles classified as N-particles *(N for nuclear-active)* that interact strongly with nuclear matter. The preponderance of electrons and photons (the *Electro-Magnetic*, or *EM*, component) gives the impression that these particles are the most important components of the shower and led to the early interpretation of AS as a pure electromagnetic cascade. Now, instead, it is well proved that practically no *EM* particles are present among the most energetic primary Cosmic Rays, and that the *EM* component is all of secondary origin. The same is true for the μ-component. The N-particles, instead, constitute the backbone of the showers, and though relatively few in number these particles are the most energetic, and keep supplying secondaries around the core.

For these reasons it is now felt that the study of the number, distribution, and energy of the N-particles near the core can give the closest insight into the details of the ultra-relativistic interactions that occur along the axis of the showers. Though the work done thus far in interpreting AS is highly consistent and clarifying, the central problem of the ultra-relativistic interactions is still quite open and much more experimental and theoretical skill will be required before the true solution is satisfactorily cornered.

2. Survey of essential processes. Before starting a detailed discussion we want to give here a qualitative description of the processes involved in the formation of a typical AS. This will help to put in the proper perspective the various problems that will be discussed separately later.

Let us start, say, with a proton of 10^{15} ev hitting vertically the upper atmosphere, and suffering a nuclear collision in a mean free path of air (~ 70 g cm^{-2}

---

[1] KERST et al.: Phys. Rev. **102**, 590 (1956).

or about $\frac{1}{13}$ of the atmosphere). This means that the primary will give rise to its first bunch of secondaries at a height of about 16 km, on the average, though the fluctuations around the average can be very large (the distribution is nearly flat from 30 to 10 km above sea level). The determinations of the nature, number, energy and angular distribution of these secondaries constitute problems yet to be solved.

Extrapolating the information obtained by studying interactions at smaller energies (10^{11} to 10^{13} ev) observed in photographic plates exposed at high altitudes, we must expect that the most abundant particles produced in the first collision are π-mesons, both charged and neutral, but that also heavy mesons, as well as nucleons (protons and neutrons) and antinucleons are present. The average total number of secondaries seems to depend both on the energy, U, of the primary and on the impact parameter, i.e., the minimum distance of approach between the primary nucleon and the nucleons in the air molecule. Let us assume here that the total number of secondaries is 30. Their angular distribution in the laboratory system is, of course, extremely peaked forward, because of the large velocity of the center of mass. In the case of one proton hitting another nucleon, the characteristic angle is $\Theta \approx 5 \times 10^{-4}$ radians. However, there are indications that the angles of emission are even smaller, on the average, than those predicted by pure kinematics; which means that in the center-of-mass system there is a preferential emission of the secondaries in the forward and backward directions.

As far as the energy distribution of the secondaries is concerned, the experimental evidence seems to favor now a model where, on the average, the total energy of the secondaries is smaller by about a factor of 3 than the total energy of the primary, i.e., a model where the collision is not strongly inelastic, so that the primary nucleon maintains, so to speak, its individuality and proceeds with still a good fraction of the energy that it had originally. In our example, after the first interaction we are left with a nucleon of $\sim 7 \times 10^{14}$ ev and with 15 secondaries, each of $\sim 2 \cdot 10^{13}$ ev.

The subsequent behavior in the air of the particles produced in the first interaction depends essentially on the value of their lifetimes; otherwise, all of them are capable of giving rise to other nuclear interactions. It seems in fact that for all the particles produced in nuclear interactions the interaction mean free path is about geometrical, i.e., $\lambda \approx 70$ g cm^{-2} in air. If then τ is the lifetime of the particle at rest and $\gamma = U/mc^2$ its Lorentz factor ($\gamma = 10^4$ to 10^5 for the secondaries in our example), the probability of decay will be small and that of interaction practically unity whenever $\dfrac{\gamma \tau c}{\lambda / \bar{\varrho}} \gg 1$ ($\bar{\varrho} = $ average density of air). In our example $\bar{\varrho} \approx \frac{1}{10} \varrho_0 \approx 10^{-4}$ g cm^{-2}, and if $\tau > 10^{-9}$ sec the particle will in general interact before decaying. Nucleons, antinucleons, π^+-mesons ($\tau = 2 \times 10^{-8}$ sec), K^+-mesons ($\tau = 10^{-8}$ sec), and perhaps some other particles, will then produce new interactions in the air and propagate a cascade of nuclear interactions (the N-cascade). The central part of the N-cascade, the core of the shower, contains the most energetic of these particles, those most responsible for the propagation of the shower, and its axis corresponds to the direction of propagation of the primary proton. The π^0-mesons ($\tau \approx 10^{-15}$ sec), some K^0 mesons and hyperons ($\tau \approx 10^{-10}$ sec) will instead preferentially decay in flight and thus transfer into their decay products the total of their energy.

Outstanding in two respects is the importance of the π^0-mesons: first, they constitute a sizable fraction (20 to 30%) of the particles created both in the first and in all subsequent N-interactions; second, the two gamma rays to which they give rise in decaying multiply in air in a completely different way than the strongly interacting particles.

The γ-rays generate what is known as the electro-magnetic cascade, a successive alternate production of electron pairs and of gamma rays of bremsstrahlung. The *EM* cascade is much more prolific than the *N*-cascade and at the surface of the earth the electrons and the photons in the shower are numerically more abundant, by a factor of ~ 20, than all the other particles and are responsible for all the main features of the completely developed AS.

Another important class of particles present in AS consists of the μ-mesons. All the $\pi^{\pm}$-mesons and a good fraction of the $K^{\pm}$-mesons that decay in flight before interacting give rise to μ-mesons having practically the same Lorentz factor; the interaction cross section of the μ-meson is so small and their lifetime (2×10^{-6} sec) so long that they can practically cross the rest of the atmosphere undisturbed. Eventually, their number at the surface of the earth amounts to $\sim 10\%$ of the electrons present in AS. It is evident that the μ-mesons are intimately correlated to the development of the *N*-cascade.

The product of all these processes is the extensive air shower. In it all the various kinds of particles are present in the lower atmosphere, and experimentally it is not always easy to distinguish one from the others, when they all arrive simultaneously on the detectors.

One can expect, however, that the μ-mesons attain the largest distances from the core of the shower; some have been detected 1 km far from its. The electrons and photons are relatively closer to the core and the strongly interacting particles are to be found mostly within some 20 meters from it.

Of the two cascades, electromagnetic and nucleonic, the first is by far the best known and a theory has been developed that interprets well most of its features. At present most of the effort, both theoretical and experimental, is devoted to the interpretation of the *N*-cascade. Our knowledge of it is still very qualitative and the picture given above for the nucleon-nucleon collisions contain already most of the details we can presently give.

The preponderance of the *EM* component in the showers and the solid theoretical basis for its interpretation make it advisable to start a description of AS with a discussion of this component. This will be done in the next chapter.

Chap. C is dedicated to the *N*-cascade, and Chap. D to other problems that at this moment seem the most challenging.

B. The electro-magnetic cascade.

3. All through this chapter our attention will be focussed on the numerically predominant electronic and photonic components of the AS, and their behavior will be interpreted and described in terms of the *EM* cascade.

The nuclear process that actually originates the shower is here left aside and the AS is pictured as an event at whose origin there is a *single* very energetic photon.

In the first place we shall summarize the most significant results of the theory of the *EM* cascade. For a detailed treatment of the problem as well as for an exhaustive discussion of the approximations and limitations of the theory, we refer the reader to the article by G. MOLIÈRE in this volume.

Here we just want to emphysize that the elementary phenomena on which the theory is based have been experimentally verified up to very high energies with sufficient accuracy, hence that the theoretical formulae are substantially sound and trustworthy.

However, not necessarily accurate and reliable are the consequences that one derives from application of the pure *EM* theory to the interpretation of the complex phenomena occuring in the AS; a pure *EM* air shower does not exist and the real showers hardly resemble it in first approximation.

In the second part we shall present those experimental results that shed light mostly on the features of the *EM* component of AS. We shall compare them with the theory and underline their importance in bringing into focus the inadequacies of the pure *EM* treatment for the description of the behavior of the AS.

I. Results of the electro-magnetic cascade theory.

a) The longitudinal development.

4. In describing the development of an *EM* shower in the atmosphere it is convenient to treat separately the longitudinal and the lateral development. This simplification does not introduce appreciable errors, because the thickness of the atmosphere in which the showers develop extends over more than 10 km, while the lateral displacements of most of the particles are smaller than some 100 meters. The deviations from a straight path are thus very small.

The characteristic unit used to express the thickness of matter in the *EM* processes is the *radiation unit*, x_0, (r.u.). In air

$$x_0 = 37.7 \, \text{g cm}^{-2} = 2.92 \, \frac{1}{P} \, \frac{T}{273} \, 10^2 \, \text{m}$$

(P = pressure in atmospheres, T = absolute temperature of the air).

The average energy lost through collisions by an electron in the thickness x_0 is called the *critical energy*, ε_0; in air:

$$\varepsilon_0 = 84.2 \, \text{Mev}.$$

The fact that the density of the atmosphere changes with height can be forgotten here because electrons (positive and negative) and photons, i.e., all the particles involved in the *EM* cascade are stable, and their multiplication depends only on the amount of matter traversed. This is not the case when one comes to treat the lateral displacement, where the geometrical distances traversed by the particles are determinant, and when one studies the nucleon cascade, where unstable particles are present, whose probability of decay depends on the time spent in the medium. This is one of the advantages of separating the treatments of the longitudinal and lateral development.

The results reported here are those obtained under the so-called approximation *B*. This means that the calculations take into account only the three processes of pair production by the photons, bremsstrahlung by the electrons, and collision energy losses suffered by the electrons, and utilize for the pair production and bremsstrahlung cross sections the expressions valid at extremely high energies, the so-called asymptotic formulae, rather than the exact expressions[1].

The total number of electrons of energy $E \geqq 0$ calculated as a function of the thickness of air traversed, for showers initiated by a single photon of energy W_0 is plotted in Fig. 1, for various values of W_0. The calculations used are those of SNYDER[2], but the results of other authors do not differ appreciably. These curves

---

[1] These approximations are really not bad. Of the processes not taken into account (nuclear interaction, Compton effect, direct pair production by the electrons) only the last could be of importance, if its cross section should turn out to be much larger than calculated. However, thus far this does not appear to be the case: see, e.g., J. E. NAUGLE and P. S. FREIER: Phys. Rev. **104**, 804 (1956).

[2] H. S. SNYDER: Phys. Rev. **76**, 1563 (1949).

can be be represented by the equation[1]

$$N(W_0, t) = \frac{0.31}{\beta_0^{\frac{1}{2}}} \exp\left[t\left(1 - \frac{3}{2}\log s\right)\right] \tag{4.1}$$

where $t = \dfrac{x}{x_0}$ is the thickness of air in radiation units, $\beta_0 = \log\dfrac{W_0}{\varepsilon_0}$, and $s = \dfrac{3t}{t + 2\beta_0}$.

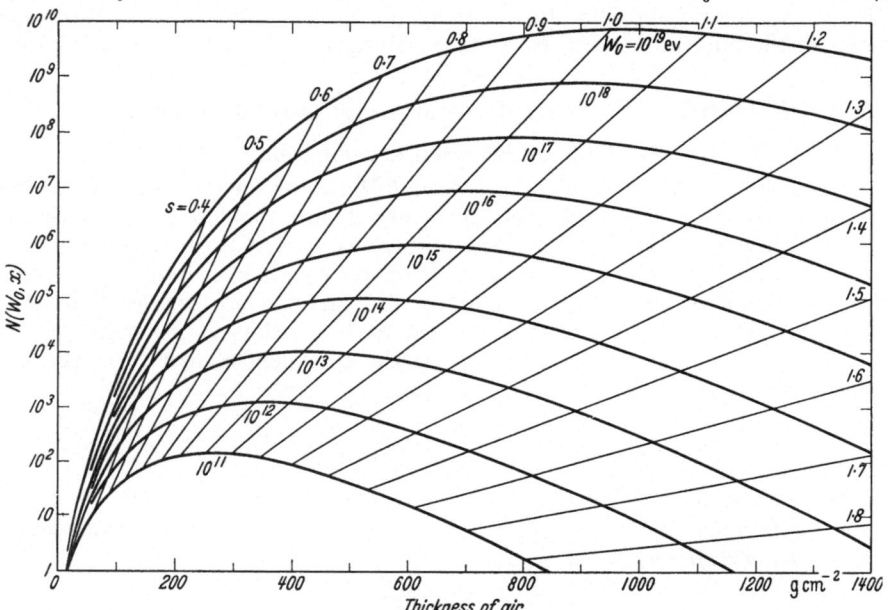

Fig. 1. The total number of electrons, as a function of the thickness (g cm⁻²) of air crossed, produced by photons of various energies, W_0, in ev. The parameter s is the age of the shower at different stages of its development.

For $E > \varepsilon_0$, the integral energy spectrum of all the electrons in the shower is given by the expression

$$N(W_0, E, t) = \frac{0.135}{\beta^{\frac{1}{2}}} \exp\left[t\left(1 - \frac{3}{2}\log s\right)\right] \tag{4.2}$$

where

$$\beta = \log\frac{W_0}{E} \quad \text{and} \quad s = \frac{3t}{t + 2\beta}.$$

Exact handling of the electrons and photons below the critical energy allows the determination of the integral spectrum for energies $E < \varepsilon_0$ (Richards and Nordheim)[2].

In Fig. 2 we have plotted the spectrum in the neighborhood of the maximum development of the shower.

The parameter s is known as the "age" parameter and is simply related to most of the characteristic properties of the shower: its behavior is plotted in Fig. 1. For instance, from Eq. (4.1) it follows that the maximum development of the shower is reached when $s = 1$, i.e., at the thickness

$$t_{\max} = \beta_0 = \log\frac{W_0}{\varepsilon_0}. \tag{4.3}$$

[1] K. Greisen: Progress in Cosmic Ray Physics, Vol. III, Amsterdam: North Holland Publishing Co. 1956.
[2] J. A. Richards and L. W. Nordheim: Phys. Rev. **74**, 1106 (1948).

At the beginning of the shower, $s \ll 1$; before the maximum $s < 1$; after the maximum $s > 1$. The relation between the number of particles in the shower at any level and the energy of the primary particle is, again from Eq. (4.1),

$$\frac{d \log N}{d \log W_0} = s - \frac{1}{2\beta_0} \approx s , \qquad (4.4)$$

the approximation being good for high energy primary particles. Analogously

$$\frac{d \log N}{dt} = \frac{1}{2} (s - 1 - 3 \log s) . \qquad (4.5)$$

Expressions (4.2) to (4.5) are very useful for the analysis of experimental data. However, one must keep in mind that these formulae refer to average values. The fluctuations around these averages are mostly due to the fluctuations of the multiplication in the first stages of the shower, where the number of photons and electrons is small; one, to start with. To a standard fluctuation of ~ 1 radiation unit for the materialization of the initial photon one must add approximately another r.u. for the rest of the cascade, so that the development of the shower under a certain thickness can eventually have a standard fluctuation corresponding to a variation of total thickness traversed of ~ 2 r.u. From (4.5), the root-mean-square deviation of $\log N$ is thus given by the expression

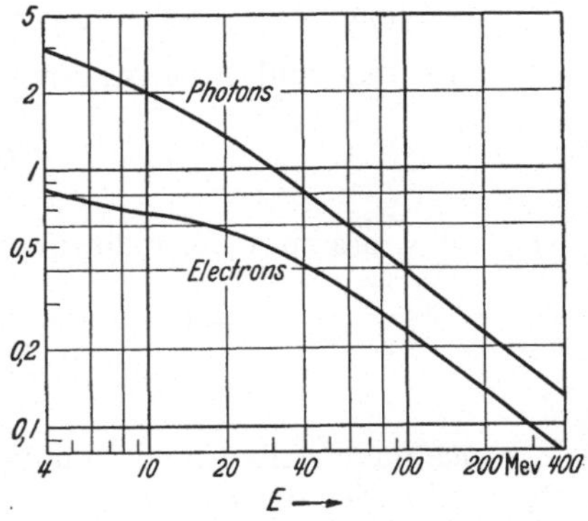

Fig. 2. Integral energy spectrum in air of all the electrons and photons in a shower near the maximum, according to RICHARDS and NORDHEIM. The ordinates are normalized to one electron of energy $\geqq 0$.

$$\langle \delta \log N \rangle \approx (s - 1 - 3 \log s) .$$

Of course, we are talking here only about the fluctuations correlated with the development of the *EM* cascade, under the assumption that a single photon is at the origin of the shower.

In real life, such a situation will probably never occur: the initiator of the *EM* cascade will, e.g., be one of the two photons produced by the decay of a π^0-meson, which in its turn will be one of the secondary particles generated in the first interaction of the primary proton with an air nucleus.

The fluctuations due to the *EM* cascade are apt to be somewhat smaller than those correlated with the nucleonic processes.

b) The lateral distribution.

5. Coulomb scattering. By far the main contribution to the displacement of the electrons and the photons from the axis of the shower, i.e., from the direction of the primary photon, comes from the Coulomb scattering suffered by the electrons in passing near the air nuclei; only this effect has been taken into account in the calculations of the lateral distributions of the *EM* cascade.

The characteristic expression that gives the angle of multiple Coulomb scattering, $\delta \Theta$, suffered by a single charged particle of energy E ($E \gg m c^2$) crossing a thickness $\delta t = \delta x / x_0$ is:

$$\langle \delta \Theta^2 \rangle = \left(\frac{E_s}{E} \right)^2 \delta t ,$$

where $E_s = 21.2$ Mev. The natural unit of length of lateral displacement in air is thus

$$r_1 = \frac{E_s}{\varepsilon_0} x_0 = 9.50 \text{ g cm}^{-2} = \frac{73.5}{P} \frac{T}{273} \text{ m} \tag{5.1}$$

(P in atmosphere and T in °K).

Referring to MOLIÈRE's article for an analysis of the methods of solution, here we give only the results of the calculations performed by NISHIMURA and KAMATA[1], that probably are the most extended and accurate thus far. These results can be considered as reliable as those obtained for the longitutdinal development under approximation B. In Fig. 3 the distribution function $f(r/r_1)$ for all electrons of energy $E \geq 0$ and for an *homogeneous* atmosphere is plotted for several values of the age parameter, s [2].

The following expression, given by GREISEN[3], represents quite well the results of NISHIMURA and KAMATA for $0.6 \leq s \leq 1.8$ and $0.01 \leq \frac{r}{r_1} \leq 10$;

$$f\left(\frac{r}{r_1}\right) = C(s) \left(\frac{r}{r_1}\right)^{s-2} \left(\frac{r}{r_1} + 1\right)^{s-4.5}, \tag{5.2}$$

where $C(s)$ assumes the following values:

$$s = 0.6 \quad 0.8 \quad 1.0 \quad 1.2 \quad 1.4 \quad 1.6 \quad 1.8,$$

$$C(s) = 0.22 \quad 0.31 \quad 0.40 \quad 0.44 \quad 0.43 \quad 0.36 \quad 0.25.$$

Since the distribution does not depend strongly on s and since the s of showers observed under the same thickness of matter changes slowly as a function of the primary energy W_0, it is justified to assume, in first approximation, that all the showers within a not too great energy range, observed, e.g., at sea level, have the same radial distribution of electrons around the core. This conclusion is convenient for the interpretation of many experimental results.

The curves of Fig. 3 have been evaluated for an homogeneous atmosphere and could be considered valid for the real atmosphere if the equilibrium between the shower and the medium were very good. In this case the variation of the atmospheric pressure, when considering experiments done at different altitudes above sea level, would change only the geometrical length of r_1 and showers of the same age would cover areas proportional to P^{-2}. Actually the equilibrium with the medium is not perfect and the lateral distribution at a certain level is influenced by that at higher levels, where the pressure is smaller. An evaluation of this effect has led GREISEN[3] to formulate the rule that in the lower atmosphere the lateral spread is equal to that observed in a uniform atmosphere having the pressure P_2 of the level two radiation units higher i.e.

$$P_2 = (P - 0.07) \text{ atm.}$$

---

[1] J. NISHIMURA and K. KAMATA: Prog;. Theo;. Phys. η, 185 (1952). — KAMATA and NISHIMURA [Suppl. Progr. Theor. Phys. **6**, 93 (1958)] give a three-parameter expression that fits their curves somewhat better than that given below (5.2).

[2] Actually the age parameter s as defined by NISHIMURA and KAMATA differs slightly from that of Eq. (4.1). Here $s = \dfrac{3t}{t + 2\beta_0 + 2 \log (r/r_1)}$, and is a slowly varying function of r/r_1 that coincides with our s only for $r/r_1 = 1$. In the lower atmosphere ($t \approx 25$, $\beta_0 \approx 15$) s changes by about 15% when r/r_1 changes by a factor of 10.

[3] K. GREISEN: Progress in Cosmic Ray Physics, Vol. III, Amsterdam: North Holland Publishing 1956.

In the analysis of data taken in the atmosphere at a pressure P and temperature T, the unit of lateral displacement must then be not the r_1 defined by (5.1), but

$$r_2 = \frac{73.5}{P - 0.07} \left(\frac{T}{273} \right) \text{m} . \tag{5.3}$$

The dependence of r_2 on the height in a "normal" atmosphere is plotted in Fig. 4. The normalization of the curves of Fig. 3 is such that

$$\int\limits_0^\infty 2\pi \frac{r}{r_1} f\left(\frac{r}{r_1}\right) d\left(\frac{r}{r_1}\right) = 1 .$$

Thus $f(r/r_1)$ represents the probability that one electron belonging to the shower falls at the distance r from the core of the shower within the surface, r_1^2. If N

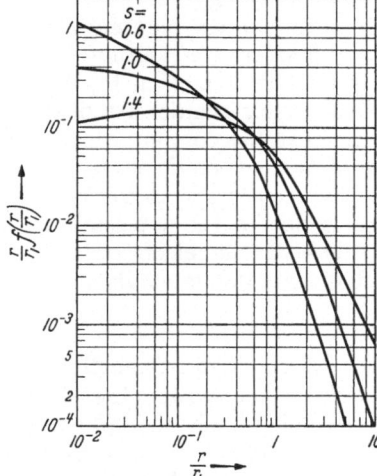

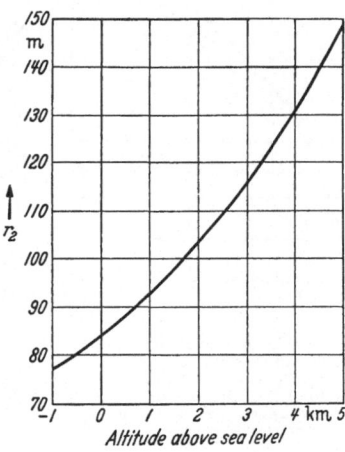

Fig. 3. Distribution function for showers of various ages according to Nishimura and Kamata. The curves have the following normalization: $2\pi \int\limits_0^\infty f(x)\, x\, dx = 1$.

Fig. 4. The unit of lateral displacement, r_2, as a function of the altitude above sea level, for the International Normal Atmosphere (INT).

is the total number of particles in the shower, the average number of particles falling in air on a unit surface in a plane normal to the axis of the shower, at the distance r from the axis is:

$$\Lambda(r) = N \frac{f(r/r_1)}{r_1^2} . \tag{5.4}$$

$\Lambda(r)$ is called the average (surface) density of the shower and is the quantity most easily measured in the experiments.

6. Energy dependence. The expressions given thus far refer to all the electrons in the shower with energy larger than zero, and can be used without modifications in the interpretation of experiments done, e.g., with unshielded GM counters. However, sometimes it is necessary to know what is the lateral distribution of the particles in the shower as a function of their energy E. In this case the most useful parameter is perhaps the root-mean-square radial displacement of all particles with energy E as calculated by Borsellino[1]: it is given in Fig. 5 as a function of the age s of the shower [see Eq. (4.2)].

[1] A. Borsellino: Nuovo Cim. **7**, 323, 601 (1950).

It may appear strange at first sight that the photons, which do not suffer Coulomb scattering, are spread farther from the core than the electrons. The angular spread of the photons is in fact smaller than that of the electrons, but this is more than offset by the fact that photons do not lose energy by ionization and can travel farther than the electrons. At very large distances from the core the presence of electrons can only be justified by admitting that low energy photons generate them locally. This effect, and the neglecting in the calculations of the single large angle scatterings limit the validity of the curves of Nishimura and Kamata to $r/r_1 < 10$.

Finally, it must be pointed out that under a fixed thickness of matter the age s of the shower is a function of E, the energy considered. In fact, the maximum development for electrons of energy E is reached in the shower under a smaller thickness than for electrons of energy smaller than E. This is the reason why the age parameter used by Nishimura and Kamata (see footnote on p. 222) is different from that defined in Eq. (4.1). For the same reason, while the density distribution of all electrons near the core is proportional to $(r/r_1)^{s-2}$, that of monoenergetic electrons is much flatter. The divergent peak at the core for all the electrons is due to the superimposition of the distributions of the electrons of various energies, each distribution much flatter but with

$$\langle r^2 \rangle \propto \sim \frac{1}{E}.$$

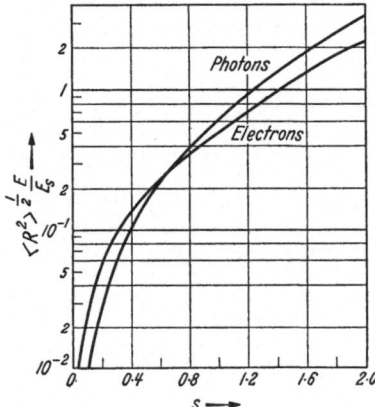

Fig. 5. Energy dependence of the root mean square radial displacement of electrons and photons, as a function of the shower age, s, according to Borsellino. $R = r/r_2$; $E(> 2\,\varepsilon_0)$ = energy considered; $E_s = 21.2$ Mev = scattering energy.

7. Influence of the earth's magnetic field. The charged particles in the showers that develop in the atmosphere, being under the influence of the earth's magnetic field, suffer an additional displacement in the East und West directions that introduces an azimuthal asymmetry with respect to the core. For vertical showers the electrons ($+$ and $-$) are displaced more in the East and the West directions than in the South-North directions by the factor[1]

$$M_0 \approx \left[1 + 0.05 \left(\frac{\cos \lambda}{P}\right)^2\right]^{\frac{1}{2}} \tag{7.1}$$

where λ is the geomagnetic latitude and P the pressure in atmospheres of the place of observation, practically independent of where the EM shower started in the atmosphere. If follows that if the density of the EM component is measured at equal distances from the core of the showers and at various azimuths, the density in the East and West directions is larger than that in the North and South directions by the factor [see Eqs. (7.1) and (5.2)]:

$$\left. \begin{aligned} M_1 &= 1 + \frac{r_2(M_0 - 1)}{\Delta(r)} \frac{d(\Delta(r))}{dr_2} \\ &= 1 + 0.025 \left(\frac{\cos \lambda}{P}\right)^2 \left[(4.5 - s)\left(1 - \frac{1}{(r/r_2) + 1}\right) - s\right]. \end{aligned} \right\} \tag{7.2}$$

However a more detailed calculation be Chaloupka[2] gives for $(M_1 - 1)$ values 2 to 3 times larger than those predicted by Eq. (7.2).

---

[1] G. Cocconi: Phys. Rev. **93**, 646; **95**, 1705 (1954).

[2] P. Chaloupka: Proceedings of the International Conference on Cosmic Radiation, p. 110. Budapest 1957.

Measurements of this effect done by Nikolsky and Satzevich[1] on the Pamirs at 100 m from the cores gave

$$M_1 - 1 = 0.12 \pm 0.03$$

in reasonable agreement with the predictions of Chaloupka.

The magnetic deflections are always of the same order as those due to the Coulomb scattering, for particles moving in the atmosphere. Hence it is expected that they influence the lateral distribution not only of the EM component but also of the other components, especially the low energy μ-mesons, since these particles can travel long distances in air without losing much energy.

As this subject will not be discussed later, we give here the expression for the magnetic displacement of vertical μ-mesons produced at the height $h\,(m)$ and arriving at sea level with kinetic energy E_0 (ev).

$$D_M = \frac{h^2}{2\,(\varrho_0 + \varrho_H)} \left\{ 1 + 2\left(\frac{H}{h}\right)\log\left(1 + \varepsilon\,\mathrm{e}^{-\frac{h}{H}}\right) + \left. + 2\left(\frac{H}{h}\right)^2 \left[\varepsilon\left(1 - \mathrm{e}^{-\frac{h}{H}}\right) + \frac{\varepsilon^2}{4}\left(1 - \mathrm{e}^{-\frac{2h}{H}}\right) + \frac{\varepsilon^3}{9}\left(1 + \mathrm{e}^{-\frac{3h}{H}}\right) + \cdots\right]\right\} \tag{7.3}$$

where

$$\varrho_0 = \frac{1.08 \times 10^{-4}}{\cos\lambda}\,E_0\,\mathrm{m}, \qquad \varrho_H = \frac{2.16 \times 10^5}{\cos\lambda}\,\mathrm{m}, \qquad \varepsilon = \frac{\varrho_H}{\varrho_0 + \varrho_H},$$

and $H = 8 \times 10^3$ m is the height of the standard atmosphere[2].

As an example, a μ-meson produced at 8 km and arriving at sea level with 1 Gev kinetic energy is displaced by the magnetic field ($\lambda = 45°$) by $D_M \approx 70$ m, while the Coulomb scattering produces a lateral displacement $\langle D_s^2\rangle^{\frac{1}{2}} \approx 110$ m. However in the case of μ-mesons in $A\,S$ the main contribution to the lateral displacement comes not from the Coulomb scattering, but from the angle of emission, and the magnetic effect is expected to be relatively small.

c) The time distribution.

8. The presence in the EM component of particles of different energy and nature, scattered at various distances from the axis, gives rise to relative delays in the time of arrival of the particles on a plane perpendicular to the direction of propagation of the shower. This must be true also for the particles of the N-component and for the μ-mesons in the shower.

However, the delays expected for the particles that trigger the usual detectors are always very small since all these particles travel in the shower with velocities close to they velocity of light, c, and the total distance they cross from the instant of creation is only a few kilometers.

The velocity differences between the electrons and the photons do not contribute, *per se*, to any appreciable delay. In fact the delay between an electron of critical energy $\left(\gamma_0 = \frac{\varepsilon_0}{m\,c^2} \approx 150\right)$ and a photon traveling parallel to it over a

---

[1] S. I. Nikolsky and I. E. Satzevich: J. Exp. Theor. Phys. **31**, 714 (1956).

[2] The projected lateral displacement due to Coulomb scattering in the atmosphere for μ-mesons produced h meters above is:

$$\langle D_s^2\rangle^{\frac{1}{2}} = \frac{E_s}{E}\left(\frac{H^3}{x_0}\right)^{\frac{1}{2}}\left[1 - \mathrm{e}^{-\frac{h}{H}}\left(1 + \frac{h}{H} + \frac{1}{2}\left(\frac{h}{H}\right)^2\right)\right]^{\frac{1}{2}}, \tag{7.4}$$

where E is the average energy of the meson along its path. The unprojected lateral displacement is $\sqrt{2}$ times larger.

distance corresponding to one radiation length, $h_0 = 300$ m at sea level, is

$$\Delta t = \frac{h_0}{c} (1 - \beta) = \frac{h_0}{c} \frac{1}{2\gamma_0^2} \approx 10^{-11} \text{ sec}.$$

The delays accumulated at higher altitudes are even smaller, because the parents are either photons or electrons of much higher energies.

Only differences of path length caused by scattering can introduce measurable delays for the electrons. The major contribution still comes from the last stage of the cascade; in one radiation length an electron at critical energy is scattered through an angle $\langle \Theta^2 \rangle^{\frac{1}{2}} \approx 0.2$ radians, and the corresponding delay with respect to a particle that travels straight is

$$\Delta t \approx \frac{h_0}{c} \frac{\langle \Theta^2 \rangle}{2} \approx 2 \times 10^{-8} \text{ sec} \equiv 6 \text{ m}/c.$$

Measurements of delays in the arrival of electrons in AS[1] are in agreement with this evaluation.

In these experiments the delays between the pulses recorded by scintillators hit by AS were found to be of the order of a few 10^{-9} sec; this figure, representing not the r.m.s. but an average delay between photons and electrons, is expected to be somewhat smaller than the figure calculated before.

The sharp definition of the front of the showers has been cleverly utilized by the MIT group for the determination of the direction of propagation of the shower (see Sect. 13).

The delays expected for particles heavier than the electrons (μ-mesons, N-particles) are not due to differences in path length, since their scattering is too small, but mostly to differences in their velocities.

For μ-mesons, assuming constant energy losses, the delay $\Delta \tau$ with respect to the front of the shower is $(\gamma_2 > 2)$

$$\Delta t = \frac{h}{2c(\gamma_2 + 20)^2} \left\{ 1 + \frac{H}{h} \left(\log \frac{\gamma_2}{\gamma_1} + \frac{(\gamma_2 + 20)(\gamma_1 - \gamma_2)}{\gamma_1 \gamma_2} \right) \right\} \text{ sec},$$

where $U_1 = \gamma_1 \mu c^2$ and $U_2 = \gamma_2 \mu c^2 (\mu c^2 = 106$ Mev) are the total energies of the μ-meson at the origin and at sea level; h is the height of the producing level, and $H = 8 \times 10^3$ m is the height of the homogeneous atmosphere.

A meson generated at 10 km and reaching sea level with 300 Mev kinetic energy would arrive $\sim 2 \times 10^{-7}$ sec after the front.

The protons cannot come from very far, their interaction mean free path being equivalent to ~ 0.5 km of air, at sea level.

For $h < 1$ km the collision losses can be neglected and

$$\Delta t \approx \frac{h}{2c\gamma^2}$$

provided $\gamma = \frac{U}{Mc^2} > 1.5$. Delays of 0.5×10^{-6} sec can thus be expected for 500 Mev protons generated some 500 m away.

The neutrons, not suffering collision losses, can have small energies and still cross large distances. Delays of several microseconds can thus be expected for them.

However, μ-mesons and protons represent a small fraction of the particles present in the shower, and even smaller is the number of them having low energy.

---

[1] P. BASSI, G. CLARK and B. ROSSI: Phys. Rev. **92**, 441 (1953). — R. SUGARMAN and S. DEBENEDETTI: Phys. Rev. **102**, 857 (1956).

It is thus difficult to detect their delays. In fact the measurements conducted by the two groups quoted failed to show delays larger than a few 10^{-9} sec. Delays of some microseconds, due to neutrons, have instead been detected by the Cornell group while measuring Air Showers with an extended grid of scintillators[1].

II. Experimental results.

a) The density spectrum.

9. The simplest method of detecting AS consists in recording the coincidences of three or four *GM* counters or groups of counters placed over a horizontal surface, some meters apart. Every time an AS strikes in the neighborhood, there is a certain probability that enough ionizing particles hit simultaneously all the counters and generate a coincidence. The overall surface covered by the system being small in comparison to the cross sectional area of the shower, in general the density Δ of particles in the shower can be considered as a constant over the system, and the probability of a coincidence can be related to Δ.

Let S be the effective surface of each counter or group of counters; the particles of the shower hitting the counter being independent, the probability that at least one particle hits S is

$$p = (1 - e^{-\Delta S})$$

and the probability that the n units in coincidence are all hit is

$$P = (1 - e^{-\Delta S})^n.$$

When $n \geq 3$, $P(\Delta)$ approaches unity rapidly as soon as $\Delta \geq 1/S$, and a measurement of the rate of coincidences $R(S)$ as a function of S is directly related to the frequency of the showers hitting the system with density $\geq 1/S$.

This frequency is called the integral density spectrum of AS, $F(\Delta)$. It has been shown experimentally that this density spectrum has approximately the form:

$$F(\Delta) = K \Delta^{-\gamma} \quad (\text{sec}^{-1}). \tag{9.1}$$

The rate of coincidences measured is therefore:

$$R(S, n) = \int_0^\infty P(\Delta) \frac{dF(\Delta)}{d\Delta} d\Delta = K \gamma S^\gamma I(n, \gamma), \tag{9.2}$$

where

$$I(n, \gamma) = -(-\gamma - 1)! \left[\binom{n}{1} - \binom{n}{2} 2^\gamma + \binom{n}{3} 3^\gamma - \cdots \right], \quad (n > \gamma > 0).$$

Both K and γ can be determined by measuring R either with counters of various surfaces, S, or by varying the order of the coincidences, n. In fact:

$$\frac{R(S_1, n)}{R(S_2, n)} = \left(\frac{S_1}{S_2} \right)^\gamma,$$

i.e., in a plot of $R(S, n)$ versus S,

$$\gamma = \frac{d \log R}{d \log S}. \tag{9.3}$$

On the other hand

$$\frac{R(S, n)}{R(S, n-1)} = \frac{I(n, \gamma)}{I(n^{-1}, \gamma)} = \frac{\binom{n}{1} - \binom{n}{2} 2^\gamma + \binom{n}{3} 3^\gamma - \cdots}{\binom{n-1}{1} - \binom{n-1}{2} 2^\gamma + \binom{n-1}{3} 3^\gamma - \cdots}. \tag{9.4}$$

Once γ is known, K is determined from (9.2).

---

[1] K. GREISEN: Private communication.

Thus far it has been assumed that R depends only on S and n and not on the average distance, D, between the counters[1]. Experimentally, this was found true only in first approximation: actually, over a limited range of D,

$$R \propto D^{-d}, \tag{9.5}$$

with d of the order of 0.1 to 0.2, depending on the geometry of the system. When absolute rates are compared it is necessary to extrapolate to $D = 0$ the results obtained with different geometries[2].

In the comparison of rates obtained with equal geometries but at different altitudes above sea level (pressures P_1 and P_2) a further correction must be introduced because the characteristic dimension, r_2, of the showers changes and [see Eq. (5.2)]:

$$D(P_2) = D(P_1) \frac{P_2 - 0.07}{P_1 - 0.07}.$$

This correction can be avoided by changing D when changing the level.

Measurements of γ with both methods have been made by several authors, at sea level as well as at mountain and airplane altitudes; also cloud chambers have occasionally been employed. Some of the results are summarized in Fig. 6. It is well established that γ is not a constant but increases slowly when Δ increases. The two straight lines in Fig. 6 correspond to the equations:

$$\left.\begin{array}{l} \gamma = a + b \log \Delta, \\ 1 < \Delta < 10^4 \, \text{m}^{-2}, \end{array}\right\} \tag{9.6}$$

with $a = 1.2$ and $b = 0.048$ at mountain altitudes, and with $a = 1.32$ and $b = 0.038$ at sea level[3].

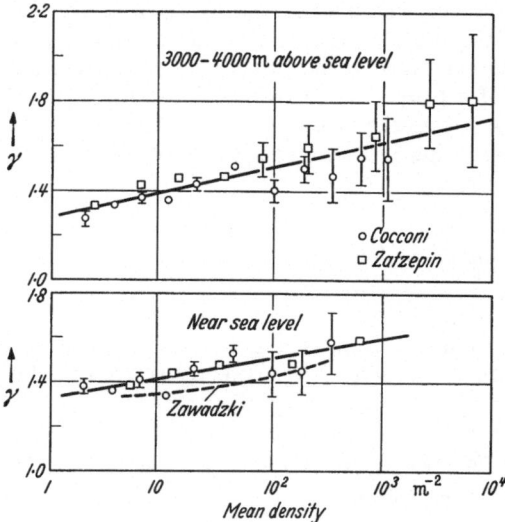

Fig. 6. Dependence of the exponent γ in the integral density spectrum on the mean shower density detected, at sea level and at mountain altitudes. The data are from: G. Cocconi and V. Tongiorgi: Phys. Rev. **72**, 964 (1941); G. T. Zatzepin, V. V. Miller, I. L. Rosenthal, and K. L. Eidus: J. Exp. Theor. Phys. **17**, 1125 (1947); G. T. Zatzepin, I. L. Rosenthal, L. I. Saritcheva, G. B. Kristiansen and L. K. Eidus: Isv. Akad. Nauk. SSSR. (Phys.) **17**, 39 (1953). At sea level the dotted line represents the best fit to the results of Zawadzki.

10. The exponent γ of the density spectrum is directly correlated with the exponent γ' of the integral number spectrum, i.e., of the frequency $F(N)$ of the showers containing N or more ionizing particles. Assume that

$$F(N) = K' N^{-\gamma'} \quad \text{m}^{-2} \sec^{-1} \tag{10.1}$$

---

[1] In these experiments the counters are generally arranged at the vertices and at the center of a symmetrical figure, e.g., and equilateral triangle.

[2] It is not possible to measure directly these rates because if the counters are very near one another small local showers created in the roofs or in the air directly above the apparatus give spurious coincidences. These are eliminated when $D > 2 - 3$ m.

[3] Zawadzki, as a result of precise measurements conducted at sea level (A. Zawadzki, Proceedings of the International Conference on Cosmic Radiation, p. 96, Budapest 1957) finds that a three parameter equation fits more closely the data; however, his measurements are still limited to a small range of densities and the less precise measurements existing thus far are more conveniently represented by a two parameter equation.

and that all the showers come from near the zenith, all with the same lateral distribution[1]. Then the radius $r(\Delta)$ of the circle centered at the core within which a shower of N particles has density $\geq \Delta$ is only a function of the ratio N/Δ,

$$r(\Delta) = r\left(\frac{N}{\Delta}\right)$$

and

$$F(\Delta) = \pi \int_0^\infty \frac{dF(N)}{dN}\, r^2\left(\frac{N}{\Delta}\right) dN \left.\begin{array}{c}\\[2em]\\\end{array}\right\}$$

$$= \left[\pi K' \gamma' \int_0^\infty \left(\frac{N}{\Delta}\right)^{-(\gamma'+1)} r^2\left(\frac{N}{\Delta}\right) d\left(\frac{N}{\Delta}\right)\right] \Delta^{-\gamma'} = K\,\Delta^{-\gamma}. \tag{10.2}$$

This shows that $\gamma' = \gamma$: an important conclusion, since γ can be measured easily with good precision, while the direct measurement of the number spectrum is much more elaborate.

The total number of particles present in the showers that contribute most to the densities around Δ can be found from Eq. (10.2) either utilizing the expressions (5.1) for the lateral distribution, or, more realistically, using the experimental information (see Fig. 9 in Sect. 16). It turns out to be[2]:

$$N \approx \Delta\, r_2^2. \tag{10.3}$$

The value of γ found for showers with $\Delta = 10^3\ \mathrm{m}^{-2}$ thus gives the exponent of the number spectrum for $N \approx 6 \times 10^6$ at sea level. If these were pure EM showers initiated near the top of the atmosphere the energy of the primary photons would be $W_0 = 10^{16}$ to 10^{17} ev. Actually, the situation is not so simple, since photons produced all along the N-cascade contribute to the development of the EM cascade, and since the electrons, though preponderant, are not the only ionizing particles in the shower. However, these figures are indicative[3].

As already mentioned, the integral of Eq. (10.2) can be performed numerically utilizing the experimental lateral distribution given in Fig. 9 (Sect. 15). It is thus possible to derive the value of K'. The result thus obtained obviously refers to all showers detected, independently of their directions of arrival; it can be reduced to those arriving in unit vertical solid angle by assuming that the zenith dependence, at sea level, is proportional to $(\cos \Theta)^{8.3}$ [see Eq. (12.1)], and that the detectors have a sensitive surface proportional to $(\cos \Theta)^{\gamma a}$ with $a \approx 0.35$.

---

[1] The last assumption is justified approximately by the cascade theory (see Sect. 5), but even more accurately by the results of the experiments that will be discussed in the next sections.

[2] This corresponds to the maximum of the distribution; showers larger and smaller, up to a factor of about ten, than those defined by (10.3) contribute appreciably to the density Δ.

[3] It is possible at this point to take another step and correlate the number spectrum with the energy spectrum of the hypothetical primary photon. In fact, from (10.1)

$$-\gamma = \frac{d \log F(N)}{d \log N},$$

and remembering (4.4) it follows that

$$\frac{d \log F(N)}{d \log N}\, \frac{d \log N}{d \log W_0} = \frac{d \log F(W_0)}{d \log W_0} = -\gamma s.$$

The product γs has been identified by some authors with the exponent Γ of the power law expressing the primary integral spectrum. This procedure is now unjustified in view of the hybrid nature of the shower: it is remarkable, however, that it gives an acceptable value for Γ when one uses for s the value ($s \approx 1.3$) estimated from the absorption of the showers near sea level. ($\Gamma \approx 1.9$ for $E = 10^{16}$ to 10^{17} ev.)

The final result for sea level showers is:

$$F(N) = 6.70\,N^{-1.32} \quad \text{m}^{-2}\sec^{-1}\text{sterad}^{-1} \quad 10^3 < N < 10^5,$$
$$F(N) = 167\,N^{-1.58} \quad \text{m}^{-2}\sec^{-1}\text{sterad}^{-1} \quad 10^6 < N < 10^8. \qquad (10.4)$$

The probable error is difficult to evaluate, but can reasonably be estimated around 30%. From these integral spectra we have derived the differential spectra plotted in Fig. 10, Sect. 17.

b) Development in the atmosphere.

11. Some important properties that characterize the behavior of AS can be deduced from the results of measurements of the density spectrum performed at various altitudes above sea level. In Fig. 7 we have plotted some points

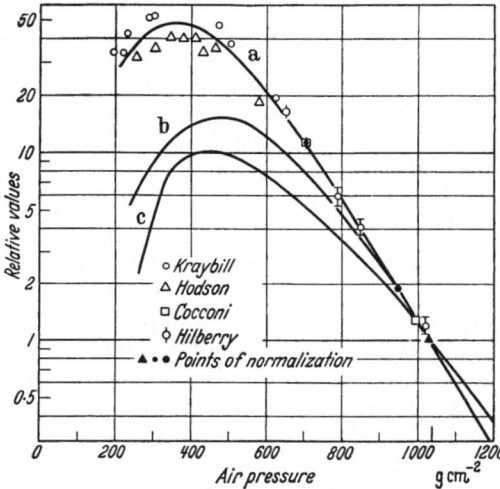

obtained up to airplane altitudes for showers of density $\Delta \geqq 50\ \text{m}^{-2}$: curve a is a plausible interpolation. Analogous curves for the lower atmosphere, $(x > 700\,\text{g cm}^{-2})$, can be drawn for showers of densities ranging from 1 to 10000 m^{-2}, but they all would practically coincide in a normalized plot with the curve of Fig. 7. So, in the lower atmosphere, the shower rates descrease exponentially with increasing pressure with a practically unique absorption coefficient

$$-\mu = \frac{d\log R}{dx}$$
$$\approx \frac{1}{135}\,(\text{g cm}^{-2})^{-1} \qquad (11.1)$$
$$= 0.28\,(\text{r.u.})^{-1}.$$

Fig. 7. Curve a: relative intensity as a function of air pressure of showers of density ~ 50 m^{-2}, $::$ measured with counter. H. L. Kraybill: Phys. Rev. **76**, 1092 (1949); A. L. Hodson: Proc. Phys. Soc. Lond. A **66**, 65 (1953); G. Cocconi and V. Tongiorgi: Phys. Rev. **75**, 1058 (1949); N. Hilberry: Phys. Rev. **60**, 1 (1941). Curve b: the result of a Gross transformation applied to curve a. It gives the vertical intensity of the showers of $\Delta \sim 50$ m^{-2}. Curve c: relative total number of particles $N(x)/N$ (sea level) in the showers that at sea level contribute mostly to $\Delta \sim 50$ m^{-2}, i.e., with N (sea level) $\approx 3.5 \times 10^5$ particles.

This conclusion is supported also by the results of measurements of the barometric coefficient, i.e., of the variation of the shower rates in a fixed locality as a function of the atmospheric pressure; both at sea level and at mountain altitudes, and for all showers $(10 \leqq \Delta \leqq 1000\ \text{m}^{-2})$

$$\frac{dR}{R\,dP} \approx 10\%\ (\text{cm Hg})^{-1},$$

again equivalent to $-\mu = 0.28\,(\text{r.u.})^{-1}$ [1].

With simple arguments, starting from these results, it is possible to reach a conclusion of fundamental importance for the understanding of AS.

If the showers are interpreted in terms of a pure *EM* cascade, it is expected that, with the usual notations (t = thickness of the absorber in radiation units),

$$\left(\frac{d\log R}{dt}\right)_{N=\text{const}} = \left(-\frac{d\log R}{d\log N}\right)\left(\frac{d\log N}{dt}\right) = \gamma\frac{1}{2}\,(s - 1 - 3\log s). \qquad (11.2)$$

---

[1] More details about the barometric effect are given in Sect. 29.

Eq. (11.2) is a consequence of Eq. (4.5) and of the fact that the number spectrum is proportional to the density spectrum (see Sect. 10). By introducing for γ the values of Fig. 6 and equating (11.2) to the experimental value of μ, the following values are found for the age, s, of the showers of various densities (third row of Table 1).

Table 1.

Δ (m^{-2}) . .	2	50	1000
N	10^4	3×10^5	5×10^6
s	1.29	1.24	1.20
x (g cm^{-2}) .	700	870	950

The values of N, appearing in the second row of Table 1, express the approximate number of particles present in the showers and are derived from Eq. (10.3).

If now the three pairs of numbers (N, s) are plotted on Fig. 1, the thickness, x, that they characterize are those of the fourth row of Table 1. The fact that x changes so drastically when Δ changes indicates that *the hypothesis that the A S are all initiated at the same height is wrong, and that a pure EM theory is not adequate to interpret the evolution of AS*. This discrepancy consists in the fact that, while the largest showers appear of about the right age and at the right depth, the smaller ones are much too "young".

The obvious explanation is that the small AS are the remnants of showers not initiated near the top of the atmosphere but much lower.

To the same interpretation points the fact that the absorption coefficient levels off with increasing atmospheric depth at around 0.3 (r.u.)$^{-1}$ while, according to the *EM* cascade theory, it has to increase steadily with increasing thickness up to the value $-\mu_{max} = \gamma \times 0.7 \approx 1.0$ (r.u.)$^{-1}$, which corresponds to the absorption minimum for gamma rays in air. The leveling of μ at a smaller value indicates that there is another source providing energy to the *EM* cascade.

As already stated several times, the N-particles, both primaries and secondaries, are the real source of the E.M. component, and the anomalous behavior here considered must be ascribed to them. However, the quantitative treatment of the problem strongly depends on the way the energy of the primary particle is distributed among its secondaries in the interactions with the air nuclei.

Two extreme points of view can be considered: the first has been recently emphasized by S. MIYAKI[1], the second, more widely accepted in the past few years, was sponsored especially by the Russian workers and by GREISEN[2].

Following MIYAKI, let us assume that the extreme relativistic nucleon-nucleon interactions are rather *inelastic*, i.e., that, when an interaction takes place, most of the energy of the primary nucleon is shared among the secondaries. In this case, once the primary Cosmic Ray has interacted, the shower will fully develop and its "age" will be determined by the thickness of air crossed after the first interaction. The probability that the "first" interaction takes place at the depth x (g cm^{-2}) from the top of the atmosphere is then:

$$P(x)\, dx \propto e^{-\frac{x}{A}}\, dx \qquad (11.3)$$

[1] S. MIYAKI: Private communication; see also H.L. GRIGOROV and V.Y. SHESTOPEROV: J. Exp. Theor. Phys. **34**, 1539 (1958) where a similar point of view is discussed.
[2] See the review article by K. GREISEN in Progress of Cosmic Ray Physics, Vol. III, Amsterdam: North Holland Publishing Co. 1956; and the summary article by N. A. DOBROTIN, G. I. ZATZEPIN, S. I. NIKOLSKY and G. B. KRISTIANSEN: Nuovo Cim. Suppl. **3**, 635 (1956).

where Λ is the effective mean free path for the "first" interaction. The shower thus initiated will not be, of course, purely electromagnetic, since strong interacting secondary particles, in their further interactions, will supply new photons, mostly near the core. However, the N interactions being inelastic, the energy degradation of the N component will be rapid, and the multiplication of the photons produced, via π^0, in the "first" interaction will eventually dictate the property of the shower at larger depths in the atmosphere. The number of particles in each shower will thus change with thickness according to Eq. (4.5). However, the showers are not initiated all at the same height, since the primaries have the first interaction governed by Eq. (11.3). If the primary spectrum has a steep energy dependence, (in fact the integral energy spectrum is proportional to $\sim E^{-1.8}$, see Sect. 28) most of the smallest showers obersved in the lower atmosphere are those produced by primaries that had the chance of crossing a good fraction of the atmosphere without interacting; their number being governed by Eq. (11.3), the observed μ of Eq. (11.1) is essentially equal to $1/\Lambda$, i.e., $\Lambda \approx$ 135 g cm^{-2}.

According to this interpretation, the constancy of μ in the lower atmosphere indicates that most of the showers observed there are produced by primaries that escaped interaction in the highest strata. At higher energies the situation would change, because the contribution of showers produced near the top of the atmosphere begins to be felt, and $1/\mu$ is expected to become larger than Λ. In the extreme case, where the thickness of the atmosphere is not sufficient to develop the shower completely, μ is expected to change sign.

Coming now to the other point of view, that we shall call Zatzepin's-Greisen's, assume that the ultrarelativistic nucleon-nucleon interaction is somewhat *elastic*, i.e., that in the interaction between the primary Cosmic Ray and the air nuclei, only a small fraction of the primary energy is shared among the secondaries, the bulk remaining in the primary. The primary will then continue to penetrate the atmosphere, generating along its path other interactions: the air shower can thus be compared to the column of knock-on secondaries left behind by a charged particle crossing a gas. The properties of each individual shower would then not change rapidly with depth inside the atmosphere, since the secondaries produced in the interactions along the core would maintain the cortège of EM and N particles at a nearly constant level. Of course, fluctuations in the inelasticity of the "primary" collisions could produce fluctuations in the number of particles at various depths, but here we assume fluctuations to be small.

The observed attenuation [Eq. (11.1)] is in this case essentially due to the attenuation of the number of secondaries in each shower, along its path,

$$- \frac{d \log N}{d x} = \frac{1}{\lambda}$$

and to the shape of the number spectrum [see Eq. (11.2)]

$$- \frac{d \log R}{d \log N} = \gamma$$

and $\lambda = \gamma/\mu = 1.45 \times 135 = 195$ g cm$^{-2} \approx 3 \lambda_{\text{geometrical}}$. This numerical value of λ corresponds to an average inelasticity, $(1 - f)$, of $\sim \frac{1}{3}$ per collision, assuming that the primaries are mostly nucleons with geometrical mean free path. The constancy of μ with depth and with shower size reflects the constancy of γ and of λ. At larger energies the situation is not expected to change much until values are reached at which the atmosphere is not thick enough to establish equilibrium conditions. At these energies one expects $d \log N/d x$ to decrease and eventually change sign.

It is clear that from the single observation of μ in the lower atmosphere one cannot decide between the two interpretations. However, in our opinion, the Zatzepin-Greisen point of view must be favored, if one takes into account other evidence derived both from air shower experiments and from the observation of ultrarelativistic interactions in photographic plates. We shall give here a quantitative interpretation of the curve a of Fig. 7 assuming that this point of view is valid.

12. A less qualitative description of the properties of the N-cascade that keeps alive the EM component in the lower atmosphere can be achieved by evaluating the fraction of the total energy spent by the showers of $\Delta \geqq 50$ m^{-2} in the lowest third of the atmosphere. This energy, being practically all supplied in these showers by the N-component, according to the ZATZEPIN-GREISEN point of view, gives an estimate of the degradation of energy in the N-cascade[1]. It is necessary first to derive from the experimental curve (a in Fig. 7) which refers to showers arriving to the detectors from all directions, the altitude dependence of the showers arriving vertically within the unit solid angle. This is achieved by applying a Gross transformation.

Let $I(x, \Theta)$ be the intensity (sec^{-1} ster^{-1}) of AS recorded with density $\geqq \Delta$ under the thickness x (g cm^{-2}) of atmosphere, and arriving with zenith angle Θ. If absorption in the atmosphere is the only process that modifies the showers,

$$I(x, \Theta) = (\cos \Theta)^n I\left(\frac{x}{\cos \Theta}, 0\right),$$

where $I\left(\frac{x}{\cos \Theta}, 0\right)$ is the intensity of the showers if the atmosphere in the vertical direction had the same distribution of pressure and temperature as that traversed in the inclined direction, and if the apparatus had isotropic sensitivity. The factor $(\cos \Theta)^n$ represents the effect of all the geometrical changes that actually occur because these conditions are not met. In the lower atmosphere $n \approx 1.0$ [2]. The Gross transformation then gives

$$I(x, 0) = \frac{R(x)}{2\pi}\left(n + 1 - x\frac{\partial \log R(x)}{\partial x}\right).$$

Curve b of Fig. 7 has been deduced from curve a by means of this transformation[3].

It is then possible to evaluate the ratio between the number of particles present at x in the vertical showers that at sea level contribute mostly to the densities $\geqq \Delta$, $N(x, \Delta)$, and the number of particles present in the same showers at sea level, $N($s.l.$, \Delta)$, by applying Eq. (11.1) to every point of curve b of Fig. 7, i.e., using the transformation:

$$\frac{d \log N}{d x} = \frac{1}{\gamma}\left[\frac{d \log I}{d x} - 0.81\frac{[2(\gamma - 1) - d]}{x - 75 \text{ g cm}^{-2}}\right].$$

The second term in the square brackets represents the usual geometrical corrections for the dependence of the counting rates on the air density (see Sect. 29).

---

[1] The procedure followed here is essentially that given by K. GREISEN in the article already quoted.

[2] $n \approx 0.81 \ [2(\gamma - 1) - d] + (0.35\gamma) - (d/s)$. The first term represents the effect of the variation of atmospheric density (see Sect. 29); the second the effect of the variation of the sensitive area of the counters, and the third the effect of the apparent change of the dimensions of the apparatus in one direction.

[3] It is also possible to deduce in the same manner $I(x, \Theta)$, the zenith angle dependence, and compare it with those measured directly at various angles. The agreement is good though the experimental curves are in general not very precise. At sea level, with good approximation,

$$I(x, \Theta) = I(x, 0) (\cos \Theta)^{8.3}.$$

Curve c of Fig. 7 gives the result of this operation for showers of $\Delta \geq 50 \,\mathrm{m}^{-2}$ at sea level[1]. Curve c has been extended beyond sea level assuming for curve a a constant absorption coefficient of 0.28 (r.u.)$^{-1}$.

The ratio

$$
\varrho(\Delta) = \frac{\displaystyle\int_{700}^{\infty} \frac{N(x, \Delta)}{N(\mathrm{s.l.}, \Delta)} \, dx}{\displaystyle\int_{0}^{\infty} \frac{N(x, \Delta)}{N(\mathrm{s.l.}, \Delta)} \, dx} \approx 0.22,
$$

is then the ratio between the track length of all the ionizing particles present in the lower atmosphere and the track length of all the ionizing particles in the shower in an infinite atmosphere[2].

The ratio $\varrho(\Delta)$ represents the fraction of the primary energy still present in the core of the shower after a thickness of air corresponding to about 10 interaction mean free paths for the N-component. At face value, this result indicates an average inelasticity parameter in each collision

$$
1 - f = \frac{\text{energy in secondaries}}{\text{primary energy}} = \frac{1}{10} \log \frac{1}{0.22} = 0.15.
$$

A more detailed treatment of the experimental evidence according to the Zatzepin-Greisen point of view indicates average values of $1 - f$ around 0.3 [3]. The condition is stringent enough to cast serious doubts that it can be met by models of the N-interactions such as that proposed by FERMI[4], where the energy of the colliding particles is shared quite evenly among the secondaries, or that proposed by LANDAU[5], where the energy is concentrated in a small fraction of secondaries. In both cases, in fact, most of the secondaries are π-mesons, and one-third of them, being π^0-mesons, decay immediately into photons and are lost for the propagation of the N-cascade.

Both these models predict that the fraction of energy left in the showers after 10 mean free paths is $< (\frac{2}{3})^{10} = 0.02$, at least an order of magnitude smaller than observed.

These questions are of fundamental importance for the interpretation of AS, as well as for the understanding of the ultrarelativistic interactions and will be more fully discussed in Sect. 26.

Coming back to the discussion of the behavior of AS in the lower atmosphere, another result of general interest is contained in curve c of Fig. 7. The value

---

[1] The behavior of curve c at small thickness, before the maximum, was guessed from the behavior of the EM cascade.

[2] The contribution to the total track length of the part absorbed beyond sea level is

$$
\frac{\displaystyle\int_{1030}^{\infty} \frac{N(x, \Delta)}{N(\mathrm{s.l.}, \Delta)} \, dx}{\displaystyle\int_{0}^{\infty} \frac{N(x, \Delta)}{N(\mathrm{s.l.}, \Delta)} \, dx} = 0.04.
$$

[3] See, e.g., A. T. ABROSIMOV, V. I. ZATZEPIN, V. I. SOLOVEKA, G. B. KRISTIANSEN and P. S. CHININ: Istv. Akad. Nauk SSSR. (Phys.) **19**, 677 (1955). — O. A. GUZHAVIN, V. V. GUZHAVIN and G. T. ZATZEPIN: J. Exp. Theor. Phys. **31**, 819 (1956). — V. V. GUZHAVIN and G. T. ZATZEPIN: J. Exp. Theor. Phys. **32**, 365 (1957).

[4] E. FERMI: Progr. Theor. Phys. **5**, 570 (1950). — Phys. Rev. **81**, 683 (1951).

[5] S. Z. BELENKII and L. D. LANDAU: Fortschr. Phys. **3**, 536 (1955).

of the integral:

$$\int_0^\infty \frac{N(x, \Delta)}{N(\text{s.l.}, \Delta)}\, dx = 4.3 \times 10^3\,\text{g cm}^{-2}$$

multiplied by the value of the average collision energy losses per unit thickness of air,

$$\frac{dE}{dx} = \frac{\varepsilon_0}{x_0} = 2.23 \times 10^6\,\text{ev (g cm}^{-2})^{-1},$$

gives the total energy spent in the shower for each ionizing particle arriving at sea level,

$$\frac{W_0(\text{s.l.}, \Delta)}{N(\text{s.l.}, \Delta)} = 9.7 \times 10^9\,\text{ev}.$$

Since for $\Delta = 50$ m^{-2} $N(\text{s.l.}, \Delta) = 3.5 \times 10^5$,

$$W_0(\text{s.l.}, \Delta = 50) = 3.4 \times 10^{15}\,\text{ev}.$$

This is the total energy lost by a shower in ionization, hence the major part of the total energy of the primary particle that originated it. If this procedure could be repeated for showers of all sizes, it would produce the best determination of the energy spectrum of the primary radiation. Unfortunately, the required experimental information at high altitudes is still lacking and the energy spectrum at higher energies has to be deduced less directly from the low altitude data with the help of a model for the N-cascade.

c) Localization of the shower core.

13. The study of the density spectrum allows one to reach several important conclusions about the *average* behavior of AS; there are, however, numerous problems for which a more powerful method of observation is needed. One of these problems is the determination of the fluctuations, which requires a measurement for each shower, not of averages; another is the measurement of properties that depend on the distance from the core, or on the size of the shower, or on both. Another still is the measurement of the *same* shower at different altitudes, to observe its development in the atmosphere[1]. All these questions can be answered only if it is possible to locate the core and measure the size of *each* shower. In the last years great effort has been spent toward this goal.

Experimental localization of the shower core can be achieved by measuring the surface density, Δ, of the particles in the shower and identifying the core with the point where Δ is maximum, or by measuring other properties of the particles in the shower, e.g., energy, nature, angle, timing of arrival, which can be related to the distance from the core. Let us begin with the first method, which has the advantage of giving at the same time a measurement of all the ionizing particles present in the shower, N_{tot}.

It consists in covering a large horizontal area with detectors capable of measuring Δ and, when a shower lands in the surface thus patrolled, in identifying the

---

[1] A measurement of this kind could be made, either keeping afloat some counters with anchored balloons, some kilometer above ground, or utilizing some natural cliff and detect those showers arriving with such a direction as to have half of them measured at the top of the cliff and the other half at the bottom. Though these measurements seem now rather difficult to realize, probably they alone could give a good answer to the problem of the development of AS in the atmosphere.

point where Δ is maximum. Batteries of *GM* counters, ion chambers, scintillators, cloud chambers, are all detectors with well defined sensitive surface that can measure the number of particles falling on them, hence Δ [1]. However, if the core must be located with a resolution of d meters, the average distance between neighboring detectors must be of the order of d. The overall number of showers recorded being proportional to the total area covered by the detection grid, this area must be of the order of a square kilometer if the most energetic showers (10^{17} to 10^{19} ev) are studied, and the number of detectors can easily run into the thousands.

The Harwell group [2] has spread ~ 500 GM counters over a triangle of 900 m on a side ($d \approx 90$ m), and the Russian group operating in Moscow and in the Parmirs [3] has used up to 2500 counters in various arrangements. High resolution ($d \approx 0.1$ m) has been reached in some cases by the Russian workers by assembling their counters within a circle of only 12 m radius [4].

However, if one does not want to investigate the details of the structure near the core, and once it is proved that $\Delta (r)$ monotonically decreases when r increases, one can avoid such full-force methods, and obtain the location of the core more easily by making good measurements of Δ in fewer places, and computing where the core must have hit in order to give the distribution of Δ observed, case by case. This method has been first used by WILLIAMS [5] who measured Δ with a few ion chambers, and more recently by the MIT group [6] who use as detectors 11

---

[1] The measurement of Δ with GM counters is based on the expression of the probability that a counter of effective surface S is activated when hit by a shower containing Δ ionizing independent particles per unit surface (see Sect. 9). Usually many counters of various sizes are used, connected to neon bulbs that flash when the counter is triggered by an AS; this indicating system is called "hodoscope". The number of counters hit determines Δ with a precision that depends on their surface and on the total number of counters available. Precautions are taken, in arranging the counters, that the same particle cannot trigger more than one counter (separation of a few inches, light roofs, etc.). The determination of Δ with scintillators is less traight-forward; the measurement gives the amount of energy, I, lost by the particles of the shower in a slab of scintillator of surface S. The average number of *ionizing* particles hitting S is then deduced by assuming that each ionizing particle contributes the same energy, i, that a minimum ionizing particle is supposed to lose in the slab. Thus $\Delta_{sci} = I/Si$. The procedure is incorrect because in the finite thickness of the scintillator the energy lost by the low energy electrons is smaller than i, and the photons can generate electrons. Both kinds of particles are abundant in a developed shower and their relative amount is a function of the distance from the core. The importance of these effects has been demonstrated experimentally by M. H. BREMMAN: Nuovo Cim. **6**, 216 (1957), who showed differences of some 10% between the two methods. However, when the density is measured over the *whole* shower with scintillators, the value of the integral

$$N_{tot} = \int_0^\infty 2\pi r \, \Delta_{sci} \, dr$$

is equal to that deduced from measurements with counters. This is so because the scinlitlator and the air have about the same atomic number, and at any depth inside the scintillators the number of electrons present is in equilibrium with the photons. In conclusion, while the functions $\Delta (r)$ determined with GM counters and with scintillators in principle are not the same, their integrals over a plane are equal.

[2] T. E. CRANSHAW and W. GALBRAITH: Phil. Mag. **2**, 797 (1957).

[3] A summary of the work performed in Russia until 1955 is given by N. A. DOBROTIN, G. T. ZATZEPIN, S. I. NIKOLSKY and G. B. KRISTIANSEN: Nuovo Cim. Suppl. **3**, 635 (1956).

[4] See, e.g., O. I. DOVCHENKO and S. I. NIKOLSKY: Dokl. Akad. Nauk SSSR. **102**, 241 (1955). — S. P. DOBROVOLSKY, S. I. NIKOLSKY, E. I. TUKISH and V. I. IAKOVLIEV: J. Exp. Theor. Phys. **31**, 959 (1956).

[5] R. W. WILLIAMS: Phys. Rev. **74**, 1689 (1948).

[6] G. CLARK, J. EARL, W. KRAUSHAAR, J. LINSLEY, B. ROSSI and F. SCHERB: Nature, Lond. **180**, 353, 406 (1957).

scintillators of ~ 1 m² surface each, spread over an area of 0.2 km². The resolution reached is $d = 10 - 20$ m [1].

In the MIT experiment an important improvement has been achieved by measuring at each detector, besides $\varDelta$, also the time of arrival of the particles. Since the front of the shower coincides well with a plane normal to the axis of the shower [2] the time of arrival of the particles on a horizontal plane is the same for all detectors if the axis is vertical, but different if the axis is inclined. A precise measurement of the timing allows the determination of the propagation direction of the shower [3]. This possibility of measuring for each shower the total number of particles present and the direction of propagation of the core permits an analysis of the showers somewhat more accurate than that possible with the previous methods, where the angle of arrival could not be determined [4]; this method is also very useful in the analysis of the time dependence of the shower intensity (see Sect. 31).

Cloud chambers in association with counters can be utilized for the localization of the core; besides the particle density, they can give the average direction of movement of the secondaries, which is essentially the axis direction [5]. However, the operation of cloud chambers being more delicate than that of scintillators, this method seems in general inferior to that described before.

Measurements made by RELF [6] with a diffusion cloud chamber of 3 m² surface, operated in coincidence with counters identifying the AS, deserve special consideration. The behavior of $\varDelta$ near the core of some showers was measured with the best resolution obtained thus far; however the relative smallness of the sensitive area is revealed by the fact that in 22 days of operation, no shower with more than $\sim 10^5$ particles landed with its core on the chamber and only a dozen within 5 meters.

14. A different approach to the problem of the core localization consists in selecting among the showers arriving at the ground only those whose cores land within a distance d from a particular detector, which in this case takes the name of "core selector".

A characteristic of the core that can be exploited is the presence in it of high energy electrons and N-component. A system that detects local showers under some inches of lead, accompanied by an AS, constitutes a core detector [7] with $d = 10 - 30$ m depending on the AS size. The local showers are mostly produced by high energy electrons and photons, but occasionally also by the N-component;

---

[1] Larger scintillators, of many square meters sensitive area, are going to be realized in the near future. It will then be possible to detect their scintillation light by means of image tubes and thus measure directly the density distribution of the shower particles in the region covered by the scintillator. Likely this method will be very useful for accurate measurements near the core of the showers.

[2] The MIT group has given evidence that the front of the shower has a curvature of some kilometers, as expected from the genesis of the cascade.

[3] With detectors separated by 300 m a timing within ~ 0.1 μsec corresponds to an angular resolution of $\sim 5°$.

[4] Actually when no timing is taken the situation is not as bad as it could appear at first sight. The number of AS decreases strongly as the zenith angle increases, and most of them arrive at sea level within a zenith angle of $\sim 30°$. When the direction of the shower is not measured, the uncertainty in the thickness of atmosphere crossed is thus only of the order of 10%.

[5] P. ROTHWELL, B. WADE and A. GOODINGS: Proc. Phys. Soc. Lond., Ser. A **69**, 902 (1956).

[6] K. E. RELF: Phys. Rev. **97**, 172 (1955).

[7] See, e.g., G. COCCONI, V. TONGIORGI and K. GREISEN: Phys. Rev. **76**, 1020 (1949). — E. W. KELLERMAN, T. SHAW and N. DICKINSON: Proc. Phys. Soc. Lond. A **70**, 452 (1957).

since also this component is concentrated near the core, its presence does not modify the resolving power of the system.

Better resolving power could be obtained with this method by making the detectors more selective for electrons of still higher energies or by making them insensitive, with opportune anticoincidence systems, to the largest showers.

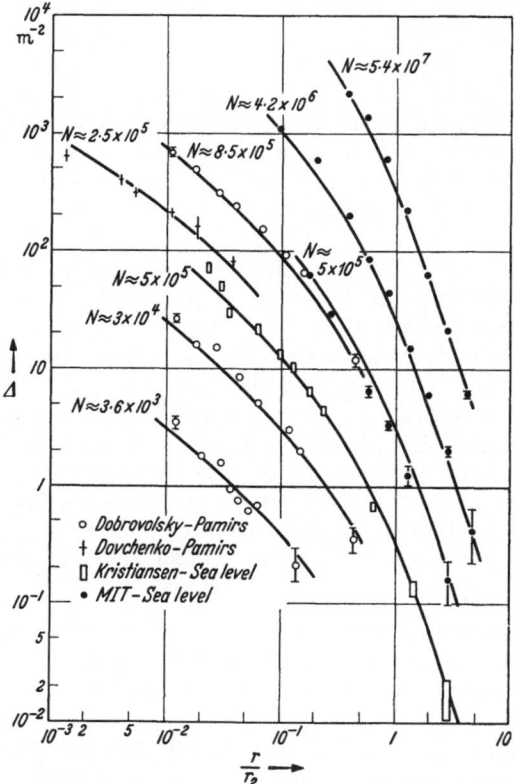

However this could be achieved only at the cost of complicating the system, and the advantage of simplicity would be lost.

Also high energy μ-mesons are concentrated around the core and could be used to locate it. However, in order to reach a resolution of $d \approx$ 10 m, only those mesons of energy $\geqq 10^{12}$ ev should be used, and their detection at sea level would be painful, requiring, e.g., both a magnetic selector and a system to measure curvatures.

d) The lateral distribution.

15. The most complete measurements of the average particle distribution in AS around the core are those performed in the last years by the Russian workers. With GM counters they have measured $\Delta(r)$ for values of r varying from 0.5 to 1000 m.

In Fig. 8 we have plotted some of their results obtained in the Pamirs and in Moscow. In the same figure we have plotted also some of the data obtained by the MIT group with scintillators at sea level. The curves drawn through the experimental points all represent the same distribution function multiplied by the factor that gives the total number of particles N_{tot}, as quoted by the authors. It appears that at least in

Fig. 8. Lateral distribution of the density of all ionizing particles for showers of various sizes, measured near sea level and at mountain altitudes with counters (Russian authors) and with scintillators (MIT). All the curves drawn through the experimental points correspond to the distribution function plotted in Fig. 9, shifted vertically to fit the total number, N, of particles in the shower. S. P. Dobrovolsky, S. I. Nikolsky, I. I. Tukish and V. I. Iakovliev: J. Exper. Theor. Phys. **31**, 939 (1956); O. I. Dovchenko and S. I. Nikolsky: Dokl. Akad. Nauk SSSR. **102**, 241 (1955); G. B. Kristiansen: Proc. Oxford Conference, April 1956; G. Clark, J. Earl, W. Kraushaar, J. Linsley, B. Rossi and F. Scherb: Nature, Lond. **180**, 353, 406 (1957).

first approximation the fit is good for all the showers, independent of their size and of the altitudes at which they were measured, provided that the distances from the core are measured in r_2 units.

The data plotted in Fig. 8 represent averages. However it is notable that the fluctuations seem not to be very large, since the lateral distributions of practically all the individual showers oberved by the MIT group follow quite well the average distribution, at least for $r > 10$ m.

The function $f(r/r_2)$ which was used to calculate the curves of Fig. 8 is given in Fig. 9, together with the curves predicted by the EM theory for various values of the age parameter s (see Sects. 5 and 6). The experimental curve is close to that predicted for pure electromagnetic showers of age $s = 1.4$. This result must be considered as mostly due to chance. In fact, the electromagnetic theory cannot justify the fact that the distribution function is independent of the shower size, and besides, most of the ionizing particles present at the largest distances ($r/r_2 > 5$) are μ-mesons and not electrons.

If the μ-meson contribution is subtracted (see Sect. 21), the distribution of the electrons is better represented by the Nishimura and Kamata curve for $s = 1.3$. The distance $r_{\frac{1}{2}}$ within which half of the particles fall, turns out to be about equal to the characteristic distance, r_2.

The experimental curve of Fig. 9 can be represented with good approximation by the equation

$$f\left(\frac{r}{r_2}\right) = 0.25 \left(\frac{r}{r_2}\right)^{-0.8} \left(\frac{r}{r_2} + 0.87\right)^{-2.64}. \quad (15.1)$$

The normalization, as usual, is such that

$$2\pi \int_0^\infty f(x)\, x\, dx = 1.$$

16. The fact that a unique lateral distribution fits all the showers, though at variance with what is predicted by the theory of the *EM* cascade, does not furnish much insight in the details of the phenomenon. Qualitatively, it indicates that the age of the shower is not related to its size, i.e. that the electrons found in the lower atmosphere are not the remnants of the multiplication of a single photon or of a few photons produced in the highest strata, but both the Miyaki and the Zatzepin-Greisen points of view are in agreement with the observations[1].

A quantitative justification of the observed distribution would require a model for the nucleonic cascade, and probably more than one model could justify the experimental evidence. However the condition that $f(r/r_2)$ be independent of shower size is rather

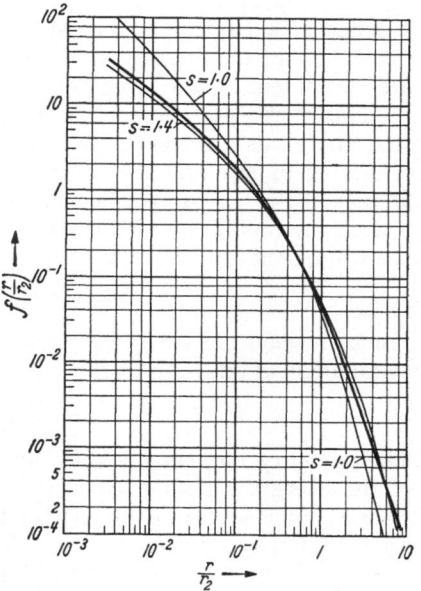

Fig. 9. The heavy line represents the lateral distribution of all ionizing particles that fits reasonably well all the experimental points of Fig. 8. For comparison we have also plotted the lateral distribution functions calculated by Nishimura and Kamata for the electrons in a pure electromagnetic shower of age $s = 1$ and $s = 1.4$. The density at a distance r from the core of a shower containing N particles is given by the expression

$$\Delta(r) = N \frac{f(r/r_2)}{r_2}.$$

stringent, since it implies that, at least in first approximation, also the energy spectrum of the N-component is independent of shower size.

Further information about the nucleonic cascade is contained in the observation that the density of the particles keeps increasing in the vicinity of the core, about in inverse proportion to r, down to distances of ~ 1 m.

At distances smaller than 1 m, $\Delta(r)$ seems to increase less steeply, but still no plateau has been found even with resolving power of ~ 0.1 m²[2].

It is however doubtful that the core can be identified with such an accuracy when the detectors are sensitive to all ionizing particles. As will be discussed in the next chapter, N-particles of high energy are present at 1 to 2 m from

---

[1] See the discussion by A. Lieda and N. Oyita in Progr. Theor. Phys. **19**, 582 (1958).
[2] See the results obtained by O. I. Dovchenko and S. I. Nikolsky (Fig. 8), as well as those obtained with a cloud chamber by W. E. Hazen: Phys. Rev. **85**, 455 (1952); see also W. E. Hazen, R. W. Williams and C. A. Randall: Phys. Rev. **93**, 578 (1954).

the core, and they can create large fluctuations in the electron and photon density that can lead to wrong identification of the shower core. Very educational in this respect are the pictures taken with a large chamber by HAZEN.

Anyway, the lack of success in finding multiple cores, or even a flattening of $\Delta(r)$ near the center of the showers, indicates that the central region of the shower is much more a "core" than it was thought to be when the word was first used to identify it. As far as it can be checked, the core is a real singular point.

If the N-component is responsible for the propagation of the shower, the absence of a flattening must be interpreted as an indication that most of the nucleons and the mesons that are responsible for the propagation of the shower are contained, at sea level, within a circle of ~ 1 m in diameter.

This is another evidence in favor of the conclusions reached in Sect. 12, i.e., that the statistical models for the ultra-relativistic interactions which postulate a sharing of the primary energy among several secondaries is not satisfactory. In fact, if the core dimensions are ~ 1 m, the angle of emission of the secondaries that in the first interaction carried away most of the available energy must satisfy the condition:

$$\Theta < \frac{0.5 \text{ m}}{15 \text{ km}} = \frac{1}{3 \times 10^4}.$$

For a nucleon of 10^{14} ev (shower with $N_{tot} \approx 10^4$ at sea level) this corresponds to an angle of emission in the center of mass

$$\vartheta = (2\gamma)^{\frac{1}{2}} \Theta < 2 \times 10^{-2}.$$

The model of Landau predicts, at these energies, an angle $\vartheta_{\frac{1}{2}}$ within which half of the energy available is emitted, about five times larger. The situation is, of course, much worse in the case of the Fermi model. This argument is perhaps, by itself, not as stringent as those discussed in Sect. 12. However, added to them, it strentghens considerably the plausibility of a model for the ultra-relativistic interaction in which the primary energy is not all subdivided among some secondaries.

e) The number spectrum.

17. The measurements of the size of the showers and of their directions have been used by the MIT group[1] for a direct determination of the differential number spectrum near sea level. The frequency with which the showers of each size land on unit surface with directions in unit solid angle has been evaluated after correcting the size of the inclined showers for the effect of atmospheric absorption, The results are collected in Fig. 10. The two straight lines (A and B) represent instead the average differential spectra deduced from the integral number spectra evaluated in Sect. 10 [Eq. (10.4)].

The agreement between the results of the two completely independent methods is very good, and the spectra of Fig. 10 can be accepted with confidence. It seems well established that at sea level showers arrive with N_{tot} as high as 10^9, and that their frequency does not suggest any large increase in the slope of the spectrum.

A rough estimate of the total energy of the primaries that generate these showers is:

$$U \approx 10^{10} N_{tot} \text{ ev},$$

i.e. around 10^{19} ev for the largest showers observed thus far.

---

[1] G. CLARK, J. EARL, W. KRAUSHAAR, J. LINSLEY, B. ROSSI and F. SCHERB: Nature, Lond. **180**, 353, 406 (1957).

From the differential spectrum of Fig. 10, we have deduced the integral number spectrum plotted in the same figure; it corresponds to the expression:

$$F(\geqq N) = 0.60\, N^{-(0.815+0.060\, \text{Log}\, N)} \quad \text{m}^{-2}\,\text{sec}^{-1}\,\text{sterad}^{-1}. \tag{17.1}$$

The slope of this integral spectrum,

$$-\frac{d\,\text{Log}\,F}{d\,\text{Log}\,N} = 0.815 + 0.12\,\text{Log}\,N$$

assumes the values 1.3 for $N \approx 10^4$ and 1.9 for $N \approx 10^9$.

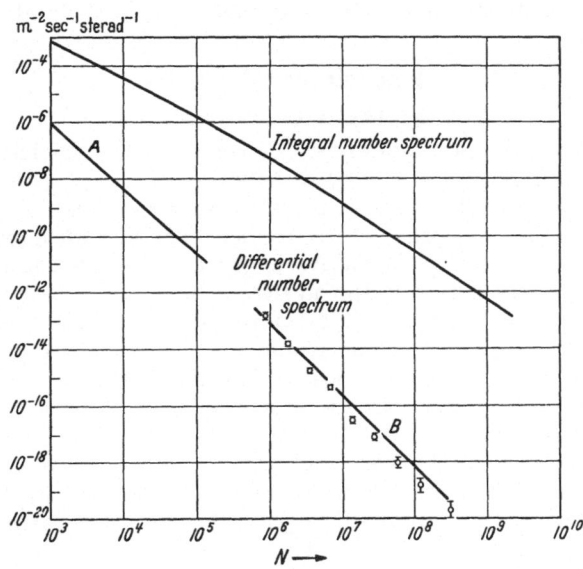

Fig. 10. In the lower part of the figure is given the differential number spectrum, $\dfrac{dF(N)}{dN}$, at sea level. Lines A and B are deduced from the density spectrum [Eq. (10.4)]. The points have been obtained by the MIT group. The integral number spectrum, $F(N)$, given in the upper part of the figure is obtained from integration of the differential spectrum.

The relation between the number spectrum and the energy spectrum will be discussed in Sect. 28.

C. The meson-nucleon cascade.

18. In the last years a great deal of experimental effort has been directed toward a quantitative interpretation of the behavior of the N-component of AS, i.e. of those particles that are capable of producing nuclear interactions (nucleons, $\pi^{\pm}$-mesons, K-mesons and hyperons). Unfortunately, the situation is as yet far from satisfactory. Experimentally it is much more difficult to obtain direct information on the N-component than on the electromagnetic component, the former being about one hundred times less abundant than the second. Furthermore, we still have no reliable quantitative theory of the N-collisions, even for energies much smaller than those involved in AS, to use as a guide in the interpretation of the experimental data. Likely, substantial advances in the interpretation of N-interactions at extremely high energies ($>10^{14}$ ev) will only be achieved when the investigations now in progress and to be made in the near future with the aid of particle accelerators have clarified the nature and characteristics of such processes at smaller energies (10^9 to 10^{11} ev).

Whatever information we have about the N-component in AS is obtained either from the direct detection of the nucleonic particles themselves or indirectly from the study of the *EM* component, as discussed in the previous chapter, and of the μ-mesons present in the showers. The μ-mesons in fact are believed to be produced from the decay in flight of the unstable N-particles (π^+-mesons and K-mesons). Their number, energy and distribution are thus correlated with those of their parents.

A lack of guiding theory in the interpretation of the phenomena makes it advisable to proceed in this chapter with an order opposite to that followed in the discussion of the electromagnetic component. The experimental results will be discussed first and an analysis of the plausible interpretations will follow.

I. Experimental results.

a) The N-component.

19. The study of the properties of the N-component of AS has as its aim the knowledge of the density and energy distribution of these particles as function of the distance, r, from the cores of the showers, as well as of the shower size and of the atmospheric depth. The N-particles are identified by "nucleonic" detectors, while their position in the AS is generally deduced from the response of a number of normal detectors spread over a large area (see Sect. 13 and 14).

The most obvious N-detector is a multiplated cloud chamber. The plates are of heavy material and the cloud chamber is possibly located under a thick layer of absorber that eliminates most of the electromagnetic component. In a cloud chamber the N-interactions are recognizable, and the energy of the secondaries can be roughly estimated.

However, in order to observe a sufficient number of events within a reasonable time it is necessary to use chambers with cross sectional areas of at least a half square meter and with a plate thickness equivalent to about one mean free path (160 g cm^{-2} of Pb). The experiment, especially when several cloud chambers must operate simultaneously, becomes rather painful so that the method has not been extensively employed.

The apparatus most commonly used thus far as N-detector consists instead of several layers of GM counters sunk in a large mass of heavy material (lead, iron) and connected to a hodoscope. With a proper arrangement of counters and absorbers the simultaneous discharge of a suitable number of counters in appropriate positions can be taken as an indication of the occurrence of an N-interaction. *EM* particles from outside are eliminated if the counters are protected on all sides by at least 20 cm Pb. The possibility that μ-mesons and their knock-on electrons simulate an N-interaction is eliminated by requiring that several counters located in the same layer and separated by some cm of Pb are simultaneously discharged. The energy of the particle that triggers this kind of detectors can be estimated only very crudely, since one can guess only roughly at the number of secondaries with range above a certain minimum produced in the interaction. The geometry of the apparatus determines the "minimum" detectable energy of the ionizing N-particles. This is usually several times 10^9 ev, and most of the events recorded correspond to N-particles with energies close to the threshold. The sensitive surface of these detectors is hard to estimate; its value is somewhat bigger than the surface of the counter layers, but smaller than the cross sectional area of the absorber; the effective surface, i.e. the product of the sensitive surface times the probability than an N-partcle gives a recordable shower is even harder to evaluate. Effective surfaces up to one square meter are commonly used.

The density of *N*-particles in AS, as determined from the frequency of recorded interactions, cannot thus be determined with good precision. One must be cautious in comparing results obtained with different detectors.

Another property of the nuclear reactions induced by *N*-particles that can be utilized to make a detector is the fact that neutrons are produced, most of them with energies in the Mev region. These neutrons can be detected, e.g. by surrounding with paraffin and BF_3 proportional counters the block of material in which the *N*-particles interact. These "neutron" *N*-detectors have thresholds of some 10^8 ev, efficiencies small and loosely dependent on energy, so they have thus far provided only qualitative information.

An ideal though cumbersome *N*-detector that avoids some of the defects of those previously described consists of a mass of material thick enough to absorb completely the *N*-particles and all their secondaries (required thickness >800 g cm^{-2}, equivalent to at least 30 radiation units), in which many layers of ionization chambers (narrow and long tubes are most suitable) are embedded, that can measure the number of ions present at the various depths. The integral number of ions generated in the material is directly correlated to the total energy of all the *N*-particles striking the detector simultaneously[1]. In principle, the method is equivalent to that used in Sect. 12 for evaluating, from the integration of the ionization produced by AS in the whole atmosphere, the energy of the primary particle.

In the arrangement so far used, the number of layers of ion chambers has been limited to one, with above it, first a layer of Pb to absorb most of the *EM* component already present in the shower, then a slab of carbon to provide material for the *N*-particles to interact, and finally some inches of lead, just above the chambers, to multiply the *EM* component generated in the *N*-collision.

The energy involved in the interaction is deduced by multiplying by a factor of 3 the estimated energy of the *EM* component, the underlying idea being that in *N*-interactions about $\frac{1}{3}$ of the energy is given to π^0-mesons on the average. The energy thus deduced represents on the average a lower limit of the energy of the *N*-particle. However, even in this imperfect realization, the method is certainly better than that utilizing GM counters. With a larger number of chambers it will be possible to obtain reliable quantitative results.

It must finally be mentioned that the distinction between ionizing and non-ionizing *N*-particles has been achieved in some experiments by covering the detecting apparatus with a layer of counters and recording the ions in which the detector was activated while none of the covering counters was struck. This "anticoincidence tray" must of course be heavily shielded from the *EM* component in the air; and corrections must be made for the loss of sensitivity to incident neutral particles, owing to interactions in this shield and also to back-projected secondaries from the underlying layers of matter.

20. Until a few years ago the information available on the *N*-component was essentially qualitative. We shall quote here a few characteristic results.

Several measurements have been performed to investigate the value of the ratio, R_N, between the density of the *N*-component and that of all the ionizing particles in the shower, as well as the variation of this ratio with the altitude above sea level. What was generally measured was a not well definable average density of *N*-particles around the core, hence comparisons among the absolute

---

[1] The distribution of the pulse amplitudes in the various layers of ion chambers can be utilized to check whether the apparatus was hit by one or more than one *N*-particles.

values obtained with different apparatus do not make much sense. Relative values obtained with the same apparatus at various altitudes are more reliable, but one must keep in mind that the N-showers can change their structure with altitude and shower size, and corrections for these effects were not calculable. For showers with total number of particles $N = 10^4 - 10^5$ it was found that R_N is about 0.02 when neutron detectors were used[1], about 0.005 with GM counter detectors[2], and about 0.003 for N-particles of energies $\geqq 10$ Gev observed with cloud chambers[3].

The energy spectrum that one can inver from these results appears much less steep than that observed for the N-particles not associated with AS, and indicates that the N-component carries a large portion of the energy present in the shower.

The measurements at various altitudes show that from sea level up to mountain altitudes R_N remains constant or decreases slightly, while at higher altitudes it increases sharply[4]. The nearly constant value of R_N between 1000 and 650 g cm^{-2} of atmosphere shows that the N-component in AS is much less absorbed in the lower atmosphere than the N-component not accompanied by AS (a factor of 5 for the first as against a factor of 15 for the second). This again points to the fact that among the N-particles in a shower there are some with very high energy, and that the behavior of the electromagnetic shower is dictated by the nucleonic cascade, a conclusion in agreement with the Zatzepin-Greisen point of view.

It was also found that the N-particles are relatively more abundant in small showers than in large ones[5], i.e., that the total number of N-particles increases less than in proportion to the number of electrons in AS. This confirms that the N-component is responsible for the propagation of the shower in the lower atmosphere, since the smaller showers are those in which the electromagnetic cascade is more exhausted.

The lateral distribution around the core at mountain altitudes was shown to be steeper than that of the electrons[6] (the opposite is true, as we shall see, for the μ-meson component), indicating again large energies concentrated in the N-component.

The ratio $\dfrac{\text{not ionizing}}{\text{ionizing}}$ N-particles was measured both with counters[7] and with cloud chambers[8] and found to be ~ 0.7 at mountain altitudes. This last result suggests that in the nucleonic cascade the energy degradation is small. In fact, the neutral particles must be mostly neutrons and since their number is so large, the energy must remain in the nucleons and not be evenly shared with the mesons, though these are produced abundantly in nucleonic interactions.

In the last years some of the measurements described above were repeated with improved techniques and results detailed enough to warrant quantitative

[1] G. Cocconi and V. Tongiori: Phys. Rev. **79**, 730 (1950). — P. L. Marsden: Proc. Oxford Conference, April 1956.

[2] G. Fujioka: J. Phys. Soc. Japan **10**, 245 (1955).

[3] B. Choudhuri, R. C. Saxena and A. Subramanian: Proc. Ind. Acad. Sci. A **36**, 457 (1952).

[4] R. L. Cool and O. Piccioni: Phys. Rev. **82**, 306 (1951).

[5] H. L. Kasnitz and K. Sitte: Phys. Rev. **94**, 977 (1954).

[6] G. Cocconi and V. Tongiorgi: Phys. Rev. **79**, 730 (1950). — K. Greisen, W. D. Walker and S. P. Walker: Phys. Rev. **80**, 535 (1950).

[7] K. Greisen and W. D. Walker: Phys. Rev. **90**, 915 (1953).

[8] H. L. Kasnitz and K. Sitte: Phys. Rev. **94**, 977 (1954).

analysis began to be available. In these experiments, the shower core was located and the total number of ionizing particles was measured for each shower (see Sect. 13), while the probability of detecting the N-component was measured simultaneously at various distances from the core.

In Fig. 11 are plotted the results obtained with GM counters by the Russian workers in Moscow and on the Pamirs.

In these measurements the energy reso-lution of the N-detectors is still very poor and the densities measured refer to all inter-acting particles of energy above $\sim 10^{10}$ ev. The straight lines drawn through the experi-mental points correspond to the equation

$$\Delta_N(r) = A\, r^{-\alpha}. \qquad (20.1)$$

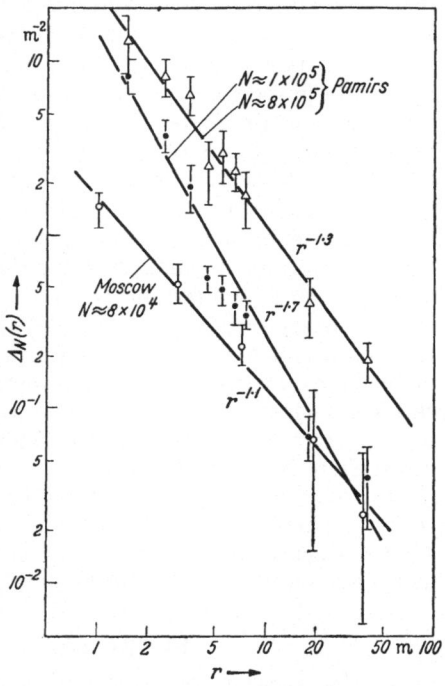

The values of α are given in the figure. Though the variations of α with N and altitude cannot be considered yet real, owing to the poor statistics and to the differences between the detectors used in the two stations, the general trend confirms most of the results obtained with the less detailed measurements discussed above.

The steady increase of $\Delta_N(r)$ from large r down to $r \approx 1$ m confirms the results ob-tained for $\Delta_{\text{total}}(r)$, hence the conclusion of the analysis conducted in Sect. 16. How-ever the value of α in (20.1) cannot remain constant when r increases, or elase the total number of N-particles would diverge. Somewhere, α must become larger than 2, if the r dependence is expressed by an equation similar to (20.1) over a limited range of r. As this does not seem to happen within the distances explored, it is then impossible to estimate the total number of the N-particles present in a shower. A

Fig. 11. Lateral distribution of the density of the N-component. The measurements at Moscow (100 m above sea level) are due to: A. I. Abrosimov, A. A. Bedniakov, V. I. Zatzepin, I. A. Nechin, V. I. Sloveva, G. B. Kristiansen and P. S. Chikin: J. Exp. Theor. Phys. **29**, 693 (1955). Those on the Pamirs (3860 m above sea level) to: S. I. Nikolsky, Y. N. Vavilov and V. V. Batov; Dokl. Akad. Nauk SSSR. **111**, 71 (1956).

rough evaluation can be made by assuming that $\Delta_N(r)$ suddenly drops to zero for $r > 100$ m. The number of N-particles in the shower is then:

$$N_N = 2\pi A \int_0^{100} r^{(1-\alpha)}\, dr.$$

Thus the ratio $R_N = N_N/N_{\text{tot}}$ turns out to be about 10^{-2}.

Measurements of the distribution near the core and of the energy of the N-particles have been made with ion chambers by two groups of experimenters[1], the first on the Pamirs (3860 m above sea leve) and the second in Moscow. Both groups found that the most energetic N-particles are confined within 1 or 2 meters from the core, and that a good fraction of the energy can be concentrated in one or a few particles.

[1] N. Dobrotin, O. Dovchenko, V. Zatzepin, E. Murzina, S. Nikolsky, I. Rakobolo-kaya and E. Tukish: Nuovo Cim. Suppl. **8**, 612 (1957). — V. A. Dimitrev, G. V. Kulikov and G. B. Kristiansen: Nuovo Cim. Suppl. **8**, 587 (1957).

Dobrotin *et al.* estimated that for showers of $\sim 10^5$ particles the total number of N-particles is $\sim 10^3$, and that the total energy they carry is about 10% of the energy of the primary particle that gave rise to the shower (4×10^{14} ev). The energy is not evenly shared: a few particles carry up to 10% of the entire energy of the N-component, i.e., 1% of the total energy of the shower. This was confirmed by Dimitrev *et al.* who found N-particles of $> 10^{12}$ ev in showers of $N_{tot} \approx 3 \times 10^4$ particles, i.e., produced by primaries of energy $U \approx 3 \times 10^{14}$ ev.

Dobrotin *et al.* also found that the integral energy spectrum of the N-particles is roughly proportional to E^{-1}, while the energy spectrum of the EM component near the core (measured by means of a cloud chamber) remains quite soft, as expected for showers of age $s = 1$ to 1.5, thus confirming the continuous supply of EM particles by the nucleons in the core.

These results are very important, because, by proving directly the existence of N-particles of such a high energy in the core of normal size showers, they establish very stringent conditions for the interpretation of the multiplication processes of the N-component.

It is necessary, however, to be aware of the fact that in these measurements the number of high energy N-particles detected in each shower is one, in most of the cases, and that the quoted distance from the core is the distance between the N-detector and the point where the detected density of all ionizing particles in the shower is maximum. The fluctuations in the multiplication in the lower atmosphere make it somewhat doubtful whether the axis of the shower should be identified with the point of maximum density or rather with the point where the N-particle itself is located, and especially so when the N-particle contains a good fraction of the energy present in the shower. Thus in some cases the identification of the core can be ambiguous.

b) The μ-component above ground.

21. The measurements on the μ-component of AS are somewhat simpler and more precise than those on the N-component both because the density of the particles far from the core is not much smaller than that of the electrons and because the energy measurements simply reduce to range measurements; the μ-mesons have in fact very small nuclear interaction cross section and are stopped mostly by the collision losses.

Both cloud chamber and counter telescopes as described in the previous section are used for μ-meson detection, the second being more commonly employed. The μ-mesons are identified with those particles that go through the absorbers without scattering and without producing penetrating secondaries. Of course an N-particle not interacting in the absorber may exhibit the same behavior, but the probability of this occuring can be evaluated and shown to be small in comparison with the number of μ-particles present at each point of a given shower. The energy resolution of the μ-meson counter detectors is good, but the energy range that can be explored with them cannot be large: from a minimum of ~ 300 Mev (20 cm Pb) to a maximum of ~ 1 Gev (1 m Pb). Higher energies, however, can be selected by placing the detector underground. In many experiments no energy analysis is made, the detector accepting all the μ-mesons with energies above, say, 300 Mev.

We shall divide the experimental evidence about the μ-component into two parts. In the first we will discuss mostly the results obtained above the surface of the earth, while in the second we will focus our attention of the experiments performed underground.

22. The radial distribution of the μ-meson component has been measured in several experiments either using μ-meson detectors located at different distances from a core selector or by means of μ-detectors spread over a large area also patrolled by detectors of all the ionizing particles in the shower (see Sect. 13 and 14).

A result confirmed by various authors is that $\varrho_\mu(r) = \dfrac{\varDelta_\mu(r)}{\varDelta_{\text{tot}}(r)}$ is independent of shower size, at least in first approximation, but depends on r, going from a minimum of less than 10^{-2} near the core to about unity at 1 km from it. Over this distance the absolute value of the density $\varDelta_\mu(r)$ decreases by a factor of about one thousand. The independence of $\varrho_\mu(r)$ of the shower size has been confirmed recently by VAVILOV *et al.* in the Pamirs for showers of $N \approx 10^6$; at sea level the MIT group finds instead that ϱ_μ slightly increases when N decreases, for values of $N \approx 3 \times 10^6$.

In Fig. 12 and 13 we have collected the values of $\varrho_\mu(r)$ obtained by various authors at mountain altitudes and at sea level.

By integrating the density of the μ-component over the whole shower we evaluate the total number of μ-particles in the shower. At sea level it turns out that, utilizing the curves of Fig. 13 and of Fig. 9,

$$R_\mu = \frac{N_\mu}{N_{\text{tot}}} = 0.10,$$

independently of shower size. At mountain altitude $R_\mu = 5 \times 10^{-2}$.

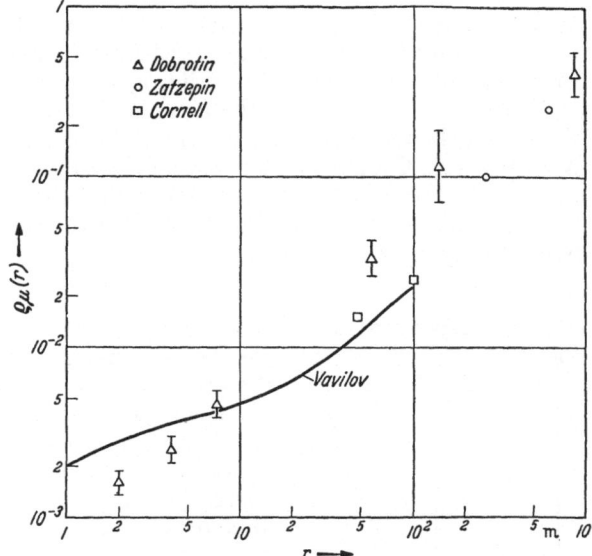

Fig. 12. The ratio $\varrho_\mu = \dfrac{\mu\text{-meson density}}{\text{total density}}$ at various distances from the shower core at mountain altitudes (3000 to 4000 m). The results utilized are from: N. A. DOBROTIN, G. T. ZATZEPIN, S. I. NIKOLSKY, and G. B. KRISTIANSEN: Nuovo Cim. Suppl. **3**, 635 (1956); G. COCCONI, V. TONGIORGI and K. GREISEN: Phys. Rev. **76**, 1020 (1949); G. T. ZATZEPIN, I. L. ROSENTHAL, L. I. SARITCHEVA, G. B. KRISTIANSEN and L. K. EIDUS: Isv. Akad. Nauk SSSR (Phys.) **17**, 39 (1953); Y. N. VAVILOV, Y. F. EVSTIGNEEV and S. I. NIKOLSKY: J. Exp. Theor. Phys. **32**, 1319 (1957).

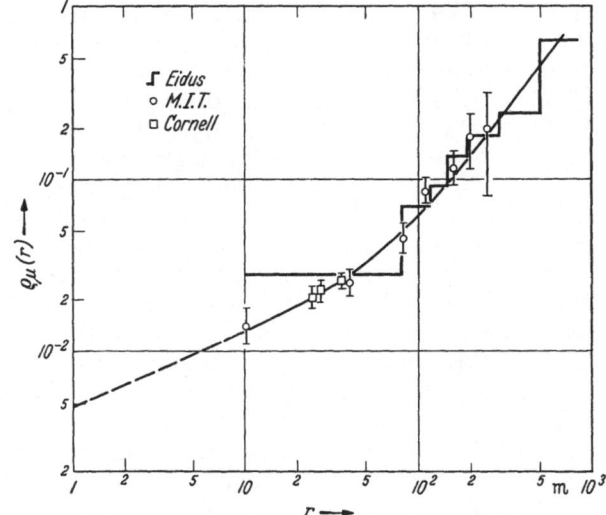

Fig. 13. The ratio $\varrho_\mu = \dfrac{\mu\text{-meson density}}{\text{total density}}$ for showers observed near sea level. The results utilized are from: K. L. EIDUS, M. I. ADAMOVITCH, I. A. IVANOVSKAYA, V. S. NIKOLAEV and M. S. TULYANKINA: J. Exp. Theor. Phys. **22**, 440 (1952); W. KRAUSHAAR: Nuovo Cim. Suppl. **8**, 631 (1957); G. COCCONI, V. TONGIORGI and K. GREISEN: Phys. Rev. **76**, 1063 (1949).

At both altitudes a large contribution to the total number of μ-mesons comes from the large distances from the cores, where the densities are small, but the

μ-particles are relatively more numerous. In fact, while 50% of all the particles in the showers are found at sea level inside a circle of radius $r_{\frac{1}{2}} = 70$ m, 50% of the μ-mesons are in a circle of radius $r_{\frac{1}{2}} = 300$ m.

The energy spectrum of the μ-component has been measured recently in the Pamirs by DOVCHENKO et al.[1] at three distances from the core of showers with $N = 10^5 - 10^6$. Counter trays of 2.4 m² placed under various thickness of absorbers (up to 10 m of earth) detected simultaneously particles of energy larger than 0.44, 1.7 and 3.5 Gev. Within this range of energies the results are consistent with an exponential spectrum

$$N(\geqq E) = B E^{-b} \tag{22.1}$$

with b independent, at least in first approximation, of shower size. In Fig. 14 the values found for b are plotted as a function of r. As expected, near the core the spectra are flatter than far from it. However the exponent b cannot remain indefinitely smaller than 1, or else the total energy would diverge.

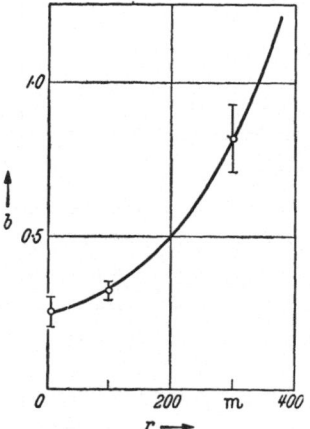

It is not possible, at this stage, to evaluate the total energy contained in the μ-component. We shall be able to partially overcome this difficulty using the data obtained deeper underground.

c) The μ-component underground.

23. The experiments performed underground on the penetrating particles associated with AS can be considered as extensions of those performed above ground with shielded detectors.

When the correlation between AS above ground and μ-mesons underground is studied, the detectors underground are located vertically below those above ground, so that the showers coming from directions near the zenith can trigger both systems. Of course, the larger the difference in level, the smaller is the solid angle of acceptance. This limitation, as well as the strong absorption undergone by the particles, cause the number of coincidences between the two groups of detectors to be so small as to make these experiments very time consuming.

Fig. 14. The dependence of the exponent, b, of the μ-meson energy spectrum on the distance, r, from the shower core, according to DOVCHENKO, NELEPO and NIKOLSKY.

The most extreme experimental conditions thus far have been realized by BARRETT et al.[2]. The underground station was 593 m below the AS detectors; only showers coming within $\sim 10°$ from the vertical could give coincidences between the two levels. The experiment had to run for about two years.

However, once it is granted that all the showers of parallel μ-mesons underground are part of AS observable above ground, the record of the correlation above and below ground may become unnecessary and one can study only the showers of μ-mesons underground.

The energy that the μ-mesons observed underground had at sea level must be at least as large as that required to cross the layer of interposed earth, E_{min}, but more likely it is equal to

$$\overline{E} = E_{min} + E_\mu,$$

[1] O. I. DOVCHENKO, B. A. NELEPO and S. I. NIKOLSKY: J. Exp. Theor. Phys. **32**, 463 (1957).
[2] P. H. BARRETT, L. M. BOLLINGER, G. COCCONI, Y. EISENBERG and K. GREISEN: Rev. Mod. Phys. **24**, 133 (1952).

where E_μ is the average energy of the μ-mesons underground[1]. The scattering of the μ-mesons along their path to the underground station is such that, at the arrival, the μ-meson is practically always less than one meter off from its initial direction. Hence the observed spreading can be considered as mostly due to the angular divergence of the mesons at their origin.

The number of μ-mesons reaching great depths is so small that thus far it has been impossible to locate the core of the μ-meson showers and study the meson distribution around it, as has been done above ground. However, information to that effect has been gathered by measuring coincidences between two detectors and varying their horizontal separation (decoherence curve); at the same time the total number of particles in the showers has been deduced from the number of cases where two, three, four, etc. μ-mesons belonging to the same shower were observed.

24. Especially from the series of experiments performed by BARRETT *et al.* one can deduce that the particles underground observed in coincidence with AS above ground are *all* μ-mesons, the few particles of different nature being secondaries of the μ-mesons produced in the absorbers. At the same time it has been

Table 2.

Location	Total absorber besides atm. (m water equivalent)	E_{min} (ev)	E_{median} (ev)	Median dist. from core (m)	Number spectrum $m^{-2} sec^{-1} ster^{-1}$	Validity range for N	Authors
Sea Level	0	3×10^8	2×10^9	300	$1.6\ N^{-1.5}$	10^3 to 10^7	See Sect. 22
30 m Underground	60	1.25×10^{10}	10^{10}	40	$0.8\ N^{-2.2}$	10^3 to 10^4	GEORGE *et al.*
30 m Underground	65.5	1.4×10^{10}	10^{10}	35			ANDRONIKASHVILI *et al.*
593 m Underground	1580	5×10^{11}	5×10^{11}	9	$0.007\ N^{-2.4}$	1 to 10^3	BARRETT *et al.*

The measurements of ANDRONIKASHVILI *et al.* refer to AS of $N_{tot} = 1 - 5 \times 10^5$ at sea level.

shown that *all* AS with energy above a certain limit contain μ-mesons capable of reaching a certain depth, i.e., mesons with energy above a certain limit, and vice-versa that *all* showers of parallel μ-mesons underground are part of AS.

The main characteristics of the μ-meson showers underground are summarized in Table 2. The median distance from the core given in column 5 has been measured at sea level directly, as described in Sect. 22; at 65 m [2] the density of the μ-mesons at various distances from the cores was measured by determining the position of the shower cores above ground, and, for distances from 15 to 70 m, was found to be well represented by the following Gaussian distribution (showers of $N_{total} \approx 2 \times 10^5$):

$$\Delta_\mu(r) = 0.66\, e^{-\frac{r^2}{2 \times (30)^2}} \qquad (r \text{ in meters})$$

---

[1] In some cases instead of the average energy, the median energy or the average of the inverse of the energy can be more significant, especially when the energy spectrum is quite flat.

[2] E. L. ANDRONIKASHVILI and M. F. BIBILASHVILI: J. Exp. Theor. Phys. **32**, 403 (1957).

with errors of $\sim 15\%$ for both constants. The corresponding median distance of 35 m is in good agreement with that deduced by George et al. [1] from a measurement of the decoherence curve at the same depth.

The median distance is deduced from a decoherence curve by assuming that μ-meson showers have a constant density up to a distance R from the core, independent of the number of particles in them, and density zero beyond R [2]. A justification for this assumption, at least at the largest depths, lies in the fact that the μ-mesons underground, being of high energy, are not created all along the shower axis, but preferentially in the first stages of the N-cascade, high in the atmosphere. The geometrical factor that creates the core singularity is thus not present, and the distribution is uniform up to distances that correspond to the maximum angle of emission of the parent π-meson. The values quoted in column 5 of Table 2 correspond to $R/\sqrt{2}$.

The number spectrum, i.e., the frequency of the showers with N_μ or more μ-mesons in them has been deduced from the frequency of cases in which several μ-mesons were observed simultaneously. A power law describes the result, as for the other AS components [3]:

$$F(N) = D\,N^{-\delta}.$$

The results of these evaluations are given in column 6 of Table 2 (integral spectra). In column 7 we have indicated the ranges of values for which the formulae can be considered as valid. From the number spectra of Table 2 it follows that, e.g., out of 10^5 μ-mesons present in a shower at sea level about 2000 survive at 60 m water equivalent and 140 reach 1600 m. Starting instead with 10^4 μ-mesons, 400 reach 60 m and 30 reach 1600 m water equivalent.

From these numbers it is possible to have an idea of the range spectrum of all μ-mesons in the showers; when the ranges are translated into energies, one obtains for the energy spectrum of all the μ-mesons present in AS at sea level the following expression:

$$N(\geq E) = B(E + b)^{-\beta}$$

with $b \approx 10^9$ ev and $\beta \approx 1.20$.

If this spectrum were taken seriously, the average energy of the μ-mesons in AS at sea level would be found to be around 10^{10} ev. The underground measurements are still too inaccurate to place much confidence in this figure, since it is too sensitively dependent on the precise value of β, and we like better to quote in column 4 of Table 2 the median energy at each level.

Since the μ-component contributes at sea level 10% of all ionizing particles in AS, its contribution to the average energy present in the shower corresponds to some 10^9 ev per ionizing particle observed at sea level. If all μ-mesons are decay products of π^+-mesons, the neutrinos created with the μ-mesons have an energy equivalent to $\sim\frac{1}{4}$ of that of the μ-mesons; their energy should also be taken into account in evaluating the total energy of a shower.

As a conclusion, for showers observed at sea level the μ-meson and neutrino components carry an amount of energy corresponding to

$$\frac{U_\mu}{N_{\text{tot}}} = 1 \text{ to } 2 \times 10^9 \text{ ev.} \tag{24.1}$$

[1] E. D. George, J. W. MacAnuff and J. W. Sturgess: Proc. Phys. Soc. Lond. A **66**, 345 (1953).

[2] If instead of a boxlike distribution one would use a Gaussian the final result would be practically identical.

[3] The method of evaluation of D and δ is in this case somewhat different from that described in Sect. 9. See, e.g., the article of Barrett et al.

The data discussed above can give information also about the energy spectrum of the μ-mesons at various distances from the core.

For instance, the μ-mesons present at sea level within 13 m from the core in a shower of 10^6 ionizing particles are ~ 1500 (see Fig. 16); at 600 m underground they are reduced to ~ 140. If the power spectrum of formula (22.1) is assumed, $b \approx 0.3$.

Within 60 m from the core, instead, the same showers have 9000 μ-mesons at sea level, 2000 at 30 m and again 140 at 600 m underground, but in the deepest station the μ-mesons are all concentrated within ~ 13 m from the core. So, at 60 m from the core $b \approx 0.4$ for energies smaller than some 10^{10} ev, but beyond this limit b must increase substantially.

These values of b compare reasonably well with those of DOVCHENKO (see Sect. 22).

However the underground data allow to determine, as a function of the distance from the core, the upper limit of the energy range within which a power law can hold. For the μ-mesons nearest to the axis this limit corresponds to an energy somewhat smaller than that of the primary particle.

25. Special mention must be made of the experiment performed by HIGASHI et al.[1], 13 m underground (30 m water equivalent). In 1080 hours of operation with a multiplate cloud chamber (sensitive surface $\sim \frac{1}{3}$ m²), they found 16 events in which 4 to 30 particles behaving like μ-mesons and parallel within a few degrees crossed the apparatus simultaneously. The near parallelism excluded a common origin within 6 m from the chamber, but any point from 6 m to infinity could be accepted. The non-parallelism of the tracks ($\Delta \Theta \approx 2°$), interpreted as due to scattering in the earth, indicated an average energy of some 10^9 ev per particle.

These μ-meson showers can hardly be considered of the same nature as those described before. Their average energy is in fact somewhat smaller than that expected for μ-mesons coming from the air and the frequency of their occurrence is too high. The μ-mesons coming from the air have a flat distribution up to distances from the core, at 30 m water equivalent, of ~ 70 m, and a density of ~ 15 m^{-2} as observed by HIGASHI et al. would correspond to a total number of μ-mesons $N_\mu = 2.5 \times 10^5$. From the expression given in the previous section, the frequency of showers of 2.5×10^5 μ-mesons at 30 m water equivalent is from 10 to 100 times too small to justify the frequencies obtained by HIGASHI et al. A peaked distribution near the core could make the situation less unfavorable, but then the expected average energy of the μ-mesons would be much larger than some 10^9 ev.

Though the evidence is based thus far only on few cases, it seems more likely that these "narrow showers" are produced in the earth above the apparatus. It is appealing to assume that they are the remnants of the showers created in the first meters of ground by what is left of N-component in the core of AS. The probability that, say, a 5×10^9 ev π-meson decays into a μ-meson in the ground is $\sim 2 \times 10^{-3}$, and the 4 to 30 μ-mesons observed require the production in the earth of at least some 10^3 π-mesons, i.e., the multiplication of N-particles of total energy $\sim 10^{13}$ ev. The frequency of occurrence of the phenomenon ($< 10^{-5}$ m^{-2} sec^{-1}) corresponds to that of AS of 10^{15} to 10^{16} ev total energy, and a residual energy in the core of 10^{13} is just about what is expected for these showers.

[1] S. HIGASHI, T. OSHIO, H. SHIBATA, K. WATANABE and Y. WATASE: Nuovo Cim. **5**, 597 (1957) and private communications.

If this interpretation is correct, the "narrow" μ-meson showers underground contribute another evidence in favor of the Zatzepin-Greisen point of view. A further study of these narrow μ-meson showers could be very illuminating. In fact if, as assumed here, most of the energy of the N-component of each shower is concentrated in one nucleon, the μ-mesons under 10 to 20 m of rock must be within a region not larger than 1 m² and must be correlated above ground with well developed showers. If instead the Miyaki point of view is correct *many* N-particles share about equally the energy present in the core; the μ-mesons must be spread over some 10 m² and must be correlated with "young" showers.

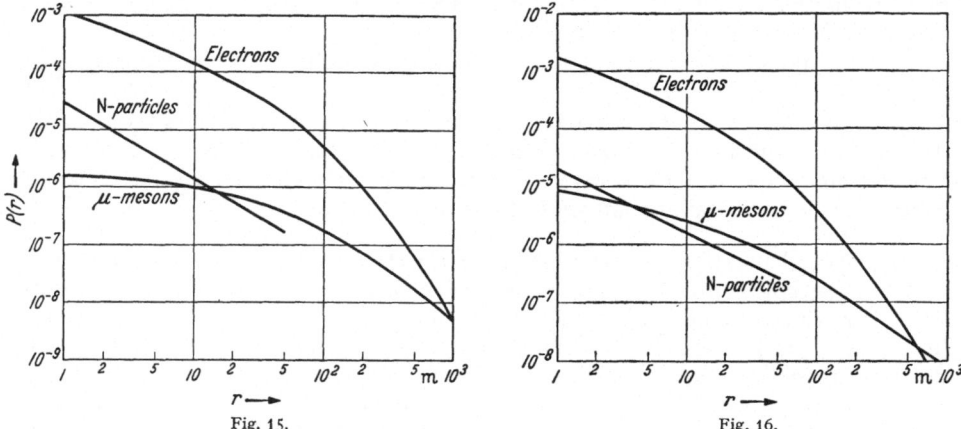

Fig. 15.   Fig. 16.

Fig. 15. Probability, $P(r)$, that one particle in a shower observed at mountain altitudes is an electron (of energy $\gtrless 0$), or a μ-meson (of energy $> 3 \times 10^8$ ev), or a N-particle, (of energy $> 10^{10}$ ev), and lands on 1 m² at a distance r from the core. The sum of the three probabilities is evaluated from Fig. 9, the μ-meson fraction from Fig. 12, and the N-particle fraction from Fig. 11. The density, $\Delta(r)$, of each component in a shower containing N ionizing particles is given by the relation $\Delta(r) = N P(r)$ m⁻².

Fig. 16. The same as Fig. 15, but for showers observed near sea level. The μ-meson fraction is deduced from Fig. 13; the N-particle fraction from Fig. 11.

As a summary of the description of the various components in AS, in Fig. 15 and 16 we have given, for a typical AS at mountain altitudes and at sea level respectively, the probabilities that one ionizing particle belonging to the shower is either an electron, or a μ-meson, or an N-particle and lands at a distance r from the core on a square meter. The sum of the three partial probabilities is, of course

$$\frac{f(r/r_2)}{r_2^2}$$

[see Eq. (5.4)] where $f(r/r_2)$ is the function plotted in Fig. 9.

II. Interpretation of the results.

26. Two phenomena concur in shaping the spectrum of the μ-mesons and in spreading them farther from the axis of the showers than the particles of the other components. One is the long range in air that, together with the long lifetime, allows the μ-mesons created in the first layers of the atmosphere to reach the ground, while the other particles cannot. The second is the larger probability for the μ-mesons to arise from decay of the π-mesons emitted at large angles with respect to the shower axis, since these mesons have, on the average, smaller energies than those emitted at small angles.

Besides, the contribution of Coulomb scattering is by no means negligible. A μ-meson arriving at sea level with 300 Mev energy, produced in the atmosphere

5 km above, is displaced by scattering at a distance $\langle r^2 \rangle^{\frac{1}{2}} \approx 300$ m. The magnetic deflections are of the same order of magnitude (see Sect. 7).

The interpretation of the properties of the μ-meson component observed at sea level is then far from simple. Furthermore it rests on the knowledge of properties of the N-cascade (such as the energy and the angular distribution of the less energetic secondaries) that are not strongly correlated with the properties of the interactions that are believed to be principally responsible for the shower development. For these reasons no quantitative interpretation of the observational data has been attempted thus far.

The situation is somewhat different for the μ-meson showers observed deep underground. There the energies are so high that only the first stages of the shower development can be involved, and the scattering cannot modify the original distribution. Their analysis leads to conclusions of some importance for the understanding of the ultra relativistic interactions[1].

For sake of simplicity, let us assume that all the primary Cosmic Rays are protons. Several independent factors contribute then to determine within relatively small limits the most probable values of the energy and of the formation height of the π^+-mesons that generate the underground showers[2]. For the energy of the π-mesons responsible for the particles observed at 1600 m water equivalent a minimum value of $\sim 6 \times 10^{11}$ ev is set by the overall thickness of the absorber. The upper limit is instead determined by the ratio:

$$\eta = \frac{\text{prob. of decay in air}}{\text{prob. of interaction in air}} = \frac{\lambda}{\tau_0 \, c \gamma \, \varrho_0 \, P} \propto \frac{1}{E_\pi P},$$

($\tau_0 =$ lifetime of $\pi^{\pm}$-mesons at rest; $\gamma = E_\pi/\mu c^2$; $\varrho_0 =$ air density at NTP; $P =$ air pressure at the production level; $\lambda =$ interaction mean free path of π-mesons in air), and by the fact that the number of particles decreases rapidly with increasing energy. These limitations make it reasonable to choose a value around 1×10^{12} ev for the energy of the parent π-meson. For $P = 0.15$ atm and $E_\pi = 1 \times 10^{12}$ ev, $\eta = \frac{1}{12}$ i.e., only one out of ~ 10 charged π-mesons decays into a μ-meson. Since the most probable number of μ-mesons present in a shower detected underground is ~ 10, the parent π-mesons must be some hundreds, hence the formation level cannot be the upper-most atmosphere, but that corresponding to $\sim 2\lambda$ ($P \approx 150$ g cm^{-2}), i.e., an altitude of ~ 13 km above sea level. About 20 km is the upper limit to this height, as set by the thinning down of the atmosphere; a lower limit of 10 km is determined by the exponential increase of the pressure, that affects η.

In conclusion, in the interactions where π-mesons of 1×10^{12} ev are created, the maximum angular spread is

$$\Theta \approx \frac{13 \text{ m}}{13 \text{ km}} = 10^{-3} \text{ rad.}$$

Here we have a specific information that can be compared with what is predicted by the theories of high energy interactions. If U is the total energy of the

---

[1] A similar analysis of the data at smaller depths, e.g., 60 m water equivalent, is not attempted here. The number of μ-mesons at that level is so great that the parent π-mesons must be produced where the AS are already very well developed. The evaluation of the average height of formation then becomes more uncertain and definite conclusions cannot be reached.

[2] The μ-mesons produced by K-mesons likely here play a negligible role since K-mesons apparently are produced less abundantly in N-interactions.

interacting N-particle and M the proton mass, Fermi statistical theory predicts

$$\Theta \approx \frac{1}{3\bar{\gamma}}, \quad \text{where} \quad \bar{\gamma} = \sqrt{\frac{U}{2Mc^2}}^{\,1}.$$

The experimental value of Θ corresponds to an energy $U = 2 \times 10^{14}$ ev. However, the same model predicts for the number of particles created in the interaction and carrying most of the energy (the forward particles, in the center-of-mass system), the value

$$\nu = K \left(\frac{U}{Mc^2}\right)^{\frac{1}{4}} \quad \text{with} \quad K \approx 0.6,$$

hence an average energy per particle, in our case, of 2×10^{13} ev, a figure an order of magnitude larger than that found before. The discrepancy is even worse in the more realistic case of a collision against an air nucleus. The Fermi model thus predicts too large angles of emission.

Better agreement can be found with the model proposed by LANDAU [2]. At energies around 10^{14} ev the angles of emission in the center-of-mass frame are about 3 times smaller than those predicted by FERMI. The energy required for the particles that create 10^{12} ev π-mesons is thus reduced by a factor of about 10, to 2×10^{13} ev, a reasonable value. These numbers apply to the case of nucleon-nucleon collisions. The more realistic cases of nucleon-nucleus (air) collisions in the Landau scheme has been analyzed by BELENKII and MILEHIN [3] and by AMAI et al. [4] with results that again agree much better with the underground data than those deduced with the Fermi model.

In the Landau-Belenkii-Amai model a parameter that is rather independent of the energy of the particles produced in the interactions, actually independent of the reference system, is the maximum transverse momentum of the secondaries, which, according to AMAI et al. is

$$p_\perp c \approx 2 \times 10^9 \text{ ev}$$

for primaries of 10^{14} ev [5].

In our case

$$p_\perp c = \Theta \times 10^{12} = 10^9 \text{ ev},$$

in good agreement with the Amai result.

NISHIMURA [6] has shown that the near constancy of the transverse momentum is to be found not only in AS, but also in the interactions observed in photographic plates, for a large range of primary energies.

While the behavior of the μ-component could be reconciled with the Landau statistical model, the experimental results on the N-component and on the developement of the showers in the atmosphere can hardly be fitted in it.

As shown in Sect. 12, the small transfer of energy in the electromagnetic component indicates high values for the elasticity parameter f. To the same conclusion we are led from the analysis of the N-component. The most striking fact is that in showers of some 10^{14} ev strongly interacting particles reach sea

---

[1] A somewhat larger angle, $\Theta \approx \dfrac{1}{2\bar{\gamma}}$, with $\bar{\gamma} = \sqrt{\dfrac{U}{2nMc^2}}$ is expect if the nucleonic particle collides with a nucleus and meets on its way n nucleons. For air $n \approx 2$, see e.g., G. COCCONI: Phys. Rev. **93**, 1107 (1954).

[2] S. Z. BELENKII and L. D. LANDAU: Fortschr. Physik **3**, 536 (1955).

[3] S. Z. BELENKII and G. A. MILEHIN: J. Exp. Theor. Phys. **29**, 20 (1955).

[4] S. AMAI, H. FUKUDA, C. ISO and M. SATO: Progr. Theor. Phys. **17**, 241 (1957).

[5] $p_\perp c$ is also almost independent of the primary energy, becoming $\sim 4 \times 10^9$ ev for nucleons of 10^{18} ev.

[6] J. NISHIMURA: Soryushiron Kenkya **12**, 24 (1956).

level with energies of more than 10^{12} ev. Other evidence comes from the observation that these particles are concentrated within ~ 1 m from the core of the shower. In fact the μ-mesons observed deep underground and produced in the high atmosphere by π-mesons also of 10^{12} ev, have a lateral spread about one order of magnitude greater; this indicates that the 10^{12} ev N-particles found at sea level clustered near the core are created less than 1 km above the detectors. The same conclusion is arrived at by requiring that the transverse momentum of the particles is of the order of 10^{9} ev.

Other striking evidence of the fact that a nucleon can maintain most of its energy after collision with another nucleon comes from many interactions of high energy cosmic rays observed in photographic plates, and is discussed in another chapter of this book.

Finally, it must be remembered that at machine energies (1 to 6 Gev) the statistical model proved inadequate to explain the observed features of multiple meson production, and positive evidence was obtained for large values of the elasticity parameter.

The model for high energy interactions between nucleons that can interpret these experimental facts is fundamentally different from the statistical ones. In it the primary nucleon maintains its identity (its charge, of course, can be modified by charge exchanges), and in each interaction with other nucleons dissipates only a fraction, $1-f$, of its energy, a fraction that is different from case to case and that on the average is of about 0.3[1].

These requirements suggest a model similar to that used to describe the bremsstrahlung of charged particles, i.e., a model in which the virtual particles surrounding the nucleon are scattered loose in the interaction.

We will identify all the specific models that enter in this category as "bremsstrahlung" models.

The first attempt at interpreting the multiple production with a bremsstrahlung model was that of LEWIS, OPPENHEIMER and WOUTHUISEN[2]; in their calculations, made when no information existed on the behavior of the π-mesons, the assumed properties of the meson field were rather arbitrary, and the predictions they made for the average multiplicity and the average emission angle were not in agreement with the experimental evidence. However, recently, LEWIS[3] has shown that if the calculations are repeated with the modifications suggested by the interpretation of the experiments on π-mesons, both the average multiplicity and the angular distribution can be brought into agreement with the experiments. Also the values predicted for the elasticity parameter are of the right magnitude. It would be very useful if the possibilities of this specific model were further analyzed.

Besides the Lewis attempt, other schemes exist that were not presented by their authors as bremsstrahlung models but that can be thought of as entering in this category.

Two similar ones are those of HEISENBERG and of BHABHA[4], in which the interaction between the two nucleons is supposed to be limited to the region

---

[1] Arguments showing that for interpreting the behavior of AS it is necessary to admit that also the average elasticity depends on the energy of the N-particles are presented by A. UEDA and N. OGITA in Progr. Theor. Phys. **18**, 269 (1957).

[2] H. W. LEWIS, J. R. OPPENHEIMER and S. A. WOUTHUISEN: Phys. Rev. **73**, 127 (1948); see also the article by LEWIS in Rev. Mod. Phys. **24**, 241 (1952) where similar calculations made by other authors are discussed.

[3] H. W. LEWIS: Proc. of the Seventh Rochester Conference, April 1957.

[4] W. HEISENBERG: Kosmische Strahlung. Berlin 1953. — H. I. BHABHA: Proc. Roy. Soc. Lond., Ser. A **219**, 293 (1953).

where the two clouds of virtual mesons overlap. Both models predict the possibility of large values for the elasticity parameters. However Landau has shown that the statistical model used by the authors for treating the interaction violates the uncertainty principle.

Another model was proposed by Takagi and by Kraushaar and Marks[1]; according to it, in the collision between the two nucleons only a fraction of the available energy and momentum is exchanged, two excited nucleins are thus created, and they subsequently decay. Kraushaar and Marks evaluated the disintegration of the two excited states with a statistical method, hence predicted zero elasticity in the collision. However it would be very easy to modify their treatment in this respect[2].

One may wonder why bremsstrahlung models have been given so far only very limited attention, while statistical models have been widely and extensively treated. Apart from the fundamental uncertainty in the choice of the correct scheme, the main reason for this situation is to be found in the fact that such models become very complicated and difficult to handle if they have to describe all possible sorts of interactions occurring in AS. Statistical models have to their advantage the fact that they are conceptually simple, mathematically easy to handle, equally well applicable to all kinds of N-particles, and furthermore give no large fluctuations. It is a pity that we have to abandon them in favor of models less amenable to analytical treatment.

The problem of deciding which among the various approaches best explains the experimental results is unfortunately very difficult, if the experimental evidence has to come only from the air showers.

This is because, owing to their complex nature, AS can give information only about the average characteristics of the N-interactions, while a reliable discrimination would require the knowledge of the specific details of individual interactions. Aside from the possibilities of more powerful accelerators, the alternate method of attack is that of studying the high energy interactions in photographic plates flown at high altitudes. There the interactions produced by single primary Cosmic Rays can be examined individually in great detail, though difficulties arise when one tries to establish the nature of many of the relativistic secondaries. With the modern techniques of balloon flight it is feasible to keep aloft a cubic meter of emulsion ($\sim$4 tons) for a period of some days; in these conditions there is a good probability of registering primaries up to about 10^{16} ev.

It is predictable that at least up to these energies the photographic method will eventually supplant the research by means of AS. But at higher energies the AS will still remain the only investigation tool[3].

D. Selected topics.

I. The nature of the primary radiation.

27. A problem of interest both for the understanding of AS and for the origin of the Cosmic Radiation is that of the nature of the primary particles that generate AS, i.e., of the particles with energy larger than, say 10^{13} ev.

---

[1] S. Takagi: Progr. Theor. Phys. **7**, 123 (1952). — W. L. Kraushaar and L. J. Marks: Phys. Rev. **93**, 233 (1954).

[2] It must be pointed out that in all these calculations only the interaction between two nucleons has been considered. A "bremsstrahlung" meson-nucleon interaction constitutes a problem by itself and, to our knowledge, has not been treated in detail yet.

[3] An empirical model for N-N interactions at $10^{12}-10^{13}$ ev, recently derived from the observations in photographic plates, seems to satisfy the requirements discussed above; its theoretical implications, however, are not yet clear. See, e.g. G. Cocconi, Phys. Rev. **111**, 1699 (1958).

The long time that these particles must spend in space before reaching the earth makes it practically certain that they are not unstable. Therefore one is restricted to consider either photons or nuclei of stable elements, the extreme case being that of associations of atoms (dust grains). Single electrons can be discarded on the ground that their bremsstrahlung emission in the magnetic fields of the galaxy would not allow them to have energies larger than $\sim 10^{12}$ ev.

The frequency of the primary Cosmic Rays with energies lower than 10^{13} ev is so high that their nature can be determined directly by means of photographic plates flown at balloon altitudes. It was thus found that all primaries are bare nuclei, the protons and the He nuclei being the most abundant and the heavier nuclei being less and less abundant as the atomic weight increases. The relative abundances as a function of A (the mass spectrum)[1] for the primaries of total energy around 10^{10} ev are given in the third row of Table 3 (p. 260). The same mass spectrum seems to hold up to $\sim 10^{13}$ ev, though at this limit the experimental data are very meager. The photons, if present, contribute to less than 1 % of the total.

At still greater energies, the air shower energies, a direct determination of the nature of the particles is lacking; other less direct methods must be utilized and all the possibilities must be taken into account.

The hypothesis that grains of dust, i.e., grains of at least some thousand nucleons, are responsible for *all* AS has been considered several times, but seems very unlikely. In fact, the showers to which such primaries would give rise would have characteristics strongly different from those observed. For instance, if, say, grains of 10^3 nucleons generate all the showers containing 10^5 particles at sea level, these showers would contain $\sim 10^7$ particles at mountain altitudes, while the observed number is less than 10^6. Besides, the lateral distribution of the particles at sea level could not be as peaked as observed. Finally, the fraction of μ-mesons in the lower atmosphere would be much larger than observed. In fact, given a shower of fixed size at sea level, the average energy of the π^+ mesons in the higher atmosphere would be much smaller in the case of a primary dust grain than in the case of a primary composed of one or few nucleons, and the probability of their decay into μ-mesons correspondingly greater. By assuming as primaries nuclei of small atomic number and utilizing a plausible model for the N-cascade, it is possible to account for the observed fraction of μ-mesons in AS[2]. In the case of primaries composed exclusively of dust grains, the same model would predict, at sea level, a fraction of μ-mesons five to ten times too large. This not withstanding, it is difficult to prove that not even a small percentage of AS, e.g., some of the largest ones that are so rare, are generated by dust grains. However, it remains to be explained how a grain of dust could move in the galactic space with ultrarelativistic energy without being stripped of a number of electrons sufficient to cause its disintegration by Coulomb repulsion. According to HERLOFSON[3] a loss of a number of electrons corresponding to some percent of the number of nuclei present in the grain should be enough to cause the break-up.

Some of the arguments against the dust grains hypothesis hold also against the hypothesis that the primaries are exclusively heavy nuclei. However, small percentages of heavy nuclei, as well as fairly large proportions of light nuclei, like He, cannot be ruled out on this basis. An answer to this problem can come only from experiments capable of showing, e.g., what fraction, if any, of the primary radiation consists of α-particles. This requires the study not of average properties of AS, but of individual showers and of their fluctuations around the

---

[1] See, e.g., B. PETERS: Progress in Cosmic Ray Physics I. Amsterdam 1952.
[2] See, e.g., M. ODA: Nuovo Cim. 5, 615 (1957) where other literature is also quoted.
[3] N. HERLOFSON: Proc. Oxford Conference, April 1956.

average; only thus one can determine how many cases are surely outside the possible fluctuations. Thus far no experiment has been successful in investigating these details.

However, the importance of the problem makes it worth discussing what kind of information could be correlated to the atomic weight, A, of the primary particles.

The EM component at large distances from the core is certainly very insensitive to A, and will not be taken into consideration. The behavior of the density function, $\Delta(r)$, near the core can instead be useful. Assume, for instance, that the nucleon-nucleon interaction is describable with a bremsstrahlung model; then the A nucleons present in the primary particle will give rise to A independent showers, and at sea level their cores will be separated because of the scattering that each of them suffers in the successive interactions against air nuclei.

The transverse momentum gained by each primary nucleon after each interaction being $p_\perp \approx 10^9$ ev/c[1], the average lateral displacement, R, is

$$R = \left[\sum_i \left(h_i \frac{p_\perp\, c}{U_i} \right)^2 \right]^{\frac{1}{2}},$$

where h_i is the height in the atmosphere at which the i-th interaction takes place. The energy U_i can be estimated by assuming, e.g., that after each interaction the energy of every primary nucleon is reduced by a factor 0.7. The numerical result for showers observed at sea level is:

$$R \approx A\, \frac{2 \times 10^{14}}{U}\, \mathrm{m},$$

where U, in ev, is the total energy of the primary nucleons of atomic number A.

For an average sea level shower of 10^6 particles, $U = 10^{15}$ ev, and if the primary is a Fe nucleus[2] $R \approx 10$ m, a value that can be excluded for the majority of the showers. For a He nucleus of the same energy, $R \approx 1$ m, a value too small to be resolved. If the primary energy is ten times smaller, $R \approx 10$ m, which can be excluded for most of the showers. As a conclusion, it could be argued that there is already evidence against a primary radiation, at these energies, all composed of heavy nuclei.

Another method for studying the nature of the primary radiation has been suggested by SITTE, who remarked that the mean free path of the primary particles that create AS can be measured, in principle, by selecting at various altitudes showers having always the same characteristics, e.g., equal total number of ionizing particles, N, *and* equal energy spectra, *and/or* equal amount of N-component, etc. These showers are by definition at the same stage of development at all altitudes (except for the difference in the π-μ equilibrium that depends on the absolute value of the atmospheric pressure), and their relative frequencies at the various altitudes is determined solely by the absorption of the primary radiation that creates them. The practical realization of this method is difficult, because all of the AS characteristics vary with altitude very much in the same manner; anyway SITTE[3] and collaborators working between 3200 and 5200 m (Chacaltaya, Bolivia) found that the primaries of $\sim 10^{15}$ ev showers have a

---

[1] To this momentum one must add, in principle, a momentum of ~ 200 Mev/c that corresponds to the Fermi noise of the nucleons in air nuclei; this contribution, however, is negligible in our case.

[2] Special attention must be focussed on Fe because it is abnormally abundant in the Universe and it has been actually identified among the heaviest primaries at low energies.

[3] K. SITTE, I.L. KOFSKY, D.L. STIERWALT, S. LEHRER, K. TAUSKAND and A. WATAGHIN: Nuovo Cim. Suppl. **8**, 684 (1957).

mean free path of (87 ± 7) g cm^{-2}; this again favors single protons as the main primary component.

Still another method of detecting the presence of heavy nuclei in the primaries of AS consists in measuring the relative abundance of μ-mesons at large distances from the core. In showers of equal size the μ-mesons must be more abundant if the primary has high A, the energy per primary nucleon being smaller, hence the probability of π-μ decay of the secondaries larger.

The possibility that photons are responsible for AS has never been seriously considered. At smaller energies the photons contribute less than 0.1% of the primary particles, besides their secondaries belong overwhelmingly to the electromagnetic component. However, so little is known about the processes that 10^{15} to 10^{20} ev photons could give rise to, that this possibility, though a priori remote, cannot be completely ruled out.

Thus far only primaries constituted of ordinary matter were taken into consideration. There might be the possibility that some primaries are formed of anti-matter and their identification would be most interesting; it is in fact conceivable that the abundance of anti-particles increases with increasing energy, since the most energetic primaries have a better chance of coming from farther away. The discrimination between matter and anti-matter as the primary radiation of AS appears difficult. Energetically the showers produced by anti-particles are indistinguishable from the normal ones; only differences in interaction behavior and cross sections could in principle be revealed by the "age" of the showers, but probably at ultrarelativistic energies also these quantities become for anti-particles practically equal to these of ordinary particles.

As a conclusion, the most plausible assumption about the mass spectrum of the primaries is that at all energies it continues to resemble the mass spectrum observed at smaller energies.

It becomes then instructive to evaluate what is the fraction of the showers observed at sea level that are produced by primaries of various mass, assuming that the mass spectrum is equal to that given in line 3 of Table 3 (p. 260) (relative number of particles of equal total energy U).

We now make the assumption that if all primaries interacted at the very top of the atmosphere $(x = 0)$ the number, N_0, of secondaries observed at sea level would depend only on U, irrespective of the atomic number A of the primaries. This is plausible, since the ratio U/N is a slowly varying function of U, in the lower atmosphere. It then may be shown that, since the heavy elements have larger cross sections, the contribution of the heavy elements to the showers observed at sea level is smaller than their relative abundance at the origin. In fact, let

$$F_A(N_0) = C_A N_0^{-\gamma}$$

be the integral number spectrum of the showers at sea level if all the primaries had interacted at the very top of the atmosphere. The number spectrum observed at sea level is then:

$$-\frac{dF_A(N)}{dN}\, dN = dN \int_{N_{\min}}^{N} -\frac{dF_A(N_0)}{dN_0}\, P_A(N_0, N)\, dN_0,$$

where $P_A(N_0, N)$ is the probability that a primary of atomic number A capable of giving a shower of N_0 particles *actually* produced at sea level a shower of N particles. $P_A(N_0, N)$ is determined by the probability that a primary interacts after a thickness of atmosphere, x:

$$p_A(x)\, dx = e^{-\frac{x}{\lambda_A}} \frac{dx}{\lambda_A}$$

17*

(λ_A = interaction mean free path in the atmosphere), and by the experimental result (see Sect. 11) that in the lower atmosphere the number of particles in a shower decreases exponentially when the thickness of atmosphere increases. So

$$N = N_0 e^{\mu x}$$

where, from curve c of Fig. 7, $\mu \approx \frac{1}{180}$ (g cm^{-2})$^{-1}$. It follows that:

$$P_A(N_0, N)\, dN = p_A(x)\, \frac{dx}{dN}\, dN = \frac{1}{\lambda_A \mu N_0} \left(\frac{N_0}{N}\right)^{1+\frac{1}{\lambda_A \mu}} dN,$$

and substituting in the first equation

$$-\frac{dF_A(N)}{dN}\, dN = \frac{C_A \gamma}{1 - \gamma \mu \lambda_A}\, N^{-\left(1+\frac{1}{\mu \lambda_A}\right)} \left[N^{\frac{1}{\mu \lambda_A} - \gamma} - N_{\min}^{\frac{1}{\mu \lambda_A} - \gamma}\right].$$

$N_{\min}$, the smallest value of N_0 that can contributes showers of size N at sea level, is smaller that $\frac{1}{10}\, N$; since $\frac{1}{\mu \lambda_A} - \gamma > 1$, the second term in parentheses can thus be neglected, and finally;

$$F_A(N) = \frac{C_A}{1 - \gamma \mu \lambda_A}\, N^{-\gamma} = C_A' N^{-\gamma}.$$

Table 3.

Elements	H	He	CNO	Ne–Si	A–Fe
$\overline{A}$	1	4	14	24	52
$C_A(\%)$. . . .	56	17	12	5	9
λ_A (g cm^{-2}) . .	70	45	31	21	13
$C_A'(\%)$. . . .	69	14	8	3	5

Experimentally, $\gamma = 1.5$ for the showers under consideration (see Sect. 17), and assuming for λ_A the values[1] given in Table 3, the values of C_A' turn out to be those given in the last row of the table. The proportion of proton showers is thus increased to about 70%[2].

II. The energy spectrum of the primary radiation.

28. The analysis of AS provides the only information thus far available from which it is possible to infer the energy spectrum of the primary cosmic radiation at energies higher than 10^{14} ev. The result of this analysis, as already mentioned several times, is an integral energy spectrum that can be expressed by the equation:

$$F(\geq U) = C\, U^{-\Gamma}, \tag{28.1}$$

---

[1] The actual values of λ_A are still controversial; when the heavy primary nucleus interacts for the first time with air, not all of its nucleons actually interact, and those left penetrate further in the atmosphere. In our calculation the values assumed for λ_A are those for the first interaction and represent a lower limit. This is compensated by the assumption that is independent of A.

[2] From the above considerations it follows also that the probability that a shower observed at sea level is initiated under a thickness x of atmosphere is proportional to

$$e^{-x\left(\frac{1}{\lambda_A} - \gamma \mu\right)} = e^{-x\left(\frac{1}{\lambda_A} - \frac{1}{120}\right)}.$$

For protons, the part of the exponential between parentheses is $\sim \frac{1}{170}$ (g cm^{-2})$^{-1}$, hence the proton showers have a good probability of being initiated at altitudes even smaller than 10 km above sea level. According to MIYAKI, arguments of this kind show that the behaviour of AS in the lower atmosphere is dictated by those primaries that escaped interaction in the highest starta. This notwithstanding, we have followed in this chapter a different point of view (the Zatzepin-Greisen point of view), because other experimental evidence emphasizes still more the role of the elasticity in ultrarelativistic interactions.

with C and Γ slowly varying functions of U[1]. However, one must be aware of the fact that possibly the primary differential spectrum is not as smooth as pretended by Eq. (28.1), but presents instead a fine structure; such a structure could hardly be revealed by any analysis of AS. The energies deduced for AS primaries are only average energies and the spectrum thence deduced must be considered as representing only the average behavior of the real spectrum.

The backbone of the energy spectrum deduced from AS measurements is the integral number spectrum evaluated in Sect. 17 and plotted in Fig. 10. The transformation from $F(N)$ to $F(U)$ can be accomplished, in principle, by determining for the showers of various sizes the total energy spent in the atmosphere and underground, as was done in Sect. 12 for showers of $N \approx 3.5 \times 10^5$ particles. However, the information on the behavior of the showers at high altitudes is so scanty, that this evaluation is possible, for the time being, just about only for showers of one size. Only one point of the energy spectrum can thus be determined; the rest must be extrapolated from it. For the showers with $N \approx 3.5 \times 10^5$ the radio η between the total energy of the primary particle and the total number of ionizing particles present at sea level is

$$\eta (N) = \frac{U}{N} = 1.2 \times 10^{10} \text{ ev}.$$

From the frequency of the showers with $N \geqq 3.5 \times 10^5$ (Fig. 10), it follows that the coordinates of this point in the integral energy spectrum are:

$$\left. \begin{array}{l} U = \eta N = 4.2 \times 10^{15} \text{ ev}, \\ F(U) = 2.5 \times 10^{-7} \text{ m}^{-2} \text{ sec}^{-1} \text{ ster}^{-1}. \end{array} \right\} \qquad (28.2)$$

The value adopted for $\eta (N)$ represents the sum of the energies given by the shower to the EM cascade [Eq. (12.6)], plus that given to the μ-meson and neutrino components [Eq. (24.1)], plus a term equivalent to $\sim 0.5 \times 10^9$ ev that represents the energy spent by the N-component in nuclear excitations[2]. The uncertainties in estimating the development of the showers in the highest atmosphere are the main responsible for our estimated error of $\pm 15\%$ in the evaluation of η. The behavior of $F(U)$ when U varies depends on the properties of the showers. If $\eta (N)$ is approximately constant, then $F(U)$ has the same behavior as $F(N)$. It is instead steeper or flatter than $F(N)$, if $\eta (N)$ decreases or increases when N increases.

If the showers were the product of a pure electromagnetic cascade, one would have:

$$\frac{d \log F(U)}{d \log U} = s \frac{d \log F(N)}{d \log N}$$

as shown in the note in Sect. 10. The slope of the energy spectrum would then be ~ 1.3 times that of the number spectrum, at energies from 10^{15} to 10^{16} ev. As the size of the showers increases, the region of maximum development would approach sea level and s would decrease.

However, the behavior of AS is strongly determined by that of the N-component in the core, and the final answer depends essentially on the model used

---

[1] It is tacitly assumed here that the primary Cosmic Rays arrive at the Earth with the same intensity from all directions, independent of time. This is certainly true, at AS energies, within better than some percent. The possible anisotropies of the primary radiation are discussed in Sect. 32.

[2] The N-component constitutes some percent of the particles in the average shower and has a mean free path equivalent to ~ 2 r.u. If in each interaction the nuclear excitation is of some 100 Mev $\approx 2\varepsilon_0$, the total energy spent will be some percent of that spent by the electrons.

for calculating the behavior of the ultra-relativistic interactions. If a statistical model is followed (Fermi, Landau), most of the primary energy is transferred to the EM component in the high atmosphere and the behavior of $\frac{d \log N}{d \log U}$ resembles that of s. For instance, the calculations of Oda[1] with a Fermi-like model give:

$$\frac{d \log N}{d \log U} = 1 + \frac{d \log \eta}{d \log U} = 1.52 - 0.045 \log N. \tag{28.3}$$

If a bremsstrahlung model is considered instead, a good fraction of the energy remains in the core, and η depends less on N.

In Fig. 17 we have given the two integral energy spectra obtained by keeping fixed the point whose coordinates are given in (28.2) and assuming for the larger energies once $\eta(N) =$ constant and once the dependence found by Oda [Eq. (28.3)]. These are two extreme cases and certainly the true spectrum must be between the two curves.

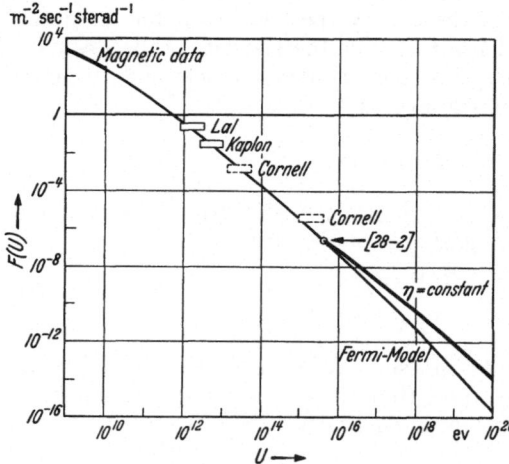

m^{-2}sec^{-1}sterad^{-1}

Fig. 17. The integral energy spectrum of the primary Cosmic Radiation. Up to total energy $U \approx 10^{16}$ ev the spectrum is directly determined (see discussion in the text). For larger energies the two curves, "$\eta =$ constant" and "Fermi model", represent two limiting possibilities. Eq. (28.4) corresponds to a spectrum that, for $U > 10^{16}$ ev, runs between the two limiting curves somewhat nearer to the lowest. The experimental evidence thus far is limited to energies up to $\sim 10^{19}$ ev.

At smaller energies the spectrum has been traced by using the geomagnetic data and the points obtained by Lal and by Kaplon et al.[2] (who observed the interactions of the primary radiation in photographic plates flown at balloon altitudes). The two points marked "Cornell" are the result of much less direct determinations of $F(U)$ obtained from the analysis of μ-meson showers underground together with the associated air showers.

Assuming that the real spectrum lies somewhere between the two curves, the following expression can considered as a good fit, for be $U > 10^{10}$ ev:

$$F(U) = 10^{13} \, U^{-(0.67 + 0.037 \log U)} \quad \text{sec}^{-1} \, \text{m}^{-2} \, \text{ster}^{-1} \, (U \text{ in ev}) \tag{28.4}$$

which gives the logarithmic derivative:

$$\Gamma = -\frac{d \log F(U)}{d \log U} = + (0.67 + 0.074 \log U).$$

The energy spectrum thus described extends up to energies of 10^{18} to 10^{19} ev. It is in fact clear that, no matter what model for the ultra-relativistic interactions will turn out to be the most successful one, the detection at sea level of showers with more than 10^9 particles indicates the existence of primary Cosmic Rays of at least the energies indicated above. Possibly with larger apparatus, showers with 10^{10} and more particles shall be observed, revealing the existence of primaries in the 10^{20} ev region, and beyond. However, when the frequencies become very small (extrapolating from Fig. 16, a primary of 10^{20} ev is expected over a surface

[1] M. Oda: Nuovo Cim. 5, 615 (1957).

[2] M. F. Kaplon, D. M. Ritson and E. P. Woodruff: Phys. Rev. 85, 933 (1952). — D. Lal: Proc. Ind. Acad. Sci. A 37, 93 (1953).

of 10 km² once every year) fluctuations in the multiplications of the secondaries in the atmosphere can lead to wrong conclusions, especially if the slope of the energy spectrum keeps increasing with increasing energy. Fluctuations can in fact even mask an abrupt cut-off in the primary energy spectrum, though this does not seem to be the case thus far.

Before closing this section, we want to raise the question of whether there is any *a priori* limit to the energy that a particle can have upon entering the Earth's atmosphere. Actually such a limit does exist if, as seems likely, the galactic spaces are filled with magnetic fields of the order of 10^{-6} gauss. The fractional energy lost as radiation in a field of intensity H (gauss) by a particle of total energy U, mass M, and charge Ze moving with velocities close to c across the field is

$$\frac{dU}{U\,dt} = \frac{2}{3}\frac{e^4 Z^4 H^2 U}{M^4 c^7} = 3.1 \times 10^{-19} \frac{Z^4}{A^3} H^2 \gamma \; \text{sec}^{-1},$$

where $\gamma = U/Mc^2$ is the Lorentz factor. If the most energetic cosmic rays come from outside the galaxy, to reach the earth they must cross a thickness of galactic space equivalent to a time of travel $\Delta t \approx 10^4$ light years $\approx 10^{12}$ sec, and the maximum energy they can have at the top of the Earth's atmosphere is that for which $\frac{dU}{U\,dt}\Delta t \approx 1$, i.e.

$$U_{\text{max}} = \gamma_{\text{max}} A \times 10^9 \approx (A/Z)^4 \times 3 \times 10^{27} \; \text{ev}.$$

This value becomes some orders of magnitude smaller if these particles instead of coming from outside, are confined within the galaxy for some million years, and/or if the galactic magnetic fields are substantially stronger than 10^{-6} gauss. However, it seems unlikely that U_{max} be smaller than 10^{23} to 10^{24} ev, a value still much larger than the maximum energies detected thus far[1].

III. Time variations correlated with atmospheric changes.

29. The atmosphere is the medium in which AS develop, and any variation in its properties causes variations in the properties of the shower. This interrelation is very complex, since the shower development does not depend only on the average values of the quantities characterizing the atmosphere (essentially pressure and temperature), but also on the distribution of the actual values as a function of the altitude above ground. For instance, the same pressure at sea level can be obtained with different temperature (hence, density) distributions at various heights; since the probability of π-μ decay depends on the air density, the pressure dependence of AS is correlated not only with the average temperature but also with the temperature distribution. The experimental disentangling of all the correlation coefficients is practically impossible, because the effects are small and the counting rates too low. Besides, information about the atmosphere is not known thus far with sufficient continuity.

What one generally measures are the correlations with the total air pressure (the barometric-coefficient α) and with the temperature of the air near the ground (the temperature coefficient β).

Let us assume that the shower number spectrum has the form (10.1). If the total number of particles present in the shower is measured, and the rate R of those showers that contain N or more particles is correlated with the barometric

---

[1] We do not consider here the analogous problem in the case of the propagation of Cosmic Rays in intergalactic spaces since too little is known about the magnetic fields in those regions.

pressure P and the air temperature T, the two coefficients can be expressed as follows[1]:

$$\alpha = \frac{\delta R}{R\,\delta P} = \gamma\,\frac{\partial \log N}{\partial P}, \qquad \beta = \frac{\delta R}{R\,\delta T} = \gamma\,\frac{\partial \log N}{\partial T}.$$

The term $\dfrac{\partial \log N}{\partial T} \approx 0$, since the variations of the ground temperature influences only the π-μ decay; the temperature coefficient β is thus small and not well measurable.

The barometric coefficient instead is quite large, and can give useful information. For showers with density in the range from 5 to 25 particles per square meter ($N \approx 10^5$) it has been found quite consistently $\alpha = 0.10$ (cm Hg)$^{-1}$, both at sea level and at mountain altitudes[2]. This confirms the constant absorption of the showers in the lower atmosphere and is in good agreement with the numerical value deducible from the altitude and zenith dependences. A more ambitious program has been started by CRANSHAW[3] who measures α for showers of different sizes. Though his results still have large statistical errors, it seems that α increases from ~ 0.10 (cm Hg)$^{-1}$ for showers with $N \approx 10^4$ particles to ~ 0.16 (cm Hg^{-1}) for $N \approx 10^7$ particles. From the known variations of γ with N (see Sect. 17) it is already possible to state that these results show that $\dfrac{\partial \log N}{\partial P}$ either remains constant or increases when N increases; *a priori* a decrease could be expected because the larger the shower the smaller is the altitude above sea level at which it reaches its maximum development. With better statistics these measurements will certainly provide useful information about the development of the most energetic showers in the lower atmosphere[4].

---

[1] Let the integral number spectrum of the showers be: $F(N) = K N^{-\gamma}$. The variation in the counting rate, R, of a device that detects the showers with total number of particles $\geqq N$ is related to the variation of N through the equation:

$$\frac{dR}{R} = \frac{dF}{F} = -\gamma\,\frac{dN}{N}.$$

The barometric coefficient is then:

$$\alpha = \frac{\delta R}{R\,\delta P} = \gamma\,\frac{\partial \log N}{\partial P},$$

and the temperature coefficient is

$$\beta = \frac{\delta R}{R\,\delta T} = \gamma\,\frac{\partial \log N}{\partial T}.$$

When the detector consists of a fixed set of counters, small in comparison to the dimensions of the showers, the counting rate varies also because the variation of the density ϱ of the air changes the horizontal dimensions of the shower, since $r_2 \propto 1/\varrho$. Consequently, if γ is the exponent of the density spectrum: (a) the density of particles, $\varDelta$, changes as ϱ^2, and $R \propto \varDelta^\gamma \propto \varrho^{2\gamma}$; (b) the area of the shower changes as ϱ^{-2} and $R \propto \varrho^{-2}$; (c) the distance between counters, measured in units of r_2, changes as ϱ^{+1}, and $R \propto D^{-d} \propto \varrho^{-d}$.

Summing up, $R \propto \varrho^{2(\gamma-1)-d} \propto (P/T)^{2(\gamma-1)-d}$, and the two coefficients become:

$$\alpha_1 = \alpha + \frac{2(\gamma-1)-d}{P}, \qquad \beta_1 = \beta - \frac{2(\gamma-1)-d}{P}.$$

When the geometrical factor varies because the altitude is changed, P and T are correlated; in the standard atmosphere, $T \propto P^{0.19}$ and $R \propto P^{0.81\,[2(\gamma-1)-d]}$. This equation has been used several times in the preceding chapters when discussing the measurements at various altitudes.

[2] A. DAUDIN and J. DAUDIN: J. Atm. Terr. Phys. **3**, 245 (1953). — A. CITRON: Z. Naturforsch. **7a**, 712 (1952). — C. CASTAGNOLI, A. GIGLI and S. SCUITI: Nuovo Cim. **7**, 307 (1950).

[3] T. E. CRANSHAW: Proc. Oxford Conference, April 1956.

[4] Preliminary results obtained at Cornell (K. GREISEN, private communication) do not confirm the increase of α with shower size; e.g. for $M \approx 10^7$ $\alpha = 0.09 \pm 0.03$.

30. Another variation that, if real, might be connected with atmospheric changes, is that described by the Harwell group[1] for showers of $N \approx 10^7$: a semidiurnal intensity variation with amplitude $\sim 10\%$ with maxima at 2 and 14 h. The effect is far from being definitely established and, e.g., has not been confirmed thus far by the measurements of other groups. For the time being there is no explanation for such large and regular variations, especially if they are limited, as it seems, to the largest showers only. We mention these results here not only for completeness but also because one must be aware of the possibility that points of view strongly at variance with those presently accepted could eventually be found more consistent with the experimental results. For instance, an interpretation of the Air Showers recently proposed by PETERS[2], assumes that the heaviest nuclei (Pb, U) are to be considered responsible for the most energetic showers: perhaps with a scheme of this kind it would be easier to explain an effect such as the diurnal variation described before, since the height at which these primaries interact would be more sensitive to the semidiurnal variations of the atmosphere, which involve a large motion only in the highest levels (the uppermost 1 g cm^{-2}). In this regard it must be admitted, however, that even if the primaries are all very heavy nuclei, the semidiurnal oscillation of the atmosphere could not account for such a large intensity variation as that reported by CRANSHAW, according to the current models of air shower development; some process of as yet unrecognized importance would have to be discovered.

Finally we want to emphasize the importance of reaching a deeper understanding of the variation induced by the atmospheres in connection with another problem.

As will be discussed in the next section, frequency measurements as a function of sidereal time are very important for the study of the origin of the Cosmic Rays and of the structure of the universe. To this effect, data are collected by grouping according to sidereal time the showers detected with the same apparatus over a period of years. Now, e.g., it is expected that the air temperature distribution up to the highest levels is different during the day than at night, and besides that both distributions depend on the season; the AS observed during winter nights (sidereal time around 6^h) will then be affected differently than those observed during summer days (again sidereal times around 6^h) and a spurious sidereal variation will appear if one does not apply the appropriate corrections. Unfortunately little is known about the magnitude of the correlation coefficients. It is the possibility of spurious effects of this kind that casts doubt on the meaning of variations of the order of 0.1% in the shower intensity, that otherwise are statistically significant.

IV. Time variations correlated with changes in the primary radiation.

31. In all the discussion carried on thus far about the properties of AS, it has been assumed that the intensity of the primary radiation responsible for them, i.e., of energy above 10^{13} ev, is constant in time. This assumption is quite legitimate, as in no part of the world the AS intensity, after correction for the changes due to the atmosphere, has ever been found to change in time by more than 1%, up to energies of 10^{17} ev. Since any detector is struck by showers coming mostly from directions close to the zenith, and since the Earth rotates, two conclusions can be drawn from the previous statement; one is that the primary

[1] T. E. CRANSHAW and W. GALBRAITH: Phil. Mag. **2**, 804 (1957).

[2] B. PETERS: Nuovo Cim. Suppl. **14**, 436 (1959).

Cosmic Ray flux in any direction is approximately constant in time, the second is that the Cosmic Ray flux is equal in all directions, i.e., is isotropic.

The constancy in time can be claimed only for the last few years and with a precision of only some percent, because systematic measurements of AS were started only in recent times, and the detectors are not yet stable enough to measure absolute rates with greater precision.

The isotropy instead can be studied more accurately, because even if the absolute shower rates show a drift due to slow changes in the sensitivity of the detectors, the same region of the sky is swept by each station once every day, and the drift is equally spread over all directions.

The isotropy of the primary Cosmic Rays can be checked against two different reference systems. One has for polar axis the Earth-sun direction (solar system), the other has for polar axis the direction at the Earth parallel to the galactic axis and for equatorial plane the plane of the galaxy (galactic system).

In the solar system the vertical at the site of observation has a polar angle Θ given by the expression:

$$\cos \Theta = \sin \lambda \sin \delta - \cos T \cos \lambda \cos \delta$$

where δ is the declination of the sun, λ the latitude of the site, and T the local solar time ($T = 0$ at midnight). The minimum angular distance from the sun, $\Theta_{min} = \lambda - \delta$, is reached every day at the local noon.

In the galactic system, the galactic latitude, b, and the longitude, l, of the vertical are given by the expressions:

$$\sin b = 0.470 \sin \lambda + 0.883 \cos \lambda \cos (\alpha - 12^{h} 40^{m})$$

$$\cos l = \cos \lambda \frac{\sin (\alpha - 12^{h} 40^{m})}{\cos b}$$

where α is the right ascension of the vertical, i.e., the local sidereal time.

The sidereal and the galactic grids of coordinates, together with some important galactic and extragalactic points, are plotted in Fig. 18. With its help it is easy to figure out what part of the galaxy is explored by the vertical of a station placed at any point on the Earth as a function of the sideral time.

In many of the experiments done thus far the AS were detected by means of coincidences among counters each of surface S several meters apart. In the region mapped by the counters the showers have densities $\Delta \approx 1/S$, but the average total number of particles in the shower, hence the average energy of the primary particle, depends not only on S but also on the counter separation. Counters only a few meters apart can be triggered by small showers as well as large ones, while counters at larger distances are triggered only by showers covering with density $\geq \Delta$ a bigger region; hence, on the average, by showers of greater energy. By varying altitude, S, and counter separation, showers with average primary energies between, e.g., 10^{14} and 10^{17} ev can be detected. The resolution in the primary energy achieved by the usual experimental set-ups is always quite poor (see Sect. 10) and the more so the smaller the multiplicity of the coincidences and the closer the counters; a poor energy resolution is a great drawback if the possibility exists that the effect sought is produced only by the more energetic particles[1].

About 70% of the showers recorded by the standard counter detectors come within zenith angles $\leq 30°$ and the angular resolutions of the system is ~ 0.7 steradians. In the experiments where the size and the direction of arrival of the

---

[1] For a more detailed discussion see K. SITTE: Nuovo Cim. **3**, 1145 (1956).

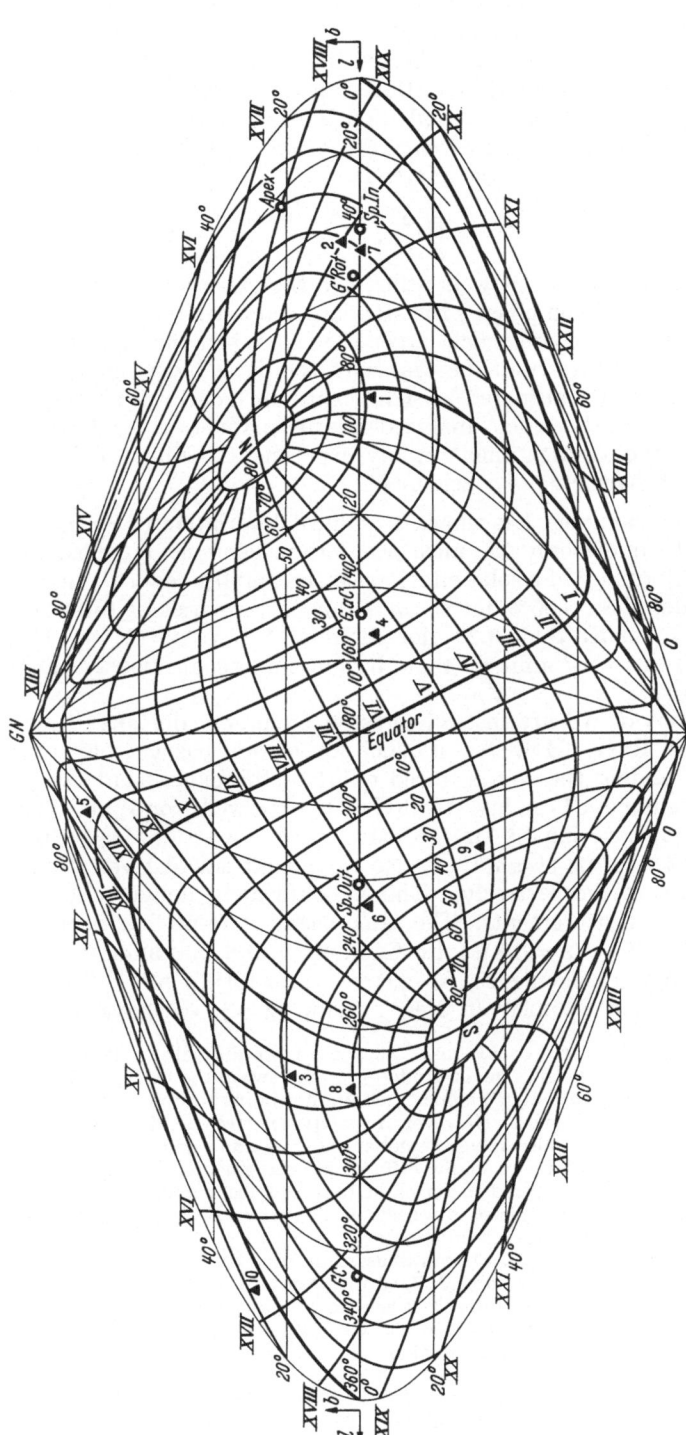

Fig. 18. The sky in Galactic (l, b) and in Sidereal (α, λ) coordinates. GM and GS corresponds to the directions in the sky of the North and South poles of the Galaxy; N and S to the terrestrial poles. The vertical of a point on the Earth at latitude λ sweeps in 24 sidereal hours the region of the sky described by a curve of constant λ. The points plotted on the map represent the following directions: $G.C.$; center of the Galaxy. $G.aC.$; anti-center of the Galaxy. $Sp.In.$; direction of the Spiral Arm containing the Sun that leads to the center of the Galaxy. $Sp.Out.$; Spiral Arm leading out of the Galaxy. $G.Rot.$; direction of the motion of the region around the Sun caused by the rotation of the Galaxy. The velocity is ~ 300 km sec^{-1}. Apex: direction toward which the Sun moves with respect to the neighboring stars. The velocity is ~ 20 km sec^{-1}. The points numbered 1 to 10 give the direction of the ten strongest radio sources, in order of decreasing intensity (J. L. Pawsey: Astrophys. Journ. $\mathbf{121}$, 1 (1955)): 1. Cassiopeia A (galactic nebulosity; remnants of supernova). 2. Cygnus A (two colliding galaxies). 3. Centaurus A (N.G.C. 5128, two colliding galaxies). 4. Taurus A (Crab Nebula, remnants of supernova A.D. 1054). 5. Virgo A (M 87, galaxy with jet of polarized light). 6. Puppis A (galactic nebulosity). 7. Cygnus X (galactic nebulosity). 8. Centaurus. 9. Pictor A (extended galactic source). 10. Hercules.

AS can be determined the showers can be grouped according to direction with angular resolution of some 0.01 steradians.

The events recorded are generally collected in hourly groups and after correction for barometric and temperature effects the data are analyzed in terms of sinusoidal variations of 24 hours period (first harmonic) and 12 hours period (second harmonic)[1]. The standard fractional errors with which these amplitudes are determined is given by the simple expression:

$$\sigma = \sqrt{\frac{2}{N}},$$

where N is the total number of showers utilized in the experiment. When there is no a priori knowledge of the phases, the standard error has the usual meaning only if the amplitudes found are much larger than σ. When the amplitudes are not very large, as has always been the case with the data under discussion, an amplitude $\leq 3\sigma$ cannot be considered as significant. In fact, even a completely random collection of events, when analyzed in this manner, gives amplitudes different from zero, the average among many experiments being equal to $\sqrt{2}\sigma$, independent of which harmonic is considered[2].

32. Thus far no experiment has shown any diurnal variation of AS intensity with solar time[3]. To quote only the most accurate measurements, those of the group at the École Normale[4], if a solar effect exists, its amplitude for the northern hemisphere is smaller than 0.06% for showers due to primaries of total energy $U \approx 3 \times 10^{14}$ ev, and smaller than 0.2% for showers of $U \approx 2 \times 10^{15}$ ev. This is expected, since it seems very unlikely that the sun emits continuously particles of such a high energy[5].

Sidereal variations of AS are instead foreseen by practically any theory of the origin of Cosmic Rays. However, no such anisotropy has yet been put in evidence not withstanding the concentrated effort of the experimenters.

The highest energies have been investigated by the Harwell group[6] at $\lambda = 52°$N with a grid of counters covering 0.5 km². About 2×10^3 showers of $U \approx 3 \times 10^{17}$ ev were recorded and 4×10^4 of $U \approx 10^{17}$ ev. No significant sidereal variation was found ($\sigma = 3\%$ and 0.7% respectively). This is confirmed, though with poorer statistics by the MIT measurements. The same negative result has been found at energies around 10^{16} ev ($\sigma = 0.5\%$) by CRANSHAW and GALBRAITH[7] with a smaller set of counters in coincidence. At energies from 10^{14} to 10^{15} ev the measurements of FARLEY and STOREY[8] in New Zealand ($\lambda = 37°$ S) gave in 1954 a

---

[1] In principle also point sources could exist. They are more difficult to detect because only the zenith of the stations at the same latitude $\pm 30°$ can sweep the useful region of the sky: besides, only apparatus with high angular resolution (timing systems) could resolve them, if their intensity is not very great. However, if galactic magnetic fields exist, no source can be a point source unless it is also monoenergetic, and this additional condition makes its existence quite improbable. A point source diffused by the galactic fields gives rise to anisotropies which can be best analyzed in terms of first and second harmonics.

[2] The situation is different if the phase is considered known a priori; the most probable amplitude for a random distribution is then zero.

[3] A possible semidiurnal variation has been discussed in Sect. 29.

[4] J. DAUDIN, P. AUGER, A. CACHON and A. DAUDIN: Nuovo Cim. **3**, 1017 (1956).

[5] The highest fields in the sun have intensities around 10^3 gauss and extend over regions of radius $< 10^7$ cm; the maximum energies that can be contained in these regions are $U \approx 300\,ZH\varrho = Z \times 3 \times 10^{14}$ ev and it seems improbable that particles can be accelerated in them to energies even approaching this limit.

[6] T. E. CRANSHAW and W. GALBRAITH: Phil. Mag. **2**, 804 (1957).

[7] T. E. CRANSHAW and W. GALBRAITH: Phil. Mag. **45**, 1109 (1954); see also J. W. CRAWSHAW and H. ELLIOT: Proc. Phys. Soc. Lond. A **49**, 102 (1956).

[8] F. J. M. FARLEY and J. R. STOREY: Proc. Phys. Soc. Lond. A **47**, 996 (1954).

first harmonic amplitude of $(1 \pm 0.26)\%$ with maximum at $\sim 20^h$ sidereal time. However the following year (Private communication from R. FARLEY) the effect dwindled to $(0.14 \pm 0.15)\%$, with different phase.

The most impressive group of measurements is that performed by the Paris group[1] at the observatory of Pic du Midi ($\lambda = 43° $ N). After five years of practically continuous measurements, 6×10^7 showers have been detected with twofold coincidences (0.236 m^2 counters at 5 m spacing; $U \approx 3 \times 10^{14}$ ev) and 7×10^6 with threefold coincidences with 80 m spacing ($U \approx 2 \times 10^{15}$ ev). After correction for the *average* barometric and temperature effects, a first harmonic amplitude of 0.15 and 0.20%, both with maxima at $\sim 20^h$ sidereal time consistently remained in the separate results of each of the five years. The standard errors for all the data taken together are $\sigma = 0.02$ and 0.05% respectively for the two shower sizes considered.

As discussed in Sect. 30, it must be remembered that the effects of periodic changes in the atmosphere are not known well enough to rule out the possibility that a residual variation correlated with time with an amplitude of 0.1% is caused by them. DAUDIN and his coworkers have taken much care in trying to disprove this interpretation; a subdivision of the data in two semiannual groups should eliminate part of such a "pseudo-sidereal" effect; even with this subdivision the sidereal variation remains, though somewhat less neat.

As a conclusion, if a sidereal effect exists, the amplitude of its first harmonic, as observed in the northern hemisphere, cannot be larger than 0.1%, at 10^{14} to 10^{15} ev[2].

Such being the situation, the following discussion of the possible interpretations of this maximum in terms of a real asymmetry in the distribution of 10^{14} to 10^{15} ev primary Cosmic Rays must be considered only as academic.

Until a few years ago the prevailing opinion was that all Cosmic Rays are accelerated and confined inside the galaxy by the galactic magnetic fields, and anisotropies were thought of as caused essentially by the asymmetries of these fields[3]. However, recently evidence has become available that a purely galactic origin of Cosmic Rays is untenable, and that likely, at least the most energetic particles come from outside the galaxy. The arguments are based on the facts that one has observed Cosmic Rays with energies up to 10^{18} ev, and that at least some of them are protons. These particles cannot be contained, and *a fortiori* cannot be accelerated, in the galaxy, if the average galactic fields are of the order of 10^{-5} to 10^{-6} gauss, as required by the astrophysical evidence[4]. Besides this, both in the galaxy and outside it, objects have been observed that emit intense continuous radiation likely due to magnetic bremsstrahlung of high energy electrons. These observations suggest that in the Universe there exist regions where high energy particles can be accelerated much more efficiently than in the average galactic space. It seems then plausible to admit that Cosmic Rays created in these regions reach our galaxy[5].

---

[1] J. DAUDIN, P. AUGER, A. CACHON and A. DAUDIN: Nuovo Cim. **3**, 1017 (1956) and private communications.

[2] The constancy in time of the showers of $\sim 10^{15}$ ev could be utilized, in future researches, as a standard for the measurement of the intensity of showers of larger size. The atmospheric variations can be expected to affect in the same manner all the showers, at least in first approximation; the real time variations would then be given directly by the variations of the relative intensities, with no meteorological corrections.

[3] See, e.g., L. DAVIS: Phys. Rev. **96**, 743 (1954).

[4] See, e.g., G. COCCONI: Nuovo Cim. **3**, 1433 (1956).

[5] See in this book the chapter on the Origin of Cosmic Rays by P. MORRISON.

These new points of view have increased the number of the possible interpretations of any anisotropy observed in the cosmic radiation. Some of the most obvious ones will be summarized below.

One of the first consequences of the existence of extragalactic Cosmic Rays is the Compton-Getting[1] effect. Since we and the stars in our vicinity rotate around the center of the galaxy with velocity $\beta c \approx 300$ km sec^{-1}, the Cosmic Rays arriving from the direction of our motion must appear more abundant than those trailing. The point of the sky toward which we rotate ($G\,rot$ in Fig. 18) has celestial coordinates $\alpha = 21^\mathrm{h}40^\mathrm{m}$ and $\lambda = 47^\circ$ N. If a detector measures AS due to primaries of energy $\geq U$, and Θ is the angle that the zenith makes with the direction $G\,rot$, the AS rate is, disregarding terms of the order of β^2,

$$R(\Theta) = R\left[1 + (3 + \Gamma)\beta \cos\Theta\right]$$
$$\approx R\left(1 + 0.005 \cos\Theta\right)$$

where $\Gamma \approx 1.8$ is the exponent of the power energy spectrum of Cosmic Rays, and R is the rate if there were no asymmetries[2].

It has been remarked by BARUT[3] that the effect found by DAUDIN et al. is in phase with the Compton-Getting effect, and that, accepted that the effect is real, this could be its explanation.

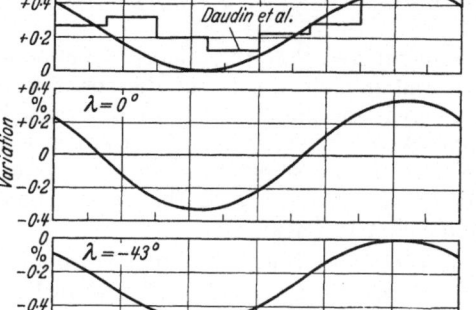

Fig. 19. The Compton-Getting effect as calculated for three latitudes on the Earth. At $\lambda = 43^\circ$ are also plotted the results of the Paris group, for $U \approx 2 \times 10^{15}$ ev (zero of ordinates arbitrary).

In Fig. 19 we have plotted the expected Compton-Getting effect at three latitudes, together with the results of DAUDIN et al. The location of $G\,rot$ is such that the Compton-Getting effect has practically the same amplitude and the same phase, when observed from any Earth latitude between 45° N and 45° S. However, while in the northern hemisphere there is an overall increase of the Cosmic Ray flux with respect to that at infinity, in the southern hemisphere there is a decrease. This asymmetric latitude effect is practically impossible to measure with the present techniques. The constant phase makes it possible to eliminate experimentally the unknown effects of the atmosphere by repeating the Daudin measurements in the southern hemisphere. Even this would be not a small task.

To admit that the Compton-Getting effect is felt with the right phase at energies around 10^{14} to 10^{15} ev entails the far reaching consequence that the Cosmic Ray particles of these energies are not strongly disturbed by the galactic fields; this requires that the storage factor of the galaxy for these particles is not much larger than unity. Such a conclusion is in disagreement with what is believed to be the situation in the galaxy, where an overall magnetic field of 10^{-5} to 10^{-6} gauss is credited to keep particles of 10^{14} to 10^{15} ev inside the galaxy for times much longer than 10^5 years, i.e., the time the light takes to cross the

[1] A. H. COMPTON and I. A. GETTING: Phys. Rev. 47, 817 (1935).

[2] The effect can be considered as the sum of three terms: (1) the change in energy, modifying the primary spectrum, is responsible for the factor $(1 + \Gamma\beta \cos\Theta)$; (2) the change in angle of arrival gives a factor $(1 + 2\beta \cos\Theta)$ in the number per unit solid angle; (3) the relative movement makes us sweep more space and gives the term $(1 + \beta \cos\Theta)$.

[3] A. O. BARUT: Nuovo Cim. 4, 661 (1956).

galaxy. On the other hand there is not yet any direct evidence that the storage factor of the galaxy for these particles is really much greater than unity.

Other explanations can be offered for the Daudin effect.

1. Our spiral arm has a direction not very different from that of $G\,rot$ (see the point, $Sp.\ In.$ in Fig. 18), and Cosmic Rays could come preferentially along it. If this is the case, the effect would be completely similar to the Compton-Getting effect. But, while the Compton-Getting effect is expected to be present at all energies, actually to increase to $\sim 0.25\%$ amplitude at the extreme energies, the effect considered now is bound to disappear at energies of 10^{17} to 10^{18} ev.

2. The plane of the galaxy is crossing the $43°$ N zenith at $\sim 20^{h}$ sidereal time, and Cosmic Rays could move preferentially in that plane. As a matter of fact, DAUDIN et al. pointed out that the intensity of the galactic radio noise that presents a maximum along the galactic equator varies in good correlation with the intensity of the AS. In this case, however, the phase of the maximum should not be a constant when the latitude is changed, and measurements at different latitudes should show phase variations. Also the energy dependence of the effect would be different from that expected for the Compton-Getting effect.

3. In locations close to $G\,rot$ are situated the Cygnus A, and the Cygnus X, two of the strongest sources of radio waves. The first is due to two colliding galaxies, 2.7×10^{8} light years away from us, and quite probably a strong source of cosmic rays; the second is a galactic object, probably the remnant of a super-nova, and likely a source of Cosmic Rays too.

The situation is very complex and only by increasing the number of stations around the world, the counting rates of each of them, and the size of the showers recorded, it could be possible to corner efficiently the correct solution.

The Hard Component of μ-Mesons in the Atmosphere.

By

G. N. FOWLER and A. W. WOLFENDALE.

With 26 Figures.

A. Introduction.

1. A general picture of the cosmic radiation. Since the discovery of the existence of the cosmic radiation at the turn of the century a large mass of data has been accumulated on its behaviour. Initially measurements were made of the intensity of the cosmic rays at different points on the earth's surface and as techniques of various kinds have improved measurements have been extended to include the momentum and mass spectrum both at sea level and at various altitudes above sea level. Although the origin of the cosmic rays remains obscure a fairly satisfactory picture of their development through the atmosphere emerges.

The cosmic radiation at great distances from the earth consists mainly of protons, with about 20% of α-particles and heavier nuclei. This primary radiation contains particles having a wide range of energies so that even at low latitudes a large fraction is able to overcome the earth's magnetic field and enter the earth's atmosphere. Here nuclear interactions occur with the disruption of nitrogen and oxygen nuclei and the emission of the secondary component of cosmic rays; high energy nucleons, unstable mesons, hyperons etc. The mesons are both charged and uncharged and consist predominantly of π- with a small percentage of K-mesons. Of the π-mesons the charged particles decay into a μ-meson and a neutrino, with a lifetime of 2.6×10^{-8} sec, and the neutral particles decay into two photons with a lifetime less than 10^{-14} sec. In several of the K-meson decays too μ-mesons are produced. Since the μ-meson has a comparatively long lifetime $(2.2 \times 10^{-6} \text{ sec})$, and a weak interaction with matter, most of those arising from the decay of the heavier mesons near the top of the atmosphere survive down to sea level. It is these μ-mesons that constitute the vast majority of the "hard" component at sea level.

2. The importance of the hard component. Measurements on μ-mesons should be useful from many points of view. In principle some, at any rate, of the properties of the basic nucleon-nucleus interaction can be inferred from the sea level μ-meson spectrum, its variation with altitude and the ratio of the fluxes of positive and negative μ-mesons. Similarly the effects of the earth's magnetic field on both the primary and secondary cosmic rays can be studied by investigating the latitude dependence of the μ-meson flux and the intensities of positive and negative particles in inclined directions, respectively.

The purpose of this article is to discuss the main experimental results on the development of the μ-meson component through the atmosphere and their relevance to the topics mentioned above.

B. Properties of the μ-meson component in the vertical direction.

I. Measurement of the momentum spectrum of μ-mesons at sea level.

Since it is nearly independent of time and position on the earth's surface, and is comparatively easily measured, the ground level spectrum (or sea level spectrum as it is often termed) of cosmic rays has become one of the "fundamental constants" of cosmic ray physics. In view of its importance a detailed discussion of its measurement will be given and an estimate made of the best momentum spectrum available at the present time.

a) A summary of the techniques of measurement.

Two main techniques are practicable for measuring the momenta of ionizing cosmic rays; these utilise

(a) measurement of the curvature of the particle's trajectory in a magnetic field and

(b) measurement of the scattering of the particle in traversing a medium of known constitution.

3. The magnetic deflection method. In method (a) the radius of curvature is found either

(i) directly from particle detectors placed in a magnetic field or

(ii) by defining the initial and final directions of the particle by detectors above and below the field region. Most of the early work was done using method (i) with cloud chambers operated in magnetic fields. Later work has used method (ii) with Geiger counters or cloud chambers as detectors. Other things being equal, method (ii) is to be preferred since the volume occupied by the magnetic field is used much more efficiently: HYAMS et al.[1] estimate that only 5% of the possible field-volume is used in a magnet-cloud chamber assembly whereas this figure is nearer 50% for an apparatus using method (ii). Another disadvantage of the cloud chamber as detector in (i) is its inability to measure high momenta, due to the masking of the small track curvature by gas turbulence. The maximum detectable momentum (the momentum corresponding to the mean spurious curvature caused wholly by turbulence) is rarely above 10 GeV/c and the fact that the velocities of the turbulent streams vary both with time and position in the chamber makes measurements of momenta approaching the maximum detectable momentum (m.d.m.) highly suspect (see WILSON[2] for a critical account of cloud chamber technique). On the other hand method (i) has the advantage that the measured spectrum refers to single particles, which may or may not have other nearby particles associated with them. In method (ii), however, accompanied particles are invariably rejected because of the difficulty of ensuring that all the particles traversed the magnetic field and the difficulty of pairing off the tracks. If the probability of a particle being accompanied varies with the momentum of the particle a biassed spectrum must result, this bias must exist at very high energies, but as yet there is little indication of the magnitude of the effect. (The lack of any great bias at low energies, $\lesssim 10$ GeV/c has been demonstrated; this will be discussed in the next section.)

---

[1] B.D. HYAMS, M.G. MYLROI, B.G. OWEN and J.G. WILSON: Proc. Phys. Soc. Lond. A **63**, 1053 (1950).

[2] J.G. WILSON: The Principles of Cloud Chamber Technique. Cambridge: Cambridge University Press 1951.

A recent method which has the advantages of (i) but none of its disadvantages is the emulsion spectrograph technique of APOSTOLAKIS and MACPHERSON[1]. Here a number of layers of nuclear emulsion act as particle detectors and most of the magnetic field volume can be used. No appreciable spectrum bias appears and the m.d.m. can be very high ($\sim$1000 GeV/c). Up to now the application of the method has been limited, however, due to the time taken in searching the emulsions for the particle tracks and in following the tracks through successive emulsions.

4. The scattering method. Method (b), using the scattering technique, is inferior in most respects to methods (a) (i) and (a) (ii) and usually consists of scattering measurements on tracks in nuclear emulsions exposed under conditions where the use of large pieces of equipment, in the form of magnets etc., is impracticable (e.g. far underground).

b) Measurements of the momentum spectrum.

5. Early measurements. ROSSI[2] has reviewed the measurements prior to 1948, most of which were taken with the magnet-cloud chamber technique, and has combined the results to give a spectrum up to about 20 GeV/c. The spectrum is shown in Fig. 1.

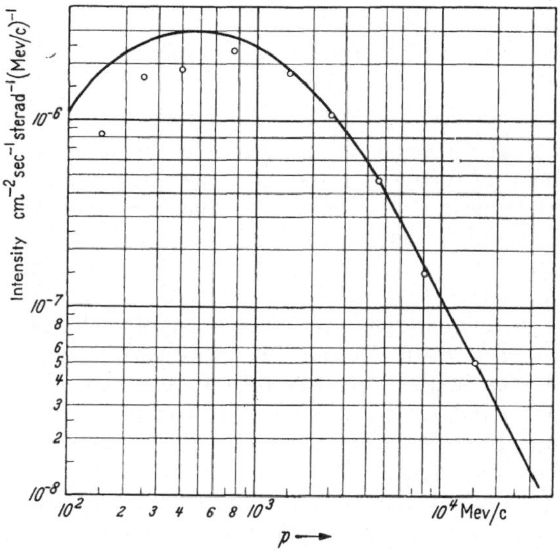

6. Measurements with cosmic-ray spectrographs. GLASER et al.[3] extended the measurements to higher momenta by the use of two cloud chambers separated by an electro-magnet having $\int B\,dl = (3.9 \pm 0.2) \times 10^5$ gauss-cm and $\int B\,dV = 1.3 \times 10^7$ gauss-cm^3. The m.d.m. was estimated to be 80 GeV/c. 1547 particles were studied and it was found that their spectrum could be written in the form kp^{-n} with $n = 1.8 \pm 0.2$ from 2.5 to 10 GeV/c, and $n = 2.1 \pm 0.6$ from 10 to 70 GeV/c. This spectrum is very nearly of the same form as that of ROSSI above about 2 GeV/c, but below it large differences occur; these arise from the difficulty in applying accurate corrections for the effect of the large magnetic deflections of the low-momentum particles.

Fig. 1. The momentum spectrum at sea level derived by ROSSI from measurements prior to 1948.
The circles represent experimental determinations of J. G. WILSON whose results agree well with those of previous authors.
Normalization was carried out to bring the number of particles with momentum above 2 GeV/c into agreement with the absolute value determined from absorption measurements.

CARO et al.[4] were the first workers to break away from the tradition of using cloud chambers as detectors and their spectrum was regarded as a standard for a number of years. In view of this, their arrangement will be described in some detail. The arrangement is shown in Fig. 2. The detecting layers consisted of

[1] A. J. APOSTOLAKIS and I. MACPHERSON: Proc. Phys. Soc. Lond. A **70**, 146, 154 (1957).
[2] B. ROSSI: Rev. Mod. Phys. **20**, 537 (1948).
[3] D. A. GLASER, B. HAMERMESH and G. SAFONOV: Phys. Rev. **80**, 625 (1950).
[4] D. E. CARO, J. K. PARRY and H. RATHGEBER: Austral. J. Sci. Res. A **4**, 16 (1950).

Geiger counters, the measuring trays 1, 2 and 3 containing two layers of overlapping counters to reduce the uncertainties in the location of the particle trajectories. The electromagnet gave

$$\int B \, dl = 7.4 \times 10^5 \text{ gauss-cm} \quad \text{and} \quad \int B \, dV = 3.3 \times 10^7 \text{ gauss-cm}^3.$$

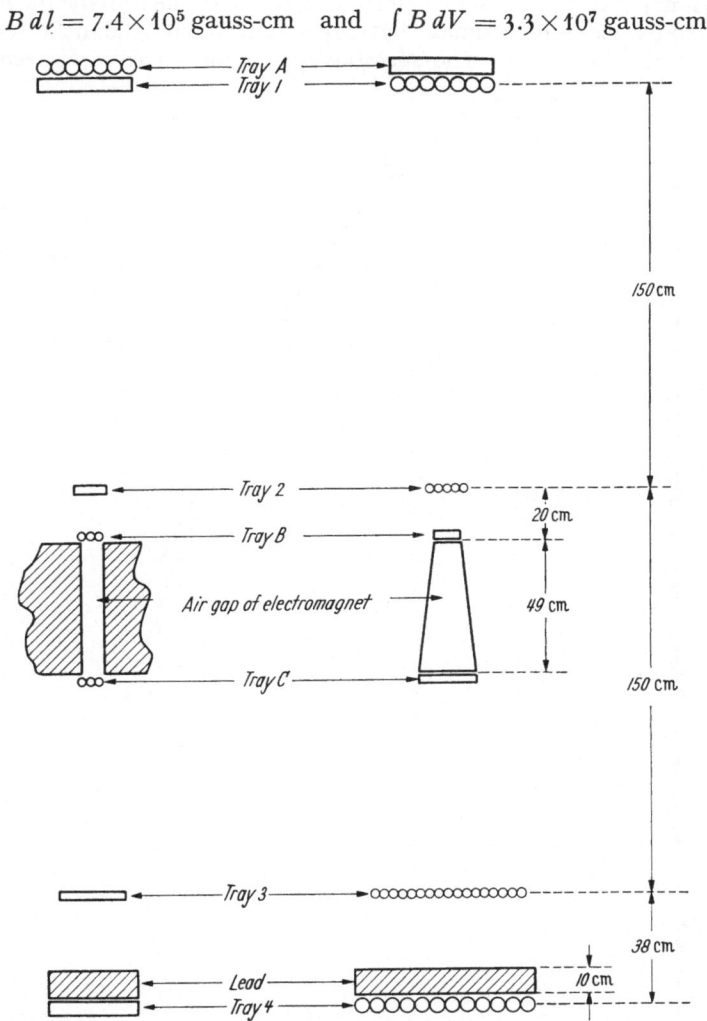

Fig. 2. The experimental arrangement of CARO *et al.* see footnote 4, p. 274. (The Melbourne cosm ic ray spectrograph)

The counters discharged by the passage of a single particle were recorded and the data analysed using a punched card system. The m.d.m. of the instrument was about 100 GeV/c: the maximum measurable momentum corresponding to a deflection of the smallest unit of measurement is quoted by the authors as 53 GeV/c at the maximum field, 13 500 gauss; 6056 particles were studied under these conditions and the resultant spectrum is shown in Fig. 3. Also shown is the spectrum deduced by ROSSI; the disagreement at momenta above about 10 GeV/c is pronounced. It is quite certain that the lower slope of the Rossi spectrum is a consequence of underestimation of the errors due to turbulence in the cloud chamber experiments contributing to ROSSI's data. The increased slope at high momenta found by CARO *et al.* is to be expected since the exponent, n, of the slope at the highest energy must, in fact, be greater than 2 for the total energy arriving in the form of cosmic ray μ-mesons to be finite.

7. Measurements by the Manchester group. These measurements have been repeated and extended by a similar but larger instrument at Manchester (Hyams *et al.*, Owen and Wilson[1]). The spectrograph employed counter recording for

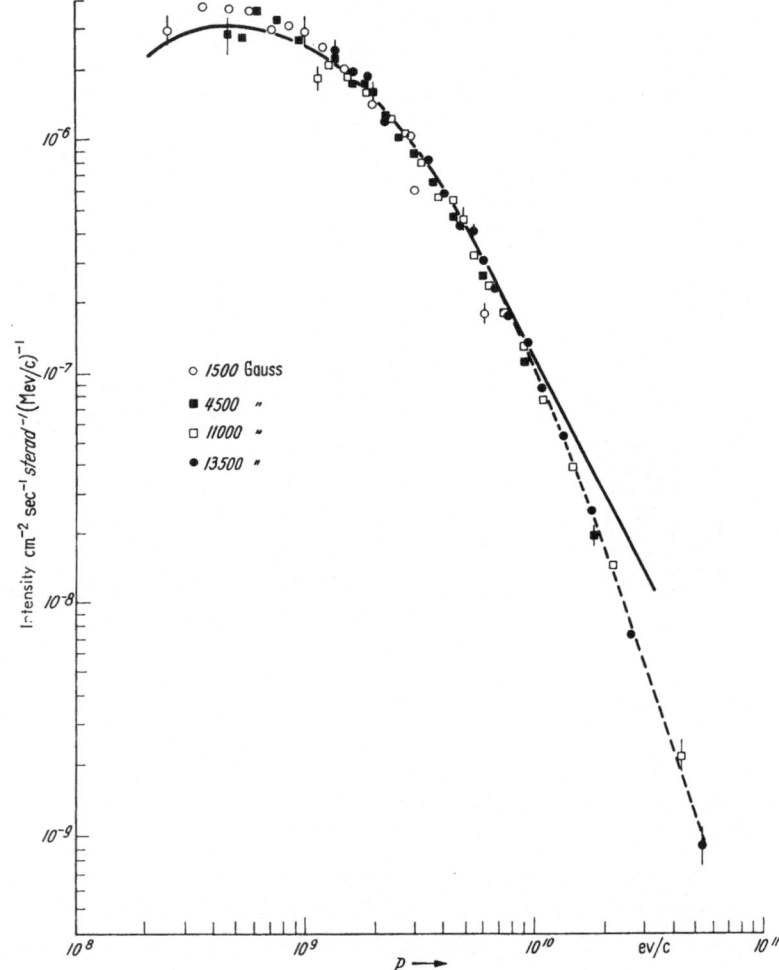

Fig. 3. A comparison of the sea level momentum spectra of Caro *et al.*, see footnote 4, p. 274, (broken line) and Rossi, see footnote 2, p. 274, (full line).

the low momentum measurements, $0.5 < p < 20$ GeV/c and consisted of two electromagnets placed vertically one above the other with three layers of counters symmetrically placed along the trajectories of the particles. The arrangement is shown in Fig. 4. Most of the accepted trajectories were within 10° of the vertical so that the spectrum is essentially that of the vertical flux at Manchester. With this instrument a very careful analysis of the deflections of 60 000 particles was carried out. The method of analysis was to compare the observed deflection spectrum with a comparison spectrum produced by modifying an assumed incident

[1] B.G. Owen and J.G. Wilson: Proc. Phys. Soc. Lond. A **68**, 409 (1955).

spectrum for the effect of instrumental bias (arising from the large magnetic deflection of slow particles and hence reduced aperture for their acceptance). A first order approximation was made to the process of formation of the sea level μ-meson component in which it was assumed effectively that production takes place at a unique altitude and that the intermediate π-meson stage is unimportant at the energies under consideration.

The comparison spectrum is then

$$S(p) = P(t, p)\, S_t(p + p_t)$$

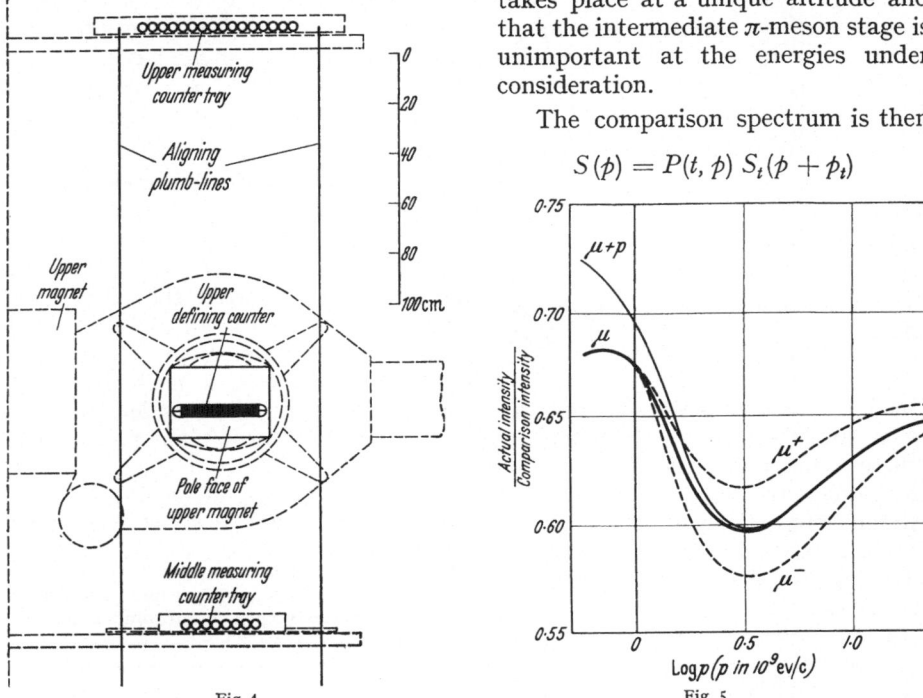

Fig. 4

Fig. 5

Fig. 4. The experimental arrangement of HYAMS et al., see footnote 1, p. 273. (The Manchester cosmic-ray spectrograph). (The upper half of the instrument, which is symmetrial about the middle counter tray, is show).

Fig. 5. The momentum spectra of OWEN and WILSON, see footnote 1, p. 276, as a function of the comparison spectrum derived from a production spectrum $(2.05 + p)^{-2.85}$, with p in GeV/c. Spectra are shown for the total hard component, $\mu + p$, for the total meson component μ, and for negative and positive μ-mesons separately.

where $S(p)$ is the sea level spectrum and is given in terms of the spectrum at formation S_t, at atmospheric depth t, and the survival probability $P(t, p)$ of a μ-meson from this depth reaching sea level with momentum p. The survival probability is

$$P(t, p) = \left\{ \frac{t_0}{t} \left(1 + \frac{p_0}{p} \right) - \frac{p_0}{p} \right\}^{-\alpha}$$

with

$$\alpha = \frac{h}{c\,\tau_\mu} \left(\frac{m_\mu c}{p + p_0} \right)$$

where t_0 is the total mass thickness of the atmosphere, p_0 is the ionization loss through the whole atmosphere, h the scale height of the atmosphere and τ_μ, m_μ the lifetime and mass of the μ-meson.

The production spectrum

$$S_t(p + p_t) \approx (p + p_t)^{-\gamma}$$

was used together with the following values of the constants:

$t/t_0 = 0.1$, $p_0 = 2.28$ GeV/c, $h = 7.02$ km, $\tau_\mu = 2.15 \times 10^{-6}$ sec, $m_\mu = 215\, m_e$

and

$$p_t = (1 - t/t_0)\, p_0 = 2.05 \text{ GeV}/c.$$

With $\gamma = 2.85$ the ratio of actual intensity to the comparison intensity is shown in Fig. 5 for both the total hard component of μ-mesons and protons, and positive and negative μ-mesons separately. The contribution of protons was taken from an earlier experiment of Mylroi and Wilson[1] using the same apparatus. The variation of the ratio with momentum is seen to be comparatively small, having a maximum variation of $\pm 7\%$. This indicates that the simple propagation model referred to above is quite a good approximation.

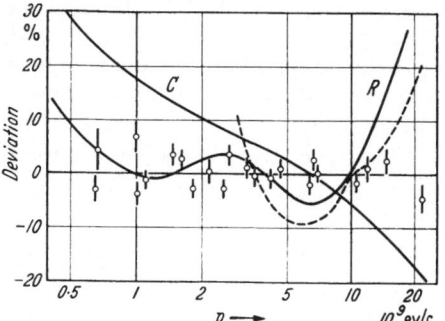

Fig. 6. Deviations of other measurements from those of Owen and Wilson[2]. C denotes Caro et al., see footnote 4, p. 274, R Rossi, see footnote 2, p. 274, broken line Glaser et al.

8. A comparison of the measurements by various workers. The various spectra are compared in Fig. 6. It will be seen that over the range $0.7\ \mathrm{GeV}/c < p < 11\ \mathrm{GeV}/c$ the agreement between the Rossi spectrum and that of Owen and Wilson is good, but at lower momenta the spectrum of the latter authors is much flatter. At the lower momenta the spectrum of Caro et al. shows a quite considerable excess over both the Rossi spectrum and that of Owen and Wilson. It is possible that these differences arise from an incomplete treatment, by some of the authors, of the bias introduced by the considerable magnetic deflection of the low momentum particles.

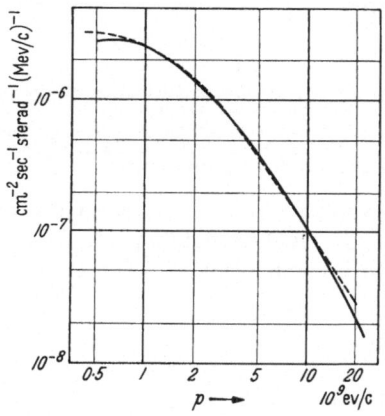

Fig. 7. The differential momentum spectra at sea level of Owen and Wilson[2] (full line) and Rossi, see footnote 2, p. 274, (dashed line). The curves have been fitted at $p = 10^9$ eV/c. The spectrum of Owen and Wilson is regarded as the standard in the momentum range 0.5 to 20 GeV/c. The statistical uncertainties on the points constituting the spectrum are about 2% but it is likely that the absolute intensities should be about 5% higher than the values indicated.

Table 1. *The vertical intensity of μ-mesons at sea level as a function of momentum, in the range 0.5 to 20 GeV/c.*

The data are taken from the experiment of Owen and Wilson[2]. As pointed out by Puppi [2] and others it is likely that the absolute intensities should be increased by 5%.

(1)	(2)	(3)	(4)	(5)
0.5	1.577	0.676	1.065	2.72
0.6	1.570	0.680	1.068	2.72
0.8	1.518	0.682	1.035	2.64
1.0	1.417	0.677	0.959	2.45 *
1.2	1.313	0.663	0.869	2.22
1.4	1.201	0.645	0.775	1.98
1.6	1.096	0.631	0.692	1.77
2.0	0.908	0.611	0.555	1.42
2.5	0.722	0.599	0.432	1.10
3	0.577	0.596	0.344	0.878
4	0.384	0.598	0.230	0.587
5	0.266	0.605	0.161	0.410
7	0.143	0.618	0.0883	0.225
10	0.0677	0.630	0.0426	0.109
15	0.0267	0.642	0.0171	0.0436
20	0.0131	0.647	0.0085	0.0220

* Normalization point.

(1) p (GeV/c); (2) $S(p)_{\mathrm{comp}}$; (3) relative correction factor; (4), (5) $S(p)_{\mathrm{incid}}$ (cm^{-2} sec^{-1} sterad^{-1} (MeV/c)$^{-1}$); (4) unnormalized, (5) normalized ($\times 10^6$).

The reason for the apparent excess of particles of high momentum in Rossi's spectrum has already been remarked upon—viz. an underestimate of the errors

[1] M.G. Mylroi and J.G. Wilson: Proc. Phys. Soc. Lond. A **64**, 404 (1951).
[2] B.G. Owen and J.G. Wilson: Proc. Phys. Soc. Lond. A **68**, 409 (1955).

due to turbulence in the cloud chambers used in the measurements. The difference between the high momentum spectra of OWEN and WILSON and CARO *et al.* probably arises from differences in treatment of the geometrical corrections applied to measurements, in the counter spectrographs, approaching the limit set by the finite size of the detecting Geiger counters.

In view of the higher statistical precision of the experiment of OWEN and WILSON, their spectrum is taken as the standard for the momentum range 0.5 to 20 GeV/c. This has been plotted in Fig. 7 where the earlier standard of ROSSI is also given. The data are also given in Table 1. Although the statistical accuracy of the experimental points is high there is some doubt as to the corresponding absolute intensities. It has been common practice to normalize to the Rossi spectrum at some low momentum point and this has been done in Fig. 7 at 1 GeV/c. However, some doubts have been cast on the accuracy of the absolute intensity, by YORK[1] and others. PUPPI [2] in his review of this subject concludes that it is likely that the Rossi spectrum at the normalization point (1 GeV/c) is too low by 5% so that the absolute values for the intensities shown in Fig. 7 and Table 1 should be increased by 5%.

9. Measurements above 100 GeV/c. Measurements of the spectrum up to momenta of several hundred GeV/c have been made by the Manchester group with the spectrograph modified by the use of flat cloud chamber as detectors, RODGERS[2], HOLMES *et al.*[3]. A cloud chamber was placed over each of the trays of Geiger counters (Fig. 4) and measurements were made of the points of intersection of the tracks with grids ruled on the top windows of the chambers. Only those combinations of counters discharged in the instrument corresponding to the passage of fast particles were selected to "trigger" the cloud chambers. The lower cut-off in momentum was chosen to give a reasonable overlap with the

Table 2. *The vertical intensity of μ-mesons at sea level as a function of momentum for the region above* 5GeV/c.

The data are from the experiment of RODGERS[2].

The spectrum is normalized to that of OWEN and WILSON[4] over the momentum range 3 to 20 GeV/c and any error in the absolute intensity in the spectrum of OWEN and WILSON is therefore carried forward (see Table 1). The errors quoted are the standard deviations (S.D.) arising from statistical fluctuations in the numbers of particles only.

Incident spectrum (normalised) cm^{-2} sec^{-1} sterad^{-1} (MeV/c)$^{-1}$	S.D. %	Mean momentum in GeV/c	Incident spectrum (normalised) cm^{-2} sec^{-1} sterad^{-1} (MeV/c)$^{-1}$	S.D. %	Mean momentum in GeV/c
4.8×10^{-13}	48	1160	3.68×10^{-8}	5	16
2.8×10^{-11}	20	271	5.90×10^{-8}	5	13
1.42×10^{-10}	18	134	7.88×10^{-8}	4	11
4.38×10^{-10}	16	89	1.35×10^{-7}	5	9.6
8.65×10^{-10}	15	67	1.83×10^{-7}	5	8.2
2.02×10^{-9}	10	49	2.24×10^{-7}	5	7.1
5.75×10^{-9}	7	36	2.44×10^{-7}	6	6.2
1.04×10^{-8}	7	28	2.98×10^{-7}	9	5.6
1.67×10^{-8}	7	23	3.98×10^{-7}	9	5.1
2.33×10^{-8}	5	19			

[1] C.M. YORK: Phys. Rev. **85**, 998 (1952).

[2] A.L. RODGERS (University of Manchester): P.D. thesis, 1957.

[3] J.E.R. HOLMES, B.G. OWEN, A.L. RODGERS and J.G. WILSON: Report on Conference on recent developments in cloud chamber and associated techniques (London Univ. Coll.) p. 177, 1956.

[4] See footnote 2, p. 278.

previous measurements of Owen and Wilson; particles below 3 GeV/c were not accepted, those between 3 GeV/c and 10 GeV/c were accepted with increasing efficiency and above 10 GeV/c the efficiency of acceptance was virtually 100%. The maximum detectable momentum of the instrument when operated at the maximum field can be written as $p_m = \dfrac{240}{\sigma_c}$ GeV/c where σ_c is the overall error in location of a track in millimeters in each of the flat cloud chambers. An accurate determination of σ_c was not possible at the time the experiments were performed but subsequent experiments by Lloyd, Rössle and Wolfendale [1] showed that σ_c is very nearly 1 mm. The knowledge of σ_c is, of course, of importance when significance is to be attached to measurements approaching this limit.

In this experiment 4566 acceptable particles were recorded of which approximately 1800 had momenta greater than 10 GeV/c and 70 exceeded 100 GeV/c. The analysis of the events followed the procedure of Owen and Wilson, viz. the use of a standard comparison spectrum derived from a production spectrum having exponent -2.85. The derived sea level spectrum was normalized to that of Owen and Wilson over the common range 3 to 20 GeV/c and the result is shown in Fig. 8 and Table 2; the histogram is derived from the ranges of deflection to which the particles are allocated and the smooth curve is the best curve through the data. A value of $\sigma_c = 1$ mm was used for the deflection error and corrections were applied to the high momentum end amounting to factors of 0.41 and 0.90 for the ultimate and penultimate groups respectively. (The correction factors assumed Gaussian errors.) It seems likely therefore that this high momentum spectrum is accurate up to nearly 200 GeV/c but above this, to $\sim$1000 GeV/c, errors of a few tens of percent may be present arising from the inherently uncertain correction for the error in track location.

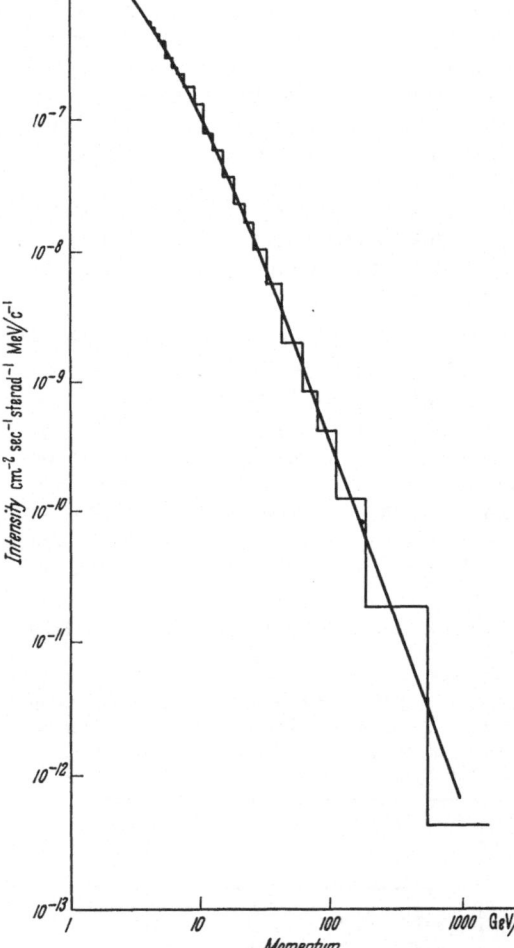

Fig. 8. The differential momentum spectrum at sea level measured by Rodgers, see footnote 2, p. 279. Normalization has been made to the spectrum of Owen and Wilson in the region 3 to 20 GeV/c. The smooth curve is the best curve through the experimental data and represents the standard in the momentum range 3 to 1000 GeV/c.

A comparison of this spectrum with that of Caro *et al.* shows that the fall-off with increasing momentum found by Rodgers is significantly less than that of

[1] J.L. Lloyd, E. Rössle and A.W. Wolfendale: Proc. Phys. Soc. Lond. A **70**, 421 (1957).

CARO *et al.* This is in agreement with the findings of OWEN and WILSON (it must be noted that a comparison of the exponent only can be made since the data of RODGERS are normalized to those of OWEN and WILSON).

In conclusion it is considered that the best estimate of the sea level spectrum at moderate latitudes is found by a combination of the data of OWEN and WILSON and RODGERS given in Figs. 7 and 8 and Tables 1 and 2.

II. The effect of instrumental bias on the spectrum measurements.

10. The emulsion spectrograph. As mentioned in Sect. 3 a spectrum measured using counter arrangements is in principle subject to bias arising from the rejection of μ-mesons accompanied by other particles within the dimensions of the counter trays (some tens of centimetres). The emulsion spectrograph is free from this bias, or at any rate the dimensions over which accompanied particles will produce ambiguity, and hence rejection, is reduced to some tens of microns. As yet the only experiment in this category is a preliminary one by APOSTOLAKIS and MACPHERSON[1]. The result is shown in Fig. 9 where the measured integral spectrum is compared with that described above due to OWEN and WILSON. It is apparent that there is no very great difference between the two spectra, but the statistics of the emulsion experiment are very poor and more data are required, particularly at high momenta.

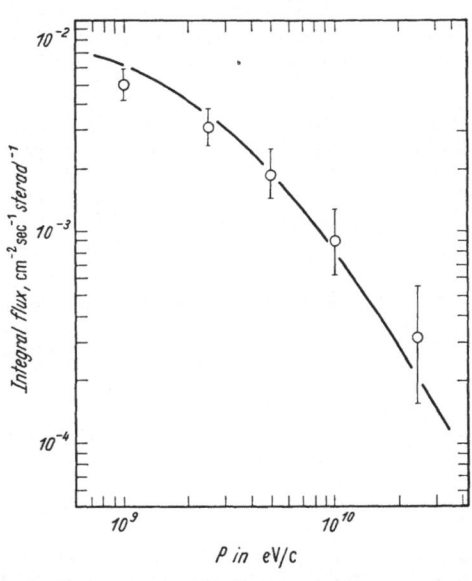

Fig. 9. A comparison of the integral μ-meson spectrum from the emulsion spectrograph and the data of OWEN and WILSON, after APOSTOLAKIS and MACPHERSON[1]. The errors quoted are the Poisson fiducial limits (REGENER[2]).

An attempt has been made by RODGERS and WOLFENDALE (unpublished) to estimate the extent of the bias from measurements on the frequency of associated tracks in the cloud chamber comprising the Manchester Spectrograph. The result is that below 50 GeV/c the bias is not expected to exceed a few percent but above this value insufficient statistics prevent a firm conclusion being reached.

III. The interpretation of the vertical momentum spectrum.

In this section we will examine in detail the connection between the observed μ-meson spectrum and the underlying processes of π- and K-meson production and decay.

Since the energy, rather than the momentum, is the operative parameter in discussing the behaviour of a particle traversing the atmosphere, it is the energy spectra of the μ- and π-mesons that will be discussed.

a) The relationships between the various spectra.

11. The diffusion equations. The most general starting point for the study of μ-meson phenomena is the pair of diffusion equations governing the π- and

---

[1] See footnote 1, p. 274.

[2] H. V. REGENER: Phys. Rev. **84**, 161 (1951).

μ-meson spectra. We will therefore start by considering these equations, and their solution, in terms of the basic π-production spectrum, in the one-dimensional approximation. With this as a basis we can then describe in detail alternative methods of deriving the spectrum which have special advantages and which have added materially to our understanding of the processes involved.

The diffusion equations referred to are the following, (Puppi [2]).

$$\frac{\partial \mu}{\partial t} = \frac{\partial}{\partial \varepsilon} \left(\beta_\mu \mu \right) - \frac{b}{t \varepsilon} \mu + \frac{B r}{t \varepsilon} \pi \left(\frac{\varepsilon}{r}, t \right), \tag{11.1}$$

$$\frac{\partial \pi}{\partial t} = \frac{\partial}{\partial E} \left(\beta_\pi \pi \right) - \left(\frac{1}{\lambda_\pi} + \frac{B}{tE} \right) \pi + \frac{1}{L_\pi} e^{-t/L_\pi} \pi_p(E). \tag{11.2}$$

Here $\mu(\varepsilon, t)$ and $\pi(E, t)$ are the μ- and π-meson spectra respectively at depth t gm cm^{-2} in the atmosphere, λ_π and L_π are the mean free paths of the π-meson and the π-meson producing components respectively and $\pi_p(E)$ is the π-meson production spectrum. The quantities β_μ and β_π are the energy losses per unit length of the μ- and π-mesons and may be taken to be

$$\beta_\mu = (0.256 \ln \varepsilon + 2) \text{ MeV gm}^{-1} \text{ cm}^2,$$

$$\beta_\pi = (0.256 \ln E r + 2) \text{ MeV gm}^{-1} \text{ cm}^2.$$

The other quantities appearing in Eqs. (11.1) and (11.2) are given by

$$b = \left(\frac{m_\mu c}{\tau_\mu} \right) \frac{x}{\varrho(x)},$$

$$B = \left(\frac{m_\pi c}{\tau_\pi} \right) \frac{x}{\varrho(x)},$$

$$r = \frac{m_\mu}{m_\pi}.$$

x is the atmospheric depth in gm cm^{-2} and $\varrho(x)$ the atmospheric density at the level x. Thus the first term on the right-hand side of (11.1) represents the energy loss of the μ-mesons, the second their decay and the third the generating spectrum. Similar interpretations apply to the terms of (11.2).

12. Validity of the approximations. The one-dimensional approximation made in (11.1) and (11.2) may be justified for energies larger than about 200 MeV by the following arguments. First of all for π-mesons produced in a given direction the maximum angle of deviation of the μ-meson, on decay, is given by

$$\tan \vartheta_{\text{max}} = \frac{1}{2} \left(\frac{m_\pi}{m_\mu} - \frac{m_\mu}{m_\pi} \right) \cdot \frac{1}{p_\pi} = \frac{0.282}{p_\pi}, \qquad p_\pi > 0.282$$

where p_π is in units of $m_\pi c$ (following Olbert [3] and Sands[1]). Thus for $p_\pi > 2$, $\vartheta_{\text{max}} < 8°$. The mesons thus retain the incident π-meson direction. For example one can show (Rossi [1], Chap. II) that the root-mean-square angle of deviation of the μ-mesons due to multiple scattering is $\sim 7°$ for a μ-meson produced at the 100 mb level arriving at sea level with momentum 300 MeV/c.

In addition it is assumed that the μ-meson energy is always r times that of the parent π-meson and that the meson production spectrum is of the special form given by the last term of (11.2).

---

[1] M. Sands: Phys. Rev. **77**, 180 (1950).

13. Solutions of the diffusion equations. From physical arguments the solution for $\mu(\varepsilon, t)$ may be written

$$\mu(\varepsilon, t) = \int_0^t \mu_p(\varepsilon', x)\, \omega(\varepsilon', t, x)\, dx \qquad (13.1)$$

where the first term represents the production spectrum of the μ-mesons at the level x and the second their probability of survival down to the level t. Making the assumptions that (i) β_μ and β_π are constants, independent of energy (this is true for relatively high energies—say $E \gtrsim 1$ GeV) and (ii) a π-meson of energy E gives a μ-meson of unique energy $\varepsilon = rE$ on decay, the solution is

$$\mu(\varepsilon, t) = \left(\frac{\varepsilon}{t}\right)^{b/(\beta_\mu t + \varepsilon)} \int_0^t \left(\frac{x}{\beta_\mu(t-x)+\varepsilon}\right)^{b/(\beta_\mu t + \varepsilon)} \frac{Br}{x(\beta_\mu \overline{t-x}+\varepsilon)}\, \pi\left(\frac{\varepsilon + \beta_\mu \overline{t-x}}{r}, x\right) dx. \quad (13.2)$$

Similarly the solution for $\pi(E, t)$ is

$$\pi(E, t) = \left(\frac{E}{t}\right)^{B/(\beta_\pi t + E)} e^{-t/\lambda_\pi} \int_0^t \left(\frac{x}{\beta_\pi \overline{t-x}+E}\right)^{B/(\beta_\pi t + E)} \times$$

$$\times \frac{e^{x\left(\frac{1}{\lambda_\pi}-\frac{1}{L_\pi}\right)}}{L_\pi}\, \pi_p(E + \beta_\pi \overline{t-x})\, dx. \qquad \Bigg\} \quad (13.3)$$

The fact that the energy of a μ-meson arising from a π-meson of energy E is not in fact unique can be taken into account by using the more accurate μ-meson energy distribution

$$p(\varepsilon)\, d\varepsilon = \frac{1}{1-r^2} \cdot \frac{d\varepsilon}{E}$$

for $r^2 E \leqq \varepsilon \leqq E$ and $v \approx c$. The spectrum of μ-mesons produced by π-decay at the level x is then

$$\mu_p(\varepsilon', x) = \frac{r}{1-r^2} \int_{\varepsilon'}^{\varepsilon'/r^2} \frac{B}{E'x}\, \pi(E', x)\, \frac{dE'}{E'}. \qquad (13.4)$$

Often, the residual range spectrum, rather than the energy spectrum, is required and this can be written

$$\mu(R, t) = \int_0^t \mu_p(R+t-x)\, \omega(R+t-x, t, x)\, dx \qquad (13.5)$$

where R is the residual range of a μ-meson of energy ε at a depth t, and $R+t-x$ the residual range when the energy is ε' at the depth x.

In order to compare these equations with experiment and to derive the connection with the process of π-meson production two procedures are possible. One may try to determine the μ-meson production spectrum and hence the π-meson production spectrum directly from the data on the observed spectrum, or one may assume a particular model for the π-meson production mechanism and compare the resulting μ-meson spectrum with that observed experimentally. In practice, both methods have been used with results which we now proceed to review, first in the low energy region in which it is assumed that all the π-mesons decay into μ-mesons, and then in the high energy region in which this assumption is not made.

b) The μ- and π-meson spectra in the low energy region.

14. Direct derivation of the μ-production spectrum. The most exhaustive study of this aspect of the problem has been given by Olbert[1] [3]. In the low energy region, π-meson interaction can be neglected and we may replace $\mu_p(\varepsilon', x)$ in (13.1) by

$$\mu_p(\varepsilon', x) = G(\varepsilon') \, e^{-x/L_\pi} \tag{14.1}$$

where L_π is, as before, the interaction mean free path of the π-meson producing component. The result is, following Sands[2] and Olbert,

$$i_v(R, t) = \int_0^t G(R + t - x) \, e^{-x/L_\pi} \, \omega(R, t, x) \, dx. \tag{14.2}$$

In this equation i_v is to be replaced by the observed spectrum at the depth t and $G(\varepsilon')$ is the production spectrum which is to be found. This equation is applicable over a restricted range of residual ranges, viz.

$$100 \text{ gm cm}^{-2} < R < 6000 \text{ gm cm}^{-2}.$$

The lower limit arises from the assumption that β_π can be taken as constant and the upper limit corresponds to the energy above which π-meson interaction is important.

The solution of (14.2) for the upper part of this range, 2000 to 6000 gm cm^{-2} is found by writing the survival probability as

$$\omega(R, t, x) = V(R) \left(\frac{x}{t}\right)^{\lambda(R)} \tag{14.3}$$

and assuming that G is a sufficiently slowly varying function of its argument, $R + t - x$. Thus for large R evidently $R + t - x$ is very insensitive to changes in x for $0 \leq x \leq t$, and we may write

$$i_v(R, t) = G(R + t - x_m) \int_0^t \exp(-x/L_\pi) \, \omega(R, t, x) \, dx \tag{14.4}$$

using the mean value theorem of integral calculus with

$$x_m = \frac{\int_0^t x \exp(-x/L_\pi) \, \omega(R, t, x) \, dx}{\int_0^t \exp(-x/L_\pi) \, \omega(R, t, x) \, dx}. \tag{14.5}$$

The remaining integral may then be evaluated to give

$$i_v = G(R + x_0 - x_m) \, L_\pi \omega(R, t, L_\pi) \, I(1 + \lambda; t/L_\pi) \tag{14.6}$$

where I is the incomplete Γ function:

$$I(z, y) = \int_0^y t^{z-1} \, e^{-t} \, dt$$

and

$$x_m = \frac{I(2 + \lambda; t/L_\pi)}{I(1 + \lambda; t/L_\pi)} \, L_\pi \approx (1 + \lambda) \, L_\pi \quad \text{if} \quad t/L_\pi \gg 1,$$

which is well satisfied for t equal to the atmospheric depth at sea level, since $L_\pi = 125$ gm/cm^2.

---

[1] S. Olbert: Phys. Rev. **96**, 1400 (1954).
[2] See footnote 1, p. 282.

Thus

$$G(R + t - x_m) = \frac{i_v(R, t)}{L_\pi \omega(R, t, L_\pi) \Gamma(1 + \lambda)} . \qquad (14.7)$$

It is found that the result may be approximated by a function of the form

$$G(R + t - x_m) = \frac{C}{(a + R')^\gamma} \quad \text{for} \quad 2000 < R < 6000 \text{ gm/cm}^2 \qquad (14.8)$$

where $\gamma = 3.58$ and $R' = R + t - x_m$ is the residual range at production. Since this range of values of R is too small to permit unambiguous determination of the other constants, recourse is made to the data due to CONVERSI[1] on the spectrum of μ-mesons of rather short residual range and its variation with atmospheric depth. This fixes the left-hand side of the equation

$$\left. \begin{aligned} i_v(100, s) = \int\limits_0^s G(100 + s - x) \times \\ \times e^{-x/L_\pi} \omega(100, s, x) \, dx \end{aligned} \right\} \qquad (14.9)$$

for various values of s between 200 and 1000 gm/cm².

Numerical evaluation of (14.9) yields for the values of the constants, at 50° N geomagnetic latitude

$$C = 7.31 \cdot 10^4 \text{ gm}^{-2+\gamma} \text{ cm}^{2-2\gamma} \text{ sec}^{-1} \text{sterad}^{-1},$$

$$a = 520 \text{ gm cm}^{-2} \text{ air equivalent},$$

$$\gamma = 3.58.$$

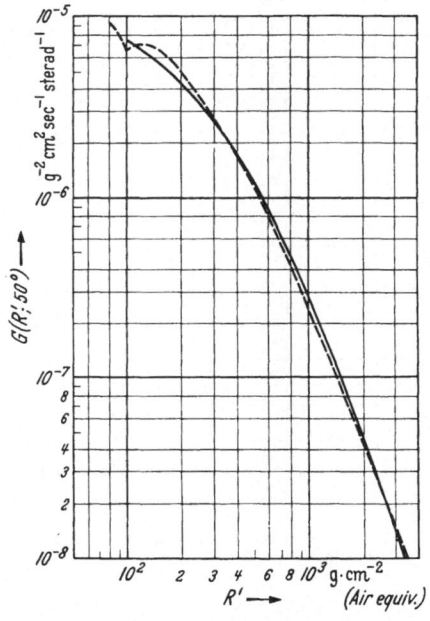

Fig. 10. A comparison of the μ-meson production spectra of OLBERT (full line) and SANDS (dashed line) (after OLBERT [3]) at 50° N.

The region between these two intervals of residual range (i.e. the region 1000 to 2000 gm cm⁻²) has also been considered by OLBERT. Working back from Eq. (14.8) with the values of C, a and γ just given, and a numerical integration of the decay probability, the expected sea level spectrum is shown to be consistent with that of CARO et al. (and hence the best sea level spectrum of Fig. 9 in the residual range region 100 to 2000 gm cm⁻²). The result is displayed in Fig. 10 and is seen to be similar to the previous result of SANDS based on earlier experimental work except for the dip at $R' \approx 100$ gm/cm² which is smoothed out in OLBERT's results.

We can conclude that the best estimate of the μ-production spectrum in the range $200 < R' < 6000$ gm cm⁻² is

$$G(R') = \frac{7.31 \times 10^4}{(520 + R')^{3.58}} \text{ gm}^{-2} \text{ cm}^2 \text{ sec}^{-1} \text{sterad}^{-1}. \qquad (14.10)$$

15. The accuracy of the relation for the production spectrum. The accuracy of the formula for G may be estimated from the standard formula for the relative error committed by using the mean value theorem. From this it is found that for $R > 2000$ gm/cm² the relative error is $\sim 1.2\%$ and for small R the chief source of error arises from errors in a. It is found by calculation that a 5% uncertainty

[1] M. CONVERSI: Phys. Rev. **79**, 749 (1950).

in $i_v(100, s)$ for $s = 3000$ gives an uncertainty in a of 2%. Thus it may be fairly asserted that the production spectrum in the region of residual range between 200 and 6000 gm/cm² is reasonably well known.

16. Direct derivation of the π-production spectrum. The π-production spectrum giving rise to the μ-production spectrum $G(R + t - x_m)$ can now be derived by inverting (13.4). A systematic procedure for doing this has been given by Ascoli[1] using the relation, exact when π-meson nuclear capture is ignored,

$$\mu_p(\varepsilon) = \frac{m_\pi m_\mu}{m_\pi^2 - m_\mu^2} \int_{E^-}^{E^+} \pi_p(E) \frac{dE_\pi}{c\, p_\pi} \qquad (16.1)$$

where

$$E_\pm = \frac{1}{2}\left(\frac{m_\pi}{m_\mu} + \frac{m_\mu}{m_\pi}\right)\left(1 \pm \frac{m_\pi^2 - m_\mu^2}{m_\pi^2 + m_\mu^2}\frac{v}{c}\right)\varepsilon$$

are the extreme energy limits of π-mesons giving rise to a μ-meson of energy ε.

This relation has been solved for $\pi_p(E)$ by Olbert and the result is shown in Fig. 11 where the relevant part of the spectrum is that below about 6 GeV.

Puppi [2] has extended the production spectrum $\pi_p(E)$ to rather higher energies (up to about 40 GeV) by assuming that the spectrum can be represented at high energies by a simple power law and solving the diffusion equations. The result is

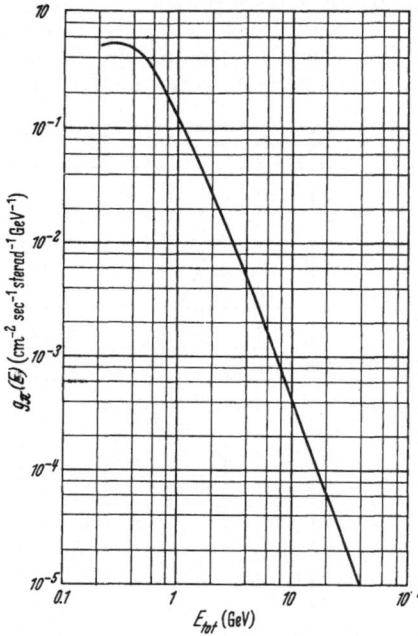

Fig. 11. The π-meson production spectrum in the atmosphere at 50° N (after Puppi [2]).

$$\pi_p(E) = \frac{0.23}{E^{2.75}} \text{GeV}^{-1}\text{cm}^{-2}\text{sec}^{-1}\text{sterad}^{-1}. \quad (16.2)$$

This production spectrum is also shown in Fig. 11 where a smooth transition has been made to energies in the region of a few GeV to the energy spectrum derived by Olbert.

The overall spectrum of Fig. 11 can be regarded as the best estimate of the π-production spectrum for energies below about 40 GeV available at the present time. A check on its validity and that of the assumed mechanism for the development of the π- and μ-components through the atmosphere is afforded by the fact that the form of $\pi_p(E)$ agrees well with that observed in nuclear emulsions exposed at high altitudes.

17. Derivation of the μ-meson and π-meson spectra from an assumed primary spectrum. A complete discussion from a theoretical point of view would involve a detailed account of the nucleon-π-meson cascade problem which is beyond the scope of the present article. We will therefore restrict ourselves to discussing some approximate solutions of the problem which are sufficient to provide a basis for further discussion of the experimental results.

18. The Budini-Molière treatment. A convenient treatment of the problem has been given by Budini and Molière [4] which may be summarised in the following way. If $g(E_0)\, dE_0$ is the differential spectrum of a "primary" cosmic ray component and $f_1^*(E, E_0)\, dE$ is the production spectrum of a "secondary"

[1] G. Ascoli: Phys. Rev. **79**, 812 (1950).

cosmic ray component (that is to say the mean number of secondary particles with energies lying between E and $E + dE$ produced by an incident particle of energy E_0) then the energy spectrum of secondary particles is given by

$$n(E)\, dE = dE \int_E^\infty g(E_0)\, f_1^*(E, E_0)\, dE_0. \tag{18.1}$$

One now assumes that the production spectrum depends only on E/E_0, that is to say that the production spectrum is homogeneous. Then,

$$f_1^*(E, E_0)\, dE = f_1\left(\frac{E}{E_0}\right) \frac{dE}{E_0} \tag{18.2}$$

and (18.1) becomes

$$n(E)\, dE = dE \int_0^1 \frac{d\eta}{\eta}\, g\left(\frac{E}{\eta}\right) f_1(\eta) \tag{18.3}$$

where $\eta = E/E_0$. This result may be used to write down the integro-differential equation satisfied by the energy spectrum $n(E, x)$ of the total penetrating component (nucleons plus π-mesons) in the one-dimensional approximation. This is

$$\frac{\partial}{\partial x} n(E, x) = -n(E, x) + \int_0^1 n\left(\frac{E}{\eta}, x\right) \left[f_p(\eta) + \frac{2}{3} \xi f_\pi(\eta) \right] \frac{d\eta}{\eta} \tag{18.4}$$

where $f_p(\eta)$ and $f_\pi(\eta)$ are the nucleon and π-meson production spectra respectively, and the assumption is made that both π-mesons and nucleons behave alike as colliding particles. The factor $\frac{2}{3}$ appears when the restriction is made to the case of charged π-mesons and ξ is the probability that the π-mesons actually do produce a nuclear interaction before decaying. This quantity is approximated by

$$\xi \approx \left(1 + \frac{B}{xE}\right)^{-1}. \tag{18.5}$$

If one now assumes that the primary spectrum has the form $n(E, 0) = A E^{-s+1}$, one may try as a solution to (18.4) the form

$$n(E, x) = A E^{-s+1} \exp\left[-x\{1 - m_p(s) - \tfrac{2}{3} \chi(E, x)\, m_\pi(s)\} \right] \tag{18.6}$$

where $m_{p,\pi}$ are given by

$$m_{p,\pi} = \int_0^1 d\eta\, \eta^s\, f_{p,\pi}(\eta). \tag{18.7}$$

Substitution in (18.4) yields, for χ

$$\chi + x \frac{d\chi}{dx} \approx \xi(E, x) \tag{18.8}$$

which gives

$$\chi(E, x) \approx 1 - \frac{\log(1 + xE/B)}{xE/B}. \tag{18.9}$$

The π-meson spectrum at production is now immediately given by

$$\pi_p(E, x) = \frac{2}{3} \int_0^{E/E_0} n(E/\eta, x)\, f_\pi(\eta) \frac{d\eta}{\eta} \tag{18.10}$$

where the upper limit E/E_0 is introduced to allow for a threshold energy of E_0 in the π-meson production process. The various integrations may be readily performed when one chooses for $f_\pi(\eta)$

$$f_\pi(\eta) = -\frac{dF(\eta)}{d\eta}$$

and

$$F(\eta) = \frac{\gamma}{\alpha!} \log^\alpha\left(\frac{1}{\eta}\right),$$

(18.11)

and a similar form for $f_p(\eta)$. Then we may use $m_\pi(s) = \frac{\gamma_\pi}{s^{\alpha_\pi}}$. If one neglects the nuclear interactions of the π-mesons the μ-meson production spectrum is given by

$$\mu_p(\varepsilon, x) = \int_0^1 \pi_p(\varepsilon/\eta, x)\, f_{\pi\to\mu}(\eta)\, d\eta/\eta$$

(18.12)

where

$$f_{\pi\to\mu}(\eta) = \begin{cases} 0 & \text{for} \quad 0 < \eta < (m_\mu/m_\pi)^2 \\ [1 - (m_\mu/m_\pi)^2]^{-1} & \text{for} \quad \left(\dfrac{m_\mu}{m_\pi}\right)^2 \le \eta < 1. \end{cases}$$

If we follow the previous practice and choose some mean energy to replace the spectrum of μ-meson energies resulting from the π-meson decay then

$$f_{\pi\to\mu}(\eta) = \delta(\eta - \eta_0)$$

and

$$\mu_p(\varepsilon, x) = \frac{1}{\eta_0} \pi_p(\varepsilon/\eta_0, x).$$

(18.13)

It is then easy to show that, with the choice (18.6) for the primary spectrum

$$\mu_p(\varepsilon, x) = \frac{2\eta_0^s \gamma_\pi\, n(\varepsilon, x)}{3\, s^{\alpha_\pi}}.$$

(18.14)

19. Comparison of the treatments of Budini-Molière and Sands-Olbert.

The result given in (18.14) cannot be directly compared with those of Sands and Olbert because the latter assume an exponential dependence on depth with a constant absorption length whereas Budini and Molière have an energy dependent absorption length. If however α_π is fixed, so that the integrals over the atmospheric depth agree, comparison may be made when a value for γ_π, the inelasticity, has been chosen. It turns out on using Sands' results that $\alpha_\pi = 2.2$ and with $\gamma_\pi = 0.2$ agreement between the two methods is quite good except at depths greater than sea level and lower energies where the Sands' spectrum departs from a pure power spectrum.

Before comparing these results with the observed μ-meson energy spectrum we must allow for ionisation loss and μ-meson decay. This may be done as before, by introducing the survival probability and assuming β_μ, β_π are constant. The result may be written

$$\mu(\varepsilon, t) = \int_0^t dx \left\{ \frac{x\,\varepsilon}{t(\varepsilon + \beta_\mu\, \overline{t - x})} \right\}^{\frac{b}{\varepsilon + \beta_\mu t}} \mu_p(\varepsilon + b_\mu\, \overline{t - x}, x).$$

(19.1)

The integral may be evaluated approximately by replacing t in the upper limit of integration by ∞ and writing $\varepsilon + \beta_\mu t$ for $\varepsilon + \beta_\mu(t - x)$. This may be justified by reference to the exponential dependence on x contained in μ_p which suppresses the contribution to the integrand arising from large x.

The result is then

$$\mu(\varepsilon, t) = \frac{A'}{(\varepsilon + \beta_\mu t)^{s+1}} \left\{ \frac{\varepsilon}{t(\varepsilon + \beta_\mu t)} \right\}^{\frac{b}{\varepsilon + \beta_\mu t}} \Gamma\left(\frac{b}{\varepsilon + \beta_\mu t} + 1 \right) \cdot \frac{1}{K^{\{1 + b/(\varepsilon + \beta_\mu t)\}}} \quad (19.2)$$

where

$$A' = \tfrac{2}{3} A \, \eta_0^s \, m_\pi(s)$$

and

$$K = 1 - m_p(s) - \frac{2}{3} m_\pi(s) \chi\left(\frac{\varepsilon}{\eta_\pi \eta_0}, t \right).$$

The quantity η_π arises from the approximate evaluation of (18.10) and may be taken to be 0.3.

20. Comparison with experiment. The main features of the experimental spectrum are now accounted for rather clearly. In particular the departure from the pure power law exhibited by the experimental results at low energies may be clearly seen from (19.2) where $\varepsilon^{-(s+1)}$ is now replaced by $(\varepsilon + \beta_\mu t)^{-(s+1)}$. In addition, at low energies it is no longer legitimate to replace $\varepsilon + \beta_\mu(t-x)$ by $\varepsilon + \beta_\mu t$ and an important contribution to the integrand comes from $t \approx x$, that is to say locally produced mesons become important (a fact already referred to by SANDS).

21. Conclusions in the low energy region. There appear to be no great inconsistencies between observation and expectation in this energy region. The development of the μ-meson component through the atmosphere is well understood and the π-production spectrum derived from observation (given in Sect. 16) is thought to be accurate.

c) The π- and μ-meson spectra in the high energy region.

22. The relation between the production spectra and the measured sea level spectrum. Interest at the high-energy end of the μ-meson spectrum centres on the information given on the mechanism of the interactions in which the parents of the μ-mesons are produced. The usual procedure is to take some primary cosmic ray spectrum and an assumed π-meson multiplicity distribution and from this calculate the π-production spectrum. The expected sea level μ-meson spectrum is then found using the diffusion equations. The expected and observed sea level spectra are then compared and divergences are attributed to some error in the initial assumptions. Although the number of uncertain parameters in the calculations is great, it is usually assumed that the most uncertain is that of the mass spectrum of the parents of the μ-mesons, and by allowing a fraction of the parents to be K-mesons, instead of π-mesons, agreement between the spectra can be guaranteed.

BARRETT et al.[1] have determined the primary spectrum from measurements on extensive showers at sea level and underground and this is usually used in the calculations.

Assuming the π-meson multiplicity distribution given by the Fermi[2] theory these authors have evaluated the exponent $\gamma_\pi(E)$ of the integral π-production spectrum to be

$$\gamma_\pi(E) \approx 1.65 + \frac{1}{6} \mathrm{Log}_{10}\left(\frac{\varepsilon}{10} \right) \quad (22.1)$$

[1] P.H. BARRETT, L.M. BOLLINGER, G. GOCCONI, Y. EISENBERG and K. GREISEN: Rev. Mod. Phys. **24**, 133 (1952).

[2] E. FERMI: Progr. Theoret. Phys. **5**, 570 (1950).

where E (in GeV) is the mean energy of π-mesons producing μ-mesons of energy ε (in GeV). (This result is for $E \gg 1$.)

It will be seen that this exponent agrees reasonably well with the value used by Puppi [Eq. (16.2)] in the common energy region, 20 to 40 GeV.

There is conflicting evidence on the extent of K-meson production required to give agreement with the sea level spectrum. Rodgers has used the primary spectrum of Barrett et al. and a simple model for propagation through the atmosphere and finds that the predicted sea level spectrum is too low at high energies (by a factor ~ 10 at 500 GeV). This would correspond to a copious production of K-mesons at primary energies of a few times 10^4 GeV.

Jakeman on the other hand has used the spectra of Barrett et al. to calculate the vertical sea level μ-meson spectrum in connexion with measurements at large zenith angles (Sects. 39 to 41). This author finds good agreement between this spectrum and that measured experimentally by Rodgers, a result indicating that K-meson production is not important at these energies.

More accurate calculations have been made by Haber-Schaim and Yekutiel-li[1], and Budini and Molière.

These authors have used the underground depth-intensity curve for comparison rather than the measured sea level spectrum, a procedure which is allowable if one considers that the range-energy relation for high-energy μ-mesons is known. The calculations will now be described.

23. The relation between the production spectra and the underground depth-intensity curve. A detailed study of the π-meson nucleon cascade has been given by Haber-Schaim and Yekutielli. These authors use the Fermi meson production model and divide the air nucleus into three layers which are treated separately. The result is given in the form

$$n'(E, x) = e^{-x/\lambda_n} \sum_{k=0}^{\infty} n_k'(E) \left(\frac{x}{\lambda_n} \right)^k \tag{23.1}$$

and

$$\pi(E, x) = e^{-x/\lambda_\pi} \sum_{k=0}^{\infty} \pi_k(E) \left(\frac{x}{\lambda_\pi} \right)^k \tag{23.2}$$

where n' and π are the nucleon and π-meson spectra respectively. n_0 is the primary spectrum and the functions n_k', π_k for $k \geq 1$ give the energy spectra of successive generations. The resulting meson spectrum is then substituted into the diffusion equation (11.1) neglecting ionisation loss and a comparison made with the μ-meson spectrum at high energy deduced from the underground observations of Barrett et al. The differential spectrum so found agrees rather well with that derived by Barrett et al. extended to 10^3 GeV, when normalised to Barrett's results for the integral spectrum for $E \geq 10^2$ GeV.

The results of Budini and Molière may be modified to allow for the competition of π-meson capture and decay by multiplying (18.13) by $1 - \xi$ where ξ is given by (18.5). This gives for the expected μ-meson spectrum

$$\mu(\varepsilon, t) = \frac{A'}{(\varepsilon + \beta_\mu t)^{s+1}} \left\{ \frac{\varepsilon}{t(\varepsilon + \beta_\mu t)} \right\}^{\frac{b}{\varepsilon + \beta_\mu t}} F(\varepsilon + \beta_\mu t) \tag{23.3}$$

where

$$F(\varepsilon + \beta_\mu t) = \Gamma \left(\frac{b}{\varepsilon + \beta_\mu t} + 1 \right) \int_0^\infty \frac{e^{-z} dz}{\left[K + \frac{z(\varepsilon + \beta_\mu t)}{\eta_0 B} \right]^{\left(1 + \frac{b}{\varepsilon + \beta_\mu t} \right)}}.$$

[1] U. Haber-Schaim and G. Yekutielli: Nuovo Cim. 11, 172, 683 (1954).

In order to estimate approximately the effect of K-meson production and decay on the μ-meson spectrum one may evaluate (23.3) for $\eta_0 = 0.4$ and $B \to \infty$ i.e. by assuming that all μ-mesons are produced from short-lived heavy K-mesons. The results derived by BUDINI and MOLIÈRE in this way, assuming $s = 1.8$, are shown in Fig. 12 and compared with the underground spectra of BARRETT et al. and the result obtained from (23.3) assuming that all μ-mesons originate in π-mesons. It will be seen that above about 500 GeV μ-mesons arise primarily from the decay of K-mesons, below 500 GeV the main contribution comes from π-mesons.

The result may also be expressed by writing for the differential energy spectrum of μ-mesons

$$\mu(E)\,dE = \frac{A}{E^{\gamma+1}}\,dE, \qquad (23.4)$$

and plotting $\gamma + 1$ as a function of E. This is shown in Fig. 13 assuming s in the primary

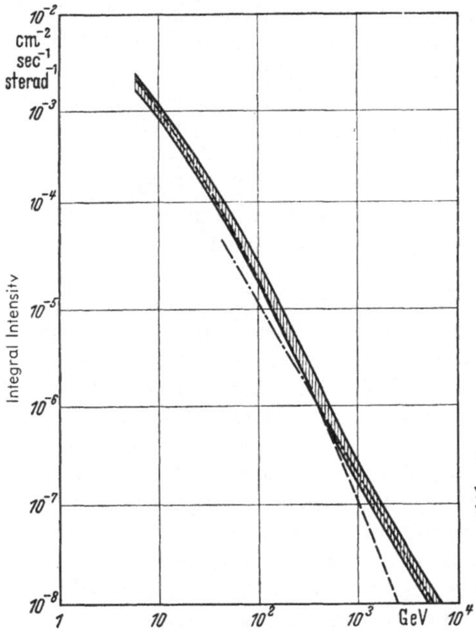

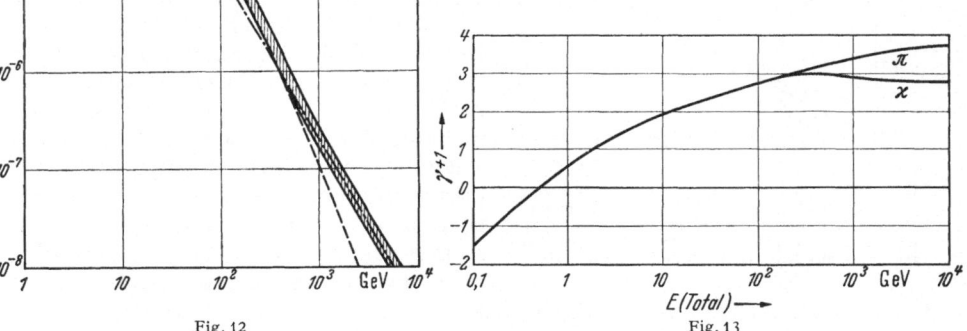

Fig. 12

Fig. 13

Fig. 12. The theoretical integral energy spectrum of vertical μ-mesons at sea level compared with experiment (after BUDINI and MOLIÈRE [4]). The shaded portion corresponds to the experimental results. ----- assumes all μ-mesons produced from π-meson decay. $-1-1-$ assumes all μ-mesons produced from K-meson decay.

Fig. 13. The exponent, $\gamma + 1$, of the μ-meson differential energy spectrum as a function of the μ-meson energy, showing the effect of production by K-mesons (after BUDINI and MOLIÈRE [4]).

spectrum to be 1.8 and including the effect of K-meson decay calculated by BUDINI and MOLIÈRE on the same basis as before.

24. Conclusions in the high-energy region. It is apparent that the onset of significant contribution to sea level μ-mesons from K-μ decay is rather uncertain but there is almost certainly some production below 1000 GeV/c. Further, precise, experiments on the high-energy spectrum of μ-mesons at sea level and underground should enable some firm conclusions to be made on this problem, particularly if experiments at various zenith angles can be made, as will be seen later in Sect. 39.

IV. Latitude effect.

The close connection between the primary cosmic ray spectrum and the μ-meson spectrum which has been described in detail above naturally leads to a latitude effect in the μ-meson spectrum. This may be examined quantitatively on the lines of the previous section.

We may, following OLBERT [3], distinguish two effects. The first, the atmospheric latitude effect, arises because the atmospheric temperature distribution which is introduced by the survival probability depends on latitude. It is found that this effect is a minimum at small atmospheric depths. In general the effect is very small and will not be discussed.

25. The effect of the earth's magnetic field. The second effect, the exclusion by the earth's magnetic field of primary protons of energies below a certain value depending on latitude, will be small for large R in (14.8). Thus if $R \sim 3000$ g/cm² $E_\pi > 8$ GeV and $E_{\text{prim}} > 12$ GeV for which energies the latitude effect is very small. It follows from this that the latitude dependence of (14.8) must be contained in the quantity a. Thus to proceed further one writes

$$G(R) = \frac{7.31 \times 10^4}{(a(\lambda) + R)^{3.58}} \quad \text{gm}^{-2}\,\text{cm}^2\,\text{sec}^{-1}\,\text{sterad}^{-1} \tag{25.1}$$

where λ is the angle of geomagnetic latitude, and by using (14.2) and the experimental results for $i_v(R, s)$ for fixed R, small s and different values of λ, one finds a functional relation between $a(\lambda)$ and $i_v(\lambda)$.

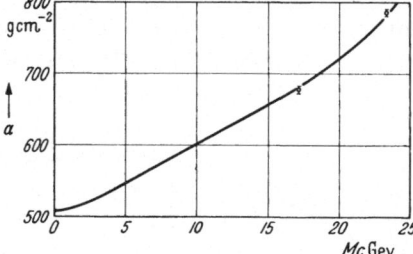

Fig. 14. The variation of the parameter a occurring in the form of the μ-meson production spectrum with geomagnetic latitude (after Harris and Escobar[2]). The experimental points refer to measurements at zenithal angles $22\frac{1}{2}°$ and $45°$ E and latitude 4.9° S.

Using the data of Conversi[1] on the variation of slow meson intensity with latitude at high altitude (30000 ft) Olbert has derived the variation of $a(\lambda)$ with λ. The value of a plotted against the geomagnetic cut-off of the primary spectrum, including the later results of Harris and Escobar[2], is shown in Fig. 14. When precise machine experiments on the variation of the π-meson production cross-section with primary energy are available it should be possible to correlate these results with those of the methods such as emulsion studies of primary protons and α-particles.

A check on the accuracy of the results in Fig. 14 is afforded by studying the variation of the low energy spectrum with latitude at sea level, and measurements on this variation will now be considered in some detail.

26. The latitude variation of the low energy spectrum. At low momenta the magnet-cloud chamber method described in Sect. 3 becomes unusable on account of the very large magnetic curvatures involved. Other methods have consequently been used, mainly (a) delayed coincidence and (b) absorption. In method (a) the differential range spectrum is found by detecting the μ-decay of those mesons penetrating one absorber and coming to rest in a further thin absorber. Method (b) is rather similar, but here an anti-coincidence array is used to detect the stopping meson. Unfortunately neither of these methods is entirely satisfactory, because a single unambiguous particle trajectory is not determined and various uncertainties arise.

The results of various experiments have been summarized recently by Fukui et al.[3]. Fig. 15 shows the differential range spectra found at various latitudes. It is apparent that the cloud chamber results of Fukui et al. are inconsistent with other results at or near the same latitude and it is likely that the geometrical corrections have been underestimated. Even among the remaining results at very

[1] See footnote 1, p. 285.

[2] F. B. Harris and I. Escobar: Phys. Rev. **100**, 255 (1955); **104**, 542 (1956).

[3] S. Fukui, T. Kitamura and Y. Murata: J. Phys. Soc. Japan **10**, 735 (1955).

similar latitudes the spread in the experimental points is large. Also shown is the variation expected according to OLBERT's relation for a latitude 24° S—Eq. (25.1)—the agreement is seen to be good. The experiments of DEL ROSARIO and DÁVILA-APONTE[1] at sea level, in particular, are the most accurate; these workers found a vertical intensity at 29° N geomagnetic latitude of $(4.25 \pm 0.13) \times 10^{-6}$ (gm sec sterad)$^{-1}$ at a residual range of 100 gm cm^{-2}. This is in very good agreement with the value of 4.17×10^{-6} (gm sec sterad)$^{-1}$ calculated by OLBERT for the conditions of this experiment.

An enhanced latitude effect at smaller residual ranges has been reported recently by SUBRAMANIAN et al.[2]. These workers measured the flux of slow mesons stopping in a cloud chamber operated at an atmospheric depth of 800 gm cm^{-2} at geomagnetic latitude 2° N. The flux of mesons in the range 47 to 64 gm cm^{-2} was found to be about 3 times that under similar conditions at 50° N. This large effect suggests that an extended study of the latitude variation of very slow mesons may be quite an accurate method of studying the latitude effect of the primaries.

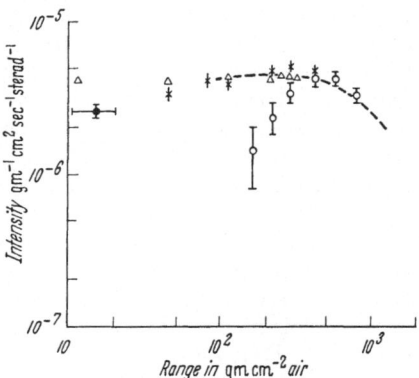

Fig. 15. The variation of the low energy spectrum of μ-mesons at sea level with latitude (after FUKUI[3] et al., see footnote 3, p. 292). × FUKUI et al.[4] — counter experiment 24° N. ○ FUKUI et al., see footnote 3, p.292. — cloud chamber experiment 24° N. ● KANEKO et al.[5] — emulsion experiment 24° N. △ DEL ROSARIO et al.[1] 29° N. — — — OLBERT[6] (theoretical spectrum for 24° N).

Returning to the theoretical aspect of the subject it may be remarked that the latitude cut-off may be introduced into the formulae of BUDINI and MOLIÈRE in a rather simple way by choosing E_0 in (18.10) to be the cut-off energy at the given latitude. This multiplies the original results by the factor

$$\left\{ 1 - \frac{I \left[\alpha_\pi - 1; s \log E_{\text{mag}}/E \right]}{(\alpha_\pi - 1)!} \right\}$$

where I is again the incomplete Γ-function. According to these authors detailed calculations at a depth of 120 gm cm^2 reproduce the "kink" in the Sands results at approximately 300 MeV which had hitherto been unexplained. The implication is that the μ-meson spectrum to the right of the "kink" includes a contribution from the first generation of incident nucleons now able to penetrate the atmosphere. Cited in confirmation of this view is some experimental work by CAMERINI et al.[7] working at the same geomagnetic latitude with photographic emulsions who also found a "kink" in the same place in the spectrum as that found by SANDS.

It should be added that OLBERT's results do not in fact show any appreciable "kink" at ~300 MeV.

The work of the other authors quoted in Sect. 23 can be easily modified to include the geomagnetic effect by choosing appropriately the lower limits of

[1] L. DEL ROSARIO and J. DÁVILA-APONTE: Phys. Rev. **88**, 998 (1952).

[2] A. SUBRAMANIAN, S. NARANAN, P. V. RAMANAMURTHY, A. B. SAHIAR and S. LAL: Nuovo Cim. **7**, 110 (1958).

[3] S. FUKUI, T. KITAMURA and Y. MURATA: Quoted by Y. WATASE et al., Ann. Rep. Sci. Works, Fac. Sci. Osaka Univ. **5**, 18 (1957).

[4] S. FUKUI, T. KITAMURA and Y. MURATA: J. Phys. Soc. Japan **12**, 854 (1957).

[5] S. KANEKO, T. KUBOZOE and M. TAKAHATA: J. Phys. Soc. Japan **10**, 915 (1955).

[6] S. OLBERT: Phys. Rev. **96**, 1400 (1954).

[7] U. CAMERINI, P. H. FOWLER, W. O. LOCK and H. MUIRHEAD: Phil. Mag. **41**, 413 (1950).

integration over the primary spectrum. No detailed calculations have as yet been made although accurate measurements at different latitudes should enable one to say something about the mechanism of π-meson production between 3 and 14 GeV. In this energy region precise information is becoming available from the accelerating machines.

V. The positive excess.

27. The origin of the excess. The presence of an excess of positive particles in the cosmic radiation at sea level has been known for some considerable time, ever since measurements on the magnetic deflection of single particles, as distinct from range, were made. This excess is still present when corrections are applied for the contribution due to protons and it is concluded that there is an excess of positive μ-mesons over negative μ-mesons at sea level. This excess is of course to be expected since the primary cosmic rays are predominantly positive (protons) and this extreme positive excess is transmitted to the π-mesons produced in the nuclear interactions of the protons and hence to their decay products—the μ-mesons. Obviously the sea level positive excess reflects the π-meson multiplicity at production and a study of the excess at high energy should give information on the variation of multiplicity with energy that is not easily available in any other way. In this section the experimental results will be reviewed critically and a best estimate made of the variation of positive excess with energy; this will then be used to study the multiplicity problem.

A review of this subject describing work done prior to 1952 has been given by Puppi and Dallaporta [5].

a) Experimental results on the positive excess.

With the exception of some early experiments of rather low statistical accuracy by Nereson[1] and Brode[2] all the sea level experiments have been carried out using the magnetic deflection of particles in air. Most of the recent experimental arrangements have already been described in Division I above and so we will confine ourselves to some general comments on these methods. By suitable reversal of the deflecting magnetic field the presence of systematic errors, which would give rise to first order errors in the positive excess, can be detected and rectified. Systematic errors in the momentum determination may occur, but these are not of great importance since it is found that the positive excess varies only slowly with momentum. Near the maximum detectable momentum of the instrument, however, serious errors may occur; this is because of the possibility of the opposite sign being attached to particles whose magnetic deflection is of the same order as the spurious or "noise" deflection.

28. The best estimate of the variation of positive excess with momentum. A summary of the results of those experiments of reasonably high statistical precision is given by Fig. 16. The quantity plotted is the positive-negative ratio, R, rather than the positive excess, δ. The two are related by (29.1). In addition to the results of direct sea level measurements those derived from the underground experiments of Filosofo et al.[3] are also given. In the latter experiments the charge ratio of slow particles at different depths underground was studied using a magnetic lens—the results then pertained to narrow energy bands at well defined energies at sea level. It will be noticed that these results have the highest precision. When

[1] N. Nereson: Phys. Rev. **73**, 565 (1948).

[2] R. B. Brode: Nuovo Cim. **6**, Suppl. 465 (1949).

[3] I. Filosofo, E. Pohl and J. Pohl-Ruling: Nuovo Cim. **12**, 809 (1954).

the range-energy relation for μ-mesons at high energies has been accurately determined this method will probably be used for future measurements of the ratio at high energies on account of its relative simplicity.

An increase in the positive-negative ratio with momentum between 1 and about 4 GeV/c followed by a slow decrease up to about 20 GeV/c is indicated from most of the measurements.

At higher momenta the apparent rapid fall in the ratio rests almost entirely on the results of RODGERS. Although, as mentioned above, the positive excess will tend to zero at the limit of measurement this limit ($\sim$240 GeV/c) is thought to be sufficiently above the highest momentum point (100 GeV/c) for the correction to the ratio to be small.

In spite of the rather wide spread in values at some momenta an attempt has been made in Fig. 16, to draw a smooth curve through the experimental points; this curve now represents the best experimental evidence on the variation of positive-negative ratio with momentum.

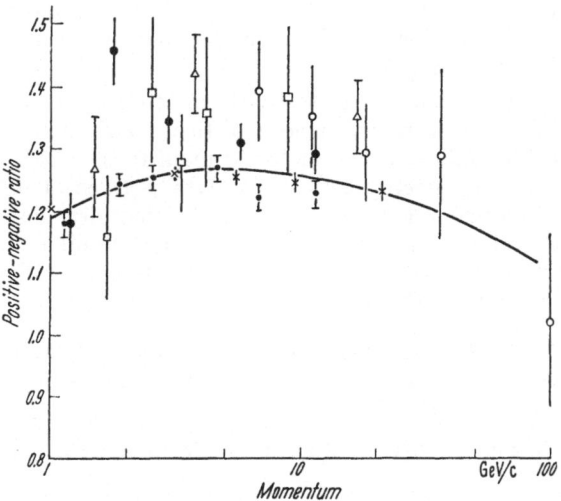

Fig. 16. The experimental results on the positive-negative ratio of μ-mesons at sea level. The curve represents an attempt to draw the best smooth curve through the experimental points. ○ RODGERS: cloud chamber arrangement (footnote 2, p. 279). ● RODGERS: counter hodoscope (same apparatus as OWEN and WILSON). ● OWEN and WILSON[1]. □ MORONEY and PARRY (footnote 1, p. 302). × FILOSOFO et al. (footnote 3, p. 294). △ CARO et al. (footnote 4, p. 274).

b) Comparison with theory.

29. The contributions from successive generations. Since the primary cosmic rays are absorbed in the upper atmosphere in more than one collision, the μ-mesons at sea level arise from π-mesons produced in several generations. However, the positive excess is reduced considerably after the first generation collision, and the positive excess for high energy particles at sea level arises almost entirely from first generation collisions. It is for this reason that a study of the multiplicity variation via the sea level positive excess is particularly useful. The positive excess rather than the positive-negative ratio is useful in the theoretical discussion and this is defined as

$$\delta = \frac{2(\mu^+ - \mu^-)}{\mu^+ + \mu^-} = 2\left(\frac{R-1}{R+1}\right). \qquad (29.1)$$

30. The treatment of BUDINI and MOLIÈRE. Following BUDINI and MOLIÈRE [4] it is convenient to introduce a quantity Δ, similar to δ, which is defined for the π-meson spectrum produced in a single collision.

Assuming that only the first generation is effective in producing the positive excess, and that the heavy primaries give no contribution, since they consist mainly of light nuclei with equal numbers of protons and neutrons, one may write

$$2\left[\pi_p^+(E, x) - \pi_p^-(E, x)\right] = \frac{2}{3} \int_0^1 \frac{d\eta}{\eta} f_\pi(\eta) \, \Delta\left(\frac{E}{\eta}\right) n\left(\frac{E}{\eta}, 0\right) e^{-x/\lambda_p}. \qquad (30.1)$$

---

[1] B. G. OWEN and J. G. WILSON: Proc. Phys. Soc. Lond. A **64**, 417 (1951).

The positive excess in the μ-meson spectrum is then obtained by the methods described in detail above and if the primary spectrum is assumed to be a pure power spectrum one finds

$$\delta = \Delta \frac{\lambda_p}{L}$$

where λ_p is the geometrical mean free path and L is the nucleon-π-meson cascade absorption length. This simple result arises from a comparison of (18.10) using (18.6) with (30.1).

The positive excess has been studied in more detail by Caldirola starting from the π-meson-nucleon diffusion equations. As a result of the assumptions made, only two parameters remain to be fixed. The collision inelasticity, taken from high-energy cosmic ray data, is fixed at 0.25 and the theoretical spectrum is normalised to give the experimentally observed number of μ-mesons with energies greater than 0.33 GeV. The result is, briefly, that up to $\sim$10 GeV the most important process is single π-meson production. This will be discussed in more detail later.

31. The treatment of Haber-Schaim, Yekutielli and Yeivin. The questions in volved may be investigated in more detail by using Haber-Schaim and Yekutielli's results[1].

The equation satisfied by the first generation of the produced π-mesons is

$$\left(1 + \frac{B}{E}\right) \pi_1(E) = \int_{E_{\min}}^{\infty} R_{\pi n}(E, E') \, n_0(E') \, dE \tag{31.1}$$

where n_0 is the primary proton spectrum and $R_{\pi n}(E, E') \, dE$ is the number of π-mesons with energies in dE, created by a primary proton of energy E'. If one assumes again that in the higher generations of nucleons equal numbers of protons and neutrons exist, all the positive excess must arise from $\pi_1(E)$. We may therefore find the positive excess in the μ-meson spectrum by substituting the positive excess in $\pi_1(E, x)$ for $\pi(E, x)$ in (13.2) and dividing by $\mu(\varepsilon, x)/2$. Since

$$\pi_1(E, x) = e^{-x/\lambda_\pi} \pi_1(E) \frac{x}{\lambda_\pi}$$

(Haber-Schaim and Yekutielli) then

$$\left.\begin{aligned}
\mu^+ - \mu^- &\approx \left(\frac{\varepsilon}{t}\right)^{b/(\beta_\mu t + \varepsilon)} \int_0^t \left(\frac{x}{\beta_\mu(t-x)+\varepsilon}\right)^{b/(\beta_\mu t + \varepsilon)} \times \\
&\qquad \times \frac{B \, e^{-x/\lambda_\pi}}{\lambda_\pi\left(\frac{\varepsilon + \beta_\mu(t-x)}{r} + B\right)} \, dx \int_{E_{\min}}^{\infty} R_{\pi n}^{+-}\left(\frac{\varepsilon}{r}, E'\right) n_0(E') \, dE'
\end{aligned}\right\} \tag{31.2}$$

where

$$R_{\pi n}^{+-} = R_{\pi n}^+ - R_{\pi n}^-.$$

We have neglected the β_μ term in the argument of $R_{\pi n}^{+-}$.

One may try to facilitate the comparison with experiment, first of all, by making certain additional assumptions which determine the form of the func-

---

[1] See footnote 1, p. 290.

tion $R_{\pi n}^{+-}$ in terms of parameters which have a direct physical significance. This has been done by YEIVIN[1]. The additional assumptions are as follows:

(a) The primary proton collides with only one nucleon in the nucleus.

(b) The average multiplicity increases as a constant power of the collision energy W' [that is $m(W') = aW'^{2\alpha}$] in the centre of mass system (or C-system).

(c) All π-mesons produced in a collision have the same energy, W, in the C-system, i.e. $W = KW'/m(W')$ where K is the inelasticity coefficient.

(d) The inelasticity and angular distribution are independent of the collision energy.

The relationship between the collision energies in the laboratory and C-systems (E' and W' respectively) is given, in the extreme relativistic region, by

$$E' = \frac{W'^2}{2M_p c^2}$$

and that between the π-meson energies in the two systems, E and W, is given by

$$E = \frac{W'}{2M_p c^2} \cdot W(1 + \omega)$$

where $\omega = \cos \vartheta$ and ϑ is the angle between the direction of the emitted π-meson and the incident proton in the C-system, and M_p is the mass of the proton.

The multiplicity as a function of collision energy in the laboratory system is then $m(E') = a(2M_p c^2)^\alpha E'^\alpha = a_L E'^\alpha$.

With assumption (c) and writing $K/a = q$ we have

$$(2M_p c^2)^\alpha E = q(1 + \omega) E'^{(1-\alpha)}. \tag{31.3}$$

We may therefore express E in terms of the primary proton energy, E', and the C-system direction, ω, in $R_{\pi n}(E', E)$ and using assumption (d) we have

$$R_{\pi n}(E', E) dE = m(E') f(\omega) d\omega, \tag{31.4}$$

where $f(\omega)$ is the π-meson angular distribution in the C-system ($\int f(\omega) d\omega = 1$).

With the assumption that the subsequent nucleon generations make no contribution to the positive excess and that the nuclear forces are charge-independent we have to a good approximation (see YEIVIN[1])

$$m^+(E') - m^-(E') = \tfrac{1}{2}$$

and

$$R^+ - R^- = \frac{1}{2} f(\omega) \frac{d\omega}{dE}.$$

On substituting in the equation for $\pi^+ - \pi^-$ derived from (31.1) and assuming that the primary differential spectrum has the form

$$n(E') dE' = s A E'^{-s-1} dE',$$

the positive-negative π-meson difference is

$$d_\pi(E) = \frac{C'}{(1 + E) E^\beta} \tag{31.5}$$

where $\beta = \dfrac{s}{1-\alpha}$ and C' is a function of the parameters K, a and α and involves the integral $\int\limits_0^1 (1+\omega)^\beta f(\omega) d\omega$. It may be shown that this integral is not sensitive

[1] Y. YEIVIN: Nuovo Cim. 2, 658 (1955) and private communication 1958.

to the form of $f(\omega)$. The passage to the μ-meson positive-negative difference is now straight-forward and one finds, using (31.5) and following Yeivin[1],

$$d_\mu(\varepsilon, \varkappa) = \frac{C\Gamma\left(1 + \dfrac{1}{g\,\varepsilon}\right) x^{-1/g\varepsilon}}{\varepsilon^{\beta+1}(1 + Y\varepsilon)}. \tag{31.6}$$

Here $g = m_\pi/\tau_\pi \cdot \tau_\mu/m_\mu$, Y is a function of r and β, C is simply related to the constant C', x is the atmospheric depth measured in units of the proton mean free path, 70 gm cm^{-2}, and ε is measured in units $m_\pi c H/\tau_\pi = 112$ GeV. Before mentioning the values of the various parameters obtained from comparison with experiment some remarks may be made with regard to the validity of the assumptions (a) to (d). It has been shown by Yeivin that (a) is not a serious restriction. More complicated collision models merely affect the physical interpretation of the constant C'. Very little ground for comment on the other assumptions exists apart from the machine results of Lindenbaum and Yuan[2] which are restricted to the low energy region. In this region of energies, ~ 3 GeV, (c) and (d) are certainly not valid, but this is probably too near to the threshold energies for multiple π-meson production to invalidate Yeivin's analysis. Finally, the mean value of $m^+(E') - m^-(E')$ has been calculated by Yeivin assuming charge independence in the collision process. It turns out that this value is quite close to $\frac{1}{2}$ at least for multiplicities > 5.

Comparison of (31.6) with experiment for energies $\lesssim 20$ GeV yield $\beta = 1.78$ and $C = 4.09 \times 10^{-6}$. After normalising the primary integral spectrum beyond 15 GeV to 0.018 particles cm^{-2} sec^{-1} sterad^{-1} and assuming that the exponent of the integral primary spectrum, s, is 1.5, one finds $\alpha = 0.16$, from the relation $\beta = \dfrac{s}{1-\alpha}$, and $\dfrac{K}{a} = 0.03$. From experimental results on jets we may take $K = 0.45$ so that

$$m(E') = 3.7\,E'^{0.16}$$

with energies in GeV.

Evidently these results for the multiplicity depend critically on the assumptions described above. In fact, as will be mentioned later, very different results have been found by various workers depending on the simplifying assumptions made.

32. An alternative derivation of the energy dependence of the meson multiplicity. One may try to determine at least the energy dependence of $m(E')$ for large E' on certain general assumptions as to the form of $R_{\pi n}(E, E')$. Thus starting from the homogeneous function f_1^* of Sect. 18 one might try

$$R_{\pi n}^{+-}(E, E') = f^{+-}(E)\,F(E'/E). \tag{32.1}$$

The integral over E' in (31.2) may now be carried out, neglecting μ-meson energy loss and using the experimental values for the meson spectrum and positive excess one may determine the form of $f^{+-}(E)$ to a normalisation constant. A best fit of the form

$$f^{+-}(E) = \frac{A}{E^\alpha} + \frac{B}{E^\beta} \tag{32.2}$$

is then found and the energy dependence of the mean π-meson multiplicity determined from

$$\int_{m_\pi c^2}^{E'} R_{\pi n}(E, E')\,dE = m(E').$$

---

[1] See footnote 1, p. 297.
[2] S. J. Lindenbaum and L. C. L. Yuan: Phys. Rev. **103**, 404 (1956).

Thus we have

$$\frac{A}{E'^{\alpha-1}} \int\limits_{1}^{E'/m_\pi c^2} z^{\alpha-2} F(z)\, dz + \frac{B}{E'^{\beta-1}} \int\limits_{1}^{E'/m_\pi c^2} z^{\beta-2} F(z)\, dz = m(E'). \tag{32.3}$$

With reasonable assumptions as to the form of $F(z)$ we may, for large E', replace the upper limit of integration by ∞. Then

$$m(E') = \frac{A'}{E'^{\alpha-1}} + \frac{B'}{E'^{\beta-1}}, \tag{32.4}$$

and the mean energy used in producing π-mesons may be written

$$\left.\begin{array}{l} h(E') = \dfrac{A}{E'^{\alpha-2}} \displaystyle\int\limits_{1}^{E'/m_\pi c^2} z^{\alpha-3} F(z)\, dz + \dfrac{B}{E'^{\beta-2}} \displaystyle\int\limits_{1}^{E'/m_\pi c^2} z^{\beta-3} F(z)\, dz \\[4mm] \sim \dfrac{A''}{E'^{\alpha-2}} + \dfrac{B''}{E'^{\beta-2}}. \end{array}\right\} \tag{32.5}$$

The values of α and β turn out to be 1.03 and 4.25 for f^+, and 1.025 and 5.55 for f^-.

Thus we find that the mean multiplicity falls for high E' as $\sim E'^{-0.03}$ but the total π-meson energy rises as $\sim E'^{0.97}$. (The fraction of energy going into π-mesons evidently also falls as $E'^{-0.03}$.) These results are certainly compatible with the most recent experimental information from cosmic ray jets, at least for energies $>10^3$ GeV[1]. Below this energy the exponent of the multiplicity is approximately 0.25 but large fluctuations in multiplicity are actually observed and it is probable that this value is not inconsistent with YEIVIN's result.

C. Comparison with π-meson production models.

33. A discussion of the results of the various treatments. It is now necessary to compare the results described in the various sections above with those to be expected on the basis of various nuclear models.

In the first place OLBERT has used the π-meson production spectrum given by the inversion of (16.1) to find the integral number of π-mesons, $\pi(E_0)$, with energies $>E_0$ as a function of geomagnetic latitude. It is then possible to derive the quantity,

$$\bar{n}(E_0; E_p) = - \frac{d\pi(E_0)}{dE_p} \Big/ \frac{dJ}{dE_p} \tag{33.1}$$

where E_p is the energy cut-off in the primary spectrum and J is the primary integral spectrum averaged over all depths. From the definition of $\pi(E_0)$, $\bar{n}$ is approximately the mean number of π-mesons of energy $>E_0$ produced in the forward cone by a nucleon of energy E_p. It is estimated that this procedure should yield results correct to a factor 2. Comparison with the Fermi theory of π-meson production is shown in Fig. 17 and it will be seen that for $2 < E_p \lesssim 6$ GeV $\bar{n} < n_{\text{Fermi}}$ and for $6 \lesssim E_p < 10$ GeV $\bar{n} > n_{\text{Fermi}}$ with $\bar{n}_{\text{max}} \approx 1.5$. The quantity $\bar{n}$ changes quite rapidly in the region of 5 GeV.

The Budini-Molière theory assumes *a priori* a π-meson production spectrum whose general form is very similar to the spectrum predicted by HEISENBERG's[2] theory of meson production. As has been seen, the two parameters appearing

---

[1] D. H. PERKINS, private communication, 1958.
[2] W. HEISENBERG: Kosmische Strahlung. Berlin-Göttingen-Heidelberg: Springer 1953.

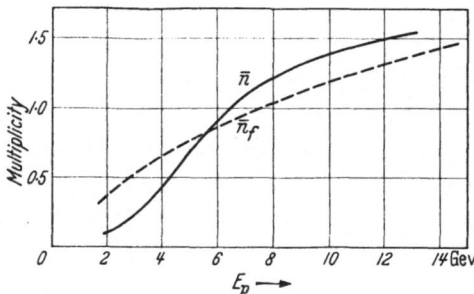

Fig. 17. Multiplicities of charged π-mesons produced in the forward cone by a primary particle of kinetic energy E_p. $\bar{n}$ is the empirical multiplicity evaluated from Eq. (33.1), $\bar{n}_f$ is the theoretical multiplicity predicted by the Fermi theory.

in the assumed spectrum can be adjusted to fit the observed data for energies beyond 10 GeV, so that for a suitable choice of parameters HEISENBERG'S model can be shown to fit the data.

Here one may also mention detailed calculations on the π-meson spectrum by GUZHAVINA et al.[1] based on LANDAU'S[2] model of meson production. About 2.5 times too many μ-mesons with $E > 1000$ GeV are predicted as compared with the experimental results of BARRETT et al. In addition, the absorption coefficient of the penetrating component at high altitudes is found to be greater than that observed. To secure agreement with experiment it is found necessary to allow for most of the collision energy to be distributed amongst nucleons and hyperons.

A more sensitive test of the π-meson production process is that afforded by a comparison of theory and experiment on the positive excess.

Reference has already been made to the work of CALDIROLA et al. These authors show that the data can be explained with an inelasticity of 0.25 and single production up to 5 GeV, with double production predominant beyond 10 GeV.

CINI and WATAGHIN[3] have calculated the positive excess assuming a production spectrum of the form $n(W) \propto p^2 W^{-4}$ where W is the energy of the π-meson in the C-system. The results agree with experiment for $E > 1$ GeV and the theory predicts a mean number of π-mesons at 7 GeV of two. The inelasticity was taken to be 0.23.

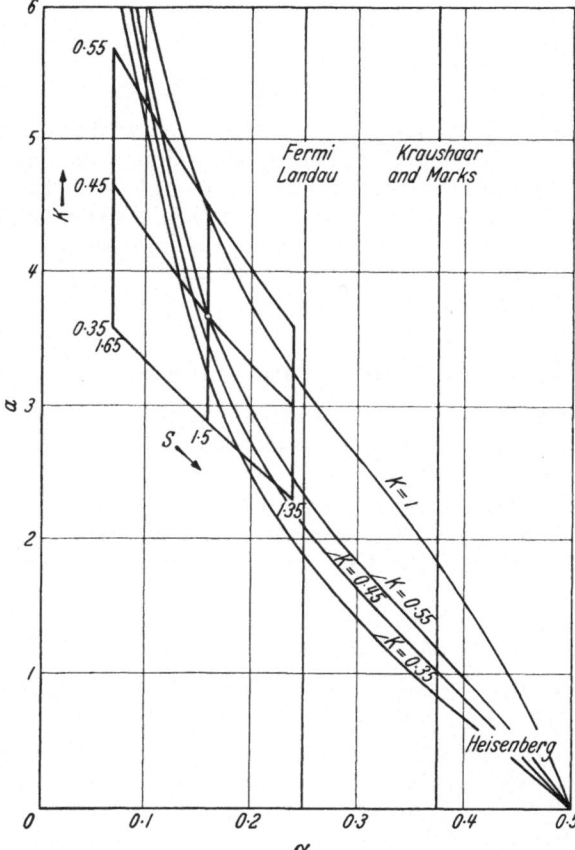

Fig. 18. Comparison of the results of the theory of YEIVIN, see footnote 1, p. 297, (and private communication) with various theories of π-meson production.

[1] O.A. GUZHAVINIA, V.V. GUZHAVIN and G.T. ZATSEPIN: Soviet Phys. J. Exp. Theor. Phys. **4**, 690 (1957).

[2] L.D. LANDAU: Dokl. Akad. Nauk SSSR **12**, 51 (1953).

[3] M. CINI and G. WATAGHIN: Nuovo Cim. **7**, 135 (1950).

The results found by Yeivin[1] are compared with various theories in Fig. 18 and it will be seen that they are consistent with Heisenberg's theory, as might be expected from assumption (b). The models due to Fermi and Landau may be acceptable but that of Kraushaar and Marks[2] is excluded. However, even at energies ~ 5 GeV the mean multiplicity is >3 which seems rather too high. A lower inelasticity consistent with the value used by Caldirola would give a reduced multiplicity, but it seems doubtful whether agreement would then be found with the Heisenberg theory for reasonable values of s.

The most reasonable conclusion from this discussion would seem to be that agreement with experiment on the positive excess can be achieved with widely different assumptions on the mean multiplicity and its energy dependence. Evidently more accurate experimental work and more detailed calculations are required before anything definite can be said about the mechanism of π-meson production from the data on μ-mesons.

D. The properties of the μ-meson component in inclined directions.

It has long been known that the flux of cosmic ray particles through a counter telescope varies markedly with the zenith angle. On account of the simplicity of the apparatus required—a simple 3-fold coincidence array—zenith angle experiments have been carried out at a variety of altitudes and depths below ground. In general it is found that the intensity I_ϑ of the hard component at a zenith angle ϑ, is related to the vertical intensity I_0 by the relation $I_\vartheta = I_0 \cos^n \vartheta$ where n is a constant for any one location and given momentum. This cosine relation is a consequence of the fact that over quite wide ranges of depth (in the atmosphere and below ground) the vertical intensity varies with depth d, such that $I_{0,d} \propto d^{-n}$ where d is measured from the top of the atmosphere. At sea level and down to moderate depths, less than ~ 300 m water equivalent, $n \sim 2$ so that the familiar cosine-squared law for the angular distribution results.

Of particular interest is the variation of the momentum spectrum with zenith angle since this is the basic phenomenon of which the zenith angle variation of the total intensity is constituted. The experimental studies of this variation will be considered first and then the variation of the total intensity.

I. The observed variation of the momentum spectrum with zenith angle for moderate angles.

a) Experimental results.

These studies can be divided into two regions:

(i) the high-energy region, where magnetic spectrographs have been operated at various zenith angles and precise spectra have been obtained at a limited number of inclinations, and

(ii) the low-energy region, where studies with magnetic lenses and delayed coincidence methods have been made.

In this study we shall be concerned with the high-energy region.

34. Measurements by the Melbourne group. Up to the time of writing, only one high-energy spectrograph has been constructed in which the apparatus is

---

[1] See footnote 1, p. 297.
[2] W.L. Kraushaar and L. J. Marks: Phys. Rev. **93**, 326 (1954).

Table 3. *Absolute intensities at various azimuth angles, after Moroney and Parry*[1].

Measurement	Number of particles recorded		Corrected intensity (sterad^{-1} cm^{-2} sec^{-1})
	1900 G	13 500 G	
Vertical	3971	2127	$(8.20 \pm 0.26) \times 10^{-3}$
30 °E.	1304	2797	$(6.22 \pm 0.27) \times 10^{-3}$
30 °W.	1523	1410	$(6.12 \pm 0.24) \times 10^{-3}$
60 °E.	1230	1234	$(1.90 \pm 0.09) \times 10^{-3}$
60 °W.	2168	1385	$(1.83 \pm 0.09) \times 10^{-3}$

rotable about a horizontal axis; this is the Melbourne instrument (CARO *et al.*[2]). The spectrograph has been operated in the vertical direction and at zenith angles of 30 and 60° in the eastern and western azimuths (MORONEY and PARRY[1]). The general form of the spectrograph has already been described in Sect. 6. In this experiment two values of the magnetic field were used, 1900 and 13 500 gauss, in order to cover a wide momentum range (0.24 to 70 GeV/c). The absolute intensities at each of the zenith angles are given in Table 3. A typical spectrum, that at 30° W, is shown in Fig. 19, illustrating the statistical precision with which the points were measured. It will be noted that the form of the spectrum above 10 GeV/c in the inclined direction is still the same as that found in the vertical direction, viz. of constant exponent, whereas as was pointed out in Sect. 9 it is likely that the errors of trajectory location were underestimated since later, more precise, measurements show an increased slope abov about 10 GeV/c. This discrepancy is not of great importance here, however, since the statistical accuracy of the points above about 20 GeV/c is too low for useful analysis to

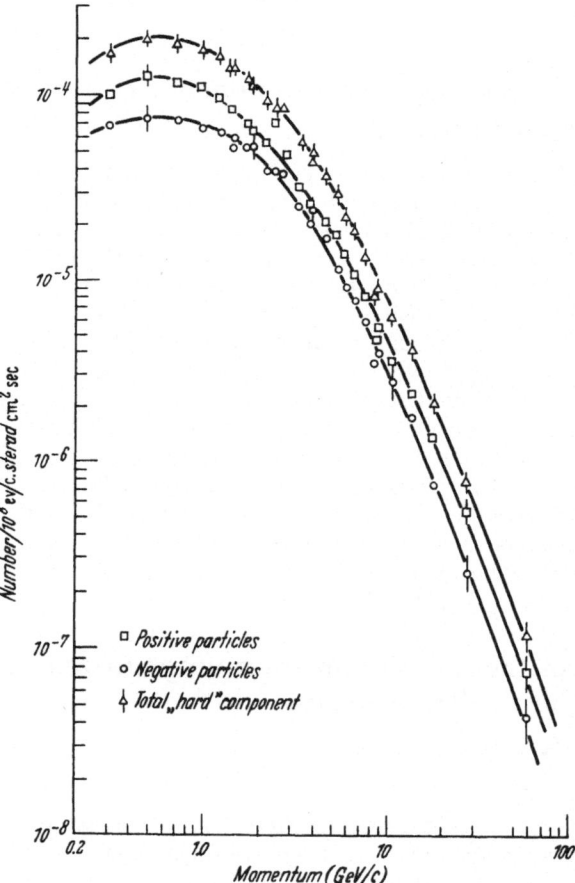

Fig. 19. The differential momentum spectra at a zenith angle of 30° W (after MORONEY and PARRY[1]).

be made in this region. The zenith angle variation found by MORONEY and PARRY is summarised in Fig. 20. In these distributions the μ-meson flux is identified with the total hard component since these authors argue that the flux of

---

[1] J.R. MORONEY and J.K. PARRY: Austral. J. Phys. **7**, 423 (1954).
[2] See footnote 4, p. 274.

protons and π-mesons is small, the proton flux being estimated at 0.5% from the results of MYLROI and WILSON[1] and the π-meson flux certainly no greater than this. It must be pointed out, however, that the proton spectrum falls off much more rapidly with momentum than the μ-meson spectrum and the proton flux at 1 GeV/c is estimated to be about 3% of the μ-meson flux under the conditions of the experiment (the particles are required to penetrate 10 cm lead below the instrument and, on emerging discharge only one of a number of counters).

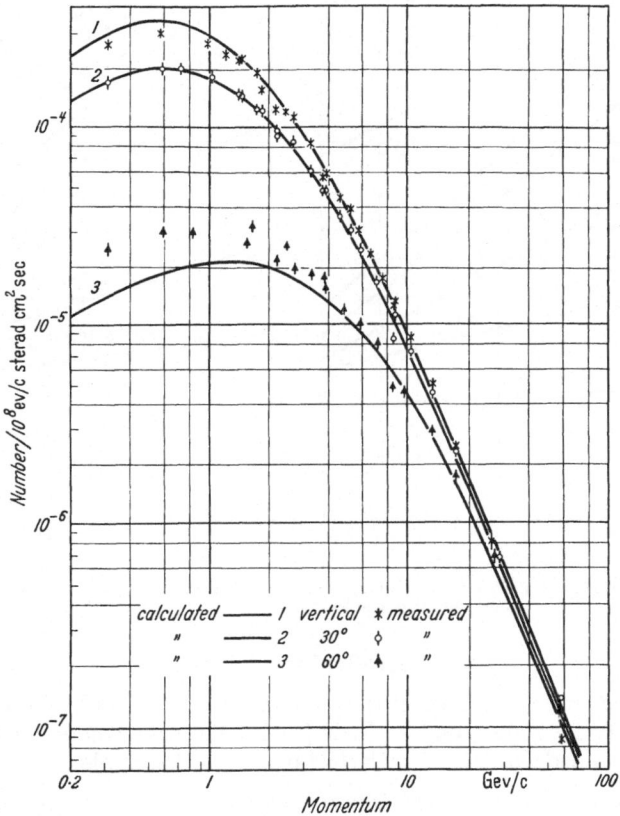

Fig. 20. Differential momentum spectra at various zenith angles (after MORONEY and PARRY ,see footnote 1, p. 302).

Although there is some small difference indicated between the spectra to the east and west, this difference is not statistically significant. Actually a small east-west asymmetry is expected at this latitude and has in fact been shown to exist by BURBURY and FENTON[2]. Its magnitude is small, 0.01 at 30° and 0.03 at 60°, and such small effects would not have been detected in the experiment of MORONEY and PARRY.

The trend of the spectrum with increasing zenith angle is as would be expected —a hardening of the spectrum arising from increased loss by decay of the low-energy mesons in travelling greater distances from the main production layer. A quantitative discussion of the results will be given later.

[1] See footnote 1, p. 278.
[2] D.W.P. BURBURY and K.B. FENTON: Austral. J. Sci. Res. A **5**, 47 (1952).

The variation of the differential intensity with zenith angle, with momentum as parameter, can be easily derived from these data. This variation is shown in Fig. 21. The decrease in n with increasing momentum is quite marked and it is interesting to observe that above about 20 GeV/c the μ-mesons are distributed almost isotropically.

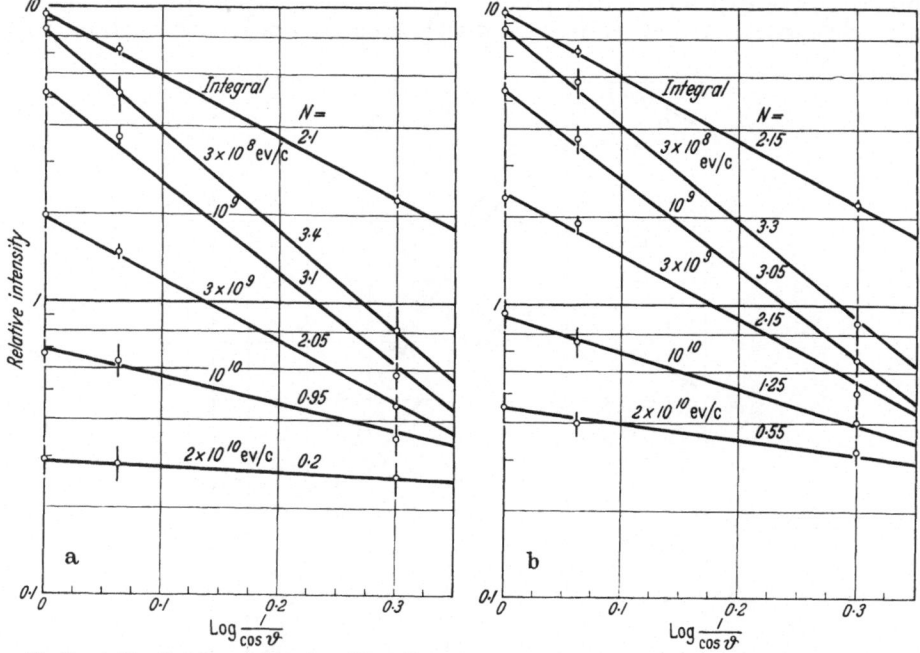

Fig. 21 a and b. Variation of intensity with zenith angle for various momenta (a) to the east (b) to the west (after Moroney and Parry, see footnote 1, p. 302).

b) Theoretical interpretation.

The methods used to interpret the μ-meson observations in the vertical direction may easily be generalised to include also the case where the observations are made in inclined directions, for arbitrary azimuthal angles.

35. Modification of the Sands-Olbert method. The Sands-Olbert method has been generalised by Harris and Escobar[1] to interpret observations on the spectrum at 45° to the vertical in the east-west plane near the geomagnetic equator. The angle of incidence is introduced by using the fact that the magnetic cut-off and hence a in (14.8) depend upon the angle of incidence of the primary particles. In addition even if the μ-mesons are collimated along the incident direction during production it is necessary to allow for the deflecting effect of the horizontal component of the earth's magnetic field during traversal of the atmosphere. This increases the length of trajectory for positive mesons incident in easterly directions and reduces it for those incident in westerly directions. The effect is of course reversed for negative mesons so that the formulae must be modified to refer to positive and negative mesons separately. Thus Eq. (14.2) is modified to read

$$i_\sigma(R,t,\Theta) = \int_0^{S'} G_\sigma(R+s+a)\left(e^{-\frac{H-Z}{L}\cdot\sec\Theta}\right)\omega_\sigma(s,R)\,ds. \qquad (35.1)$$

---

[1] See footnote 2, p. 292.

Here $\sigma = \pm 1$ according to whether positive or negative mesons are in question, H is the atmospheric depth at the point of observation, Z is the (variable) vertical height above the point of observation at which the μ-mesons are produced, and s' is the maximum track length at the angle of observation Θ s is the distance of this point measured along the trajectory. The angle Θ is the angle at which the observations are made and ϑ is the angle of incidence of the primary particles. Now ds is given by $dZ \sec \vartheta$, and ϑ may readily be found in terms of Θ, the earth's magnetic field and the sign of the charge of the incident μ-meson for a fixed value of R. The integral (35.1) may then be evaluated approximately using OLBERT's results for G with the replacement $a = a_1 + \Delta$ where Δ is a small quantity depending on Θ which is to be determined from experiment. The result is given as a power series in Δ/H up to terms of order $(\Delta/H)^2$ and involves the mean positive excess at production which must itself be deduced from the observed positive excess. The quantity a_1 is a conveniently chosen central value of a. Comparison with experiment then yields two values of a, viz. a_E and a_W corresponding to the eastern and western angles of incidence.

36. Comparison with the results of HARRIS and ESCOBAR. The values of a_E and a_W may be compared with the values of a predicted by OLBERT for the values of magnetic rigidity appropriate to the angles and geomagnetic latitude of the experiments, that is 10.6 and 23.3 GeV for the westerly and easterly directions respectively in the experiments of HARRIS and ESCOBAR. It turns out that the results fall significantly below OLBERT's values. Furthermore, since it is assumed that $\Delta = 0$ at the vertical, any change in OLBERT's values of a would lead to corresponding changes in the values of a found from this experiment; hence nothing is to be gained by adjusting the former. After an exhaustive discussion of the various factors which may explain their result, these authors conclude that the most reasonable explanation is that the π-mesons are not collimated in the direction of the incident primary but are produced in a cone whose root-mean-square angle may be determined by the condition that a_W lies on OLBERT's curve of a versus magnetic rigidity. The angle is then found to be $14° \pm 2°$. With this angle a_E is found to lie on a reasonable extrapolation of the Olbert curve. This result also leads to a correction to OLBERT's vertical spectrum of a few percent which is also given approximately by these authors. Further measurements at $22\frac{1}{2}°$ east and west were made which agreed very well with the previous measurements. Measurements at $67\frac{1}{2}°$ west were made but the experimental uncertainties were such that it was only possible to record an order of magnitude agreement between theory and experiment. The calculations on the effect of the terrestrial magnetic field on the μ-meson trajectories were however confirmed by further measurements on the east-west charge asymmetry at 45° zenith angle and latitude 48.4° N.

37. Comparison with the results of the Melbourne group. In the light of this result it is of interest to recall the results of MORONEY and PARRY mentioned above. These workers interpret their result in terms of an Olbert type spectrum. They are able to fit the vertical and 30° observations with the same production spectrum viz.,

$$\mu(p) = \begin{cases} 0.147\, p^{-3.0}\, e^{-x/125} & p > 17.6\, m_\mu c, \\ 0.147\, (17.6)^{-3}\, e^{-x/125} & p \le 17.6\, m_\mu c. \end{cases}$$

The units are $(0.1 \text{ GeV}/c)^{-1}$ cm^{-2} sec^{-1}. These spectra differ from those given by OLBERT. At 60° zenithal angle these workers observe an excess of particles with energies > 2 GeV of roughly the same magnitude as that observed by HARRIS and ESCOBAR. In addition their results at 60° show a slight westerly excess below about 3 GeV and an easterly excess at higher energies. Broadly similar results appear at 30° though in view of the experimental errors the easterly excess may not be significant in either case.

Several points are worth noting on these results. First of all, at this latitude, the main differences between the spectra should be contained in the survival probability term since the magnetic rigidity remains more or less constant for angles < 40° in both easterly and westerly directions. One should therefore expect to find a fit with experiment with the same production spectrum at 0 and 30°. That the result differs from that found by OLBERT is not significant because, as has been mentioned in Sect. 15 the production spectrum is not very sensitive to changes in the sea level spectrum. For $\vartheta = 60°$ on the other hand the increase of a in the easterly direction should reduce the production spectrum significantly for all energies below about 2 GeV. Secondly the production spectrum used by MORONEY et al. does not allow for the increased ionisation loss of particles arriving at large angles which is important in the energy region of interest. This should lead to an overestimate of the theoretical results. That the experimental results should nevertheless exceed those expected must be regarded as a rather strong confirmation of the results of HARRIS and ESCOBAR that the angle of emission of the π-meson on production should be taken into account, at least for energies ≤ 10 GeV.

38. Modification of the Budini-Molière method. One may discuss the spectrum as a function of ϑ more concisely by appropriate modification of the results

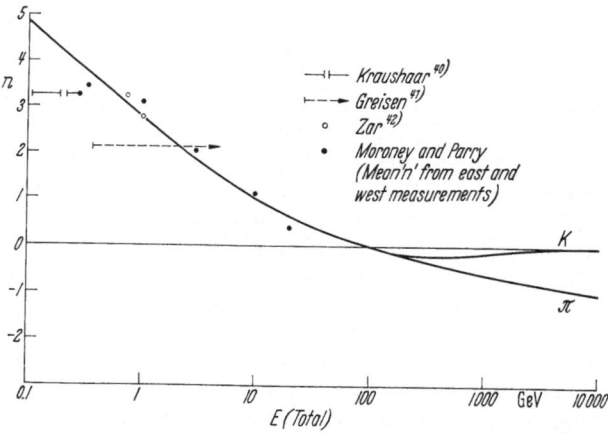

Fig. 22. The exponent $n(E)$ of the zenith angle distribution, $I_\vartheta = I_0 \cos^{n\vartheta}$, compared with experiment. The theoretical curve is due to BUDINI and MOLIÈRE [4].

of BUDINI and MOLIÈRE. One merely needs to replace b and B by $b/\cos\vartheta$ and $B/\cos\vartheta$ and t by $t/\cos\vartheta$ in (23.3). One then finds

$$\mu(\varepsilon, t, \vartheta) = [\cos\vartheta]^{s+1+\frac{b}{\varepsilon\cos\vartheta+\beta\mu t}} \cdot \mu(\varepsilon\cos\vartheta, t) \qquad (38.1)$$

and for the integral spectrum after a partial integration, neglecting the derivatives of the first term on the right-hand side

$$M(\varepsilon, t, \vartheta) = [\cos\vartheta]^{s+\frac{b}{\varepsilon\cos\vartheta+\beta\mu t}} \cdot M(\varepsilon\cos\vartheta, t). \qquad (38.2)$$

If we write

$$\mu(\varepsilon, t) \approx A \varepsilon^{-\gamma+1}$$
$$= C \cos^n \vartheta,$$

where γ is a slowly varying function of the energy, then

$$n(\varepsilon) = s + 1 + \frac{b}{\varepsilon \cos \vartheta + \beta_\mu t} - (\gamma(\varepsilon) + 1). \tag{38.3}$$

Using the results for $\gamma(\varepsilon)$ obtained from (23.3) and (23.4) MOLIÈRE and BUDINI were able to find $n(\varepsilon)$. The result is shown in Fig. 22 for $s = 1.8$. The possible effects due to K mesons are also known, calculated on the same basis as in Sect. 16. So far no experimental results are available in the very-high-energy region. In the low-energy region the experimental points below 1 GeV lie significantly below the curve. This is presumably due to the neglect of angular divergence in the π-production and decay processes.

Calculations based on the diffusion equations (11.1) and (11.2) have been carried out by MURAYAMA et al.[1] using constant β_μ, neglecting π-meson ionisation loss and neglecting also the effect on the μ-meson ionisation loss of assuming an inclined trajectory. This could evidently have serious effects on the spectra for energies $\lesssim 10$ GeV and large angles. Results are given by these authors for various angles between 40 and 83.3°.

II. The spectrum at large zenith angles.

A logical extension of the measurements referred to above is to the determination of the spectrum of particles arriving at sea level in the near-horizontal direction. Such measurements are interesting not only from the phenomenological aspect but also because they should, when combined with other measurements, give information on that fraction of sea level μ-mesons arising from particles other then π-mesons. This arises from the fact that the parents of the μ-mesons arriving in the vertical and greatly inclined directions are produced in regions of the atmosphere having very different densities. The sea-level μ-flux is thus sensitive to the mass and lifetime of the parents. Unfortunately no direct spectrum measurements at very large zenith angles have been made, but an experiment by JAKEMAN[2] with a counter array has shown the potentialities of the method and given some preliminary results on the contribution of heavier mesons to the flux of μ-mesons at sea level.

39. The expected form of the spectrum at large zenith angles. Before considering the experimental results it is useful to study the variation of the form of the sea level spectrum with zenith angle to be expected at large zenith angles.

As mentioned in Sect. 38 the problem is essentially that of solving the diffusion equations (11.1) and (11.2) for the various zenith angles. The solution for the μ-production spectrum has been given by BARRETT et al., for not too large zenith angles, as

$$\mu_p(\varepsilon, \vartheta) = \frac{F(\varepsilon/r)}{r} \frac{\lambda_\pi}{\lambda_p} \frac{B'}{\varepsilon \cos \vartheta} \times$$
$$\times \left\{ \frac{1}{1 + B'/\varepsilon \cos \vartheta} - \frac{\lambda_\pi/\lambda'}{2 + B'/\varepsilon \cos \vartheta} + \frac{(\lambda_\pi/\lambda')^2}{3 + B'/\varepsilon \cos \vartheta} - \dots \right\} \tag{39.1}$$

---

[1] T. MURAYAMA, K. MURAKAMI, K. TANAKA and S. OGAWA: Proc. Theor. Phys. **15**, 421 (1956).

[2] D. JAKEMAN: Can. J. Phys. **34**, 432 (1956).

where $\dfrac{1}{\lambda'} = \dfrac{1}{\lambda_p} - \dfrac{1}{\lambda_\pi}$, $B' = r B$, where r and B are defined in Sect. 11, and $F(\varepsilon/r)$ is the π-meson production spectrum.

Murayama *et al.* have evaluated (39.1) for $\lambda_p = 120$ gm cm^{-2} and $\lambda_\pi = 60$ gm cm^{-2} with the result shown in Fig. 23. The resulting sea-level spectra have not been given by these authors but calculations have been made by Jakeman, and extended by Allen (private communication) for the case of $\lambda_p = \lambda_\pi$ (this is prob-

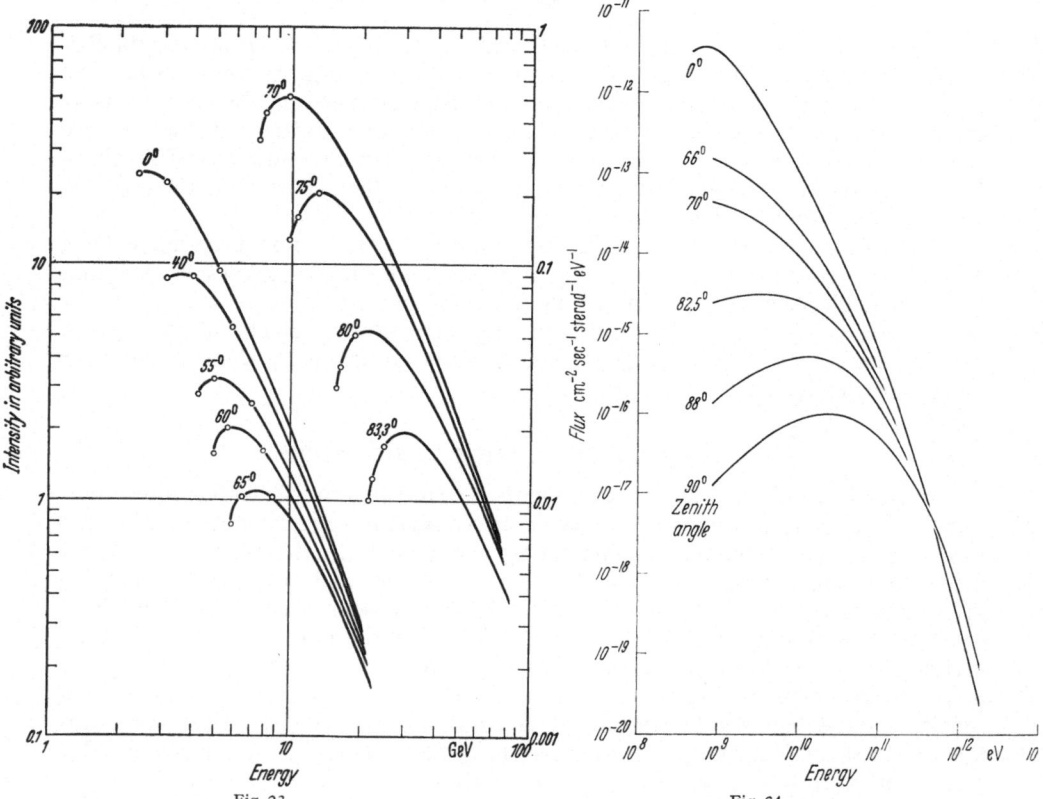

Fig. 23.   Fig. 24

Fig. 23. The zenith angle dependence of the differential energy spectrum of μ-mesons at production after Murayama *et al.*, see footnote 1, p. 307. The three points attached to each curve are the minimum energies of particles able to penetrate the absorbers of Sekido *et al.*, see footnote 1, p. 310, at sea level (211, 523 and 1588 gm cm^{-2} of iron respectively).

Fig. 24. The differential energy spectra of μ-mesons at sea level calculated by Jakeman, (at $\vartheta = 0°$, 88° and 90° and Allen (private communication). To avoid confusion only the spectra of $\vartheta = 0°$ and 90° are shown in the energy region above their intersection. The spectra have been calculated on the assumption that π-mesons alone are the source of sea level μ-mesons.

ably a better approximation to the actual case). Eq. (39.1) can then be rewritten in the general form applicable to all zenith angles

$$\mu_p(\varepsilon, \vartheta) = \frac{F(\varepsilon/r)}{r} \frac{B'}{B' + \varepsilon(\varrho/x)_\vartheta}. \tag{39.2}$$

The sea-level spectra calculated using this relation, and the survival probabilities for the μ-mesons to reach sea level, are shown in Fig. 24.

The main features of the spectra can be readily understood. The considerable reduction in intensity at low momenta arises from the large ionization loss and high decay probability in traversing the considerable atmospheric path in the horizontal direction. The increase in intensity at high energies in the greatly inclined directions compared with the vertical direction arises from the increased value of x/ϱ in these directions. In the region of the main producing layers of the atmosphere $(x/\varrho)_{90^\circ} \sim 10\,(x/\varrho)_{0^\circ}$ and the ratio of the intensities tends, therefore, to about 10.

In experiments designed to study the properties of fast μ-mesons there are obvious advantages of working at large angles where the background of slow particles is low. As yet no experiments with momentum resolution have been performed under these conditions.

40. The effect of K-meson production on the sea-level spectrum. At high energies, above about 100 GeV, it is likely that K-meson production is important. Of the various decay schemes, known at the moment, those in which a μ-meson is one of the secondaries are:

(i) $K_{\mu_2} \rightarrow \mu + \nu$

(ii) $K_{\mu_3} \rightarrow \mu + \nu + \pi^0$

The K-meson mass is near $965\,m_e$ and its lifetime is $\sim 10^{-8}$ sec (see the review article by W. D. WALKER[1] for a recent summary). The evidence from machine experiments indicates that the decay scheme (i) is 20 times more probable than (ii) but for the purposes of comparison calculations have been made for both cases.

As can be seen from Eq. (39.1) there will be two effects on the μ-meson production spectrum, arising from changes in r and B' respectively.

For a two-body decay, as in (i), in which a heavy particle decays into two much lighter particles, the mean energy of the secondaries in the laboratory system will be $\sim 50\%$ of the energy of the parent and there will be a considerable spread in the energy

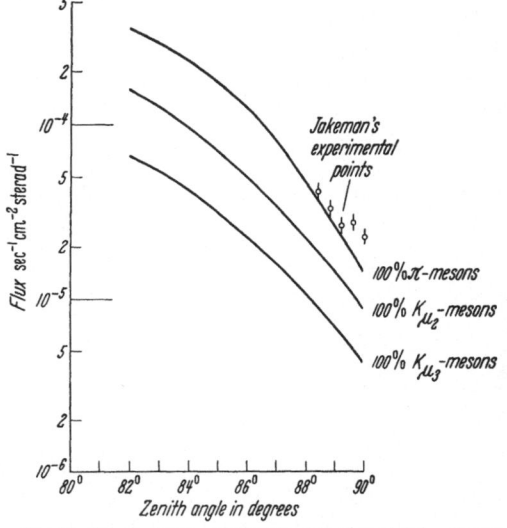

Fig. 25. The variation of absolute rate of sea level μ-mesons with zenith angle. The curves have been calculated by ALLEN (private communication) and the authors for three limiting assumptions as to the nature of the parents of the μ-mesons viz. 100% π-mesons, 100% K-mesons, decay scheme (i) and 100% K mesons, decay scheme (ii). The experimental points refer to the experiment of JAKEMAN, see footnote 2, p. 307. The curves are approximate in that they are derived using a rather uncertain primary spectrum.

distribution. Thus a K_{μ_2}-meson of energy 100 GeV will produce a μ-meson having energy between 4.6 GeV and 100 GeV, the energy corresponding to emission in the K-meson system perpendicular to the line of flight being 52 GeV. (For a π-meson the corresponding figures are 57, 100 and 78 GeV.) For a three-body decay, as in (ii) the μ-meson energy range is even greater and the mean value of $r'\,(= m_\mu/m_K)$ is $\sim\frac{1}{3}$. As yet no accurate calculations have been made including the variation in r' but approximate calculations with $r' = \frac{1}{3}$ have been made by various authors. At present it suffices to point out that if the production spectrum of the K-mesons is identical with that of π-mesons[2] the flux of μ-mesons at sea level will be reduced on account of the significantly lower value of r'.

[1] W. D. WALKER: Progress in Cosmic Ray Physics, Vol. IV. Amsterdam: North Holland Publishing Co. 1958.

[2] It has recently been pointed out, however, by ALLEN and APOSTOLAKIS (private communication) that the preceding argument regarding the effect of K-meson production is in correct. They show that the production spectra for π- and K-mesons cannot be the same if agreement is to be maintained with the vertical spectrum. They conclude that, if consistent production spectra are chosen the sea level intensity, at large zenith angles, for 100% K_{μ_2}-mesons is some 10—15% *higher* than that for 100% π-mesons.

ALLEN and the authors have calculated the variation·in ·total flux at sea level at large zenith angles under extreme assumptions as to the parents of the μ-mesons viz. 100% π-mesons, 100% K-mesons decaying according to scheme (i) and 100% K-mesons decaying according to scheme (ii). The result is shown in Fig. 25. It will be seen that the behaviour, whereby the ratio of the μ-meson fluxes is reduced at the very largest zenith angles, is explained by the discussion given above.

41. Jakeman's experiment on the near-horizontal particles. Jakeman used a horizontal counter telescope with hodoscope circuits to determine the angular distribution of the total component able to penetrate at least 30 cm of lead. Essentially the apparatus consisted of a series of telescopes having resolution $\sim 0.2°$. The angular distribution found for the total intensity as a function of zenith angle between 88° and 90° is shown in Fig. 25. Also shown is the distribution calculated by Jakeman for 100% π-mesons together with those for 100% K_{μ_2}- and 100% K_{μ_3}-mesons referred to above. It is apparent that the form of the measured variation of intensity with zenith angle is as expected but that the absolute magnitude is in error. Although Jakeman concludes from this evidence that the percentage contribution from K-mesons is small this conclusion does not seem justified from the data; it seems likely that there are further corrections to be applied to the data which will have the effect of reducing the intensities— the effect of multiple scattering in the atmosphere and deflections in the earth's magnetic field appear to be important examples.

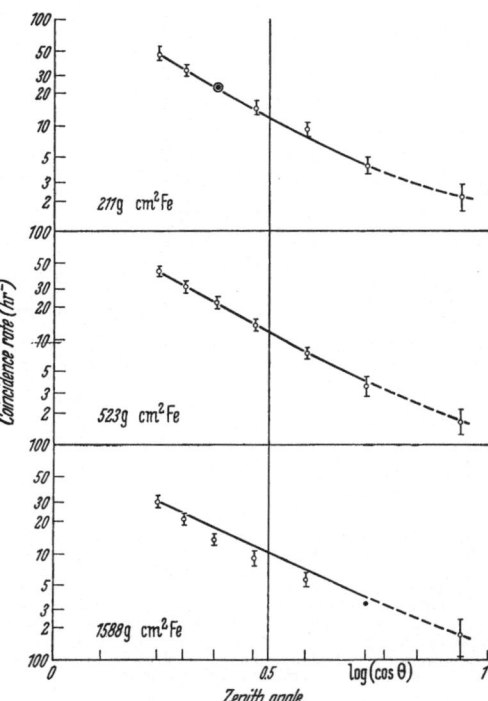

Fig. 26. The observed zenith angle distribution of cosmic ray intensity found by Sekido *et al.*[1] at sea level using a counter telescope with iron absorbers of various thickness. The minimum μ-meson energies at production for each zenith angle and absorber thickness are given in Fig. 23.

Experiments of a similar type but at smaller zenith angles have been carried out by Sekido *et al.*[1]. These authors used a narrow-angle counter telescope with a thick iron absorber. The results found for the variation of total intensity (above the low momentum cut-off imposed by the iron absorber) with zenith angle, are shown in Fig. 26.

The results are compared with the calculated rates derived from the analysis of Murayama *et al.* and it will be noted that theory and experiment are normalized at 65° for the 211 gm cm^{-2} absorber.

It is obvious that there is in general quite good agreement indicating that the method of analysis and basic assumptions are sound, at least for moderately large zenith angles. At the greatest zenith angle however the statistical accuracy of the experimental points is low and the extrapolation of the theoretical curves,

---

[1] Y. Sekido, S. Yoshida and Y. Kamiya: J. Geom. Geod. **6**, 22 (1954).

represented by the dotted lines, rather dubious. It is just in this region where the effect of K-meson production is expected to be most pronounced, and it must be concluded that this experiment, too, gives little information on the question of the contribution of high-energy K-mesons to the sea-level μ-meson flux. Further experiments and more detailed calculations on this interesting topic are awaited.

III. The latitude effect in inclined directions.

42. The contributions from the various sources. It will be seen from the discussion in Sects. 25 and 35 that the latitude effect on spectra in inclined directions is two-fold. The most important effect is the same as that observed on the vertical spectrum, i.e. magnetic cut-off of the primary spectrum, except that the magnetic cut-off of primary energies at a given latitude varies with zenithal and azimuthal angles. It is found to increase with ϑ in easterly, and decrease with ϑ in westerly directions. An additional effect not present in the vertical case is the variation with latitude in the trajectory of the mesons due to the variation in strength of the horizontal component of the earth's magnetic field. This produces an effect on the total spectrum through the positive excess. One interesting experimental quantity at inclined directions which should reflect these effects is the east-west asymmetry, A, given by $(I_W - I_E)/(I_W + I_E)$, as a function of latitude. One should find that $A > 0$ at all latitudes, becoming larger at low latitudes and energies and large angles. It is obviously rather difficult to disentangle the two effects, but, from the way in which magnetic rigidity varies with zenith angle at a given latitude, one can say that the second effect might be predominant at small angles for which the east and west rigidities are nearly the same. If one may neglect the direct dependence of the positive excess on geomagnetic latitude then comparison of A at small angles and different latitudes should yield an estimate of the second effect. Detailed calculations of the asymmetry at latitude $9°$ N, $\vartheta = 45°$ and $60°$ and latitude $42°$ N, $\vartheta = 20°$, $40°$ and $60°$ have been given by QUERCIA and RISPOLI[1,2] on the basis of CALDIROLA's model of π-meson production allowing for the second effect alone.

At latitude $9°$ N for both angles and at latitude $42°$ N for $\vartheta = 60°$ both effects may be expected to appear together and calculations of the second contribution to A at latitude $9°$ N did in fact fall short of the experimental results. At the higher latitude the experimental results were not sufficiently accurate to allow any firm conclusion to be drawn although it does appear that in some cases $A < 0$ (cf. the results of MORONEY and PARRY), a result which is rather difficult to understand.

It is noteworthy that measurement in the north-south plane of the north-south asymmetry should show only the first effect, since the effect produced by deflection of the μ-mesons is negligible for these azimuths. This has been used by QUERCIA and RISPOLI to evaluate the east-west asymmetries in an approximate way by writing

$$w^{\pm} \approx s^{\pm}\left(1 + \tfrac{1}{2}A(\varepsilon, \vartheta)\right),$$
$$e^{\pm} \approx s^{\pm}\left(1 - \tfrac{1}{2}A(\varepsilon, \vartheta)\right),$$

where $w^{\pm}$, $e^{\pm}$ and $s^{\pm}$ are the counting rates at the three azimuths. The proton energy cut-offs, at least for the western and southern azimuths, are approximately the same so that, roughly speaking, all the magnetic effects should be contained in A. Comparison of experimental values of A found in this way with those found directly showed qualitative agreement.

---

[1] L. F. QUERCIA and B. RISPOLI: Nuovo Cim. **10**, 1142 (1953).
[2] I. F. QUERCIA and B. RISPOLI: Nuovo Cim. **12**, 490 (1954).

IV. The positive excess.

43. The results of QUERCIA and RISPOLI. The positive excess in inclined directions is also discussed by QUERCIA and RISPOLI in the work referred to in the previous section.

Again the two main differences from the vertical positive excess are the variation of magnetic cut-off with zenith angle (which prevents primaries of relatively higher energies initiating a positive excess) and the differential deflection of positive and negative mesons in various azimuths. These can be separated fairly well by comparing measurements in the north-south plane with those in the east-west plane. In particular if, as is to be expected, the meson trajectories are symmetrical about the zenith, then w^+/e^-, e^+/w^- should give the same result as that observed in the vertical direction for energies, angles and latitudes such that differences in magnetic rigidity may be neglected. QUERCIA and RISPOLI find that experimentally this is not the case, though they do find that s^+/s^- is in fact less than the theoretical estimate as indeed it should be since in the latter no allowance was made for effects of magnetic rigidity.

44. The results of MORONEY and PARRY. On the other hand the measurements by MORONEY and PARRY[1] of w^+/e^- and e^+/w^- at 30° and 60° do agree with the vertical result within the experimental errors. The result at 60° is perhaps slightly smaller than that at the vertical but this is to be expected from the increase of magnetic rigidity with angle. These authors were also able to fit the observed energy dependence of the conventional positive excess (w^+/w^- etc) by assuming a positive excess at production equal to the vertical value and allowing for the different trajectories of positive and negative mesons.

In conclusion it seems that the present ideas on zenith and latitude dependence of the positive excess are correct but detailed calculations and more accurate experimental results are required before this can be confirmed quantitatively. Finally, accurate measurements in the north-south plane at various zenithal angles should enable one to deduce rather accurately the positive excess produced by protons within a definite energy interval.

E. Meteorological effects on the μ-meson component.

45. The scope of the discussion on meteorological effects. Summaries of the time variation of cosmic rays, a subject which includes the present topic, have appeared from time to time. Of these we may mention the reviews of ELLIOT[2], SIMPSON[3] and SARABHAI and NERURHAR[4]. More up to date reviews will be found in various articles in the present volume.

A brief treatment will be given of the various meteorological effects on the μ-meson component as a function of zenith angle and on the positive and negative components separately.

46. Effects on the vertical spectrum. That the μ-meson spectrum at sea level is influenced by atmospheric conditions is clear from the formulae quoted in various sections (e.g. in the survival probability term in Sect. 14). Experimentally the first observations seem to be those of MYSSOWSKI and TUWIM[5] of the baro-

---

[1] See footnote 1, p. 302.
[2] H. ELLIOT: Progress in Cosmic Ray Physics, Vol. I. Amsterdam: North Holland Publishing Co. 1952.
[3] J. A. SIMPSON: Ann. Geophys. **11**, 305 (1955).
[4] V. SARABHAI and N. W. NERURHAR: Annual Rev. Nucl. Sci. **6** (1956).
[5] L. MYSSOWSKI and L. TUWIM: Z. Physik **50**, 273 (1928).

meter effect. This implies a correlation between μ-meson intensity and the atmospheric pressure at sea level of the form

$$\delta I_v/I = A_P \, \delta x_0 \qquad (46.1)$$

where x_0 is the atmospheric depth at sea level. It was found that $A_P < 0$, as one would expect if the μ-mesons were actually absorbed in the atmosphere. Since that time the subject has been developed by DUPERIER[1-3] who introduced a correlation of I_v with the height (H) of an assumed μ-meson production layer and subsequently with its temperature (T). Eq. (46.1) thus becomes

$$\delta I_v/I = A_P \, \delta x_0 + A_H \, \delta H + A_T \, \delta T. \qquad (46.2)$$

The first term was interpreted as representing the absorption of μ-mesons, the second an effect introduced by their finite lifetime and the third the effect of competition between $\pi - \mu$ decay and π-meson capture by nuclei which should lead to $A_T > 0$ as indeed was found to be the case experimentally. However it has been shown by TREFALL[4,5] and MAEDA and WADA[6] that for any reasonable value of the π-meson absorption mean free path the theoretical values of A_T were much smaller than those found experimentally.

47. The positive and negative temperature coefficients. Accordingly detailed calculations have been made by OLBERT[7] starting from the expression (14.2) and differentiating with respect to T. Two terms appear, the first giving rise to $A_H \, \delta H$ since H depends on the atmospheric temperature, and the second produces an additional dependence on δT. This second term represents the effect of ionisation loss and depends essentially on the temperature of the troposphere. If the temperature of the troposphere falls, then some of the ionisation loss will be transferred to the latter part of the μ-meson trajectory enhancing its survival probability. The results of calculation are

$$A_H = -3.15\% \text{ per km,}$$
$$A_T = -0.059\% \text{ per °C,}$$
$$A_P = -1.79\% \text{ per cm Hg.}$$

The height H is found to be that of an average production layer at the depth $x_1 \sim 115$ gm cm^{-2}. OLBERT was able to show that these values of A_H and A_T are in agreement with the results of DOLBEAR and ELLIOT[8] on the annual and seasonal variations in T and H. So far as the positive temperature effect is concerned it is known that $(\delta T)_{\text{trop}}/(\delta T)_{\text{strat}} < 0$ and if one assumes $(\delta T)_{\text{trop}}/(\delta T)_{\text{strat}} = -2$ one recovers the results of DUPERIER that $A_T = 0.12\%$ per °C. However it is likely that this ratio for the two temperature changes is too large and indeed later experimenters have found smaller values of A_T.

More recently, TREFALL[5] has found the same effect using a method broadly similar to that of OLBERT, though somewhat less detailed. In addition this author notes that the reference level H, chosen arbitrarily by most experimentalists to be the 100 mb level, may not, and in general will not, cor-

[1] A. DUPERIER: Terr. Magn. Atmos. Electr. **49**, 1 (1944).
[2] A. DUPERIER: Proc. Phys. Soc. Lond. A **62**, 685 (1949).
[3] A. DUPERIER: J. Atmos. Terr. Phys. **1**, 296 (1951).
[4] H. TREFALL: Proc. Phys. Soc. Lond. A **68**, 625 (1955).
[5] H. TREFALL: Proc. Phys. Soc. Lond. A **68**, 893 (1955).
[6] K. MAEDA and M. WADA: J. Sci. Res. Inst. Tokyo **48**, 71 (1954).
[7] S. OLBERT: Phys. Rev. **92**, 454 (1953).
[8] D. W. N. DOLBEAR and H. ELLIOT: J. Atmos. Terr. Phys. **1**, 215 (1951).

respond with an actual production layer, supposing this to exist. This will give rise to a further term depending on stratospheric temperature since the variation of reference level and of production level with temperature need not be the same and the actual phenomena depend on the real production level. After a detailed comparison of the results with all the relevant experimental information it appears that these two terms each account for about one half of DUPERIER's observations. It will be seen that OLBERT's result takes the reference level to be the same as the production level and indeed x_1 is not very different from 100 mb. However, before detailed comparison of the two results is possible, it would be necessary to allow for the difference in assumed reference level and also for the difference between $(\delta T)_{\text{trop}}$ and $(\delta T)_{\text{strat}}$. At present it would seem that a combination of these two contributions should account for the observed positive temperature effect.

48. The pressure coefficient. So far as the coefficient A_P is concerned this has been evaluated by OLBERT by writing x_0 for t in (14.2), differentiating with respect to x_0 and then writing

$$A_P = \frac{1}{I_v(R_0)} \int\limits_{R_0}^{\infty} a_P(R)\, i_v(R)\, dR \qquad (48.1)$$

with the result quoted above, viz. -1.79%.

It is found that the principal contribution arises from the $\partial G/\partial x_0$ term. No attempt is made to ensure that the atmospheric temperature distribution is changed in such a way that H remains constant during this variation.

More recently TREFALL[1] has allowed for the dependence of R_0 (the momentum cut-off of the apparatus) on x_0, the true absorption effect, and has also calculated the contribution to A_P arising from the dependence of ω, the survival probability, on x_0 with the result $A_P = -1.6\%$ per cm Hg, for a cut-off energy of 2 GeV. After a detailed review of the experimental data he finds reasonable agreement between experiment and calculation when the restriction $\delta H = 0$ is imposed. Without this restriction $A_P = -2.5\%$ per cm Hg. Although the agreement with OLBERT's calculation is good it is not clear that the two calculations refer to the same effect since OLBERT's result arises mainly from the change in the production spectrum of the μ-mesons when the energy of the μ-mesons produced is changed. This discrepancy remains to be elucidated.

49. Effects in inclined directions. So far as the effects in inclined directions are concerned MURAYAMA et al., in the work referred to above, have evaluated the temperature correlation by replacing the expression (46.2) by

$$\delta I_\vartheta/I_\vartheta = A_1\, \delta T_1 + A_2\, \delta T_2 + A_3\, \delta T_3 + A_P\, \delta x_0. \qquad (49.1)$$

where T_1 refers to the stratosphere and T_2 and T_3 are the atmospheric temperatures above and below the 300 mb level respectively. This value was chosen in the light of meteorological observations in Japan.

They find the following values for the constants in Eq. (49.1).

$$A_1 = +0.0568\% \text{ per } °C,$$
$$A_2 = -0.305\% \text{ per } °C,$$
$$A_3 = -0.157\% \text{ per } °C,$$
$$A_P = -0.072\% \text{ per mb}.$$

[1] H. TREFALL: Proc. Phys. Soc. Lond. A **68**, 953 (1955).

The only experimental data available as yet with which to compare these results are those of DOLBEAR and ELLIOT who find a coefficient A_1 of 0.14% per °C at an inclination of 45°. This compares with a theoretical result due to TREFALL of 0.17% per °C.

50. Effects for positive and negative particles separately. Lastly the correlation coefficients A_P and A_H have been measured separately for positive and negative μ-mesons in the vertical direction by FASOLI et al.[1]. A model based on an approximate solution of the diffusion equations (11.1) and (11.2) was used in deducing the theoretical results and the difference in positive and negative μ-meson production was introduced by asuming two sources, one for the negative and one for the positive mesons. The theoretical results, with the reference level taken to be at 100 mb are

$$A_P^+ = -2.25\% \text{ per cm Hg} \qquad A_H^+ = -7.06\% \text{ per km}$$
$$A_P^- = -2.28\% \text{ per cm Hg} \qquad A_H^- = -6.87\% \text{ per km.}$$

These results were found to be in good agreement with the experimental values at least when averaged over the 18 month period of observation. It was noticed however that the coefficient A_H seemed to be rather strongly correlated with the seasonal variation of ground temperature, and the height of the reference level. The theoretical dependence of A_H on H was found to be insufficient to account for the observed results. Accordingly the ground temperature was introduced as an extra variable but this did not remove the seasonal variation in A_H. A_H was however reduced to

$$A_H^+ = (-5.00 \pm 0.7)\% \text{ per km,}$$
$$A_H^- = (-3.26 \pm 0.75)\% \text{ per km.}$$

It will be seen that the agreement between these results and the mean value, for positive and negative particles together, evaluated by OLBERT (-3.15% per km) is much improved.

Bibliography.

[1] ROSSI, B.: High Energy Particles. Prentice Hall 1952.
[2] PUPPI, G.: In: Progress in Cosmic Ray Physics, Vol. III. Amsterdam: North Holland Publishing Co. 1956.
[3] OLBERT, S.: Technical Report 61 M.I.T. 1954.
[4] BUDINI, P., and G. MOLIÈRE: In: Kosmische Strahlung, ed. by W. HEISENBERG. Berlin: Springer 1953.
[5] PUPPI, G., and N. DALLAPORTA: In: Progress in Cosmic Ray Physics, Vol. 1. Amsterdam: North Holland Publishing Co. 1952.

---

[1] U. FASOLI, C. MARONI, I. MODENA, E. POHL and J. POHL-RÜLING: Nuovo Cim. **5**, 473 (1957).

Sachverzeichnis.
(Deutsch-Englisch.)

Bei gleicher Schreibweise in beiden Sprachen sind die Stichwörter nur einmal aufgeführt.

Subject Index.
(English-German.)

Where English and German spelling of a word is identical the German version is omitted.